T0172920

Beginner's Guide to SOLIDWORKS 2019 – Level I

Parts, Assemblies, Drawings, PhotoView 360 and Simulation Xpress

Alejandro Reyes, MSME, CSWE, CSWI
Certified SOLIDWORKS Expert

SDC
Publications

SDC Publications
P.O. Box 1334
Mission, KS 66222
913-262-2664
www.SDCpublications.com
Publisher: Stephen Schroff

Copyright 2018 Alejandro Reyes

All rights reserved. This document may not be copied, photocopied, reproduced, transmitted, or translated in any form or for any purpose without the express written consent of the publisher, SDC Publications.

Examination Copies
Books received as examination copies are for review purposes only and may not be made available for student use. Resale of examination copies is prohibited.

Electronic Files
Any electronic files associated with this book are licensed to the original user only. These files may not be transferred to any other party.

Trademarks and Disclaimer
SolidWorks and its family of products are registered trademarks of Dassault Systemes. Microsoft Windows and its family products are registered trademarks of the Microsoft Corporation.

Every effort has been made to provide an accurate text. The author and the manufacturers shall not be held liable for any parts developed with this book or held responsible for any inaccuracies or errors that appear in the book.

ISBN-13: 978-1-63057-220-4
ISBN-10: 1-63057-220-9

Printed and bound in the United States of America.

Acknowledgements

Beginner's Guide to SOLIDWORKS 2019 – Level I is dedicated to my lovely wife Patricia and my kids Liz, Ale and Hector, all of whom have always been very supportive, patient and understanding during the writing of this book. To you, all my love.

Also, I wish to thank the hundreds of students, users, teachers and engineers whose great ideas and words of encouragement have helped me improve this book and make it a great success.

About the Author

Alejandro Reyes holds a BSME from the Instituto Tecnológico de Ciudad Juárez, Mexico, in electro-mechanical engineering and a Master's Degree from the University of Texas at El Paso in mechanical design, with a strong focus in Materials Science and Finite Element Analysis.

Alejandro spent more than 8 years as a SOLIDWORKS Value Added Reseller. During this time he was a Certified SOLIDWORKS Instructor and Support Technician, Simulation Support Technician, and a Certified SOLIDWORKS Expert, credentials that he still maintains. Alejandro has over 20 years of experience using CAD/CAM/FEA software and is currently the President of MechaniCAD Inc.

His professional interests include finding alternatives and improvements to existing products, FEA analysis, and new technologies. On a personal level, he enjoys bicycle riding, off-road driving and spending time with family and friends.

Table of Contents

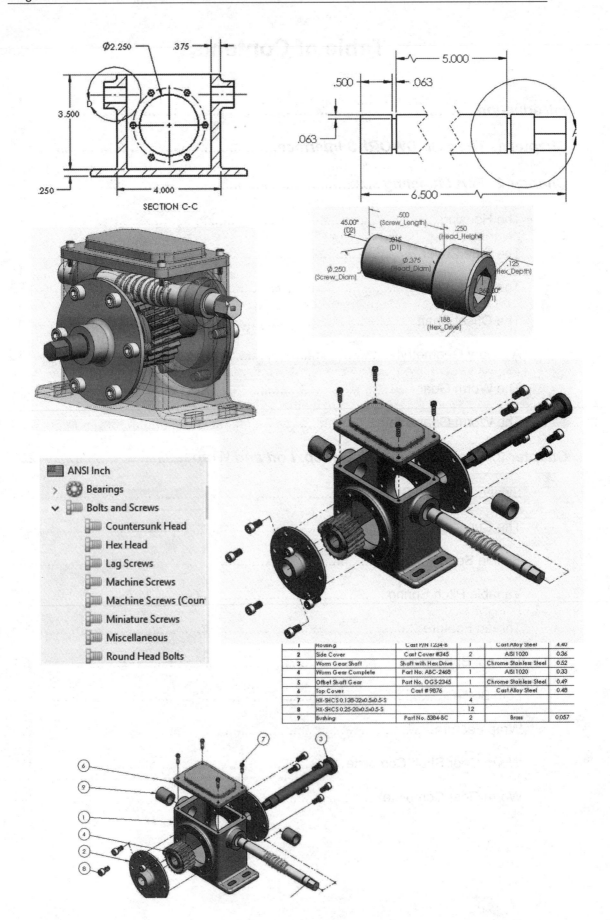

	Housing	Cast P/N 1234-8	1	Cast Alloy Steel	4.40
2	Side Cover	Cast Cover #345	2	AISI 1020	0.36
3	Worm Gear Shaft	Shaft with Hex Drive	1	Chrome Stainless Steel	0.52
4	Worm Gear Complete	Part No. ABC-2468	1	AISI 1020	0.33
5	Offset Shaft Gear	Part No. OGS-2345	1	Chrome Stainless Steel	0.49
6	Top Cover	Cast # 9876	1	Cast Alloy Steel	0.48
7	HX-SHCS 0.138-32x0.5x0.5-S		4		
8	HX-SHCS 0.25-20x0.5x0.5-S		12		
9	Bushing	Part No. 5384-BC	2	Brass	0.057

Notes:

List of commands introduced in each chapter. Note that many commands are used extensively in following chapters after they have been presented.

PART MODELING

Housing:
Part Templates
New Part
Create Sketch
Confirmation Corner
Sketch Grid
Auto-Rotate view normal
to Sketch plane
Sketch Rectangle
Sketch Centerline
Sketch Relations
Smart Dimension
Sketch Status
Extrude Boss/Base
Document Units
View Orientation
Mouse Gestures
Menu Customization
Center Rectangle
Fillet
Magnifying Glass
Extruded Cut
Through All (End
condition)
Sketch Fillet
Rename Features
Circle
Instant 3D
Mirror Features
Model Display Styles
Fly-Out FeatureManager
Dimension Tolerance
Hole Wizard
Cosmetic Threads
Sketch Point
Edit Sketch
Rebuild
Circular Pattern
Automatic Relations
Temporary Axes
Sketch Slot

Sketch Numeric Input
Linear Pattern
Edit Material
Mass Properties

Side Cover
Revolved Boss/Base
Trim Entities
Extend Entities
Construction Geometry

Top Cover
Offset Entities
Mirror Entities (Sketch)
Up to Surface (End
condition)
Shell
Measure Tool
Select Other
More Fillet options

Offset Shaft
Revolved Cut
Auxiliary Planes
Hide/Show sketch
Convert Entities
Polygon (Sketch)
Flip Side to Cut
Axis (Reference geometry)
Coordinate Systems

Worm Gear
Mid Plane (End Condition)
Chamfer
Dimension to Arc
Direction 2 (End
Condition)

Worm Gear Shaft
General Review of
previous commands

SWEEP, LOFT, WRAP

Sweep
Thin Feature
Ellipse
Auxiliary Plane at point
Sweep
Up to Next (End
condition)
Full Round Fillet
Helix
Variable pitch helix
Open Sketch Cut
Guide Curves

Loft
Offset Plane
Hide/Show Plane
Loft
Start/End Conditions
Model Section View

Wrap
Wrap
Intersection Curve

Worm Gear Shaft
Complete
Helix
Sweep
Convert Entities

Worm Gear Complete
Sweep
Circular Pattern
Fillets

DETAIL DRAWING

Housing Drawing
Part Configurations
Suppress Feature
Parent Child Relation
Unsuppress
Configure Dimensions
Change Configuration
Drawing Templates
New Drawing
Make Drawing from Part
View Palette (Drawing)
Projected View
Tangent Edge Display
Section View
Detail View
Model Items
(Import dimensions)
Drawing cleanup
Center Mark/Centerlines
Add Sheet to Drawing
Model View
   Configuration
Rename Sheet

Layers

Side Cover Drawing
Reference Dimension
Dimension Parentheses
Sheet Scale
View Scale

Top Cover Drawing
Display only Cut Surface
Display Dimension As
diameter
Ordinate Dimension
Add to Ordinate
Notes
Custom Document
   Properties
Parametric Notes
Link to Property

Offset Shaft Drawing
Broken View

Worm Gear Drawing
Display Sheet Format
Insert Model View
Crop View
Ellipse
Angle Dimension
Tolerance/Precision
Chamfer Dimension
Edit Sheet Format / Title
   Block
Edit Sheet
Create and Reuse a
   Sheet Format

Worm Gear Shaft Drawing
Broken-Out Section
Spline
Property Tab Builder

ASSEMBLY MODELING

Assembly Templates
New Assembly
Add New Component
Change Configuration
Mates
Concentric Mate
Coincident Mate
Move Component
Rotate Component
Change Appearance
Parallel Mate
Hide/Show Component
SmartMates
Mirror Components
Width Mate
Toolbox Fasteners
Design Library
Add Fasteners
Cosmetic Thread Display
Configurations using

Design Tables
Show Feature
   Dimensions
Rename Dimensions
Edit Design Table
Mate Reference
Design Library
Assembly Configurations
Interference Detection
Change dimensions in
   the assembly
Collision Detection
Isolate
Gear Mate
Component Transparency
Display Pane
Drawing update
Exploded View creation
Collapse View
Animate Exploded view

Exploded View Sketch
   Line

Assembly Drawing
Show Assembly in
   exploded view
Bill of Materials
Add Custom Properties
   to bill of Materials
Open Part from
   assembly drawing
Format Bill of Materials
Auto Balloons
Assembly Views
Selective Assembly
   Section View

ANIMATION AND RENDERING

SimulationXpress

eDrawings

Appendix

In the next book,
Beginner's Guide to SOLIDWORKS 2019 <u>Level II</u>
you'll learn:

- Multi body part techniques
- Part editing, equations and errors
- Top Down design techniques
- Design sheet metal parts
- How to build and use Libraries
- 3D Sketches
- Design welded structures
- Structural member libraries
- Surface modeling
- Mold design tools
- And more…

Introduction

This book is intended to help new users learn the basic concepts of SOLIDWORKS and good solid modeling techniques in an easy to follow guide by building the models on the cover one at a time. This book will be a great starting point for those new to SOLIDWORKS or as a teaching aid in classroom training to become familiar with the software's interface, learn commands and modeling strategies as the user completes a series of models, while learning different ways to accomplish a particular task. At the end of this book, the user will have a good understanding of the SOLIDWORKS interface and the most commonly used commands for part modeling, assembly and detailing after completing the speed reducer pictured on the cover and all detail drawings complete with Bill of Materials. Our books are primarily focused on the processes to complete a task, (Modeling of components, assembly or drawing), instead of individual commands or operations, which are learned as we progress through the project. We strived very hard to cover the majority of the commands required to pass the multiple certification tests, including **Certified SOLIDWORKS Associate, Certified SOLIDWORKS Professional,** and in the level II book many of the commands for the **Advanced Sheet Metal, Weldments, Surfacing, Mold Tools,** and **Certified Expert** tests as well.

SOLIDWORKS is an easy to use, yet powerful CAD software that includes many time saving tools that enable new and experienced users to complete design tasks in a very short time. The majority of the commands covered in this book are first introduced using simple examples, and while many of these commands also have advanced options, these options may or not be covered in this book, as it is meant to be a starting point to help new users learn the basic and most frequently used commands, instead of an extensive in-depth command tutorial which could potentially confuse a new user; the Level II book goes deeper and covers many advanced commands, options and trade specific tools for sheet metal, mold making, surfacing, weldments, etc.

SOLIDWORKS has hundreds of thousands of users ranging from one-man shops to Fortune 500 companies, as well as a very strong presence in the educational market including high schools, vocational schools and many world renowned prestigious universities.

We always love to hear from our readers about your experience with our books, questions you may have, and ideas or suggestions. As much as we'd love to hear what you think we did right, we are a lot more interested, and it is that much more important to us, to know which areas can be improved. This is how we have been able to make this the best SOLIDWORKS book available.

Please send us your questions, suggestions and comments by email to alejandro@mechanicad.com. Your message will be personally answered.

Prerequisites:

This book was written assuming the reader has knowledge of the following topics:

- The reader must be familiar with the Windows operating system, conventions and generalities (open, save, close, copy files, copy/paste etc.).
- Concepts of mechanical design, drafting (detailing), engineering graphics.
- Experience with other CAD systems is a plus.
- Principles of mechanics of materials is a must to understand Finite Element Analysis using SimulationXpress.

Chapter 1: The SOLIDWORKS Interface

The very first time we open SOLIDWORKS we are presented with the option to select our default drafting standard and units of measure. Selecting a default doesn't mean we have to use it always; we can change either one or both at any time as we see fit. Most samples and exercises in the book are presented using the ANSI dimensioning standard and inches for units of measure. If this is not the first time you load SOLIDWORKS, this option will be pre-set and you will not be presented with this option.

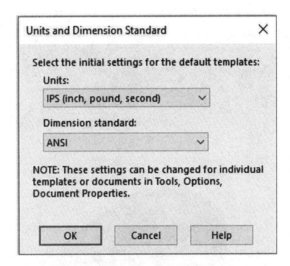

1.1. – After loading SOLIDWORKS, the Menu bar is hidden by default. It is automatically displayed when the user moves the pointer over the SOLIDWORKS banner in the upper left corner of the window and is hidden when we move away from it. In order to make the menu bar always visible for ease of clarity in the book, press the pin icon at the end of the menu bar as indicated.

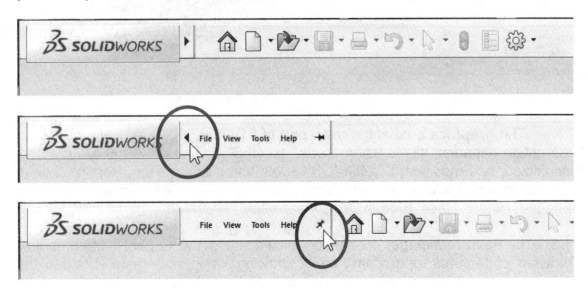

1.2. - The SOLIDWORKS interface is simple and easy to navigate. The main areas in the interface include toolbars, menus, graphics area, Feature-Manager/PropertyManager and Task Pane. SOLIDWORKS includes an intelligent system of pop-up toolbars which is automatically activated when the user selects elements in the FeatureManager or graphics area. SOLIDWORKS has icons and menus similar to those of Microsoft Office applications, and follows Windows rules like drag and drop, copy/paste, etc., typical of any Windows compliant software.

FeatureManager **Menus and Toolbars** **View Toolbar** **Graphics Area**

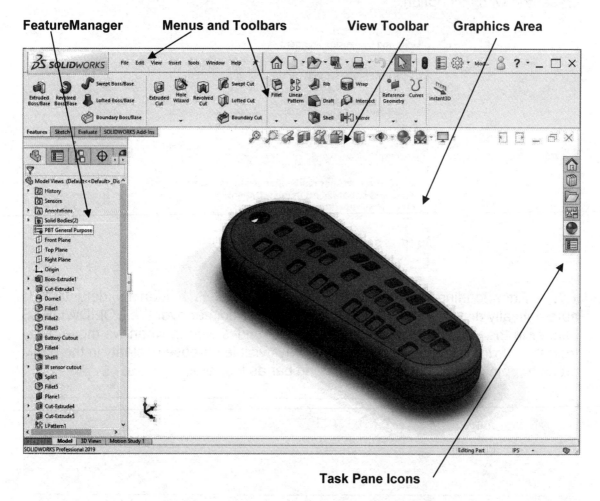

Task Pane Icons

The **graphics area** is the main part of SOLIDWORKS and where most of the action happens; this is where parts, assemblies, and drawings are created, visualized, and modified. SOLIDWORKS lets users Zoom, Pan, Rotate, change view orientations, etc., as well as change how the models are displayed, either as Shaded, Hidden Lines, Hidden Lines Visible and Wireframe to name a few.

1.3. - The **FeatureManager**, located on the left side of the screen, is the graphical browser of features, operations, parts, drawing views and more; besides the graphics area, this is where we can edit, modify, delete, or otherwise make changes.

The FeatureManager's space is also shared by the **PropertyManager** and the **ConfigurationManager**. The PropertyManager is where most of the SOLIDWORKS' command options are presented to the user; this is also where a selected entity's properties are displayed and Configurations are created. The PropertyManager is displayed automatically when needed, so the user does not need to worry about it. Later in the book we'll learn how to view both the FeatureManager and PropertyManager at the same time.

1.4. - In the PropertyManager we are presented with a consistent, common interface for most commands in SOLIDWORKS, including common controls such as check boxes, open and closed option boxes, action buttons, etc.

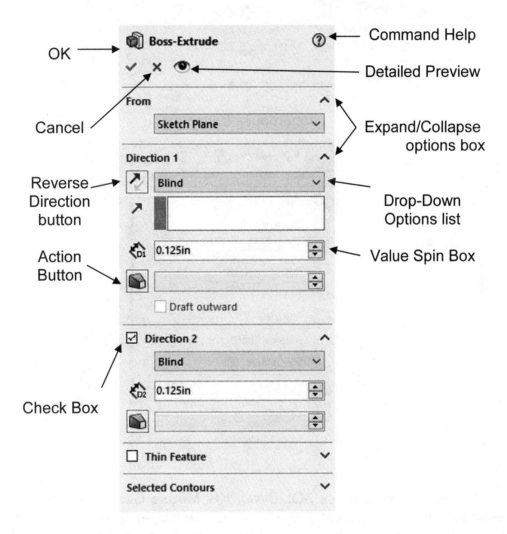

1.5. - To manipulate the models in the graphics area, a set of tools is available from the menu "**View, Modify**" or the right-mouse-button menu in the graphics screen, to Zoom, Pan, Rotate, etc. Using the menu "**File, Open**" or the "**Open**" command icon, open the *'Model Views'* file from the *'Needed Files'* folder in the accompanying book files, select a view manipulation tool, and left-click-and-drag the mouse in the graphics area to see its effect.

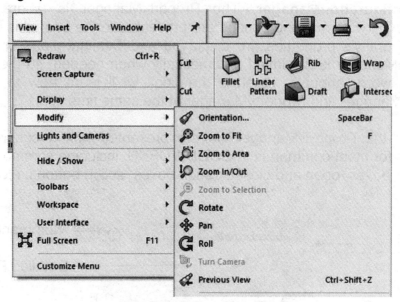

Right-mouse-button menu

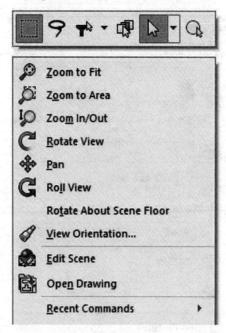

1.6. - Alternatively, we can use SOLIDWORKS' **Mouse Gestures**. Gestures are activated by right-mouse-button-dragging in the graphics area to reveal a shortcuts wheel, where we keep dragging the mouse to touch the command we want to activate. In order to modify the commands in the wheel we have to select the menu **"Tools, Customize"** and select the Mouse Gestures tab. By default, Mouse Gestures are enabled, but they can also be turned off in this tab; we can configure **Mouse Gestures** to have 2, 4, 8 or 12 gestures for each environment (Part,

Assembly, Drawing and Sketch) by assigning a command to each gesture position. Selecting this tab will show the configured commands for each gesture. In this book, we'll use the 8 gestures option with the default command settings.

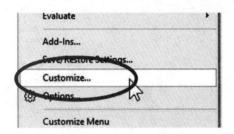
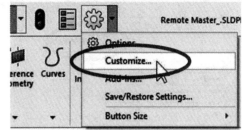

 The option to customize the interface is only available if a document is open.

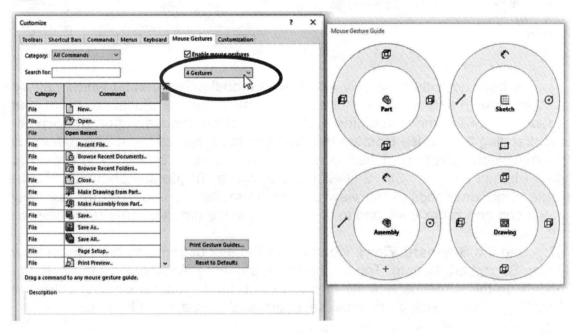

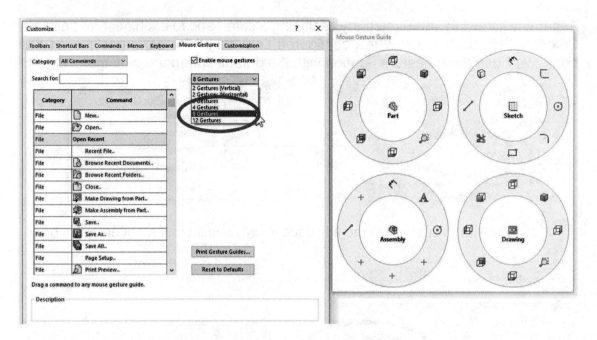

A few shortcuts to quickly manipulate models: The mouse wheel can be used to zoom in and out in the model; keep in mind that the model will zoom in at the location of the mouse pointer; clicking the middle mouse button (the wheel) and dragging the mouse **Rotates** the model in the graphics area. The rotation is automatic about the area of the model where the pointer is located. The **Previous View** (default shortcut "Ctrl+Shift+Z") and **Zoom to fit** (default shortcut "F") are single click commands in the view orientation toolbar to return to the last view orientation and zoom level, and to fit the model in the graphics area respectively.

1.7. - Use the **Standard Views** icon to view the model from any orthogonal view (Front, Back, Left, Bottom, Top, Right, Isometric, etc.). Another way to rotate the models on the screen is with the arrow keys on the keyboard. Holding down the "Shift" key while pressing the arrow keys rotates the model in 90° increments.

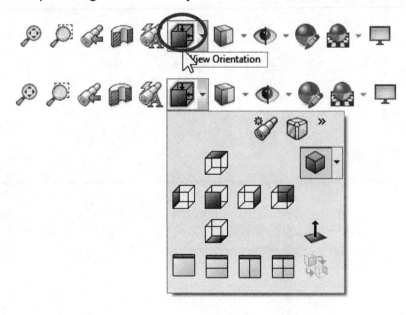

1.8. - Yet another way to change the view orientation is using the View Selector. We can activate it from the "View Orientation" drop down menu or using the shortcut "Ctrl + Spacebar." After activating the View Selector, every time we click in the "View Orientation" we'll get the View Selector until we turn it off. From here we can select the view we want in the translucent box and the model will be reoriented to that view.

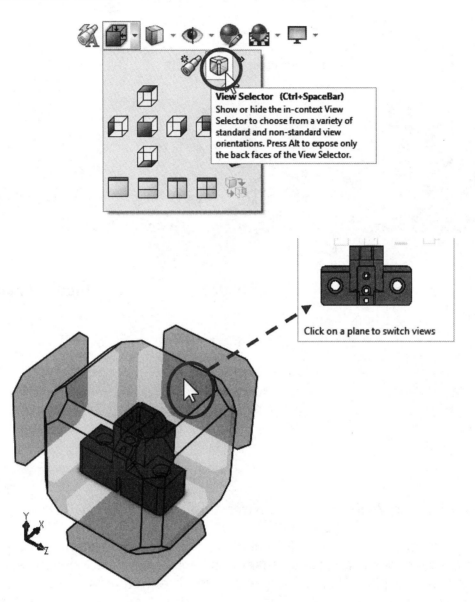

View Selector (Ctrl+SpaceBar)
Show or hide the in-context View Selector to choose from a variety of standard and non-standard view orientations. Press Alt to expose only the back faces of the View Selector.

Click on a plane to switch views

1.9. - To change the **Display Style** (the way the model looks in the graphics area), the Display Style icon can be selected in the View toolbar; the effects will be immediately visible to the user. Feel free to explore them with this model to become familiar; sometimes it's convenient to switch to a different view style for visibility or easy selection of internal or hidden entities.

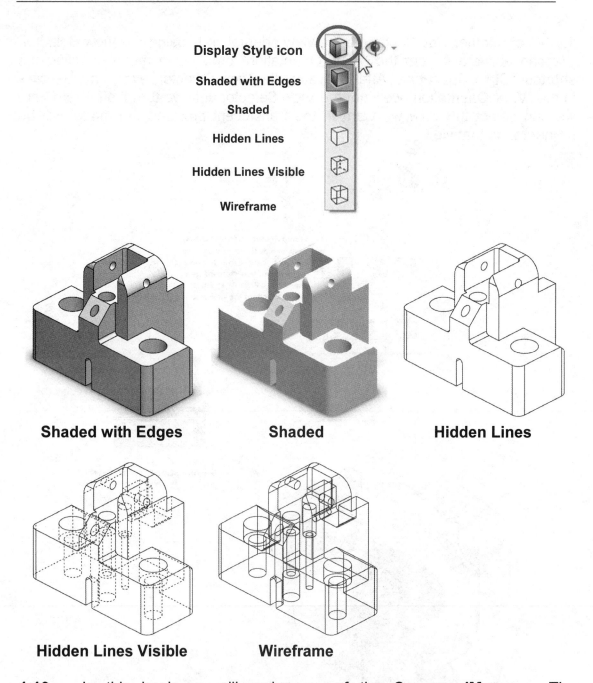

Display Style icon

Shaded with Edges

Shaded

Hidden Lines

Hidden Lines Visible

Wireframe

Shaded with Edges   **Shaded**   **Hidden Lines**

Hidden Lines Visible   **Wireframe**

1.10. - In this book we will make use of the **CommandManager**. The CommandManager consolidates many toolbars in a single location, selecting a toolbar's tab displays commands available, like Features, Sketch, Detailing, Assembly, Sheet Metal, etc. The CommandManager is a smart feature in SOLIDWORKS – depending on the task at hand, different toolbars will be available to the user – and is enabled by default.

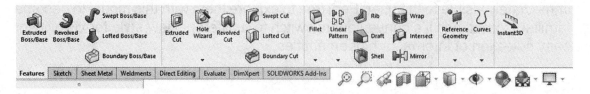

1.11. - To enable or disable the CommandManager, select the menu "**View, Toolbars, CommandManager**" or right-mouse-click any toolbar. You must have a document open to be able to turn the CommandManager on or off.

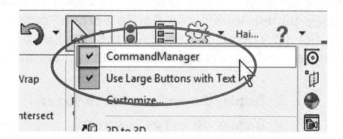

For clarity, the option "Use Large Buttons with Text" has been enabled; it can be activated by right-mouse-clicking anywhere in the CommandManager or a toolbar and selecting the option from the pop-up menu.

Models in SOLIDWORKS can be displayed as simple solid colors or with high quality images; depending on the video card (graphics accelerator) used, real time reflections and shadows can be displayed using "RealView" technology. In our book RealView graphics will be used sparsely to make images clear and easier to understand.

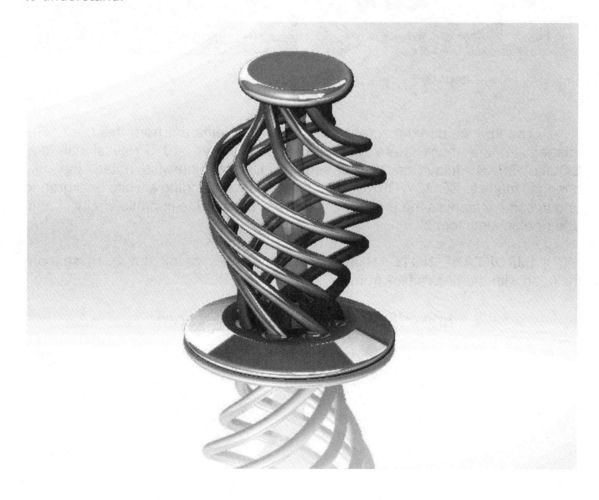

One option the user may wish to change is to display dimensions flat to screen; this way, regardless of the orientation of the part, the dimensions are easier to read. This option can be found in the menu "**Tools, Options**" under the "System Options" tab, in the "Display" section. The images in this book will use the "**Display dimensions/notes flat to screen**" option toggled ON for clarity. By default, SOLIDWORKS shows dimensions aligned to a plane, which could sometimes make it difficult to read dimensions and annotations.

☑ **Display dimensions flat to screen**
☑ **Display notes flat to screen**

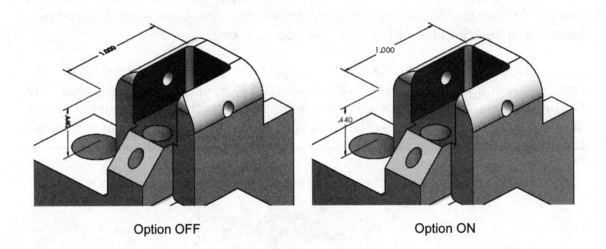

Option OFF   Option ON

The images on your screen *may* be slightly different from this book. The images in the book were made using Windows 10 Professional and SOLIDWORKS' default installation settings. Unless otherwise noted, the only changes made to SOLIDWORKS' default options were adding a white background and in some instances the preview colors were changed to improve clarity in print and/or electronic format.

IMPORTANT NOTE: High resolution images of all the exercises are included with the book's files from:

https://www.sdcpublications.com/

Document Templates

One way to be more efficient in SOLIDWORKS is by creating document templates. SOLIDWORKS' templates include many settings that are inherited when we create a part, a drawing or an assembly from them. Depending on the type of template used, it will include information about the measuring system, drafting standard, material properties, color preferences, image quality, annotations display, drawing notes display, even pre-defined geometry if needed.

When we first install SOLIDWORKS and define the system's units of measure, the default templates are created using the selected units and drafting standards. To create a new part template with different properties, first we need to create a new part from an existing template, modify its properties as needed, and finally save it as a template with a meaningful name. Optionally we can save our templates to an alternate location to better manage them.

1.12. – Select the "**New**" document icon from the main toolbar or the menu "**File, New…**". From the "New SOLIDWORKS Document" select the "Part" document and click OK to make a new part.

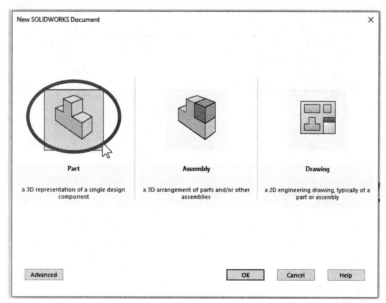

 If the "New SOLIDWORKS Document" window is not like the above image, select the "**Novice**" button in the lower left corner, or select an existing template.

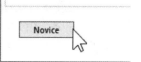

1.13. – After creating the first part, we need to change the document's units and drafting standard, then save it as a new template. Go to the menu "**Tools, Options, Document Options**" or select the "**Options**" command from the main toolbar and go to the "Document Options" tab.

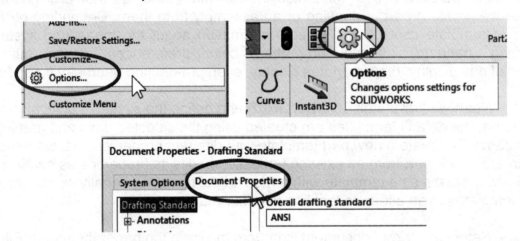

1.14. – From the "Overall drafting standard" drop down menu select "ANSI" if not already selected.

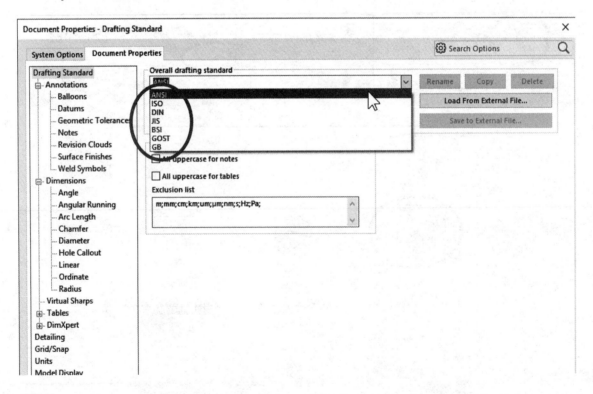

A drafting standard defines the display style of annotations in parts, drawings and assemblies including text notes, dimensions, arrow styles, tables, symbols, etc. The ANSI standard (American National Standards Institute) is predominantly used in the USA, whereas the ISO standard (International Organization for Standardization) is used in over 160-member countries.

1.15. – After setting the Drafting Standard, select the "Units" section, from the "Unit system" select "IPS (inch, pound, second) to change the units of measure to inches, pounds and seconds. In the same section, set the Length decimals to three significant numbers (.123) and click OK to finish.

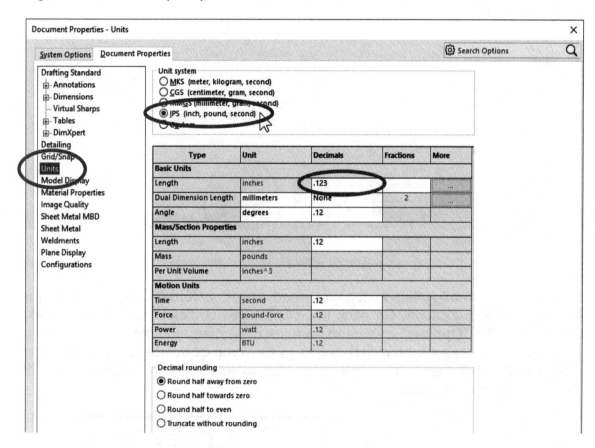

1.16. – Among other options which can be defined in a part's template is the component's material. At this point we'll assign a material to the template, later in the book material selection and mass properties will be covered in depth. In the FeatureManager, right-mouse-click in "Material" and select "Cast Alloy Steel" from the list of favorites to set the new material properties.

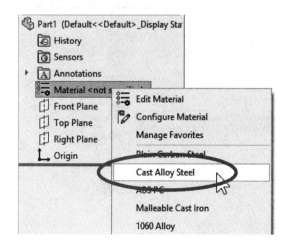 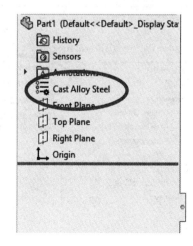

1.17. – Now that we have defined the drafting standard, units of measure and material for the part's template we need to save it, but before that we need to create a new folder for our custom templates and direct SOLIDWORKS to this folder. Using Windows Explorer create a new folder in your hard drive and rename it *'My SW Templates'*.

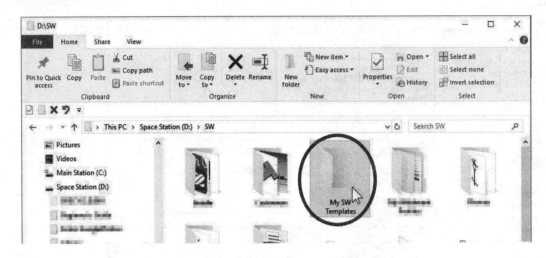

1.18. – After creating the folder, we need to tell SOLIDWORKS to look for templates in this location. Go back to SOLIDWORKS, select the menu "**Tools, Options**" or the "**Options**" command from the main menu toolbar. Go to "System Options, File Locations," the first option in the drop-down menu is "Document Templates". Click the "Add..." button and scroll to the newly created *'My SW Templates'* folder to select it.

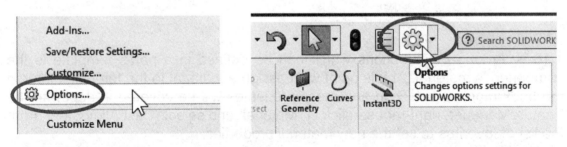

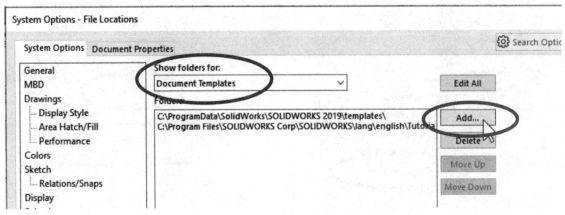

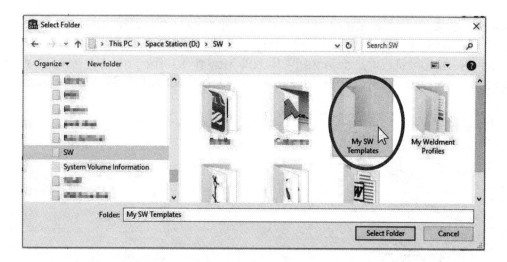

After selecting the folder click OK and confirm to add the selected folder to our templates search path. Depending on your administrative settings, Microsoft Windows may ask to confirm the change.

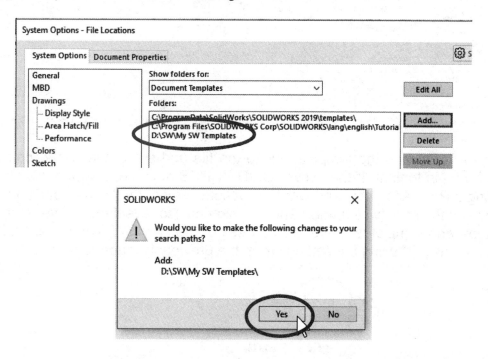

1.19. – Now we are ready to save our template. Select the menu "**File, Save As...**" or the drop-down command "**Save As**" from the "**Save**" command in the main toolbar.

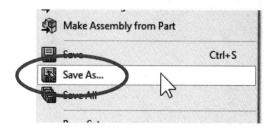

In the "**Save As**" dialog select the drop-down list in "Save as type:" and scroll to "Part Templates (*.prtdot)", immediately after selecting it the default templates location is automatically selected. Scroll down to the newly made '*My SW Templates*' and save the template as '*ANSI-Inch-Cast Alloy Steel*' in this folder.

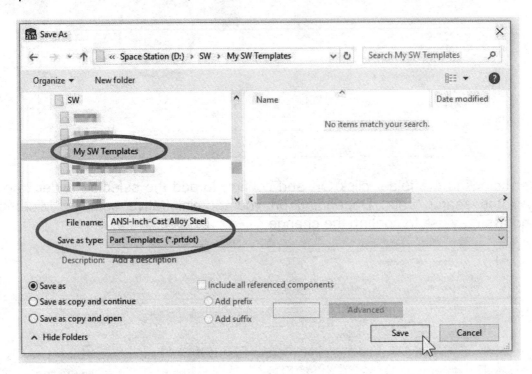

1.20. – After saving the template, close the file and select the "**New**" command from the main toolbar. If the "New SOLIDWORKS document" window is showing the single Part, Assembly and Drawing options, select the "**Advanced**" button in the lower left corner (which will change to "**Novice**") to reveal the different template locations and templates. Here we can select the "My SW Templates" tab and create a new part using the settings from the previously made template.

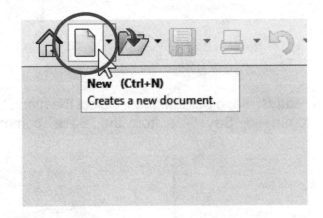

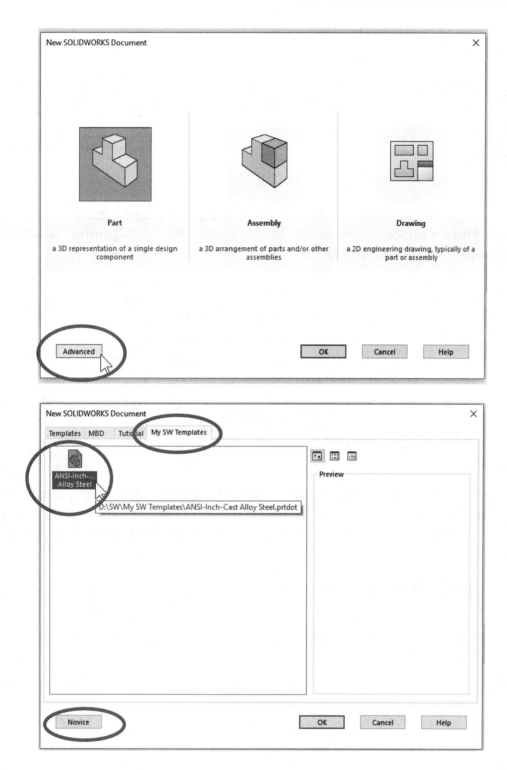

Select the 'ANSI-Inch-Cast Alloy Steel' template and click OK to make a new part based on this template. After creating it we can see the material option is already set to "Cast Alloy Steel" in the FeatureManager and the units are set to inch, pound, second in the lower right corner of the status bar, as had been defined in the template.

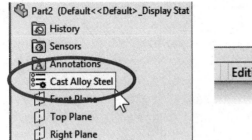

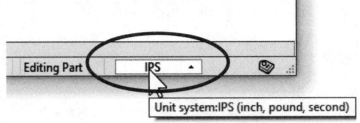

 By selecting the "IPS" button in the status bar we can quickly change to a different measurement system or go directly to the "**Options, Document Properties, Units**" section to modify other document's settings.

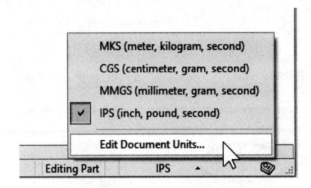

Keep in mind that we can change the unit system, material, drafting standard or any other document properties at any time with no consequences other than the modified setting, and we can always go back to the previous or a different setting as well.

With that said, let's design something...

Chapter 2: Part Modeling

The design process in SOLIDWORKS generally starts in the part modeling environment, where we create the different parts that make the product or machine being designed. These are later assembled to other parts; at that time the group of parts becomes an Assembly. In SOLIDWORKS, every component of the design will be modeled separately, and each one is a single file with the extension *.sldprt*. SOLIDWORKS is *Feature based software*; this means that the parts are created by incrementally adding features to the model. In the simplest of terms, features are operations that either add or remove material to a part; for example, extrusions, cuts, rounds, etc. There are also features that do not create geometry but are used as a construction aid, such as auxiliary planes, axes, etc.

This book will cover many different features to create parts, including the most commonly used tools and their options. Some features require a **Sketch** or profile to be created first; these are known as Sketched features. A Sketch is a 2D profile created on a plane or flat face that will be later used to generate a 3D feature. It is in the Sketch where most of the design information is added to the model, including dimensions and geometric relations between the different sketch elements and existing geometry. Examples of sketched features include Extrusions, Revolved features, Sweeps and Lofts, all of which will be covered in this section.

A 2D Sketch can be created only in a Plane or planar (flat) face. By default, every SOLIDWORKS Part and Assembly has three **default planes** (Front, Top and Right) and an Origin at their intersection. Most parts can be started in any one of these planes. It is not really critical which plane we use to start our designs; however, the plane's initial selection can potentially save us a little time when working in an assembly or when we start detailing the part in the detail drawing for manufacturing.

The initial planning that takes place before we start modeling a part is called the *Design Intent.* Basically, the Design Intent includes the general plan of how the part is going to be modeled, sort of a *"Step 1, Step 2, etc.,"* and how we anticipate (or guess) our model may change to accommodate possible design changes to fit other parts in an assembly or overall design needs. For example, we may choose to create a revolved feature instead of multiple extrusions, or the other way around, based on the particular needs of the task at hand.

SOLIDWORKS is a 3D parametric design software. *Parametric* means the models created are driven by parameters. These parameters are dimensions, geometric relations, equations, etc. When a parameter is modified, the 3D model is updated to reflect the changes. Good design practices are evident in how well the Design Intent and model integrity is maintained when parameters are modified. In other words, *the model updates predictably when we change the parameters.*

Notes:

The Housing

Notes:

When we start a new design, we must decide how we are going to model it. Remember that the parts will be made one feature or operation at a time. It takes a little practice to define the optimum feature sequence for any given part, but this is something that you will master once you learn to think of parts as a sequence of features or operations. To help you understand how to make the *'Housing'* part, we'll show a sequence of features. The order of some of these features can be changed, but always remember that sometimes we must make some features before others. For example, we cannot round a corner if there are no corners to round! A sequence will be shown at the beginning of each part, and the dimensional details will be given as we progress. In this part, some features could be combined to simplify the design, but instead we chose to repeat previously covered commands to practice those commands and improve retention.

In this lesson we will cover the following tools and features: creating various sketch elements, geometric relations and dimensions, Extrusions, Cuts, Fillets, Mirror Features, Hole Wizard, Linear, and Circular Patterns. For the *'Housing'* part, we'll follow the next sequence of features:

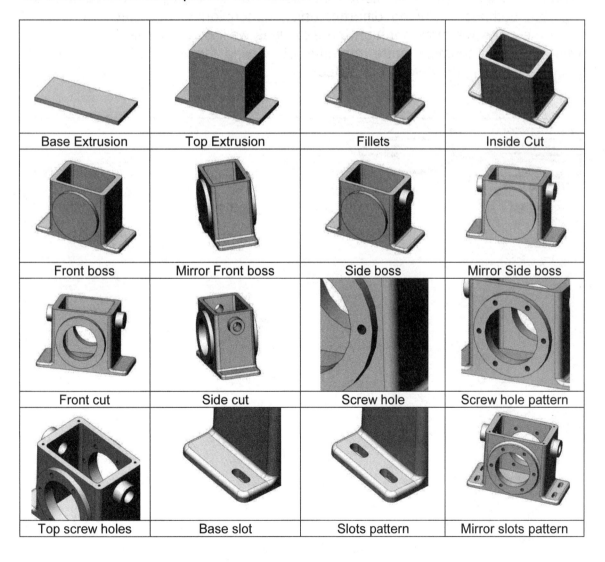

Base Extrusion	Top Extrusion	Fillets	Inside Cut
Front boss	Mirror Front boss	Side boss	Mirror Side boss
Front cut	Side cut	Screw hole	Screw hole pattern
Top screw holes	Base slot	Slots pattern	Mirror slots pattern

2.1. - The first thing we need to do after opening SOLIDWORKS is to make a **New Part** file. Go to the "New" document icon in the main toolbar and select it.

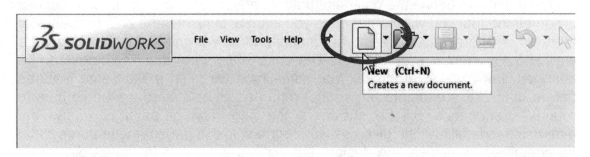

We are now presented with the New SOLIDWORKS Document dialog. For our first part we'll select the '*ANSI-Inch-Cast Alloy Steel*' template made in the previous section and click OK; if the dialog is not showing the different template tabs, select the "**Advanced**" button in the lower left corner. In this step is where we tell SOLIDWORKS that we want to create a Part using the selected template's properties. As previously mentioned, additional Part templates can be created with different settings, including different units, dimensioning standards, materials, colors, etc. As we move through the exercises we'll create additional templates with different document **units**, material, etc. Using the "Novice" option creates documents using the default templates.

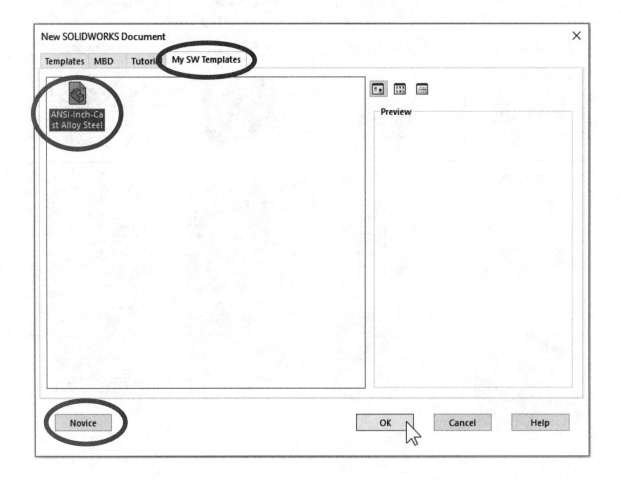

2.2. - Now that we have made a new Part file, we can start modeling the part, and the first thing we need to do is to make the extrusion for the base of the *'Housing'*. The first feature is usually one that other features can be added to or one that can be used as a starting point for our model. Select the "**Extruded Boss/Base**" icon from the CommandManager's Features tab (active by default). SOLIDWORKS will automatically start a new **Sketch**, and we will be asked to select the plane in which we want to start working. Since this is the first feature of the part, we will be shown the three standard planes (Front, Top, and Right). Remember the sketch is the 2D environment where we draw the profile before creating an extrusion, in other words, before we make it "3D."

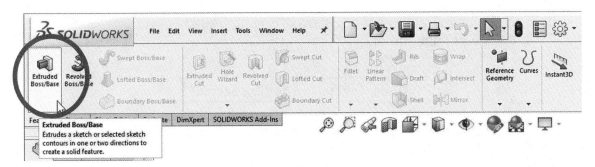

2.3. - For the *'Housing'* we'll select the *"Top Plane"* to create the first sketch. We want to select the *"Top Plane"* because we are going to start modeling the part at the base of the *'Housing'* and build it up as was shown in the sequence at the beginning of the chapter. Don't get too concerned if you can't figure out which plane to choose first when starting to model a part. At worst, what you thought would be a Front view may not be the front; this is for the most part irrelevant, as the user is able to choose the views at the time of detailing the part in the 2D drawing for manufacturing. Select the *"Top Plane"* from the screen using the left mouse button. Notice the plane is highlighted as we move the mouse to it. For the first feature, the view orientation will be automatically rotated to the selected plane.

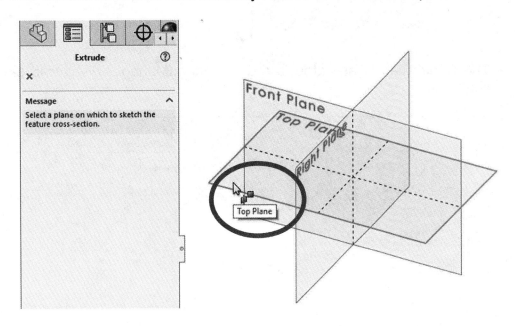

After selecting the *"Top Plane,"* a new **Sketch** is created and now we are working in the first sketch. It is in the Sketch environment where we will create the profiles that will be used to make features like Extrusions, Cuts, etc. SOLIDWORKS gives us many indications, most of them graphical, to help us know when we are working in the Sketch environment:

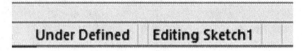

a) The **Confirmation Corner** is activated in the upper right corner and displays the Sketch icon in transparent colors.

b) The Status bar at the bottom shows "Editing Sketch" in the lower right corner.

Under Defined	Editing Sketch1

c) In the FeatureManager "*Sketch1*" is added at the bottom just under "*Origin.*"

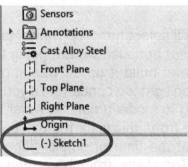

d) The part's Origin is projected in red.

e) The Sketch tab is activated in the CommandManager displaying sketch tools.

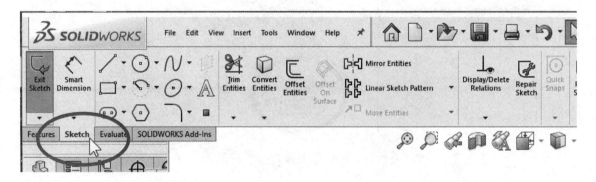

f) If the "Display Grid" option is activated, it will be displayed. This can be easily turned on or off *while in the Sketch environment*, by right-mouse-clicking in the graphics area and selecting the "Display Grid" command. This option can also be set in the template.

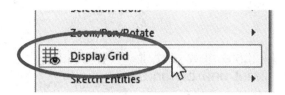

As the reader can see, SOLIDWORKS gives us plenty of clues to help us know when we are working in a sketch.

2.4. - Notice that when we make the first sketch, SOLIDWORKS rotates the view to match the plane that we selected. In this case, we are looking at the part from *above*. By default, the part is automatically oriented only when the part's first sketch is made to help the user get oriented. For subsequent sketch operations we can rotate the view manually using view orientation tools, or turn on the option to always rotate the view to be normal to the sketch plane in the menu "**Tools, Options, System Options, Sketch**" and turn on the option "**Auto-rotate view normal to sketch plane on sketch creation and sketch edit**." Turning this option ON will help new users get oriented in 3D. Feel free to turn it ON or OFF as you feel comfortable.

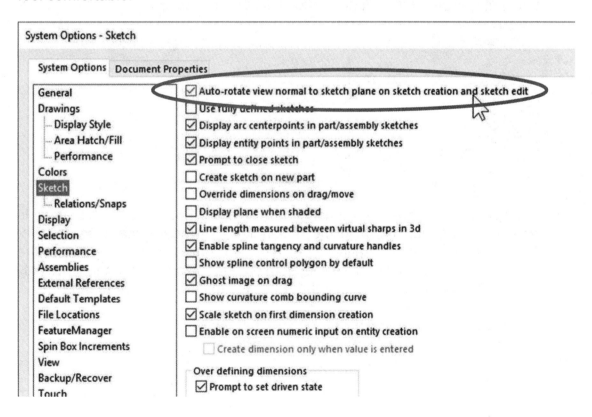

2.5. - The first thing we need to do in the first sketch is to draw a rectangle and center it about the origin; this way the part will be centered about the model's origin and planes making future operations easier.

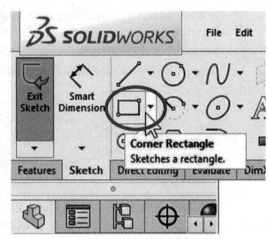

Every time we design a new part, if the part is symmetric about one or more planes, centering the part about the part's origin is a good idea, since the default planes can be used as a mirror plane simplifying the design and better maintaining the design intent.

Select the "**Rectangle**" tool from the "Sketch" tab *or* make a right-mouse-button-click in the graphics area to select it from the pop-up menu or click-and-drag with the right-mouse-button to select it in the Mouse Gestures. For this step, make sure we have the "Corner Rectangle" option selected in the PropertyManager of the rectangle's command.

Draw a Rectangle around the origin as shown. To draw it, left-mouse-click-and-drag (or click to start, move, click to end) to the opposite corner. Don't worry about the rectangle's size; we'll dimension it in a later step.

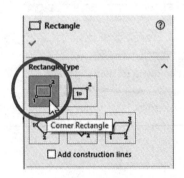
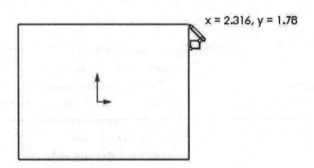

2.6. - Notice the lines are colored using the "Selected Entity" color as defined in the system options after finishing the rectangle. This means the lines are pre-selected immediately after creating them. You can unselect them by hitting the Escape (Esc) key; this will also de-select (turn off) the rectangle tool. Since we only need one rectangle in this sketch, hit the "Esc" key to finish the command.

 Default system colors can be changed in the menu "**Tools, Options, System Options, Colors**" or the **Options** icon.

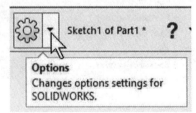

After drawing the rectangle, we may see automatically added geometric relations icons, and the rectangle may or not be shaded. These options can be toggled using the "**Hide/Show Items**" drop down menu and selecting "**View, Sketch relations**", and the "**Shaded Sketch Contours**" command from the Sketch Toolbar.

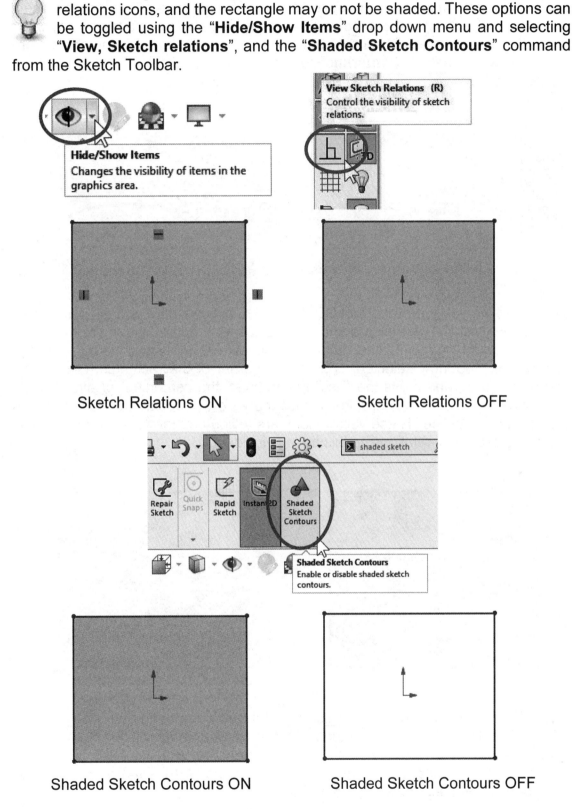

Sketch Relations ON   Sketch Relations OFF

Shaded Sketch Contours ON   Shaded Sketch Contours OFF

Throughout the book either option will be used as needed to improve visibility.

2.7. - After drawing the rectangle, we need to draw a "**Centerline**" from one corner of the rectangle to the opposite corner. The purpose of this line is to help us center the rectangle about the part's origin. (We'll also learn a faster way to do this in the next few steps.) From the Sketch tab select the "**Line**" command's drop-down arrow and select "**Centerline**" (or use the menu "**Tools, Sketch Entities, Centerline**").

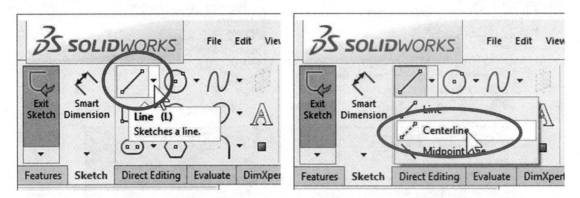

Every time we use a drawing tool SOLIDWORKS shows us feedback icons to let us know we will start or finish a line at an existing endpoint; when we move the cursor near an endpoint, line, edge, origin, etc. it will "snap" to it. With the "**Centerline**" tool now selected, click in any corner of the rectangle, click in the opposite corner, and press the "Esc" key to finish the centerline, or alternatively click-and-drag to draw a single line. Notice the yellow endpoint feedback icons as we add the centerline. These yellow icons are telling us that we are automatically capturing geometric relations, in this case coincident.

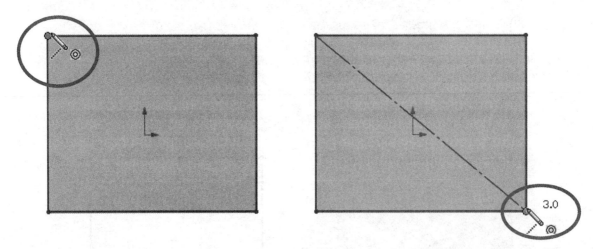

2.8. - Next we want to make the *midpoint* of the new centerline *coincident* to the part's origin; for this step we will add a "**Midpoint**" geometric relation between the centerline we just drew and the part's origin. Select the "**Add Relation**" command from the menu "**Tools, Relations, Add**," or from the "**Add Relation**" icon in the "**Display/Delete Relations**" drop-down command.

By adding this relation, the centerline's midpoint will be forced to coincide with the origin; this way the rectangle (and later the part) will be centered about the origin, and as we mentioned earlier, will better keep the design intent.

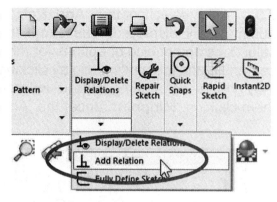

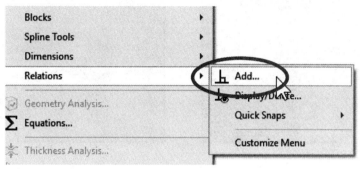

 "Add Relations" can also be configured to be available in the **Mouse Gestures** shortcuts.

 A word about the sketch right mouse button menu: The SOLIDWORKS shortcut menu can be customized to include the commands we use more frequently. To customize it, make a right-mouse-click in the graphics area, click in the double down arrow at the bottom of the menu to reveal the hidden commands, and select "Customize Menu" to reveal a checkbox next to each menu item. The checked commands will be displayed immediately; the unchecked commands will be revealed after pressing the double down arrow at the bottom.

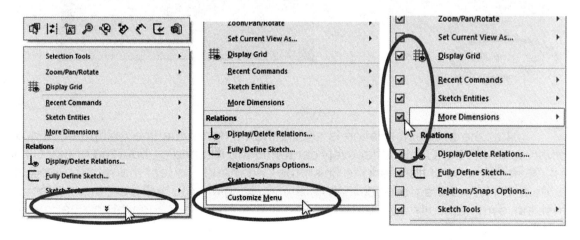

2.9. - After selecting the "**Add Relations**" command the PropertyManager is automatically displayed. The **Property-Manager** is the area where we make selections and choice of options for most commands. Select the previously made centerline and the part's origin by clicking on them in the graphics area (notice how they change color and get listed under the "Selected Entities" box). After selecting them click on "**Midpoint**" under the "Add Relations" box to add the relation.

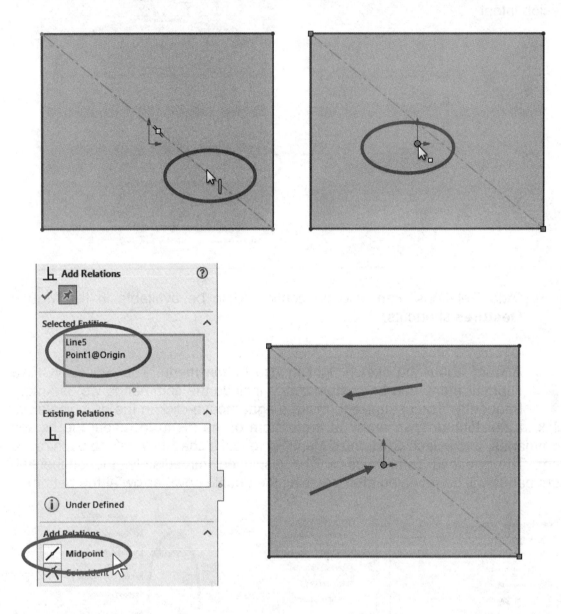

After the Midpoint relation is added the *center* of the line (midpoint) is now *coincident* with the Origin, effectively centering the rectangle about the origin. Click on OK (the green checkmark) to finish the command. To test the relation we just added, click-and-drag one corner of the rectangle. You will see the rectangle resizing symmetrically *and* centered about the origin because of the geometric relation we added.

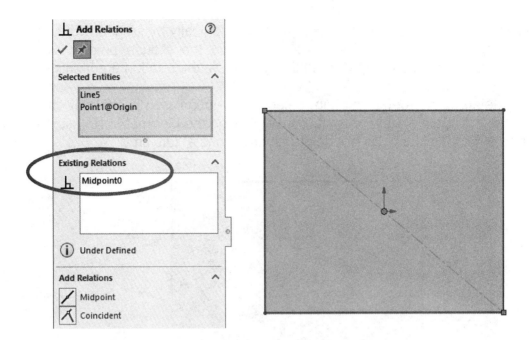

2.10. - What we just did is we *manually* added a **geometric relation**; we also added some geometric relations *automatically* when we drew the rectangle and the centerline in the previous step. SOLIDWORKS allows us to view the existing relations between sketch elements graphically by going to the menu "**View, Hide/Show, Sketch Relations,**" or from the "**Hide/Show Items**" drop down icon in the graphics area. Use this option as needed to toggle the relations' visibility ON or OFF.

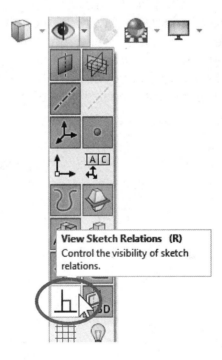

2.11. - The geometric relations are represented graphically by small icons next to each sketch element. Notice that when we move the mouse pointer over a geometric relation icon, the entity or entities that share the relation are highlighted.

 To delete a geometric relation, select the relation icon in the screen and press the "Delete" key, or right-mouse-click on the Geometric Relation icon and select "Delete." (Do not delete any relations at this time!)

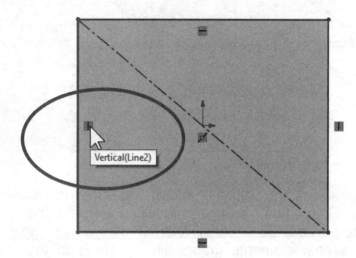

The **Sketch Origin** ⌐→ shows the sketch's Horizontal direction (short red arrow) and Vertical direction (long red arrow). It's important to be aware of the sketch's "orientation" when working in 3D, since we may be looking at the part in a different orientation, and the "sketch vertical" may not necessarily be "up" on the screen. This is a visual indicator to let us know where the vertical direction ("up") is in the sketch regardless of which way we are looking at it. In SOLIDWORKS we have the following basic types of geometric relations for sketch entities:

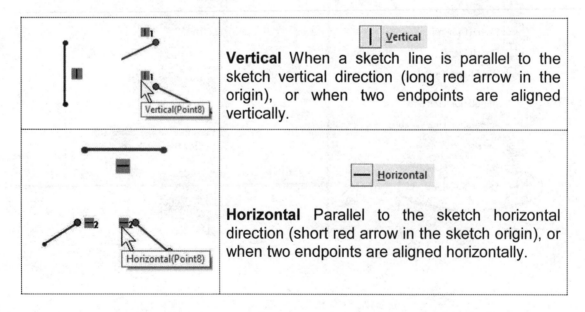

	Vertical When a sketch line is parallel to the sketch vertical direction (long red arrow in the origin), or when two endpoints are aligned vertically.
	Horizontal Parallel to the sketch horizontal direction (short red arrow in the sketch origin), or when two endpoints are aligned horizontally.

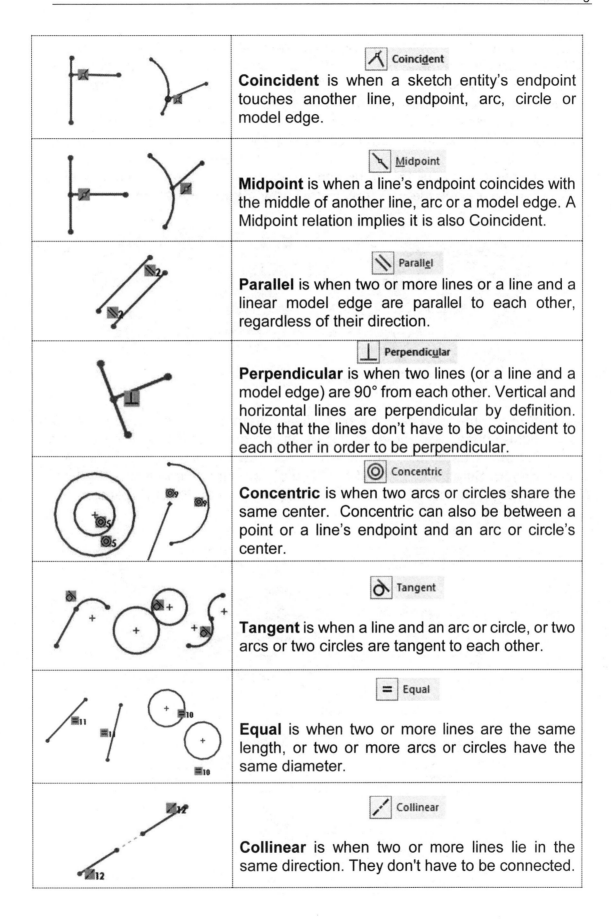

	Coincident **Coincident** is when a sketch entity's endpoint touches another line, endpoint, arc, circle or model edge.
	Midpoint **Midpoint** is when a line's endpoint coincides with the middle of another line, arc or a model edge. A Midpoint relation implies it is also Coincident.
	Parallel **Parallel** is when two or more lines or a line and a linear model edge are parallel to each other, regardless of their direction.
	Perpendicular **Perpendicular** is when two lines (or a line and a model edge) are 90° from each other. Vertical and horizontal lines are perpendicular by definition. Note that the lines don't have to be coincident to each other in order to be perpendicular.
	Concentric **Concentric** is when two arcs or circles share the same center. Concentric can also be between a point or a line's endpoint and an arc or circle's center.
	Tangent **Tangent** is when a line and an arc or circle, or two arcs or two circles are tangent to each other.
	Equal **Equal** is when two or more lines are the same length, or two or more arcs or circles have the same diameter.
	Collinear **Collinear** is when two or more lines lie in the same direction. They don't have to be connected.

2.12. - Once we have added the "Midpoint" geometric relation, the next step is to define the size of the rectangle by adding dimensions. Click with the right mouse button in the graphics area and select "**Smart Dimension**" from the pop-up menu or select the "**Smart Dimension**" icon from the Sketch toolbar. Notice the cursor changes by adding a small dimension icon next to it. This icon will let us know the Smart Dimension tool is selected.

 To avoid visual clutter in the screen we can turn off the geometric relation icons using the menu "**View, Hide/Show, Sketch Relations**."

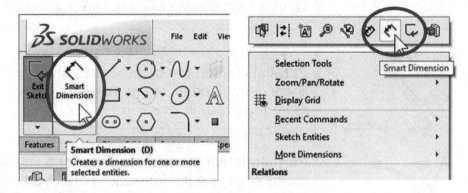

Smart Dimension can also be activated from the Mouse Gestures (default right-mouse-click-and-drag up) or assign a shortcut key using the menu "**Tools, Customize, Keyboard**." Feel free to customize a shortcut that works for you. Scroll down to find the "Tools" category, "Dimensions" and select "Smart" Type "D" in the "Shortcuts" column to assign it to Smart Dimension.

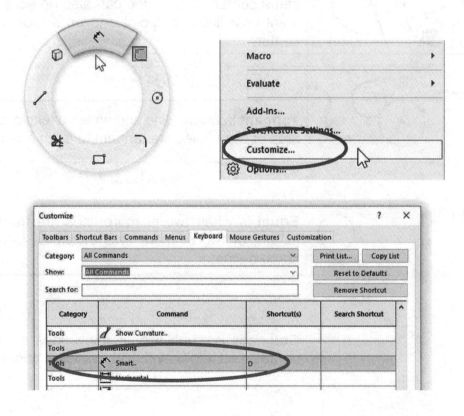

2.13. - Adding dimensions in SOLIDWORKS is simple and straightforward. After activating the Smart Dimension tool, click to select the right (or left) vertical line in the sketch and then click next to it to locate the dimension. SOLIDWORKS will show the "Modify" dialog box, where we can enter the 2.625″ dimension and press "Enter" or click the green checkmark.

Repeat with the top horizontal line and enter a dimension of 6″. As soon as the dimension value is entered, the geometry updates to reflect the correct size.

 If, for some reason, your document is set to a different measuring units system, you can override the default units by adding "*in*" or ″ at the end of the value in the Modify dialog box. If this is the case, type **2.625 in** *or* **2.625″** for the vertical dimension, and **6 in** *or* **6″** for the horizontal dimension to override the document's units to inches.

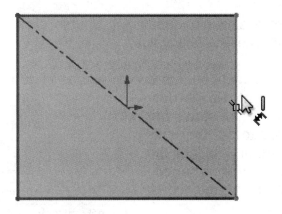

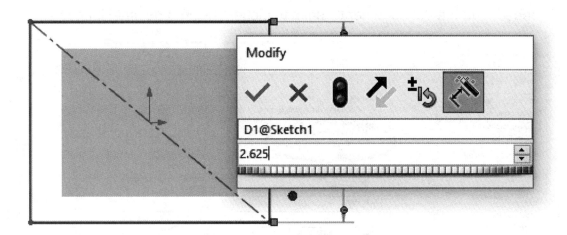

 Another way to select the units of measure is to enter the numerical value, and from the newly revealed pop-up "Units" menu directly under the value box, select the units of choice.

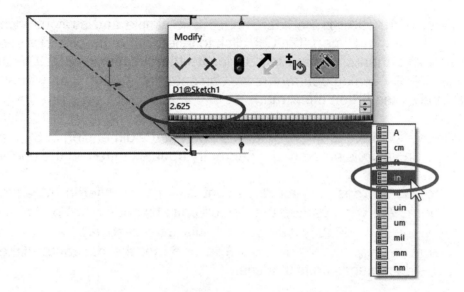

 To change a dimension after adding it, double click on it to display the "Modify" box and enter the new value.

 To change the document's units we can either 1) go to the menu "**Tools, Options, Document Properties, Units**," or 2) click in the status bar in the lower right corner to set the units from the quick pick menu, or 3) launch the "**Edit Document Units**…" options page to change units, decimal places, dual dimension units, etc.

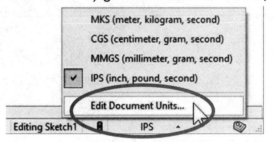

If needed, change the document's units to inches, using 3 decimal places.

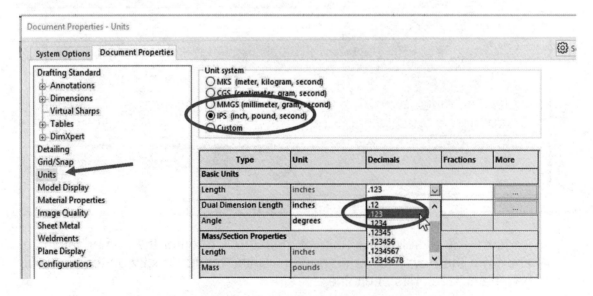

Most exercises in this book are done using inches with three decimal places unless otherwise noted.

After dimensioning the lines, notice the sketch lines changed from Blue to Black. This is one way SOLIDWORKS lets us know that the geometry is *defined*, meaning that we have added enough information (dimensions and/or geometric relations) to completely define the geometry in the sketch. The status bar also shows "Fully Defined" in the lower right corner. This is the preferred state before creating a feature, since there is no information missing and the geometry has been accurately and completely described.

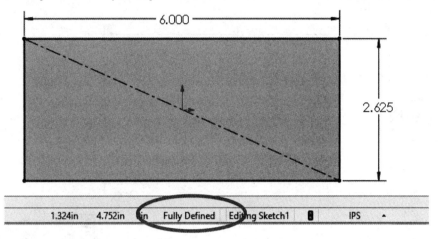

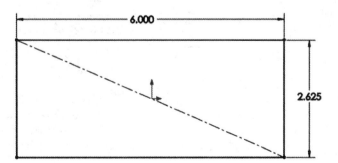

More about the sketch's state. A sketch can be in one of several states; these are the three main states:

- **Under Defined**: (BLUE) Not enough dimensions and/or geometric relations have been provided to completely define the sketch. Sketch geometry is blue and lines/endpoints can be dragged with the left mouse button.

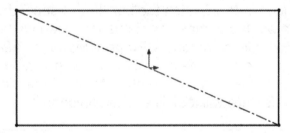

- **Fully Defined**: (BLACK) The Sketch has all the necessary dimensions and/or geometric relations to completely define it. *This is the desired state*. Fully defined geometry is black.

- **Over Defined/Unsolvable**: (RED/YELLOW) Redundant and/or conflicting dimensions and/or geometric relations have been added to the sketch. If an over-defining dimension or relation is added, SOLIDWORKS will immediately warn the user. If an over-defining geometric relation (or dimension) is added, delete it immediately or use the menu "**Edit, Undo**" or select the "Undo" icon. If an over-defining dimension is added, the user will be offered an option to cancel it.

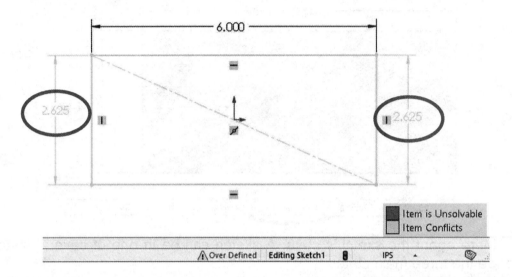

2.14. - Now that the sketch is fully defined, we will create the first solid feature of the *'Housing'*; this is where we go from the 2D Sketch to a 3D feature. Select the Features tab in the CommandManager and click in the "**Extrude**" icon or click in the "**Exit Sketch**" icon in the Sketch toolbar. In the second case, SOLIDWORKS remembers that we wanted to make an Extrusion when we first started this feature and displays the Extrude command's PropertyManager after exiting the sketch. Notice that the first time we create a 3D feature in a new part, SOLIDWORKS automatically changes the part's orientation to an Isometric view and gives us a preview of what the feature will look like when finished.

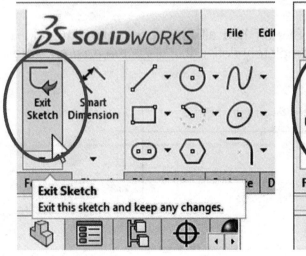
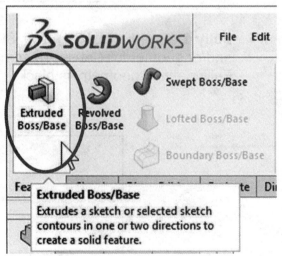

We will make the first extrusion 0.25″ thick. To do this, use the default option "Blind" for the extrusion's end condition, and enter the 0.25″ dimension indicated in the "**Extrude**" command. To create the extruded 3D feature, select the OK button or press the "Enter" key.

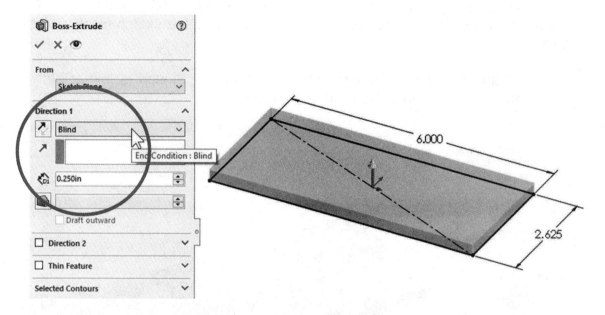

2.15. - After the first extrusion is completed, notice that the "Boss-Extrude1" 3D feature has been added to the FeatureManager. The confirmation corner is no longer showing the Sketch icon, and the status bar now reads "Editing Part," alerting us that we are now editing the 3D solid part and not a 2D sketch.

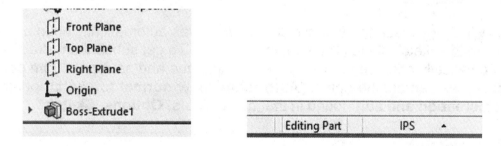

 Expanding the "Boss-Extrude1" feature in the FeatureManager by clicking on the arrow ▶ on the left side of it, we see that "Sketch1" has been absorbed by the "Boss-Extrude1" feature.

43

2.16. - The second 3D feature will be like the first one, but with different dimensions. To create the second extrusion, we need to make a new sketch. We can either select the **Extruded Boss/Base** in the Features tab or the **Sketch** icon in the Sketch tab; SOLIDWORKS gives us a message in the PropertyManager asking us to select a Plane or a planar (flat) face in the 3D model to add the sketch to it. Select the top face of the previous extrusion to add the sketch for the next feature.

 If a Plane or a flat face is pre-selected, the new Sketch is immediately created in that Plane or face without a message.

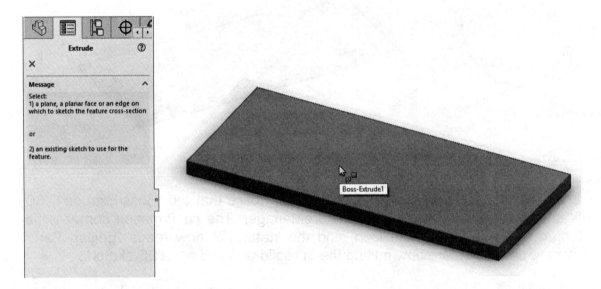

2.17. - When we made the first sketch, the view was automatically oriented to be normal to the sketch plane (Top view in this case). To get subsequent sketches to be automatically oriented normal to the sketch plane and saving us from doing it manually, we can set the option "**Auto-rotate view normal to sketch plane on sketch creation and edit**" found in the menu "**Tools, Options, Sketch**."

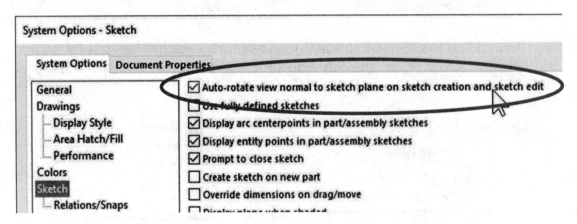

If you are not looking at the part from the Top view after selecting the top face, or chose *not* to turn on the previous option, we have to manually change to a **Top View** to see the part from the top using the "**View Orientation**" icon or the default keyboard shortcut "**Ctrl+5**," as indicated in the tooltip. Notice that when we place the mouse over a view orientation, we get a pop-up preview of it. In SOLIDWORKS we are free to work in any orientation we like as long as we can see what we are doing. Re-orienting the view helps us get used to 3D in a more familiar way by looking at our models in 2D.

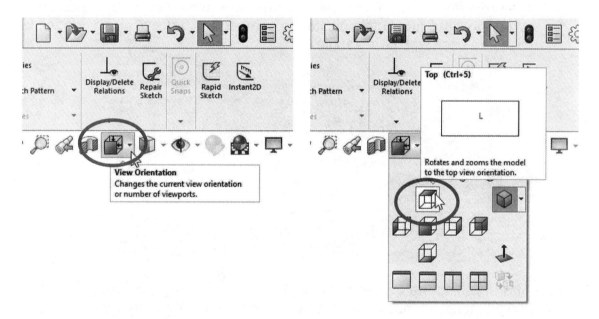

2.18. - Notice that after the second sketch is started, the Sketch tab in the CommandManager is automatically selected. For the second sketch we'll use the "**Center Rectangle**" command; by making the rectangle using this option, the rectangle will be automatically centered about the start point, in our case, the part's origin. Click in the Rectangle's tool drop down menu in the Sketch tab and select "**Center Rectangle**"; if you selected the rectangle as in the previous step, you can select the "**Center Rectangle**" option from the Rectangle's PropertyManager.

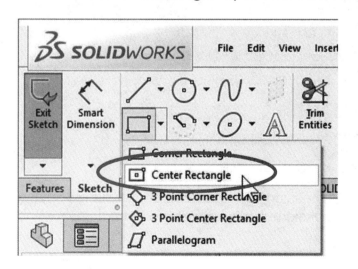
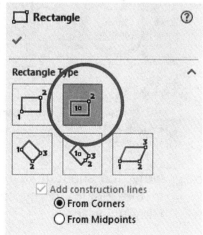

To make a Center Rectangle click in the Origin first, then on the top edge of the first extrusion (or the bottom; it really doesn't matter) to finish it. Notice the yellow Coincident icon as the pointer is in the origin and then on the model edge. By doing the rectangle this way we automatically add coincident relations to the origin and the top edge. The "**Center Rectangle**" command saves us from adding the centerlines and midpoint relations making the rectangle centered about the origin in a single operation.

 In the different types of options available to draw a rectangle, we can see numbers; these numbers indicate the number and order of clicks needed to complete each type of rectangle.

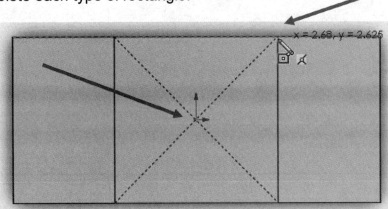

2.19. - After finishing the centered rectangle, select the "**Smart Dimension**" tool (toolbar, right mouse menu, Mouse Gesture or keyboard shortcut), and dimension the rectangle 4″ wide by selecting the top (or bottom) line and locating the dimension above. Adding this dimension will fully define the sketch since we had added automatic geometric relations to the origin and the top edge when the rectangle was created, defining its center and height.

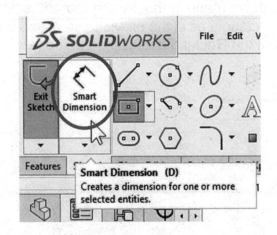 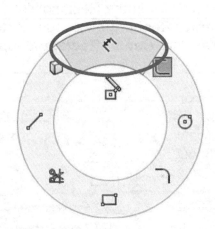

 Note: If a shortcut keyboard is assigned to a command, it will be displayed in the tooltip. In my case I assigned the keyboard shortcut ("D") to the Dimension tool.

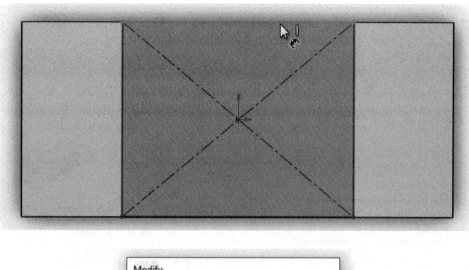

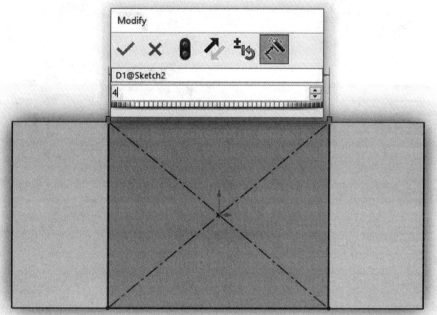

2.20. - Now that the second sketch is fully defined, we are now ready to make the second extruded feature. Select the "**Extrude**" command in the Features tab of the CommandManager (or "**Exit Sketch**" if you initially selected "**Extrude**") and make the extrusion 3.5″ high. Also, notice the part does not rotate to show an isometric view as it did in the first extrusion—we must rotate the view ourselves to see a preview. From the Standard Views icon, select the **Isometric** view ("Ctrl+7" shortcut or Mouse Gesture) to see the preview of the second extrusion. Click OK to complete the command.

 As opposed to automatically rotating the part when we add a new sketch, there is no option to change the part's orientation when making a new 3D feature.

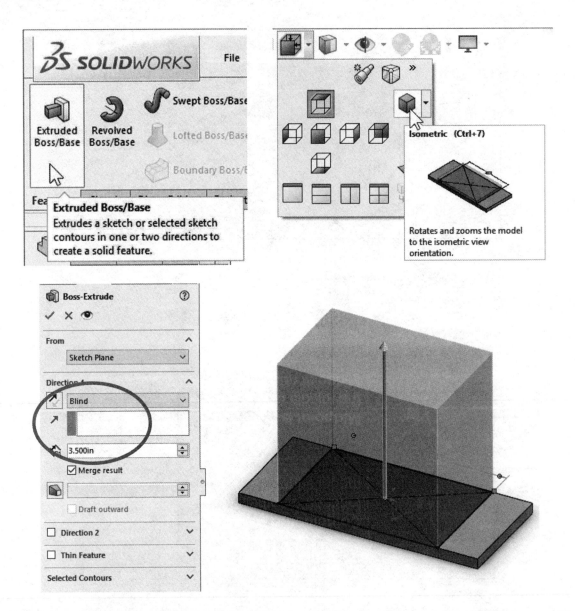

2.21. - The next step is to round the edges of the two extrusions. To do this, we will select the "**Fillet**" command. The Fillet is an applied feature, which means we don't need a sketch to create it, and it's applied directly to the solid model. From the Features Tab of the CommandManager select the "**Fillet**" icon.

Use the "Manual" mode and the "**Constant Size Fillet**" option. Change the fillet's radius to 0.25″ and select the eight vertical corners indicated in the preview. SOLIDWORKS highlights the model edges when we place the cursor on top of them to let us know what we'll be selecting. If an edge to be selected is not visible, rotate the model using the menu "**View, Modify, Rotate.**"

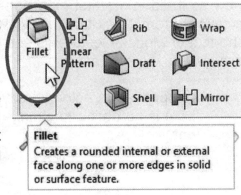

Click-and-drag in the graphics area to rotate the part. Another way to rotate the model is by holding down the middle mouse button (scroll wheel), and dragging in the graphics area, or using the arrow keys. Click OK when the eight edges are selected to complete the command.

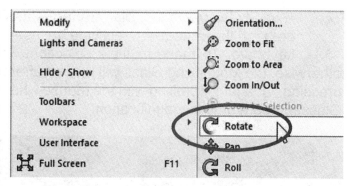

 To use the mouse wheel's button to rotate the model, you *may* have to configure the middle mouse button to change it from the default "Scroll" mode to "Middle Mouse Button."

 While adding the fillet, if a model edge or face is mistakenly selected, simply click on it again to de-select it.

 Select the "Full preview" option to see the resulting fillets in the graphics area as we select the edges.

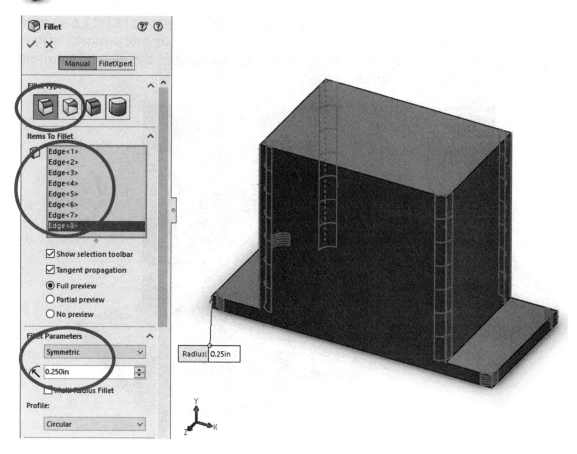

A handy feature is the "**Magnifying Glass**" to selectively zoom in only one area of the model. To activate it, use the default shortcut "G" in the keyboard. To make multiple selections with it, hold down the "Ctrl" key; otherwise, the Magnifying Glass will turn off after making the first selection or after pressing "G" again. Scrolling with the mouse wheel will zoom inside the Magnifying Glass for more or less magnification.

 Alternate Option: Instead of selecting the **Fillet** command first and then selecting the edges to round, we can select one or more model edges (using the Ctrl key while selecting) and *then* select the Fillet command from the fly-out features toolbar.

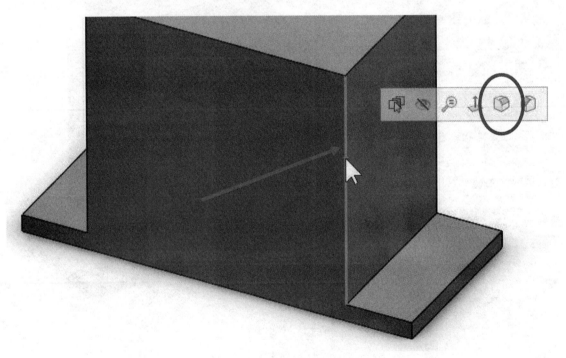

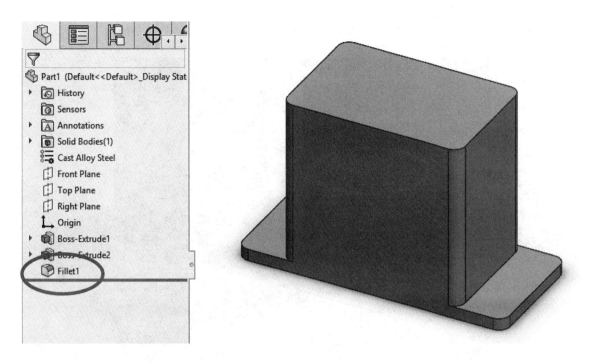

2.22. - Repeat the **Fillet** command to add a 0.125″ radius fillet at the base of the *'Housing'* but instead of edges we'll select the faces indicated. When we select a face, SOLIDWORKS rounds all the edges connected to it. Click OK to continue.

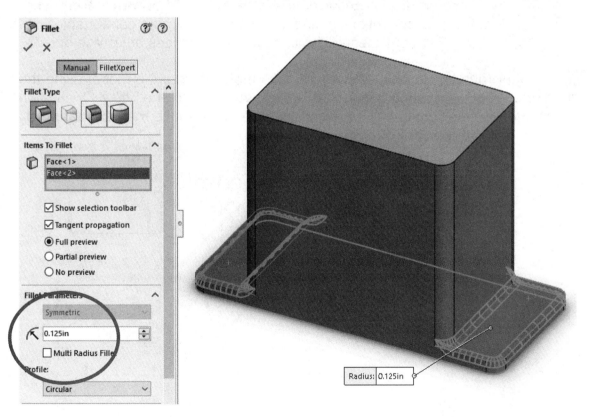

Notice how the new fillets are blended with the previous vertical fillets.

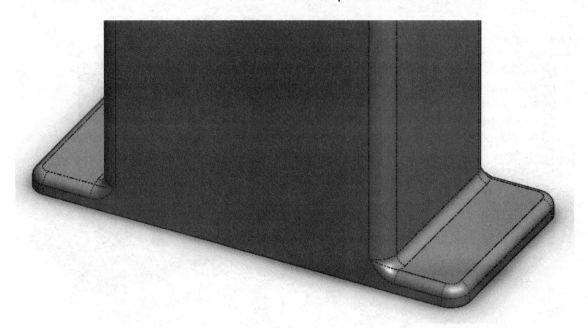

 We can change the appearance of tangent edges (the edges where two tangent faces meet) by selecting the menu "**View, Display**" and selecting the tangent edge display option desired: **Visible**, as **Phantom** or **Removed**. Explore the different options to find the one you feel more comfortable with. In this book Phantom lines will be used for clarity unless otherwise noted.

2.23. - After rounding the edges, we'll remove material from the model using the "**Extruded Cut**" command. Switch to a **Top View** using the View Orientation toolbar, Mouse Gesture or keyboard shortcut (Ctrl+5) to make the next feature.

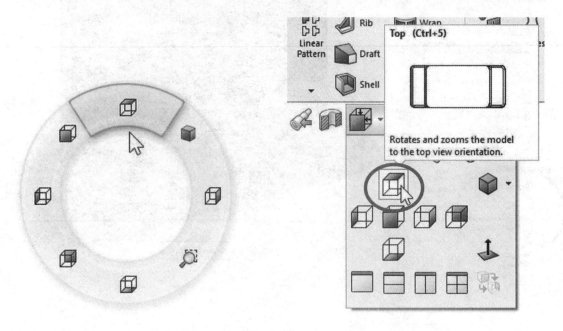

Select the "**Extruded Cut**" command from the Features tab or the menu "**Insert, Cut, Extrude…**"; you will be asked to select a plane or flat face just as with the "**Extruded Boss**." Select the topmost face to create a new Sketch in it and using the "**Corner Rectangle**" tool from the now visible Sketch toolbar, the menu "**Tools, Sketch Entities, Corner Rectangle**," or the sketch mouse gestures. Draw a rectangle *inside* the top face.

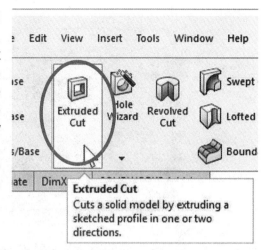

Extruded Cut
Cuts a solid model by extruding a sketched profile in one or two directions.

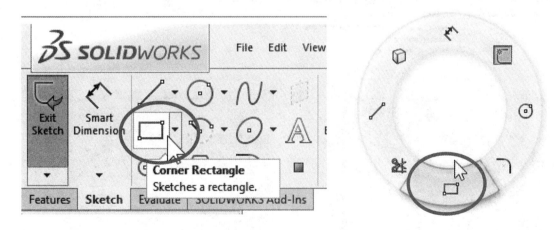

 If the option "**Auto-rotate view normal to sketch plane on sketch creation and sketch edit**" is ON, the model will automatically rotate to a Top View after selecting the top face without having to change the model's view.

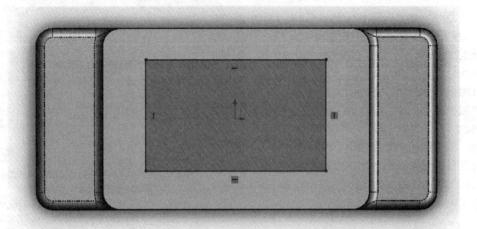

2.24. - To add the dimensions, select the "**Smart Dimension**" tool; we can add dimensions from sketch geometry to model edges simply by selecting them. Select a Sketch line, click on a model edge parallel to it, and finally click to locate the dimension in the screen. When asked, enter a 0.375″ dimension. Repeat to add the other three dimensions and make them 0.375″. Feel free to turn off the Sketch Relations.

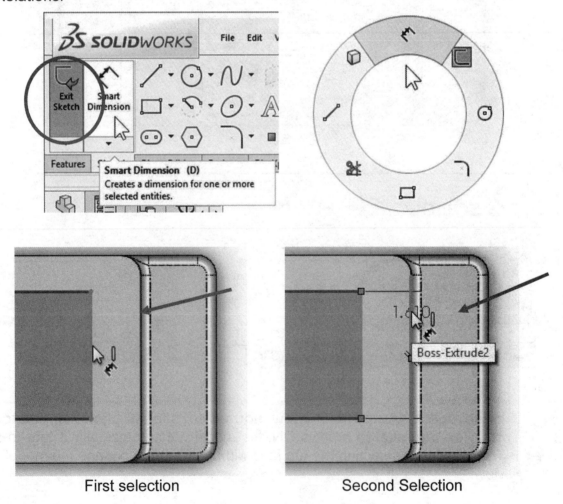

First selection   Second Selection

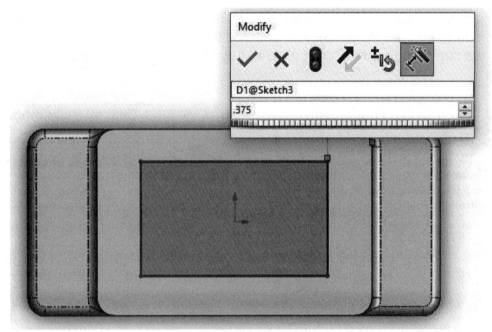

Locate and enter dimension value

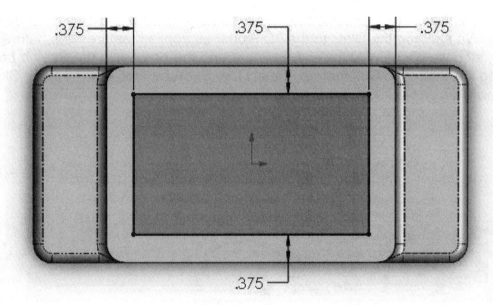

Finished dimensions between the sketch lines and the four surrounding edges

If needed, switch to a "**Hidden Lines Removed**" mode from the **View Style** icon to see the model without shading to facilitate visualization. Remember we can change the display style at any time during modeling.

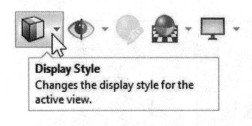

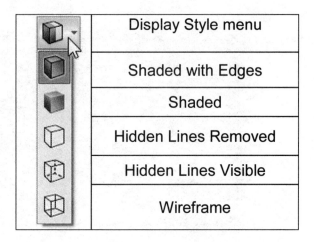

	Display Style menu
	Shaded with Edges
	Shaded
	Hidden Lines Removed
	Hidden Lines Visible
	Wireframe

2.25. - For this feature, we will round the corners in the sketch using a **Sketch Fillet**. We can add the fillets to the 3D model as applied features like before, but in this step we chose to show you how to round the corners in the Sketch *before* making the "Extruded Cut" feature. Select the "**Sketch Fillet**" icon from the Sketch toolbar.

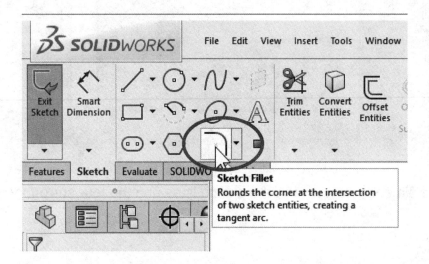

Set the fillet radius to 0.150", and click on the corners of the sketch lines as indicated to round them. Notice the preview in the screen. After clicking on all 4 corners, click OK to finish the Sketch Fillet command. Adding multiple fillets at the same time results in only one dimension being added; the reason is that SOLIDWORKS adds an equal relation from each fillet to the dimensioned one.

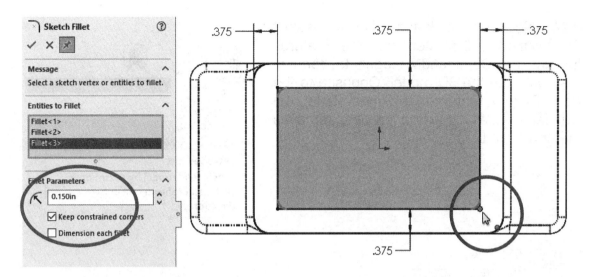

After adding the sketch fillets, if the sketch relations are visible, we can see the number of geometric relations are starting to clutter the screen. To help clear up the screen, we can turn geometric relations OFF. Select the menu **"View, Hide/Show, Sketch Relations,"** or from the **"Hide/Show Items"** drop down icon, turn off "View Sketch Relations." In this case the keyboard shortcut "R" has been added to the "View Sketch Relations" command.

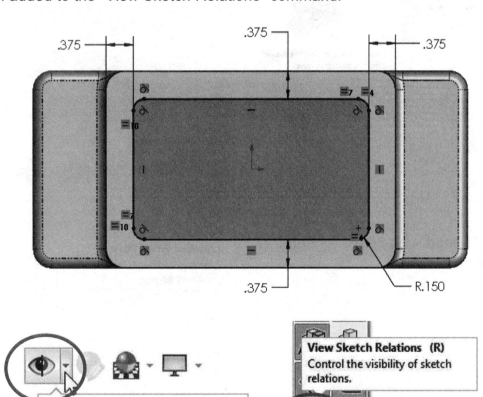

2.26. - Now that our sketch is finished, select the "**Extruded Cut**" icon from the Features tab in the CommandManager to remove material from the 3D model. Opposite to the Boss Extrude feature that adds material, the Cut feature, as its name implies, removes material from the model.

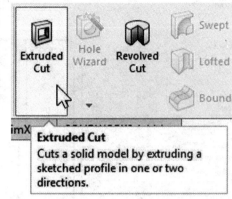

Extruded Cut
Cuts a solid model by extruding a sketched profile in one or two directions.

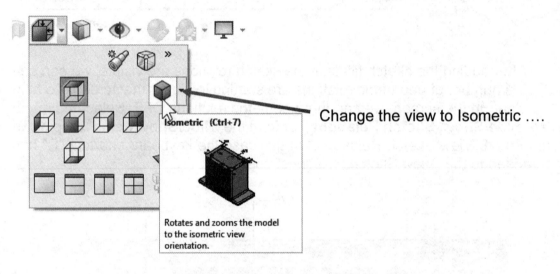

Isometric (Ctrl+7)

Rotates and zooms the model to the isometric view orientation.

Change the view to Isometric ….

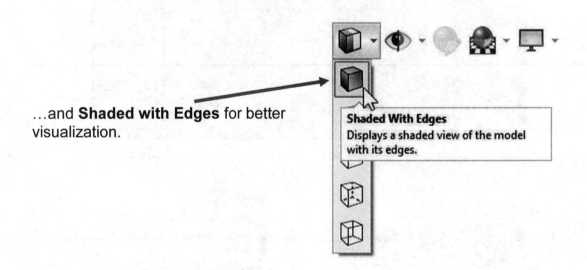

…and **Shaded with Edges** for better visualization.

Shaded With Edges
Displays a shaded view of the model with its edges.

Make the cut 3.5″ deep and click on OK to finish the cut.

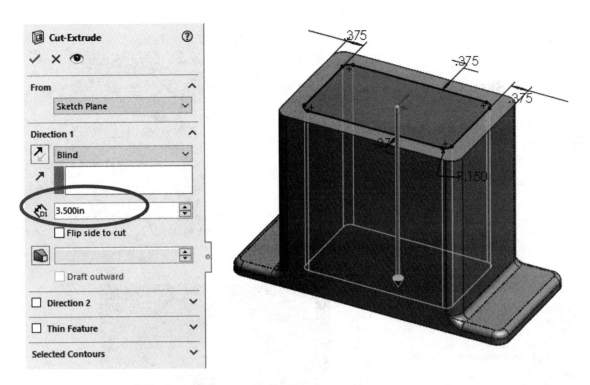

 Features can be **Renamed** in the FeatureManager for easier identification. To rename a feature, slowly double-click the feature, or select it and press F2, and type a new name (just like renaming files in Windows Explorer).

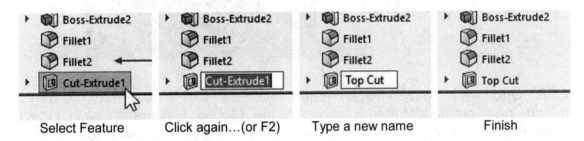

| Select Feature | Click again...(or F2) | Type a new name | Finish |

2.27. - In the next step we will add a round boss to the front of the *'Housing'*. Switch to a **Front View** using the "View Orientation" toolbar, Mouse Gestures or shortcut Ctrl+1.

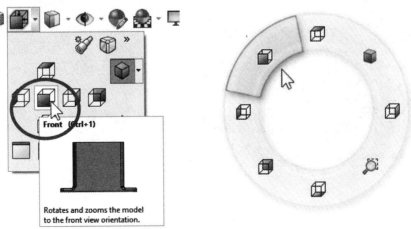

Select the "**Sketch**" icon from the Sketch tab in the CommandManager and click in the front face, or the reverse order: select the face first, and then click in the "**Sketch**" icon.

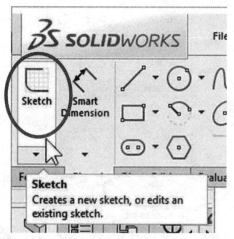

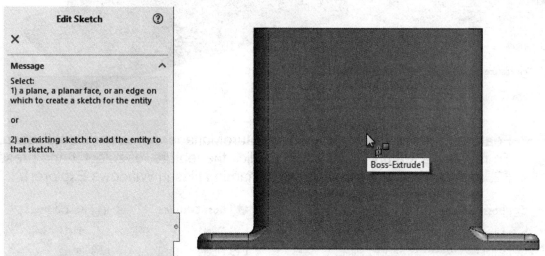

2.28. - Once we have the sketch created select the "**Circle**" tool from the Sketch tab in the CommandManager or using the Mouse Gestures. Draw a circle approximately as shown; click near the middle of the part to locate the center of the circle, and click again to set its size. (You can also click-and-drag from the center to draw the circle.) Don't worry about the size; we'll dimension it in the next step.

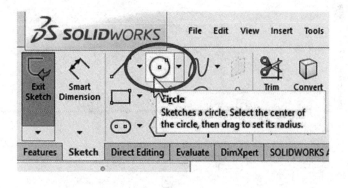

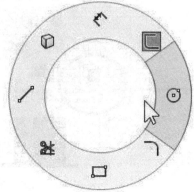

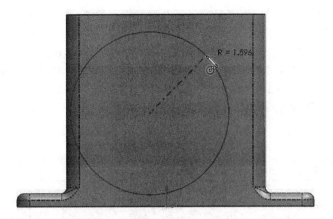

To define the circle's location, select the "**Smart Dimension**" tool. Click on either the center of the circle or its perimeter, and then on the top edge of the *'Housing'* locate the dimension and enter the value of 1.875″. For the Diameter, select the circle and locate the 3.25″ diameter dimension as shown.

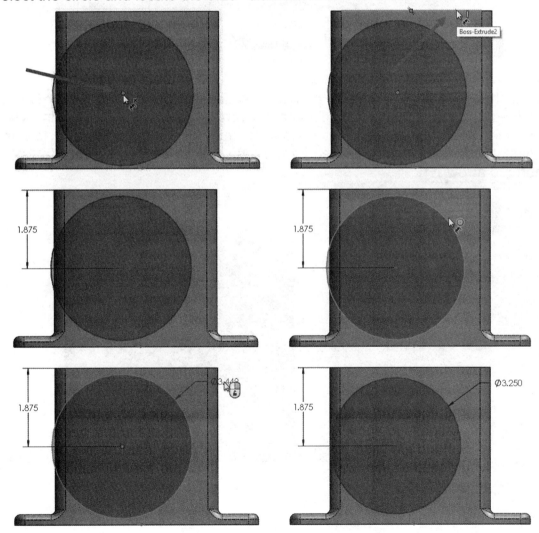

2.29. - To make the circle horizontally centered in the part, we will manually add a **Vertical Relation** between the center of the circle and the part's origin. SOLIDWORKS allows us to align sketch elements to each other or to existing model geometry (edges, faces, vertices, planes, origin, etc.). From the "**Display/Delete Relations**" drop-down icon, select "**Add Relation**" or use the menu "**Tools, Relations, Add**."

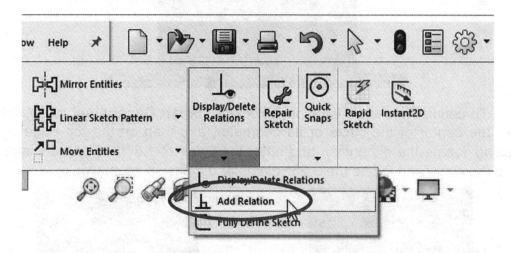

Select the circle's center (not the perimeter!) and the origin. Click on "**Vertical**" to add the relation and make the circle's center vertical in relation to the origin. When done click **OK** to finish. Adding this relation fully defines our sketch. Note the origin's Vertical direction identified by the long red arrow. (The horizontal is the short red arrow.)

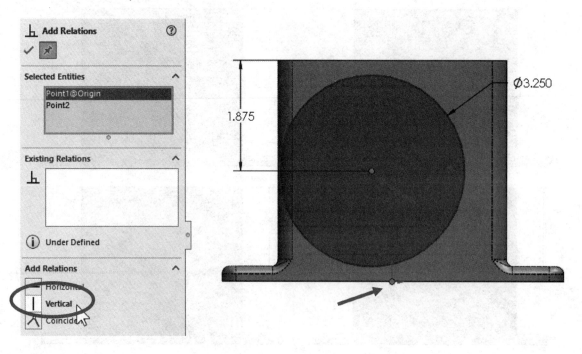

Alternatively, we can add the vertical relation by pre-selecting the origin and the circle's center, and then selecting the "Vertical" relation from the pop-up menu.

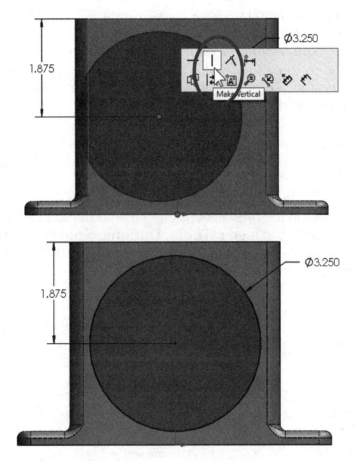

2.30. - Instead of making an extrusion like we did before, we are going to add material using a different technique. After adding the **Vertical** geometric relation, our sketch is fully defined and we need to exit the sketch. We'll use a different feature called "**Instant 3D**" to make the extrusion. It is activated in the Features tab in the CommandManager; if it is not already enabled, click to activate it.

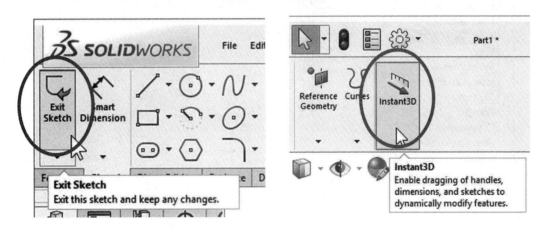

To make the extrusion, switch to an isometric view using the View Orientation toolbar and select the circle of the previously made sketch. Be aware that now we are editing the part, we left the sketch editing environment when we exit the sketch. After the sketch circle is selected with the "Instant3D" active, click-and-drag on the arrow that appears in it; this is the handle to extrude the sketch. You will immediately see a dynamic ruler that will show the size of the extrusion as you drag it. Make sure to extrude it 0.250". When you release the handle, a new extrusion will be created. To modify this (or any other) extrusion, select the front face of the extrusion and drag the handle to a new size.

You can control the size of the extrusion with more precision by dragging the handle over the ruler's marks; this way the handle will snap to the markers.

The smaller step in the ruler is controlled by the default increment in the spin box settings. Go to the menu "**Tools, Options, Spin Box Increments**." For convenience, we have set the length increment to 0.125" for English units and 2.5mm for metric units.

System Options - Spin Box Increments

System Options | Document Properties

General
Drawings
   Display Style
   Area Hatch/Fill
   Performance
Colors
Sketch
   Relations/Snaps
Display
Selection
Performance
Assemblies
External References
Default Templates
File Locations
FeatureManager
Spin Box Increments
View
Backup/Recover
Hole Wizard/Toolbox
File Explorer
Search
Collaboration
Messages/Errors/Warnings
Import
Export

Length increments

English units: 0.125in

Metric units: 2.50mm

Angle increments: 1.00°

Time increments: 0.10s

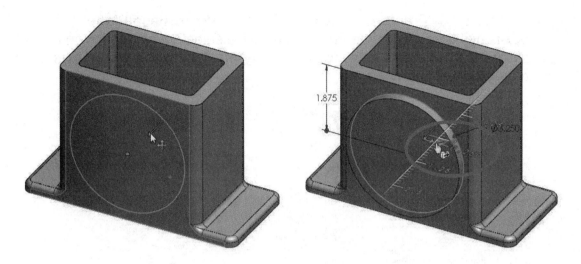

2.31. - Rename this extrusion as *"Front Boss"* by slowly double-clicking the feature's name in the FeatureManager or selecting it and then pressing F2.

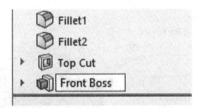

2.32. - The next step is to create an identical extrusion in the opposite side of the *'Housing'*. The "**Mirror**" command will create an identical 3D copy of the extrusion we just made. Switch to an Isometric view to help us visualize the Mirror's preview and make sure we are getting what we want. Select the "**Mirror**" command from the Features toolbar in the CommandManager.

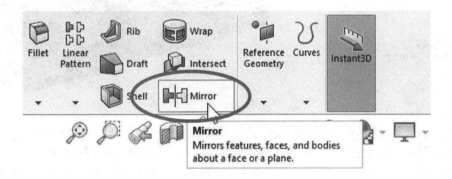

2.33. - From the Mirror's PropertyManager, we must make two selections. The first one is the Mirror Face or Plane and the second is the feature(s) we want to make a mirror of. The face or plane that will be used to mirror the feature has to be in the middle between the original feature and the desired mirrored copy. By making the first extrusion centered about the origin caused the default *"Front Plane"* of the *'Housing'* to be in the middle of the part, making it the best (and only) option for a Mirror Plane in this case.

To select the "*Front Plane*" (make sure the "**Mirror Face/Plane**" selection box is highlighted; this means it is the active selection box), click on the "▸" sign next to the part's name to reveal a **fly-out FeatureManager**, where we can select the "*Front Plane.*"

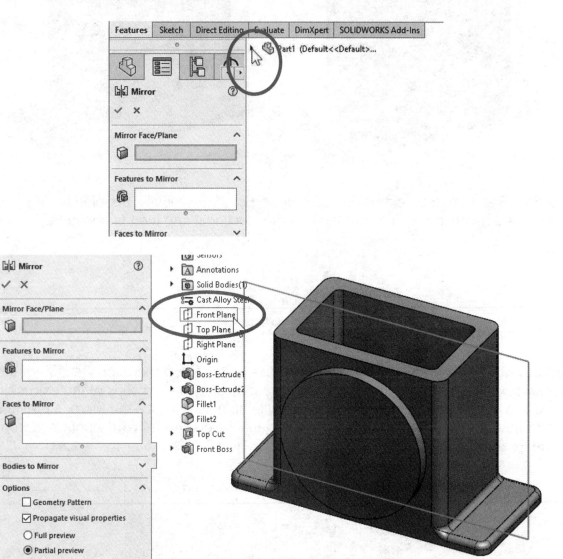

2.34. - After selecting the "*Front Plane*" from the fly-out FeatureManager, SOLIDWORKS automatically activates the "**Features to Mirror**" selection box (now highlighted) and is ready for us to select the feature(s) we want to mirror. If the *'Front Boss'* extrusion was pre-selected before activating the Mirror command, it is automatically added to the "Features to Mirror" selection box; otherwise, we need to select it either from the FeatureManager or in the graphics area.

 When selecting features from the graphics area be sure to select a face that belongs to the feature. When traversing the FeatureManager with the mouse pointer, notice how SOLIDWORKS highlights the features in the screen before selecting it.

Look at the preview after making your selections and click OK. Rotate the view to inspect the mirrored feature.

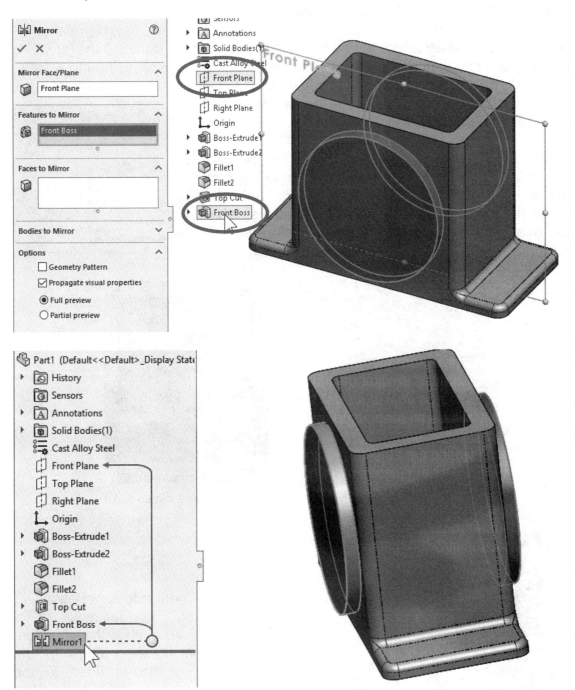

2.35. - In the next step we'll add the small boss at the right side of the *'Housing'*. Switch to a **Right view** using the View Orientation toolbar (Shortcut Ctrl+4), and select the "**Sketch**" icon from the CommandManager's Sketch tab.

When asked to select a face to add the sketch, click in the rightmost face (or select the face first and then click the Sketch command), draw a circle using the "**Circle**" tool, and add the dimensions shown. Just as we did with the front cylindrical boss, add a **Vertical Relation** between the center of the circle and the part's origin.

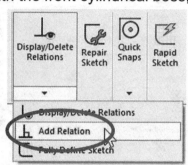

From the "**Display/Delete Relations**" drop-down icon, click "**Add Relation,**" select the circle's center and the origin, and add a "Vertical" relation between them by selecting it in the "Add Relations" box. Optionally, pre-select both and select "Make Vertical" from the pop-up menu to fully define the sketch.

Select Sketch plane

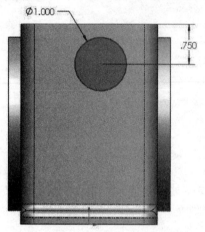

Draw circle & dimension it

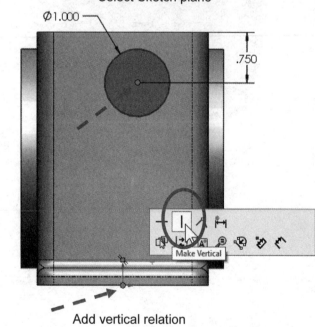

Add vertical relation

When adding relations, if you accidentally select more entities than needed, you can unselect them in the screen, or delete them in the selection box by clicking on them and then pressing the "Delete" key.

2.36. - Now that our sketch is fully defined, we are ready to extrude the sketch to make the side boss. We'll use the "**Instant 3D**" function as we did in the previous extrusion.

Exit the Sketch by selecting "**Exit Sketch**" in the CommandManager's Sketch tab or the Sketch icon in the confirmation corner (*not* the red X) and change to an Isometric view. In the graphics area select the circle of the sketch we just drew and click-and-drag the arrow along the ruler markers to make the extrusion 0.5″ long.

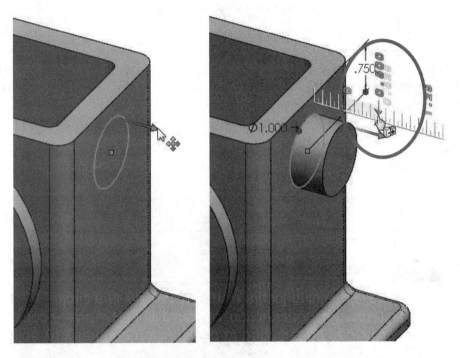

Rename this extrusion *'Side Boss'* in the FeatureManager by slow double-click or select and press the F2 key.

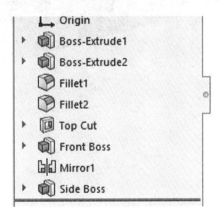

2.37. - Just like we did with the front circular boss, we'll mirror the last extrusion, but in this case, it will be mirrored about the *"Right Plane"* which is also in the middle of the part.

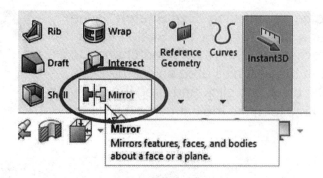

Select the "**Mirror**" command from the Features tab in the CommandManager; using the fly-out FeatureManager, select the *"Right Plane"* as the "**Mirror Face/Plane**" and the '*Side Boss*' extrusion in the "**Features to Mirror**" selection box to complete the Mirror command.

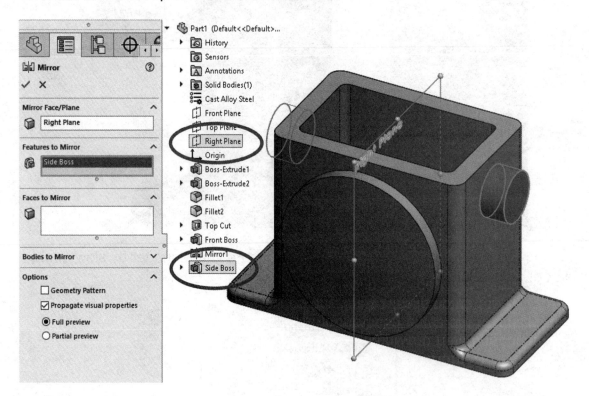

2.38. - We'll now make the circular cut in the front of the *'Housing'*. Change to a Front view (Shortcut Ctrl+1) for easier visualization. Select the "**Sketch**" icon from the Sketch tab and click in the round front face of the part, or select the face first, and then the Sketch icon.

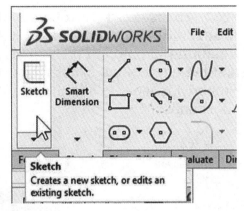

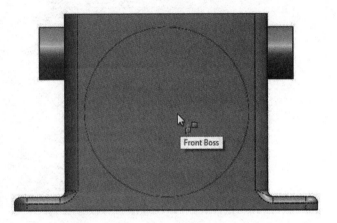

Draw a circle using the "**Circle**" tool and dimension it 2.250″ in diameter. For this step, do not start the circle at the center of the "Front Boss." We'll show how to manually add a concentric relation between the circle and the "Front Boss."

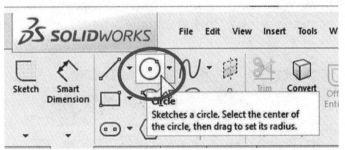

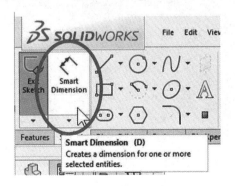

To locate the circle in the center of the circular face, we'll manually add a "**Concentric Relation**." Select the "**Add Relation**" icon from "**Display/Delete Relations**" drop-down menu; select the circle we just drew (perimeter or center) and the edge of the circular face. After selecting them click "**Concentric**" in the PropertyManager to add the relation and center the circle. Click **OK** to finish the command.

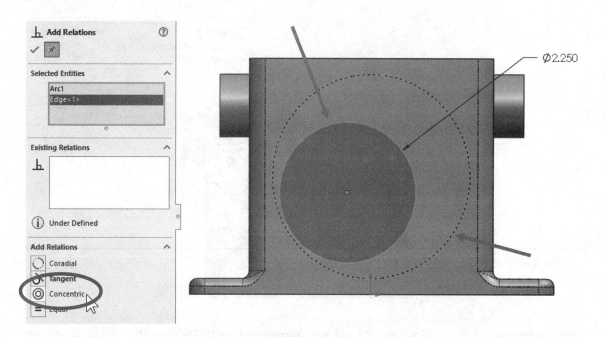

 Another way to add a concentric relation is to pre-select the circle and the edge of the "*Front Boss*" (hold down the **Ctrl** key while selecting) and select **Concentric** from the pop-up menu or the PropertyManager. When adding a single geometric relation this is usually a faster way to do it.

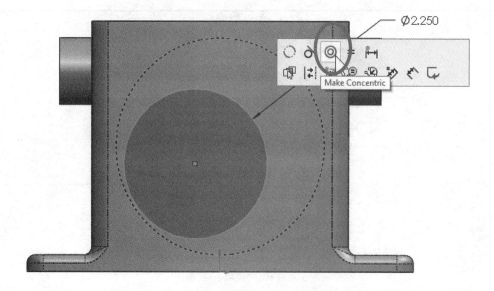

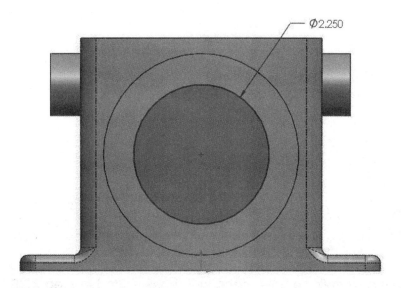

Ø2.250

2.39. - Now that the circle is concentric with the boss, we'll make the cut. Select the "**Extruded Cut**" command from the Features tab and switch to an isometric view for better visualization. In the "**Extruded Cut**" properties select the "**Through All**" option and OK to finish; by using this end condition the cut will go through the entire part regardless of its size. In other words, *if* we change the *'Housing'* to be wider, the cut will still go through all of it. Rename the new feature "Front Cut."

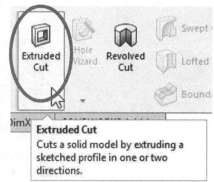

Extruded Cut
Cuts a solid model by extruding a sketched profile in one or two directions.

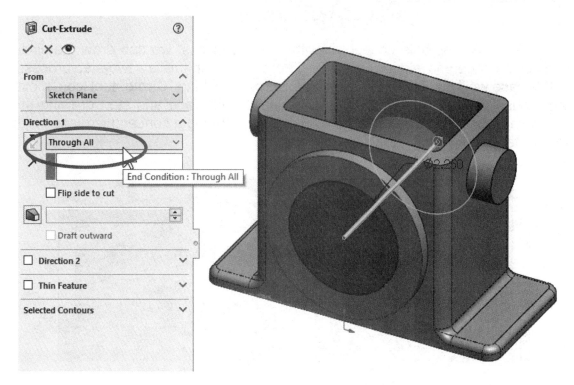

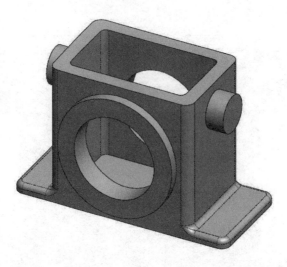

2.40. - We will now make a hole in the side boss added in step **2.36** to fit a shaft. Switch to a Right view and create a new sketch on the small circular face of the "*Side Boss*" by selecting the "**Sketch**" icon and then the circular face to locate it.

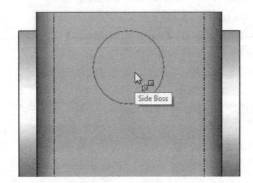

To make the hole concentric with the "Side Boss" we can draw the circle and add a concentric relation as we did before; however, this is a two-step process. In this case we'll do both in one step as follows: Select the "**Circle**" tool icon, move the mouse pointer towards the center of the circular face and start drawing the circle when its center is revealed automatically capturing a concentric relation with the center of the boss.

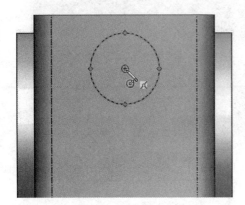
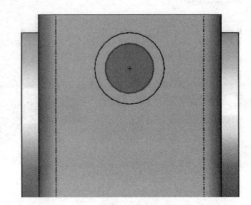

2.41. - The last step is to add a 0.575″ diameter dimension the circle, fully defining the sketch.

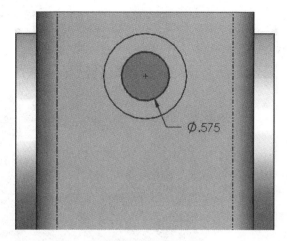

2.42. - Since this hole will be used for a shaft, we need to add a bilateral **tolerance** to the dimension. Select the 0.575″ dimension in the graphics area, and from the dimension's PropertyManager, under "Tolerance/Precision" select "Bilateral." Now we can add the tolerances. Notice that the dimension changes immediately in the graphics area. This tolerance will be transferred to the *Housing's* detail drawing later on. If needed, tolerances can also be added later in the detail drawing.

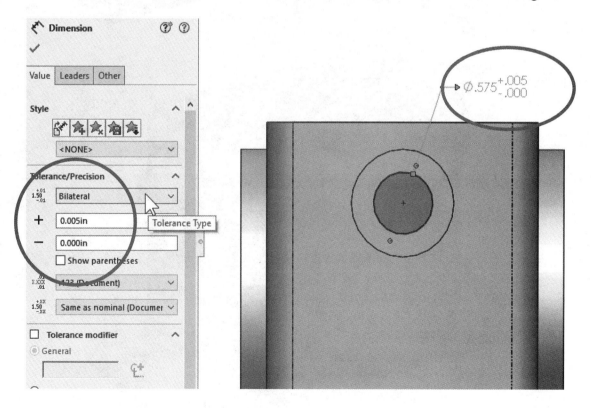

2.43. - Now we can make a Cut with the "**Through All**" option using the "**Extruded Cut**" command. Switch to an isometric view if needed. Rename the finished Feature "*Shaft Hole.*"

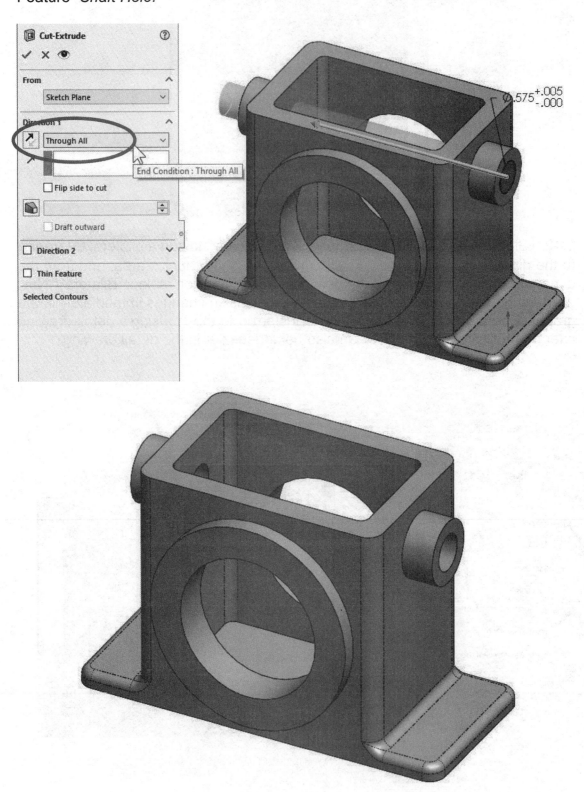

2.44. - To try a different approach using the "**Instant 3D**" command, select the "*Shaft Hole*" feature and delete it, but do not delete the sketch; to avoid deleting the hole's sketch make sure the "Delete absorbed features" checkbox is not selected.

To make an extruded cut using the "**Instant 3D**" functionality after deleting the cut extrude, select the sketch's circle and click-and-drag the handle *into* the part. You'll see how the part is cut as the arrow is dragged. The only disadvantage to making the cut using this technique is that the "Through All" option is not available, and we are only allowed to define the hole's depth as we would with a Blind end condition. Drag the cut until we cross the entire '*Housing*' part.

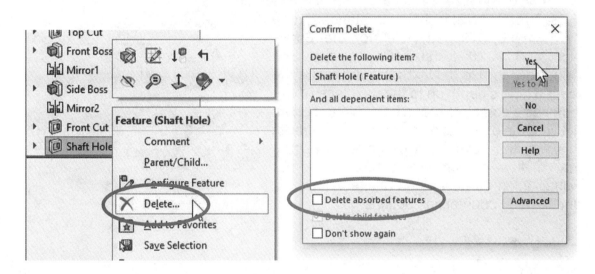

2.45. - For the next feature we'll add a ¼"-20 tapped hole in the front face of the '*Housing*'. SOLIDWORKS provides us with a tool to automate the creation of simple, Countersunk and Counterbore holes, slots, tap, Pipe taps and more by simply selecting a fastener size, depth, and location. The "**Hole Wizard**" command is a two-step process: in the first step we define the hole's type and size, and in the second step we define the location of the hole(s). The **Hole Wizard** is a special type of feature that uses 2 sketches that are automatically created, so there is no need to add a sketch first; in fact, it works very much like an applied feature.

Change to a Front View for clarity and select the "**Hole Wizard**" icon from the Features tab in the CommandManager. The first thing we'll do is to define the hole's type and size. Select the "**Tap**" icon for "Hole Type," "ANSI Inch" in the Standard drop down menu, "Tapped Hole" for the Screw type, and "¼-20" for size. Change the "End Condition" to "Up to Next"; this will make the tapped hole's depth up to the next face where it makes a complete round hole. At the bottom activate the option to add **Cosmetic Threads**, this way the hole(s) will be added a texture that makes it *look* like a real thread for visualization purposes and uncheck the near and far side countersink options.

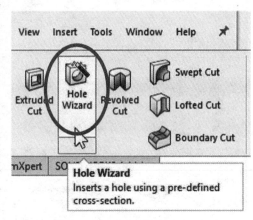

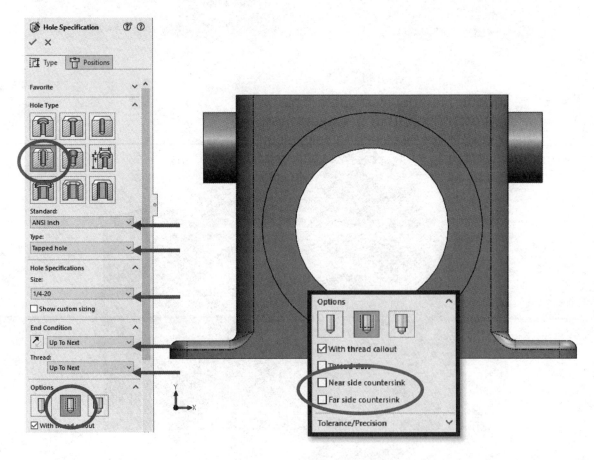

2.46. - After setting the options for the "**Hole Wizard**," the second step is to define its location. When we select the "**Positions**" tab, SOLIDWORKS asks us to select a flat face to locate the hole(s) or optionally press the "3D Sketch" button to add holes in multiple faces. For our example we'll add a single hole in the round face at the front of the part.

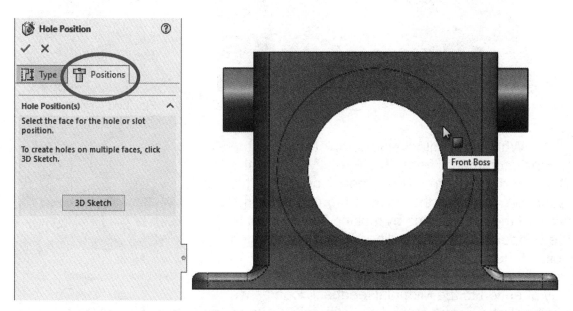

Immediately after selecting the face, SOLIDWORKS activates the "**Sketch Point**" tool, ready for us to define the hole's location. Anywhere we add a sketch point, the Hole Wizard will add a hole. For this exercise we only need to make one, and to locate it we'll use regular sketch tools including dimensions and geometric relations; notice that we are working in a Sketch. Our intent is to locate the hole in the middle of the flat face's width; to locate it, we will first draw a "**Centerline**." Select the "**Centerline**" command from the drop-down menu in the "**Line**" command, and start drawing it from the right quadrant of the outer circular edge, and finish it in the same quadrant of the inner circular edge (or vice versa.) The quadrants are automatically displayed as we approach them with a drawing tool active. Press the Esc key *once* or turn the Centerline command OFF when finished.

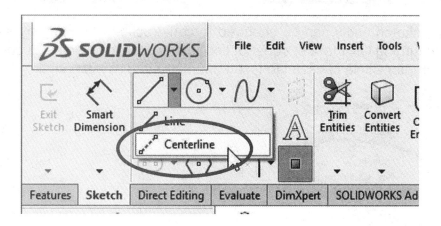

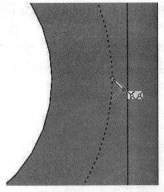

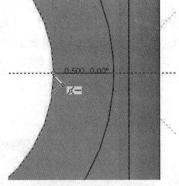

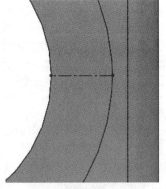

Click for Start of centerline   Click for End of centerline   Finished Centerline

When the centerline is complete, select the **"Point"** command from the Sketch tab in the CommandManager. The idea behind this technique is to add the point in the middle of the width of the circular face by making it coincident to the midpoint of the centerline. To add the sketch point, touch the centerline for a split second to reveal its midpoint, and click on it to add the point. Pay attention to the Midpoint feedback icon as we

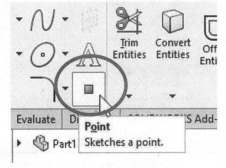

approach the center of the line. With this strategy, the hole will be centered in the face's width even if we change its size. Click OK to finish the Hole Wizard.

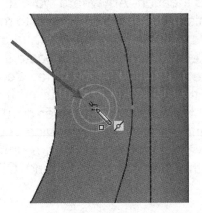

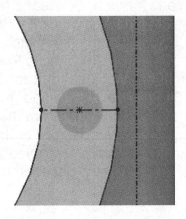

Any sketch entity can be changed to construction geometry by selecting it and turning ON the "For construction" option in its properties. A construction geometry Circle is useful to locate elements radially in a circle.

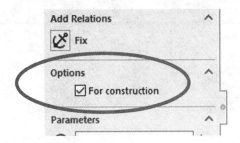

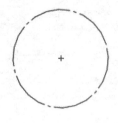

If the point had been added in a different location, we could add the midpoint relation by Window-Selecting the "**Point**" and the "**Centerline**," and from the pop-up toolbar selecting "**Make Midpoint**." Pre-selecting the face before activating the hole wizard command will add a "**Point**" where the face was selected.

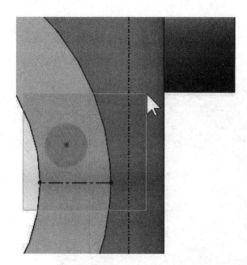 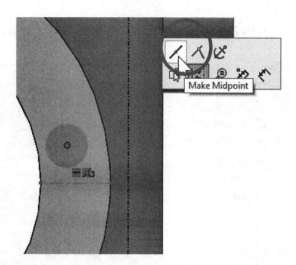

Window select entities   Add Midpoint relation from pop-up menu

 The "Cosmetic Thread" option adds a threaded texture to the holes, instead of an actual thread for looks and performance purposes. To show or hide the "Cosmetic Threads" right-mouse-click on the "*Annotations*" folder at the top of the FeatureManager, select "Details" and toggle the options "Cosmetic Threads" and "Shaded Cosmetic Threads." Remember that these are not real threads in the model. Real helical threads can be made, but it's mostly unnecessary in these cases. Later in the book we'll learn how to model real threads.

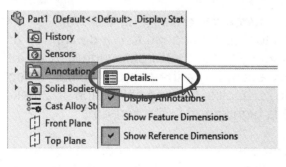 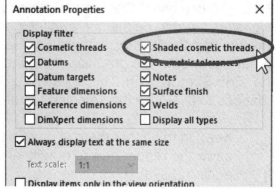

This is the difference between the two types of cosmetic threads:

1. **Cosmetic thread** is the annotation that shows up in a 2D drawing to indicate a thread needs to be made.
2. **Shaded cosmetic thread** is a texture added to the holes to give the 3D model the *appearance* of a thread and it's only for visual effects.

This is the finished ¼"-20 Tapped Hole with cosmetic threads.

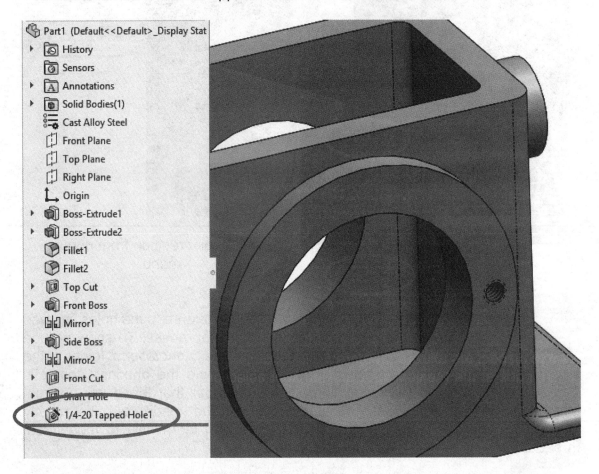

2.47. - After making the Tapped hole, we *suddenly* realize that the walls of the *'Housing'* need to be thinner, and we need to make a change to our design. To do this, we find the feature that we want to modify in the FeatureManager (*'Top Cut'* in our case) or in the graphics area and select it. From the pop-up toolbar, select the "**Edit Sketch**" icon. This will allow us to go back to the original sketch and make changes to it. Notice the selected feature is highlighted in the screen.

 Keep in mind that there is no real purpose to this dimensional change but to show the reader how to modify an existing feature's sketch if needed. When editing a sketch, its dimensions, geometry, and geometric relations can be changed, removed, etc. as needed.

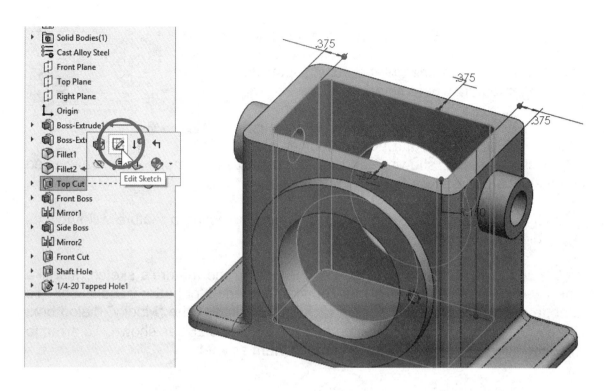

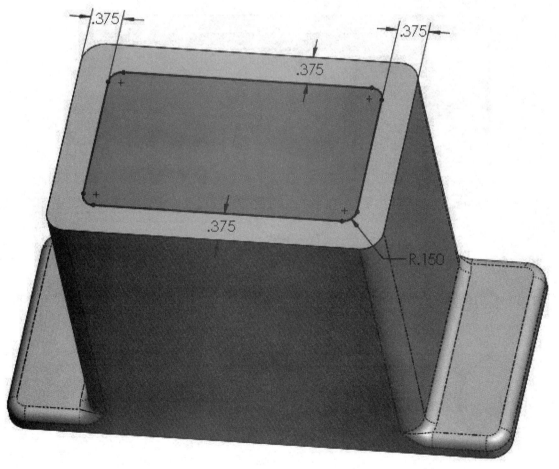

Selecting the "Edit Feature" icon will show the Cut Extrude command options; this is where we can change the cut's depth and other feature's parameters.

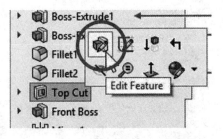

 If we select the feature with a right-mouse-click, we will see the pop-up toolbar along with a menu showing additional options. The most commonly used commands are already in the pop-up toolbar.

 If the "**Instant 3D**" command is activated, selecting a feature will show its dimensions on the screen.

2.48. - What we just did was to go back to editing the feature's sketch, just like when we first created it. Switch to a Top view if needed for better visualization. To change a dimension's value double click on it to display the "Modify" dialog box. Change the two dimensions indicated from 0.375″ to 0.25″ as shown. To arrange the dimensions just click-and-drag to move them around.

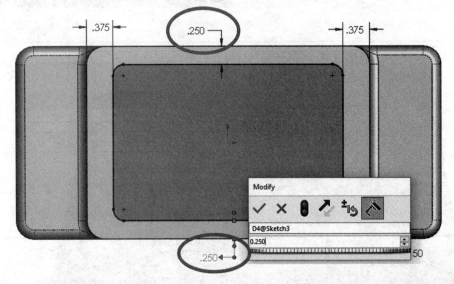

2.49. - After changing the dimensions, we cannot select "**Cut Extrude**" because we had already made a cut; what we need to do now is select "**Exit Sketch**" or "**Rebuild**" the part (Shortcut Ctrl + B) to update the model with the new dimensions.

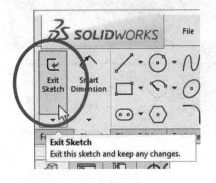

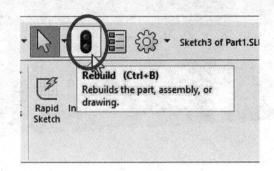

A different way to change these dimensions is using the "**Instant3D**" functionality. The way it works is very simple: instead of having to edit the sketch, we select the feature that we want to modify either in the FeatureManager or the graphics area (in this case one of the inside faces which were made with the Cut Extrude) and click-and-drag the **blue dots** at each of the dimensions that need to be modified until we get the desired value without having to edit the sketch. Dragging the mouse pointer over the ruler markers will give you values in exact increments. Depending on the speed of your PC and the feature being modified, Instant3D may be slow as the model is being dynamically updated.

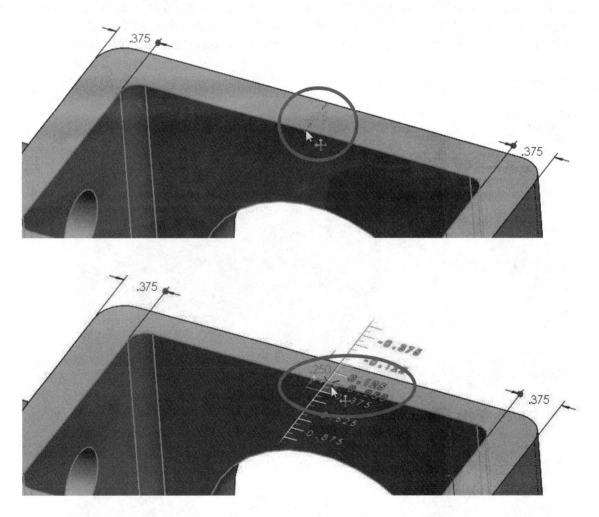

 A third option to change the dimensions is to click on the dimension's value and type a new one. If Instant3D is not active, double-click in the feature to show its dimensions, and double click a dimension to change its value; after changing the dimension's value we need to rebuild the model.

2.50. - Now we will add more tapped holes to complete the flange's mounting holes. We'll use the first hole as a "seed" to make copies of it using the "**Circular Pattern**" command. In the Features tab, select the drop-down menu under the "**Linear Pattern**" and select "**Circular Pattern**" or use the menu "**Insert, Pattern/Mirror, Circular Pattern**."

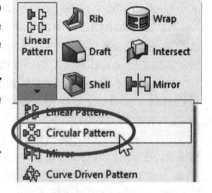

 Note that commands are grouped by similar functionality.

2.51. - To create a "**Circular Pattern**" we need to define the direction of the pattern using a circular edge, a cylindrical surface or an axis. Click inside the "Direction 1" selection box to activate it, and then select the edge indicated for the pattern axis.

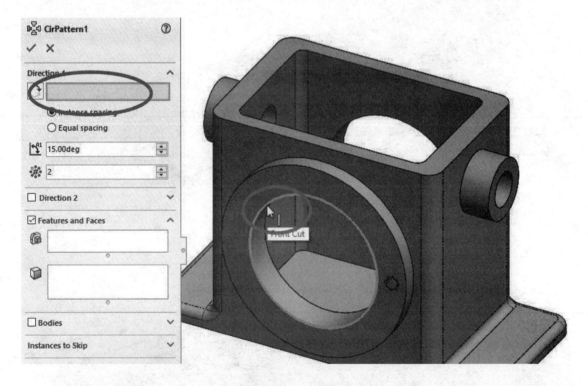

      Any circular edge or cylindrical face that shares the same axis can be used for the pattern's direction, as shown in the following images.

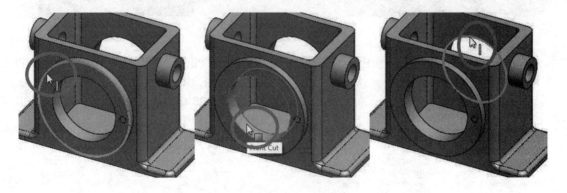

Another option for the pattern's direction is a temporary axis. Every cylindrical surface has a "Temporary Axis" that runs through its center. To see the temporary axes in a model, select the menu "**View, Hide/Show, Temporary Axes**" or turn them ON in the "**Hide/Show Items**" toolbar.

Temporary axes (and other auxiliary geometry) can be turned on or off while a command is in progress. In this picture we can see that the shortcut letter "T" has been assigned to toggle the temporary axes on and off. Go to the menu "**Tools, Customize, Keyboard**" to add a keyboard shortcut to a command.

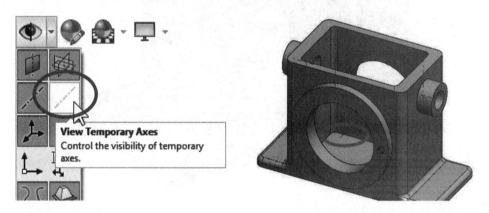

2.52. - After selecting the edge (or axis) for the pattern axis, in the "Features and Faces" section click inside the "Features to Pattern" selection box to activate it (notice it gets highlighted). Select the "*¼-20 Tapped Hole1*" feature from the fly-out FeatureManager; change the number of copies to six (this value includes the original), and make sure the "Equal spacing" option is selected to equally space the copies in 360 degrees. If the option "Instance spacing" is checked instead, we need to define the angular distance between copies in degrees. Notice the preview in the graphics area and click OK to finish the command.

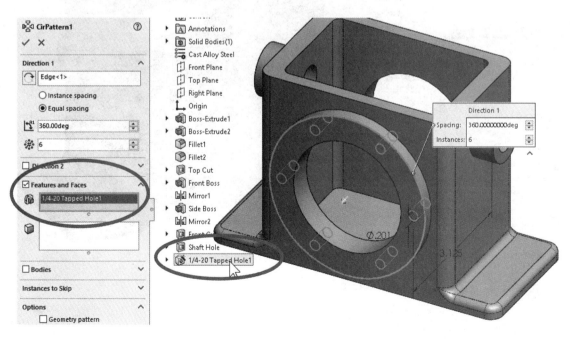

The feature to be patterned can also be selected from the graphics area; in this case a face of the feature needs to be selected. Sometimes a face can be difficult to select because it may be small, like this hole. In this case, we can use a "**Magnifying Glass**" (Default shortcut "G") to make selection easier.

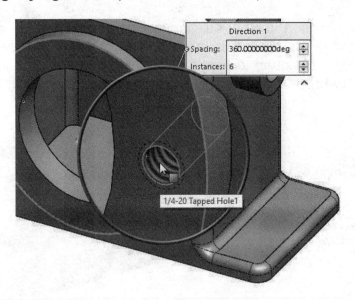

2.53. - Since we need to have the same six threaded holes in the other side of the *'Housing'*, we will use the "**Mirror**" command to copy the Circular Pattern about the "*Front Plane.*" Make this mirror about the "*Front Plane*" and mirror the "*CirPattern1*" feature created in the previous step. Click OK to finish.

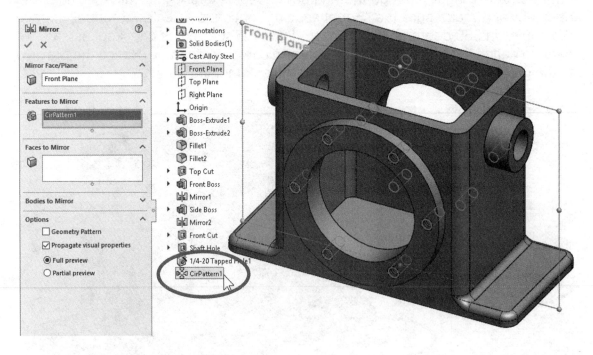

After mirroring the circular pattern our part looks like this. The Cosmetic Threads have been turned off for clarity (Right-mouse-click in Annotations, Details, uncheck "Cosmetic Threads").

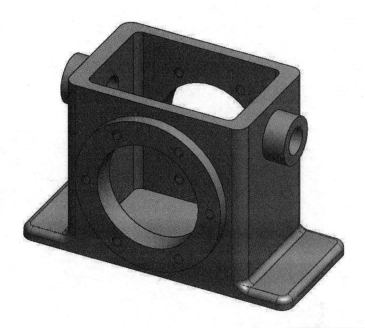

2.54. - We will now add four #6-32 tapped holes to the topmost face using the **Hole Wizard**. Switch to a Top view (shortcut Ctrl+5 or Mouse Gestures) for visibility and select the "**Hole Wizard**" icon.

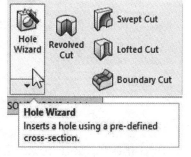

2.55. - In the Hole Wizard's PropertyManager, select the "**Tap**" Hole Specification icon, and select the options shown for a #6-32 Tapped Hole. The "Blind" condition tells SOLIDWORKS to make the hole an exact depth.

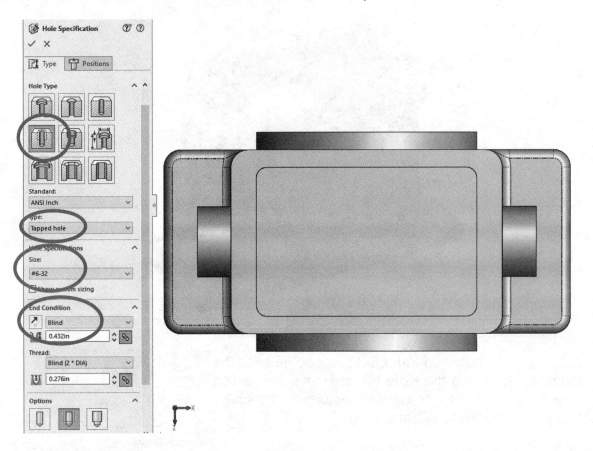

2.56. - Click in the "Positions" tab to define the hole locations. Select the top face to add the tapped holes, and notice that immediately after we select the face the "**Sketch Point**" tool is automatically selected; we are editing a sketch and the Sketch toolbar is activated.

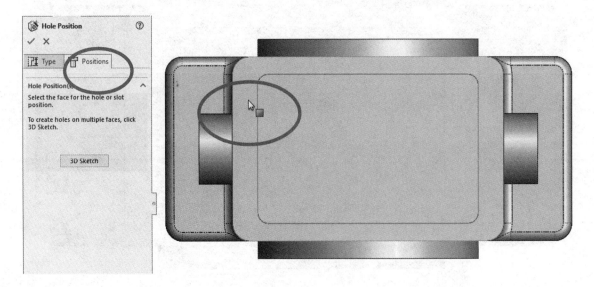

2.57. - With the "**Point**" tool active, *touch* each of the round corner edges to reveal their centers, and then click in the arc's center to add a point in each one; this way we'll make the four points concentric to each corner fillet's center. Click OK to finish the Hole Wizard.

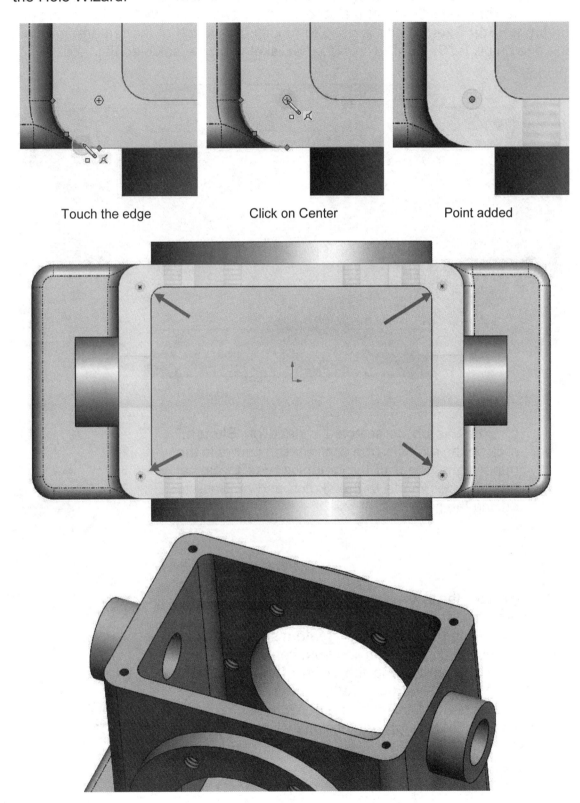

| Touch the edge | Click on Center | Point added |

2.58. - We are now ready to make the slots at the base of the *'Housing'*. For this task it will be easier to switch to a Top view. To add a new sketch, we can select the "**Sketch**" command and click on the selected face as before, but in this case, we'll learn how to use the pop-up toolbar. Select the face indicated, and from the pop-up toolbar, select "**Sketch**." Notice there are two similar looking icons: the one on top is "**Edit Sketch**" and is used to modify the sketch of the feature selected; the one below is "**Sketch**" to create a new sketch on the selected flat face.

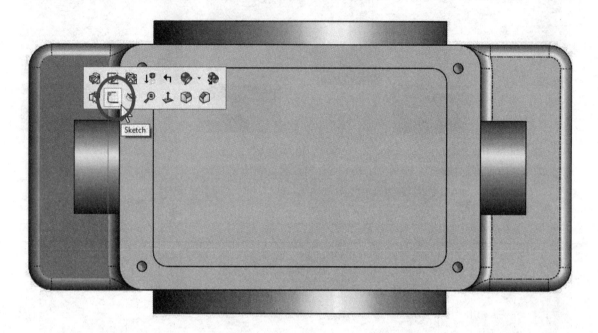

 If "**Edit Sketch**" is selected instead of "**Sketch**," click on the red "X" in the confirmation corner in the upper right corner of the graphics area to cancel any changes made to the sketch and go back to editing the model.

To make the slot, we'll use the "**Straight Slot**" command from the Sketch tab in the CommandManager. This tool will create a slot by first drawing a construction line from center to center, and then defining the width of the slot. Select the "**Straight Slot**" icon and activate the "Add Dimensions" option to automatically add the slot's dimensions when we finish.

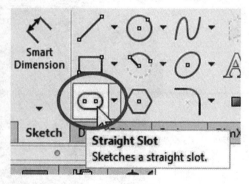

Straight Slot
Sketches a straight slot.

First click to locate the center of one arc, click to locate the center of the second arc (this is a construction line), and click a third time to define the slot's width. When finished, double click the dimensions and change them to make the slot 0.375″ long and 0.250″ wide.

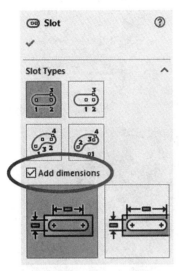

Add dimension option

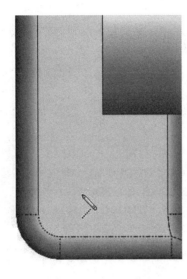

Locate first center

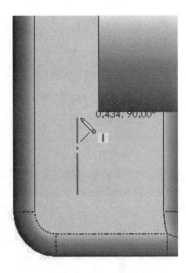

Locate second Center

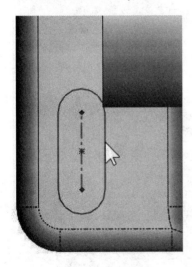

Define the slot width

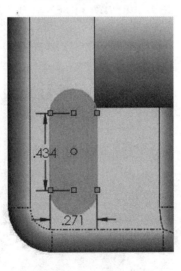

Automatic dimensions added

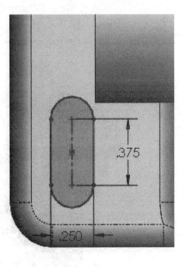

Corrected dimensions

 The "**Slot**" command has more options, including arc slots and overall slot length dimension. The numbers refer to the order of clicks to create each slot type.

 To enable auto dimension while adding other sketch elements, select the menu "**Tools, Options, System Options, Sketch**" and activate:

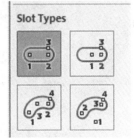

☑ Enable on screen numeric input on entity creation

Another way to activate this option is by a right-mouse-click and turning "**Sketch Numeric Input**" ON. This option will allow you to type the entity size as you sketch. To automatically add the corresponding dimension, select a sketch tool (line, arc, circle, etc.), right-mouse-click and turn on the option "**Add Dimension**."

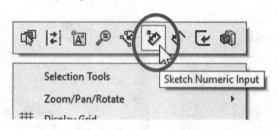

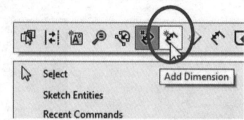

After the slot's size is defined, locate the slot by adding two 0.5″ dimensions to the lower and left edges of the base as shown. Finish the slot by making an "**Extruded Cut**" using the "Through All" end condition. Rename this feature "*Slot*."

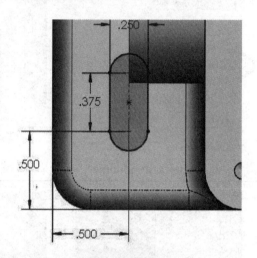

 Slots can also be made using the Hole Wizard.

2.59. - We will now create a "**Linear Pattern**" of the slot. A linear pattern allows us to make copies of one or more features along one or two directions (usually along model edges). Select the "**Linear Pattern**" command from the Features tab in the CommandManager or the menu "**Insert, Pattern/Mirror, Linear Pattern.**"

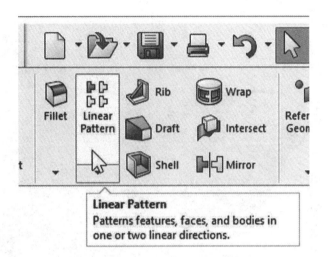

Linear Pattern
Patterns features, faces, and bodies in one or two linear directions.

 Using the Mirror command keeps the design intent better, but we chose to show the user how to use the Linear Pattern command instead.

2.60. - In the Linear Pattern's Property-Manager, the "Direction 1" selection box is active; select the indicated edge for the direction of the copies. Using the "Spacing and Instances" option we can define the number of copies and the spacing between them. Keep in mind that any linear edge or linear sketch element can be used to define the direction.

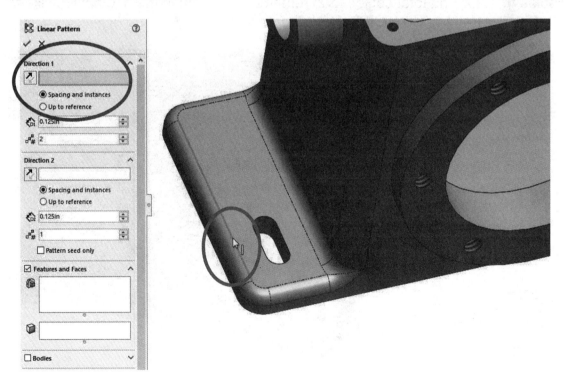

Once the edge is selected, an arrow indicates the direction in which the copies will be made. If the Direction Arrow in the graphics area is pointing in the wrong direction, click in the **"Reverse Direction"** button next to the "Direction 1" selection box.

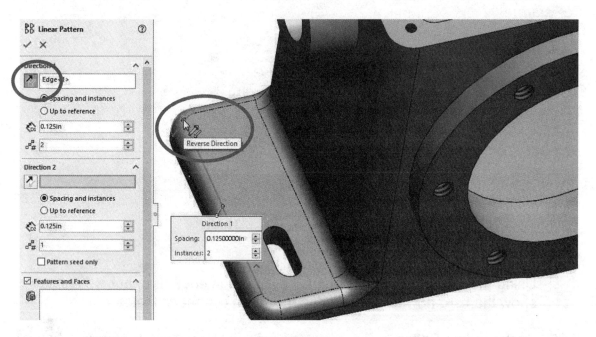

2.61. - After defining the direction, the "Direction 2" selection box is activated. In this example we will only copy the slot in one direction. In the "Features and Faces" section, activate the "Features" selection box and select the slot feature either from the fly-out FeatureManager or the graphics area. Change the spacing between the copies to 1.25″ and total copies to 2; remember it includes the original just like in the Circular Pattern. Click OK to finish the command.

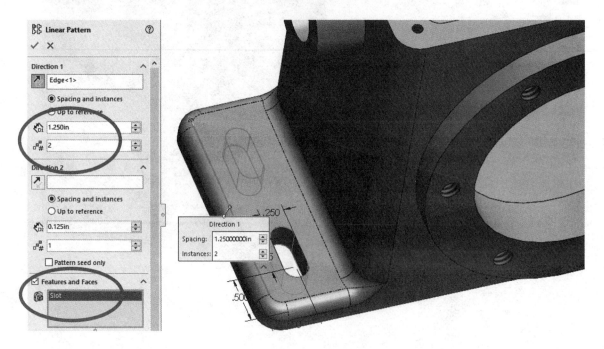

2.62. - The mounting slots are needed on both sides of the *'Housing'*, so we'll copy the previous linear pattern to the other side of the *'Housing'* using the "**Mirror**" command about the "*Right Plane*." Click on the "**Mirror**" icon in the Features tab of the CommandManager; select the "*Right Plane*" as the mirror plane and the "*LPattern1*" in the "Features to Mirror" selection box to copy the slots.

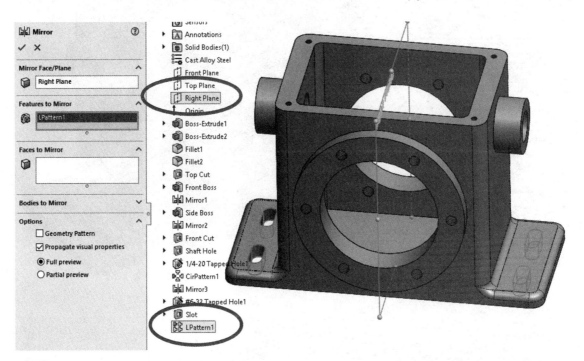

 Selecting the Linear Pattern feature for the mirror also includes the pattern's seed feature, the "*Slot*."

2.63. - Using the "**Fillet**" command from the Features tab, add a 0.125″ radius fillet to the edges indicated as a finishing touch. Rotate the model using the middle mouse button and/or change the display style to "Hidden Lines Visible" mode to make selection easier. Click OK to finish.

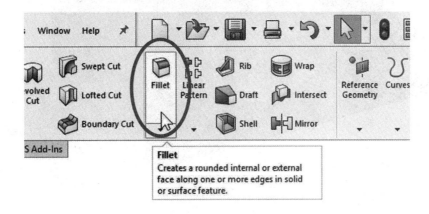

2.64. - Now that the model is finished, we can easily determine its physical properties, such as **Weight, Volume, Center of Mass and Moments of Inertia**. SOLIDWORKS includes a built-in material library with many different metals, alloys, plastics, woods, composite materials and others like air, glass and water.

The library includes these materials' mechanical and thermal properties:

- Mass density
- Elastic and Shear modulus
- Tensile, Compressive and Yield strengths
- Poisson's ratio
- Thermal expansion coefficient
- Thermal conductivity
- Specific heat

 These properties are also used by SOLIDWORKS to determine a part's weight or determine if a component will fail under a given set of loading conditions using SimulationXpress (the built in structural analysis software).

Since we made our part based on a template with a pre-defined material (Cast Alloy Steel), the material is already defined. To change the component's material, right-mouse-click on the material's name in the FeatureManager, or if a material has not been assigned, in the "**Material**" icon in the FeatureManager select "Edit Material" or pick one from the list of favorites listed. The list of favorites can be changed selecting the "Manage Favorites" option from the pop-up menu, or in the "Favorites" tab in the Materials library. Select "Edit Material" to see the material properties that are defined. Click "Apply" to accept the material and Close the library.

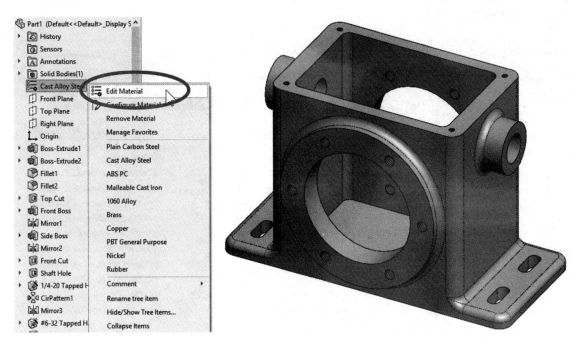

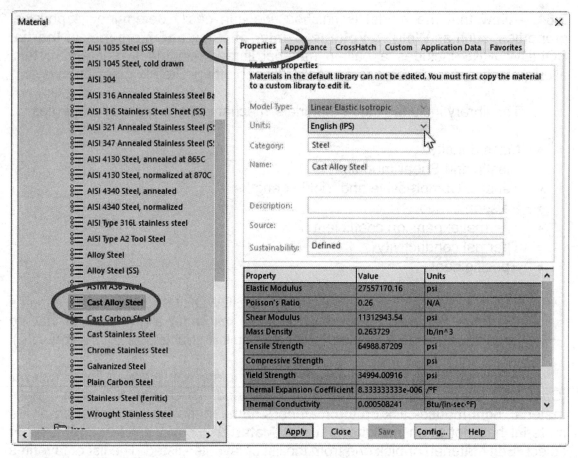

 In the Properties tab we can select the units to display the material properties.

 In the "Favorites" tab we can define which materials to show in the pop-up Material menu.

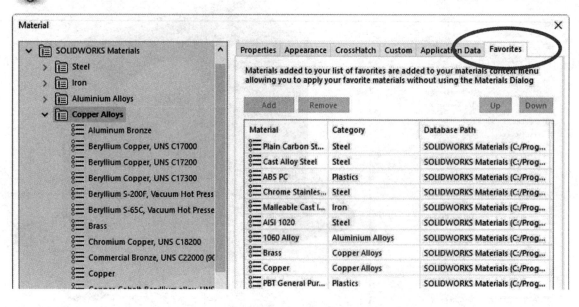

2.65. – After assigning or modifying the material the FeatureManager will show the new material's name, in our case "*Cast Alloy Steel*". Also the part's color will be changed to resemble the material's visual properties. Activate the "Evaluate" tab in the CommandManager, and select "**Mass Properties.**" The report will display the material's Density (provided by the material selection), Mass (Calculated from the part's volume and its density), Volume, Surface Area and Center of Mass coordinates relative to the origin (also indicated by a magenta triad in the graphics area), Principal Axes of inertia and Moments of inertia about the Center of Mass and the part's origin, which can later be used in reports.

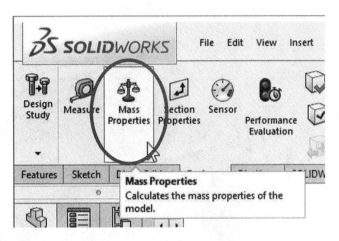

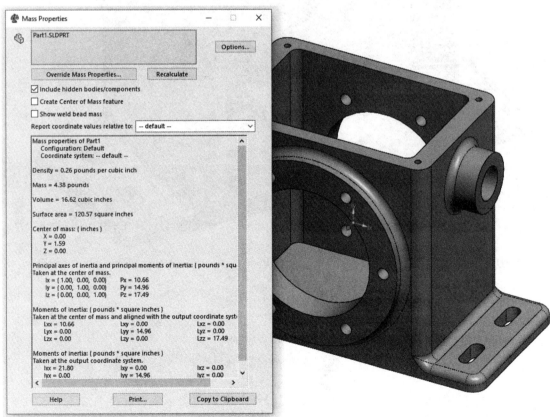

Mass properties are referenced to the part's origin by default, but they can also be referenced to a user defined coordinate system by selecting one from the "Report coordinate values relative to:" drop down list. User defined coordinate systems will be covered later in the book.

 In the "Options" button we can change the units used to display the mass properties results, which by default are calculated using the document's units of measure.

Save the finished part as *'Housing'* and close the file.

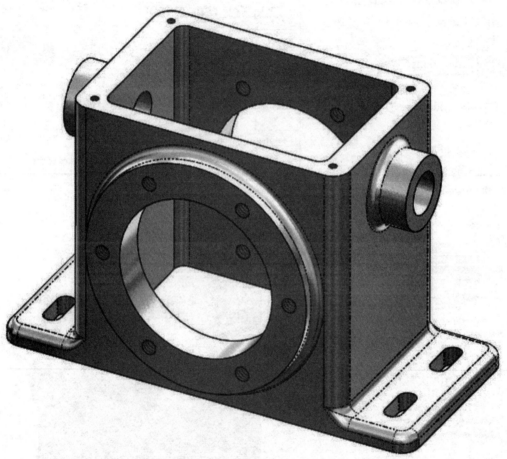

*Image made using RealView Graphics

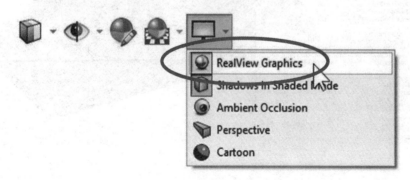

Exercises: Build the following parts using the knowledge acquired in this lesson. Make a new template or multiple templates using millimeters as the unit system and optionally define a material, then use the appropriate template for each exercise, or use an existing template and change the units of measure and material as needed. Try to use the most efficient method to complete each model.
High resolution images are included on the exercise files.

Exercise 1
DIMENSIONS: INCHES
MATERIAL: AISI 1020

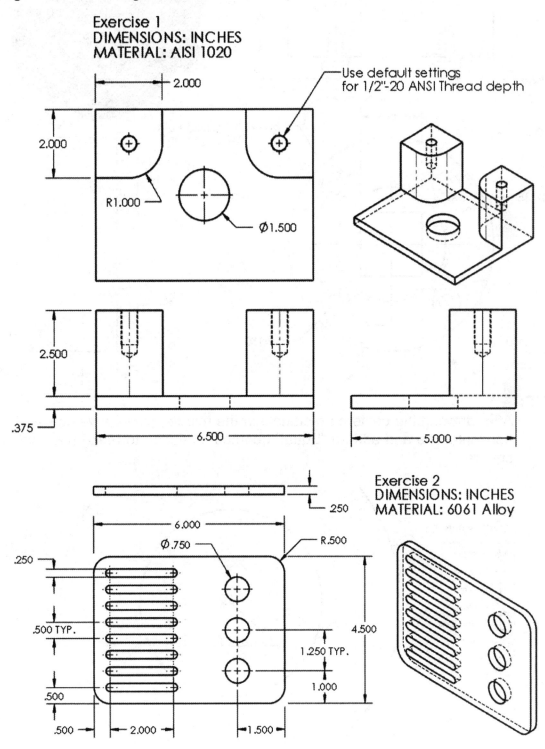

Use default settings
for 1/2"-20 ANSI Thread depth

2.000

2.000

R1.000

Ø1.500

2.500

.375

6.500

5.000

.250

Exercise 2
DIMENSIONS: INCHES
MATERIAL: 6061 Alloy

6.000

Ø.750

R.500

.250

.500 TYP.

4.500

1.250 TYP.

1.000

.500

.500

2.000

1.500

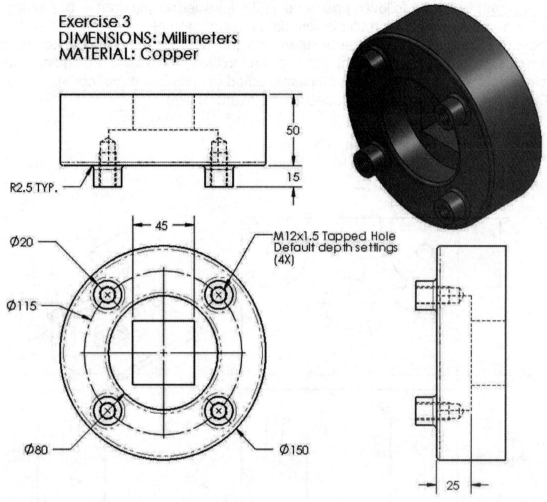

Exercise 3
DIMENSIONS: Millimeters
MATERIAL: Copper

50

15

R2.5 TYP.

45

Ø20

M12x1.5 Tapped Hole
Default depth settings
(4X)

Ø115

Ø80

Ø150

25

TIPS:

- After drawing the centered rectangle in the middle, select a vertical and a horizontal line and add an "Equal" geometric relation to make it a perfect square.

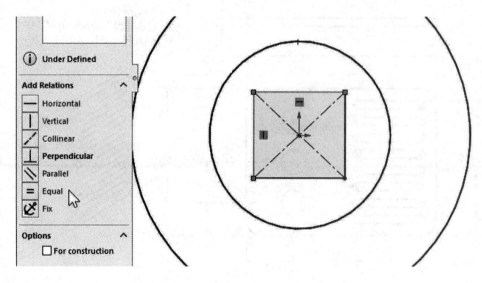

- To locate the circular bosses, draw a circle and convert it to construction geometry.

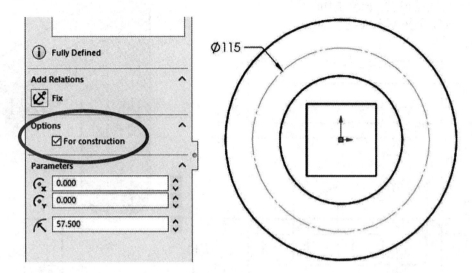

- Add a "Centerline" from the origin to the construction circle, and add a 45° dimension between the centerline and a horizontal or vertical edge. To add the circular boss, start the circle at the end of the centerline.

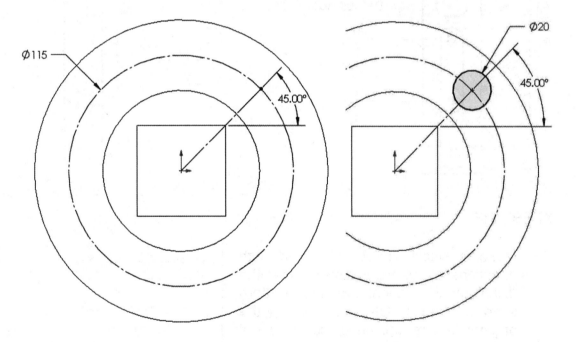

Exercise 4
DIMENSIONS: Millimeters
MATERIAL: ABS

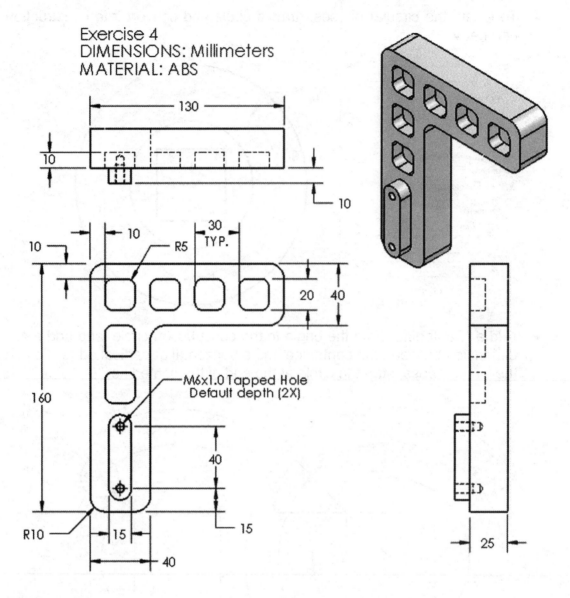

EXPERT TIP:

- Make a single linear pattern in both directions at the same time. Under the "Direction 2" section, turn the "Pattern seed only" option ON; this way only the original feature will be copied and not create a rectangular pattern.

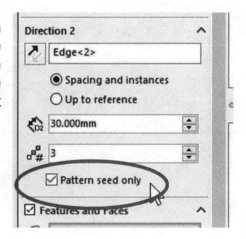

Rectangular pattern in 2 directions

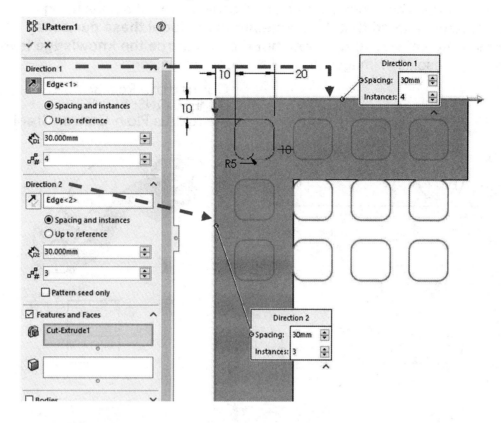

Rectangular pattern with the "Pattern seed only" option

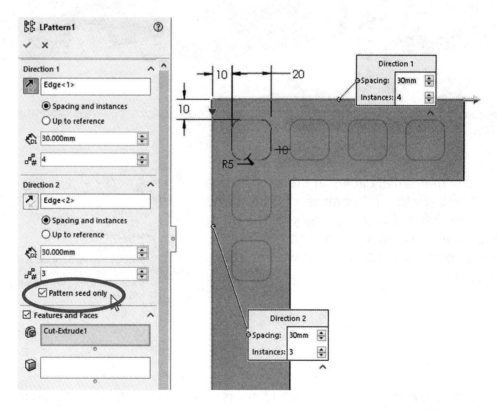

Engine Project Parts:

Make the following components to build the engine. Save the parts using the name provided. **Keep in mind that the suggestions to model these parts may not be the most efficient way to do them, but they reinforce the knowledge learned so far.** High resolution images at www.sdcpublications.com.

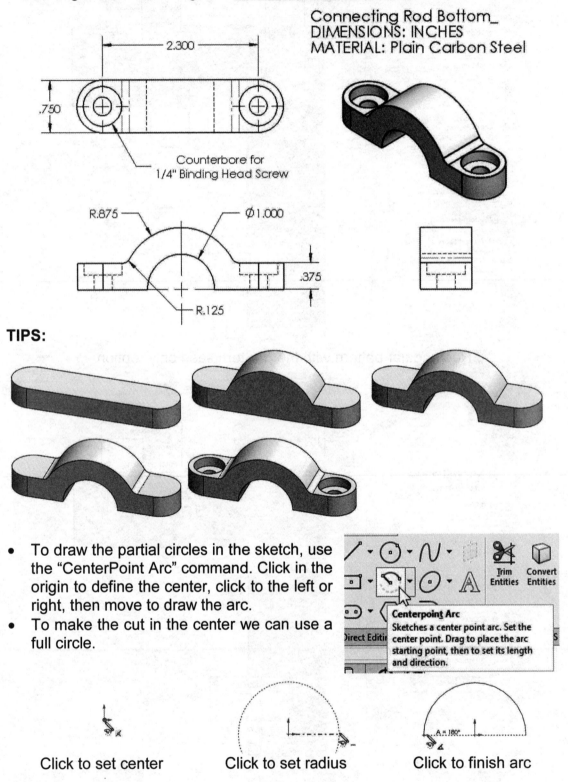

Connecting Rod Bottom_
DIMENSIONS: INCHES
MATERIAL: Plain Carbon Steel

2.300

.750

Counterbore for
1/4" Binding Head Screw

R.875   Ø1.000

.375

R.125

TIPS:

- To draw the partial circles in the sketch, use the "CenterPoint Arc" command. Click in the origin to define the center, click to the left or right, then move to draw the arc.
- To make the cut in the center we can use a full circle.

Centerpoint Arc
Sketches a center point arc. Set the center point. Drag to place the arc starting point, then to set its length and direction.

Trim Entities   Convert Entities

Click to set center   Click to set radius   Click to finish arc

A = 180°

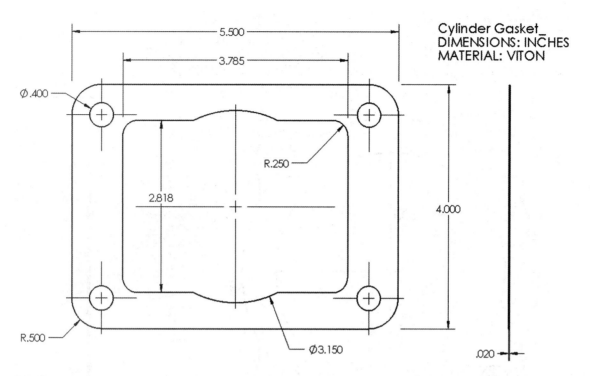

Cylinder Gasket_
DIMENSIONS: INCHES
MATERIAL: VITON

TIPS: Draw the gasket's outside and inside rectangles, add sketch fillets, holes, and extrude. Make a circular cut as a second feature (it can all be done with a single sketch, but we haven't covered the trim tool yet).

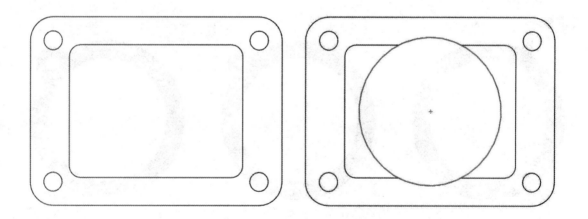

Head Gasket_
DIMENSIONS: INCHES
MATERIAL: VITON

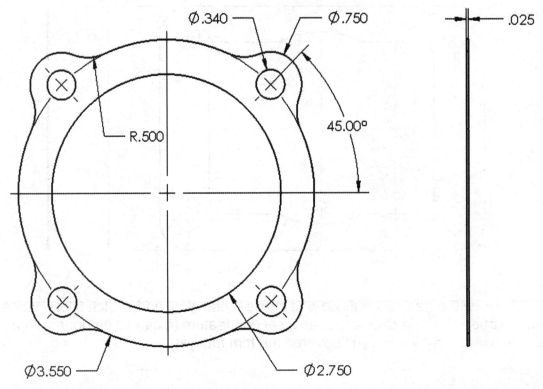

TIPS:

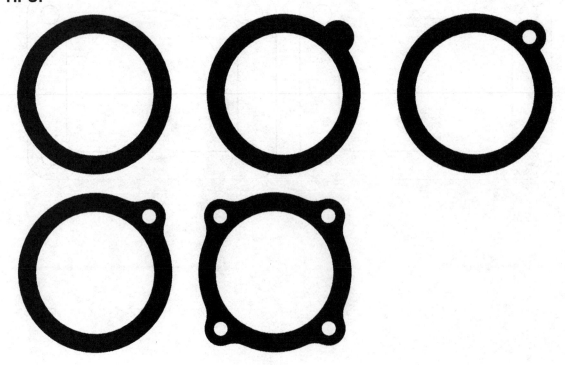

The Side Cover

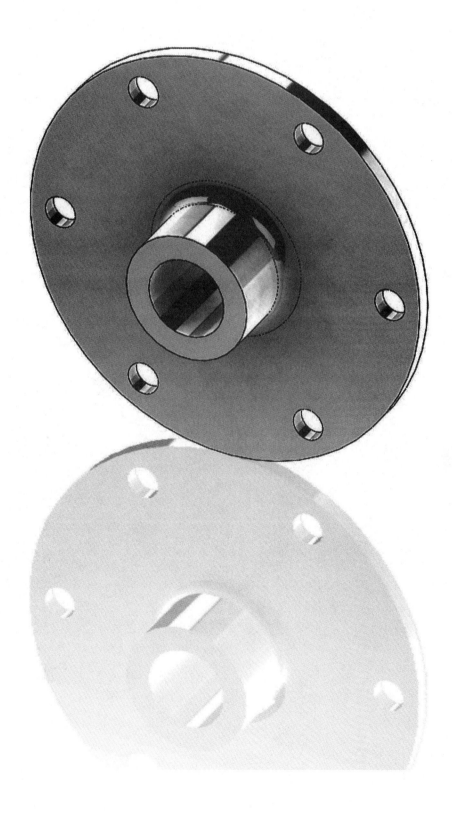

Notes:

In making the *'Side Cover'* part we will learn the following features and commands: Revolved Feature, Sketch Trim and Extend, and construction geometry. We will also review some of the commands previously learned in the *'Housing'* part. This is the sequence of features we'll follow for the *'Side Cover'*:

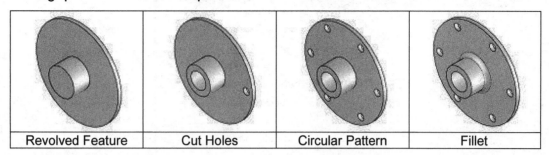

| Revolved Feature | Cut Holes | Circular Pattern | Fillet |

3.1. - Make a new part using the "ANSI-Inch-Cast Alloy Steel" template and change the material to "AISI 1020." If you use a different template be sure to set the units to Inches and three decimal places.

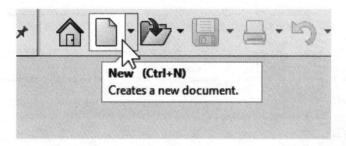

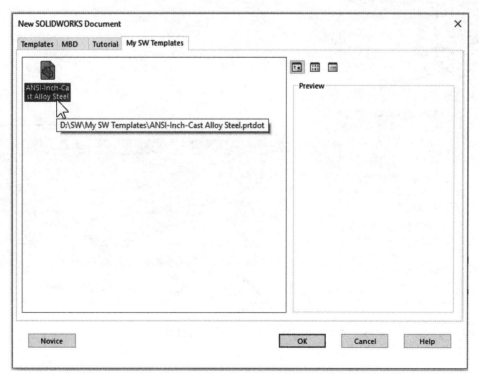

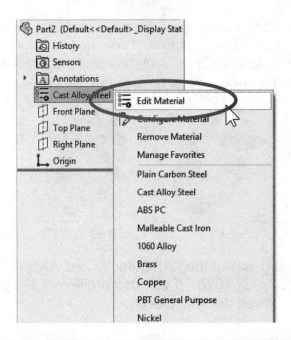

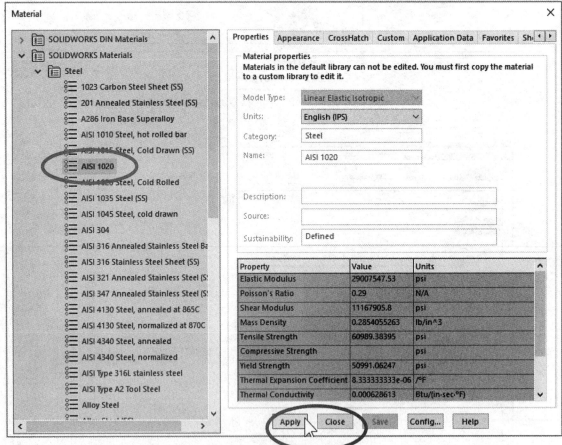

Click "Apply" and "Close" to assign the new material.

3.2. - The first feature will be a **Revolved Feature**. As its name implies, it is created by revolving a sketch about an axis. Select the "**Revolved Boss/Base**" icon from the Features tab in the CommandManager. When asked to select a plane for the sketch, select the "*Right Plane*" (no particular reason to choose this plane except to have a good looking isometric view ☺).

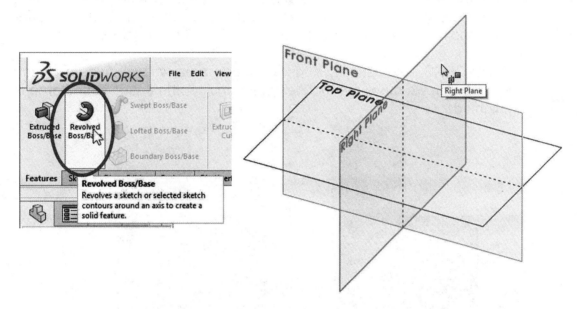

3.3. - Select the "**Corner Rectangle**" command from the Sketch toolbar and draw the following sketch using two rectangles; for the first rectangle click-and-drag starting at the origin going up to the left; for the second rectangle start at the lower left corner of the first rectangle also going up to the left (there will be two lines overlapping in the middle). Don't be too concerned with their size; we'll add the correct dimensions later.

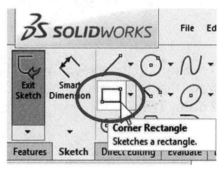

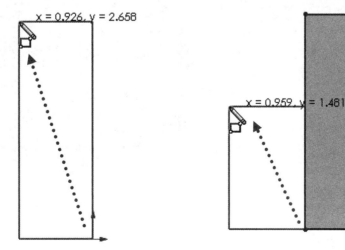

3.4. - It is a good practice to have single, non-intersecting profiles in the sketch, and no more than 2 lines sharing an endpoint. While it is possible to use a sketch with intersecting lines using a function called "**Contours**," it will be covered in the Level II book, and for now we'll continue using single contour sketch.

 In general, it is a good idea to work with single contour sketches and advance to other techniques like Contours after getting a better understanding of the use of open and closed contour sketches.

 SOLIDWORKS automatically adds a shaded area when it finds a closed contour; when we move the mouse pointer inside, the entire closed area is highlighted.

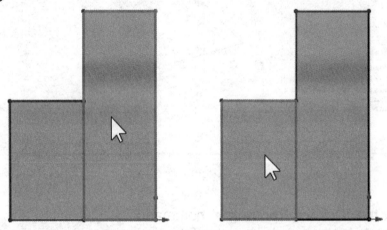

To clean up the sketch, we will use the "**Trim**" command from the Sketch toolbar, from the lower left corner in the Mouse Gestures or the menu "**Tools, Sketch Tools, Trim**."

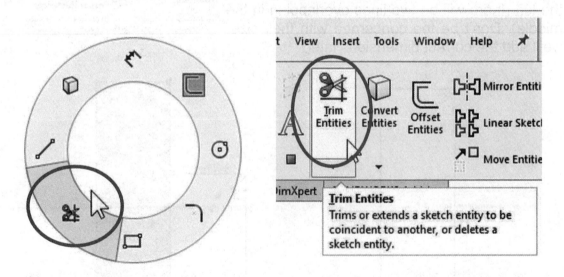

3.5. - The Trim tool allows us to cut sketch entities using other geometric elements as a trim boundary. After selecting the "**Trim Entities**" icon, select the "Power Trim" option from the PropertyManager. The Power Trim allows us to click-and-drag *across* the entities that we want to trim. Click-and-drag the cursor crossing the two lines indicated next. Notice the lines are trimmed as you cross them.

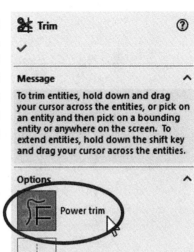

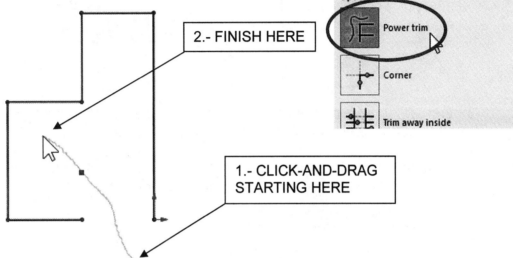

2.- FINISH HERE

1.- CLICK-AND-DRAG STARTING HERE

3.6. - The next step is to extend the short line to close the sketch and have a single closed profile. Select the "**Extend Entities**" icon from the drop-down menu under "**Trim Entities**." Click on the short line indicated; a preview will show you how the line will be extended. If the extension does not cross a line, you will not get a preview and the line will not be extended. After the line is extended and the sketch forms a closed contour, the area defined will be shaded.

 Extend Entities can also be accessed from the right mouse button menu when the **Trim Entities** command is active.

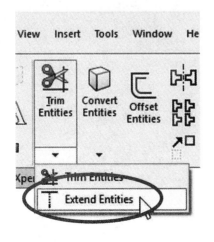

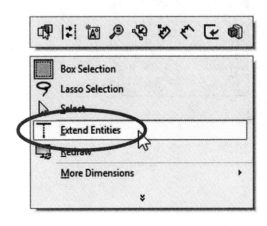

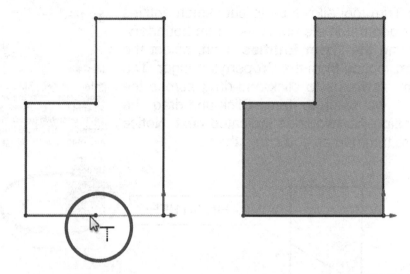

 Another way to extend the line is to turn OFF the Trim tool and click-and-drag its endpoint onto other entities, in this case the origin.

3.7. - Add the following dimensions to the sketch using the "**Smart Dimension**" tool from the Sketch tab in the CommandManager or the Mouse Gestures.

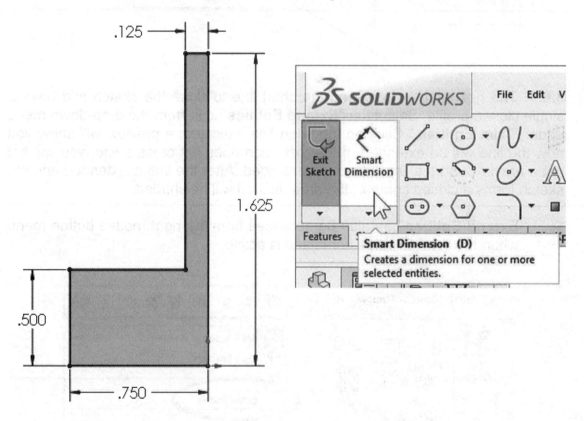

If needed, change the document's units of measure to inches with 3 decimal places.

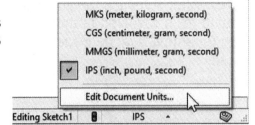

3.8. - After adding the dimensions, the Sketch is fully defined, and we can make the **Revolved Boss/Base**. Select the **Revolved Boss/Base** icon from the Features tab.

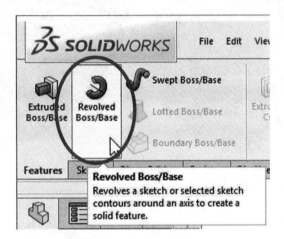

3.9. - The **Revolve** Property-Manager is presented and waits for us to select a line or centerline to make the revolved feature about it; if the sketch has a single centerline, it is automatically selected as the default axis of rotation. Select the line that we extended as the axis of rotation to make the revolved base.

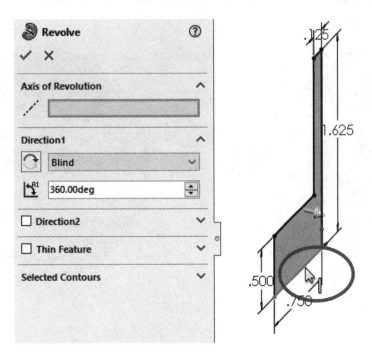

119

3.10. - After selecting the line, notice the preview in the graphics area. The default setting for a revolved feature is 360°. Click OK to complete the revolved feature and rename it "*Flange Base*."

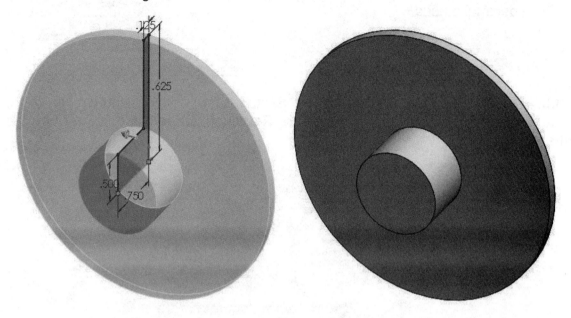

3.11. - Now switch to a Front view. We'll make a hole in the center of the cover for a shaft and a hole for screws to pass through. The two holes will be made at the same time with the same sketch.

Create a new sketch in the front most face of the cover (small round face) by selecting the small face and click in the "**Sketch**" icon from the pop-up toolbar. Draw a circle starting at the origin and dimension as indicated.

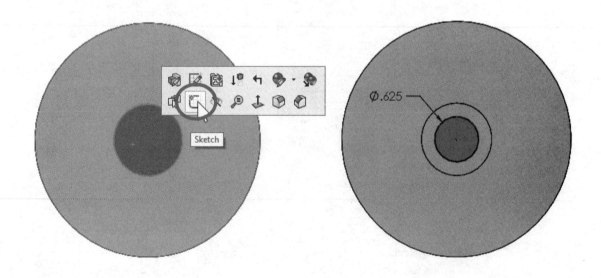

3.12. - After drawing the circle for the shaft's hole the next step is to draw the circle for the first screw. We'll make one hole, and then use a Circular Pattern to make the rest as we did in the *'Housing'*, but in this case we'll use a "**Cut Extrude**" feature to show a different approach. In this step we'll draw a circle and convert it to construction geometry to dimension the locating circle's diameter. Draw a circle as shown and convert it to "**Construction Geometry**" either by selecting it in the graphics area and clicking the icon from the pop-up toolbar or activating the "For construction" check box in the element's PropertyManager.

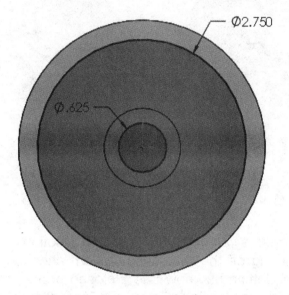

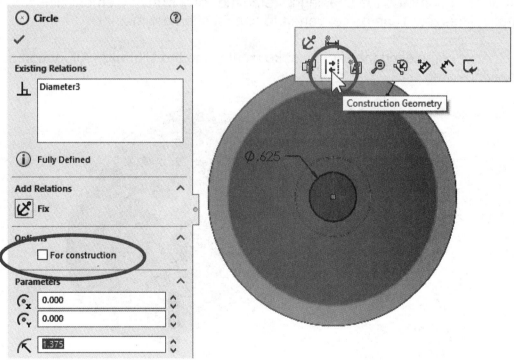

 An advantage of locating the screw's hole using this approach is that you can add a diameter dimension to the locating circle.

3.13. – After converting the large circle to construction geometry (also known as *reference geometry*), draw the circle for the hole making its center coincident with one of the quadrants of the reference circle as shown. Dimension the sketch and make a cut using the **"Extruded Cut"** command using the "Through All" option.

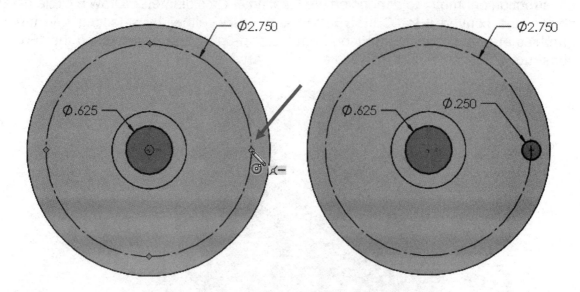

Geometric elements can be changed to construction geometry and back as needed by turning off the "For construction" checkbox. Note the circle is shaded at first, letting us know it is a closed profile; after it is changed to construction geometry, the shading is removed. Any sketch element (lines, arcs, circles, splines, etc.) can be converted to construction geometry

3.14. – After completing the sketch make a cut using the "Through All" option as in the previous step.

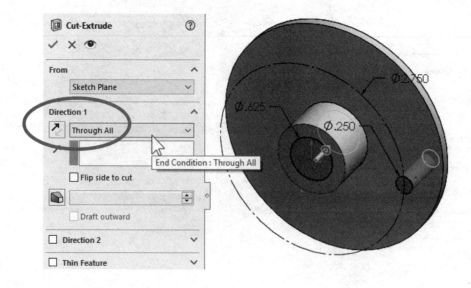

Note the Cut Extrude feature starts at the front-most face and goes through the entire part.

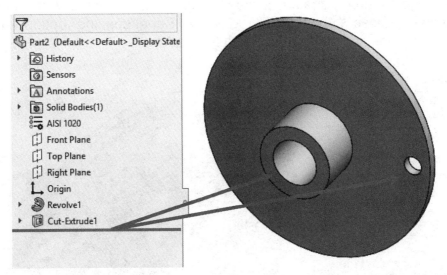

3.15. - To complete the rest of the screw holes we'll make a circular pattern. In the previous part we made a circular pattern of a feature, but in this case the feature includes two holes, one of which doesn't need to be patterned. In this step we'll make a pattern of a model's face, not a feature. Select the "**Circular Pattern**" icon from the Features tab in the CommandManager and select any circular edge for the pattern direction.

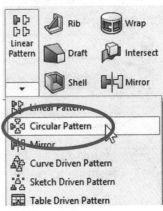

 For a face pattern (Linear, Circular or otherwise) to work, the resulting patterned faces must be geometrically the same as the original face, otherwise the pattern will fail.

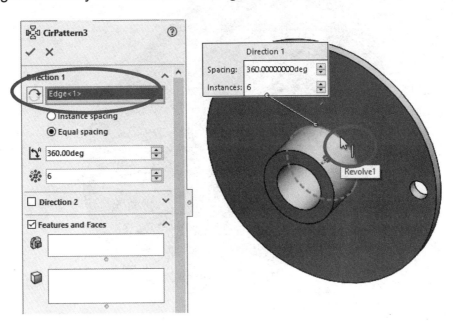

After selecting the pattern's direction click inside the "Faces to Pattern" selection box to activate it and select the screw hole's face. A face to pattern cannot be selected in the FeatureManager, it must be selected graphically. If needed, use the "Magnifying Glass" (Shortcut "G"). Use the "Equal spacing" option with a total number of 6 copies and click OK to finish the pattern.

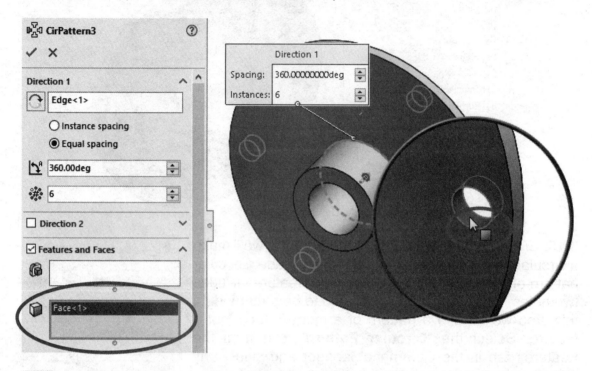

 Using faces to make patterns is particularly useful when we don't have a feature to pattern, for example when working with imported geometry.

3.16. - As a final step, select the "**Fillet**" command and round the indicated edge with a 0.125″ fillet radius.

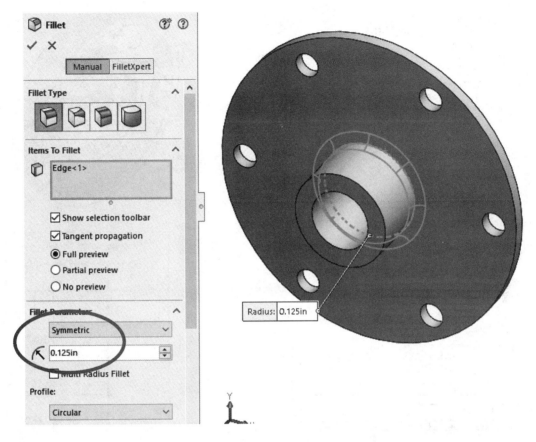

Save the finished component as *'Side Cover'* and close the file.

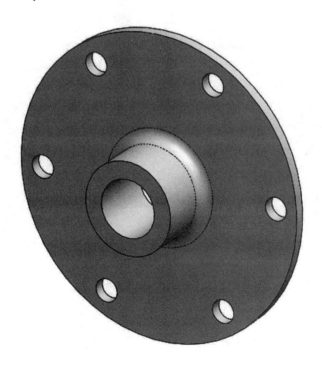

Exercises: Build the following parts using the knowledge acquired so far. Try to use the most efficient method to complete each model.

Exercise 5
DIMENSIONS: INCHES
MATERIAL:

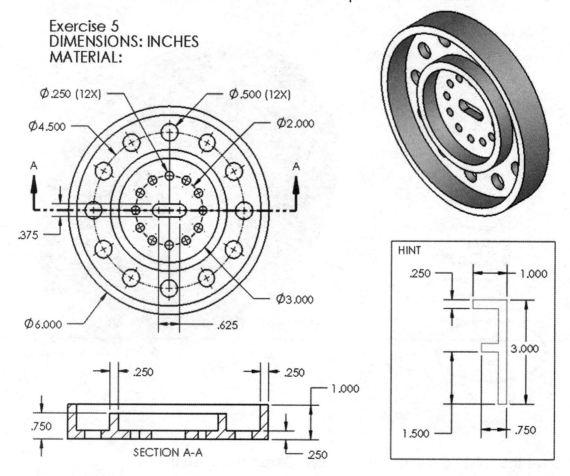

Ø.250 (12X)
Ø.500 (12X)
Ø4.500
Ø2.000
A
A
.375
Ø3.000
Ø6.000
.625

.250
.250
1.000
.750
.250
SECTION A-A

HINT
.250
1.000
3.000
1.500
.750

TIPS:
- Draw the profile and create the revolved feature.
- Add both holes with the same sketch; make one circular pattern.

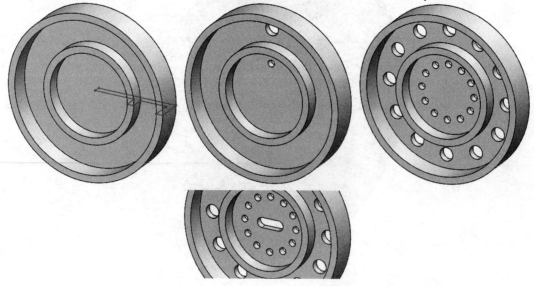

126

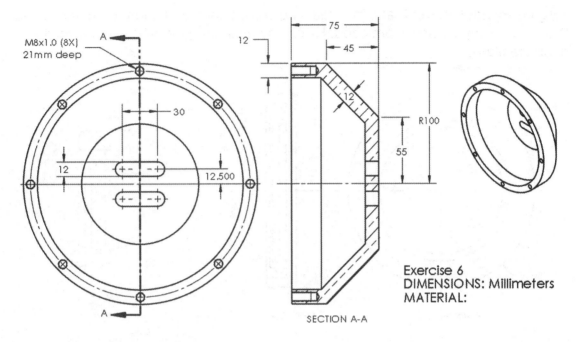

M8x1.0 (8X)
21mm deep

75

12

45

30

12

12.500

12

R100

55

Exercise 6
DIMENSIONS: Millimeters
MATERIAL:

SECTION A-A

TIPS:

- Draw the revolved profile and create the revolved feature. Using the Top or Right plane will give the same result.
- Add a slot, above or below, and mirror it about the Top Plane.
- Add the tapped hole; make the circular pattern.

Engine Project Parts: Make the following components to build the engine. Save the parts using the name provided. High resolution images are included on the exercise files.

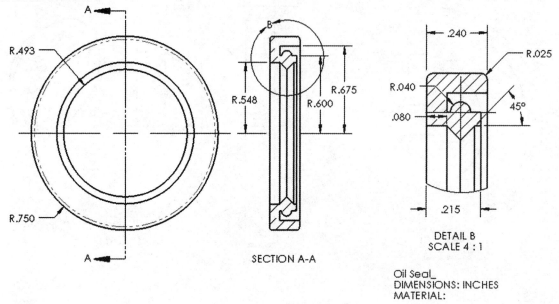

SECTION A-A

DETAIL B
SCALE 4 : 1

Oil Seal_
DIMENSIONS: INCHES
MATERIAL:

TIPS:

- The '*Oil Seal*' is made using a single feature.
- Draw the revolved feature profile, dimensions to the centerline.

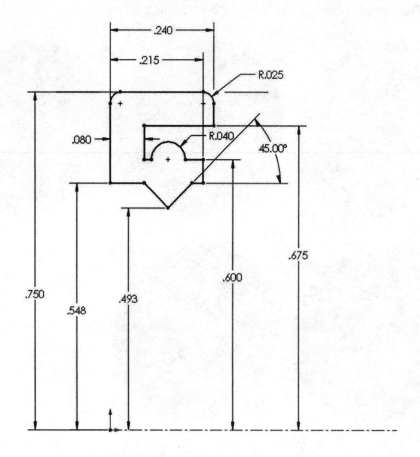

The Top Cover

Notes:

For the *'Top Cover'* part we will follow the next sequence of features. In this part we will learn a new feature called Shell, new options for Fillet, new end conditions for features (Boss, Cut, Revolve, etc.) and we'll practice some of the previously learned features and options.

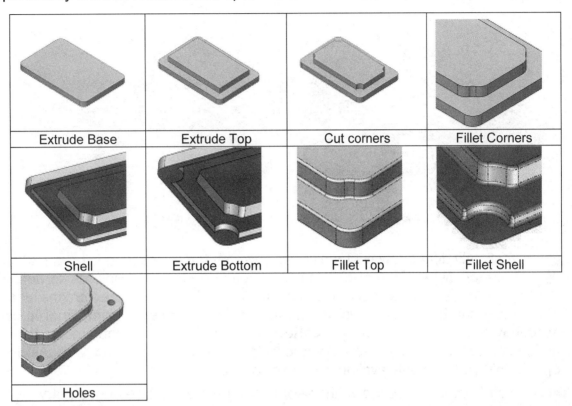

Extrude Base	Extrude Top	Cut corners	Fillet Corners
Shell	Extrude Bottom	Fillet Top	Fillet Shell
Holes			

4.1. - We'll start by making a new document using the "ANSI-Inch-Cast Alloy Steel" template from the "**New Document**" command, and just like we did with the *'Housing'*, create a new sketch. Click in the "**Extruded Boss**" command and then select the "*Top Plane.*" This part should already be set to inches with three decimal places.

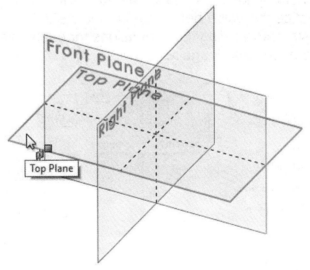

4.2. – Draw a rectangle using the "**Center Rectangle**" command; first click in the origin to locate the center and then outside to complete the rectangle. Add the dimensions shown with the "**Smart Dimension**" tool and round the corners with the "**Sketch Fillet**" command using a 0.25″ radius.

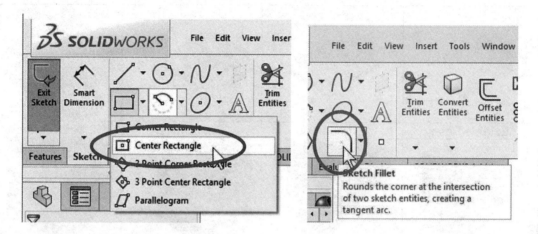

A slightly different way of making this sketch (or any sketch) is by using the "**Sketch Numeric Input**." What this option does is to allow us to define geometry's exact size as we sketch, making it the correct size from the start. This option can be turned on in the menu "**Tools, Options, Sketch, Enable on screen numeric input on entity creation**," or the right mouse button menu. If this option is enabled, it may also be a good idea to activate the "**Add Dimension**" option (this option is only available if a sketch tool is active).

- *Sketch Numeric Input* allows the user to enter the exact size of geometry.
- *Add Dimension* will add the dimensions to the geometry created.
- *Add Dimension* is only available when *Sketch Numeric Input* is activated and a sketch drawing tool is selected, as seen in the following images.

Select the "**Center Rectangle**" tool, make a right-mouse-button click and activate both options; click to locate the center of the rectangle *(do not click-and-drag)*. Notice the value boxes are immediately displayed as we move the mouse pointer. Now start typing the dimensions for each side of the rectangle followed by Enter; no need to click again.

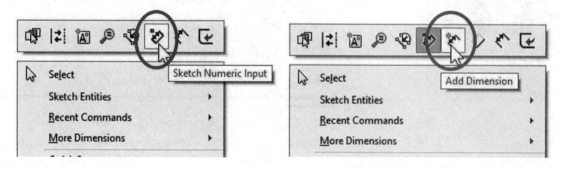

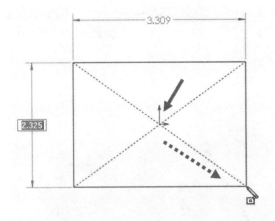

Click on center and move the mouse

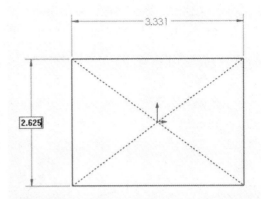

Type vertical dimension, press Enter

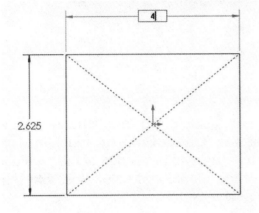

Type horizontal dimension, press Enter

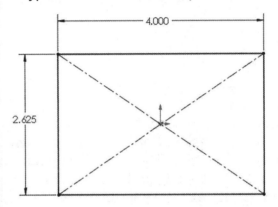

Finished rectangle with dimensions

4.3. – Using the "Sketch Fillet" command round the corners 0.25".

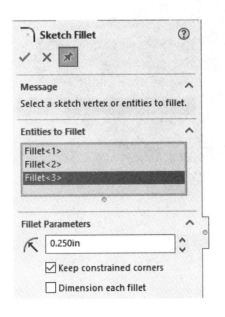

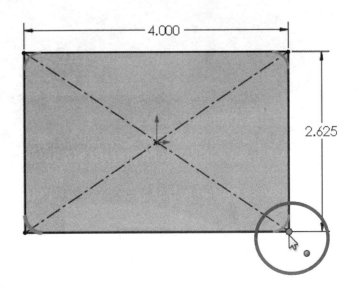

4.4. - To make the first extrusion, click on "**Exit Sketch**" or "**Extruded Boss/Base**" and set the extrusion's depth to 0.25″. Rename the extrusion *'Base'*.

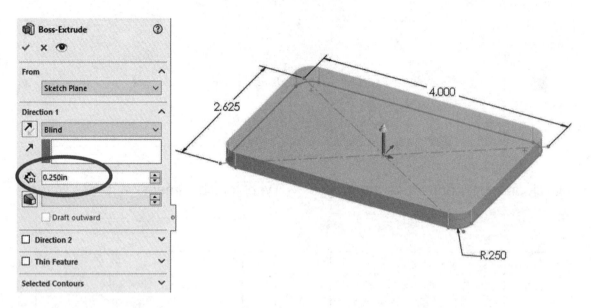

4.5. - For the second feature, we'll make an extrusion of similar shape to the first one, but smaller. For this feature we will use the "**Offset Entities**" function. This way the sketch will be created automatically by offsetting the edges of the previous feature face's edges. Click in the top face of the model and select the "**Sketch**" command from the pop-up toolbar to create a new sketch.

4.6. - After creating the sketch, notice the top face of the first feature remains selected (in highlighted color); while the face is selected, activate the "**Offset Entities**" icon in the Sketch tab of the CommandManager. A preview of the offset geometry is immediately visible.

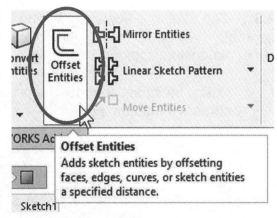

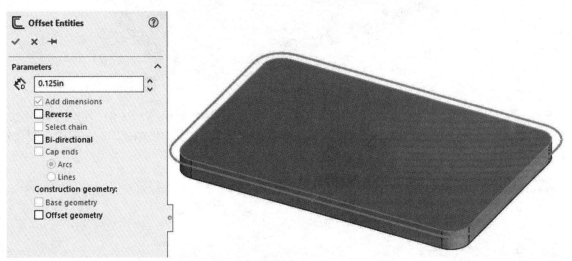

4.7. - Change the offset value to 0.375" and then click in the "Reverse" checkbox to make the offset inside, not outside. Notice that when we change the direction the preview updates accordingly. Click OK when done.

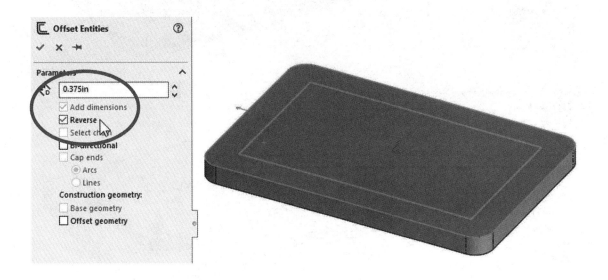

135

4.8. - The Sketch is now Fully Defined (all geometry is black) because the sketch geometry has an "Offset Entities" relation to the edges of the face and only the offset dimension is added.

 Notice the offset command is powerful enough to eliminate the rounds in the corners if needed.

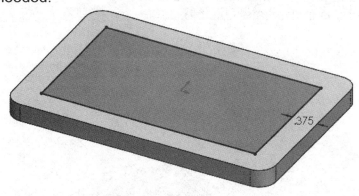

4.9. - To make the second feature select "**Extruded Boss**" from the Features toolbar in the CommandManager and extrude it 0.25″. Rename the feature "*Top Boss.*"

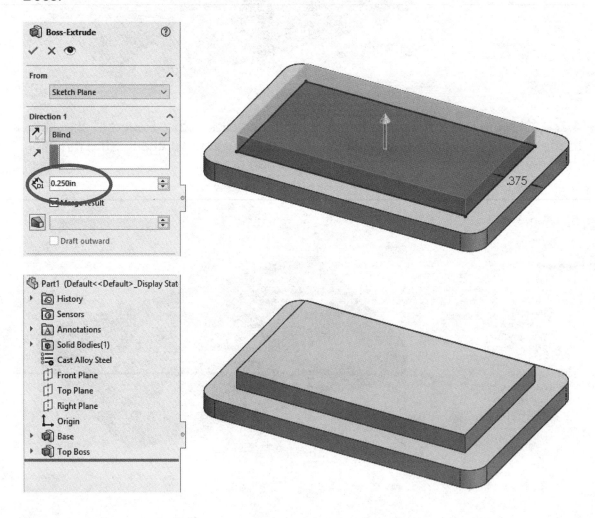

4.10. - For the next feature, we'll make round cuts in the corners of the top extrusion to allow space for a screw head, washer, and tools. Switch to a Top view and create a sketch in the topmost face. Make a circle as shown making sure the center is coincident to the corner. Feel free to use the Sketch Numeric Input in this sketch. Add two centerlines starting in the origin, one vertical, and one horizontal; they'll be used in the next step to create a **Sketch Mirror**. Change the part's view mode to "Hidden Lines Removed" for clarity.

 The "Sketch Numeric Input" option can be left on or turned off as needed. It will be turned off in the following exercises to make the explanation easier.

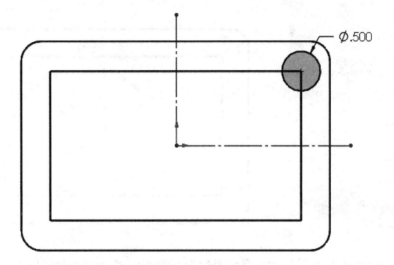

4.11. - We will now use the "**Mirror Entities**" command from the Sketch tab in the CommandManager. This tool will help us make an exact copy of any sketch entity, in this case the circle, about any straight line, edge, or for our example, the vertical centerline; then we'll copy both circles about the horizontal centerline to make a total of four equal circles. Select the "**Mirror Entities**" icon.

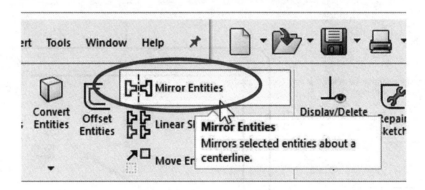

In the "Entities to Mirror" selection box, select the circle. Then click inside the "Mirror About" selection box to activate it (it will be highlighted) and select the vertical centerline. Click OK to complete the first sketch mirror.

In certain commands after making a selection the mouse pointer will change to let us know that pressing the Right Mouse Button will activate either the next selection box or finish the command, helping us reduce mouse travel and work more efficiently. If we ignore the Right Mouse Button it will be dismissed after moving the pointer.

Next Step   Finish/OK

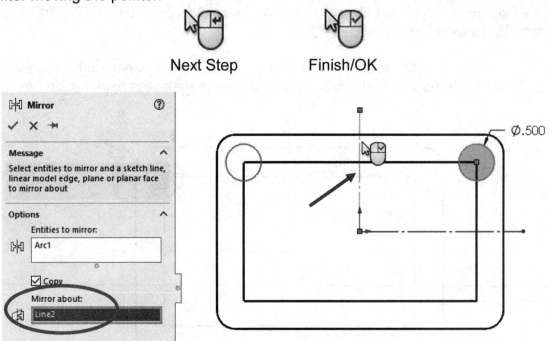

4.12. - Now repeat the "**Mirror Entities**" command selecting both circles in the "Entities to Mirror" selection box and using the Horizontal centerline in the "Mirror About" selection box. Click OK to finish the Mirror. Since the new circles are mirror copies of the original, and the original was fully defined, the sketch is therefore fully defined. The mirrored circles have an automatically generated "Symmetric" geometric relation about the mirror centerline.

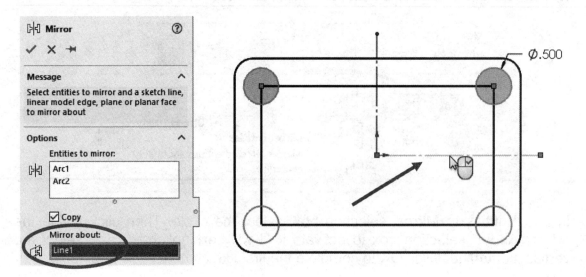

4.13. - We are now ready to make the cut. In this step we'll cut all four corners at the same time. SOLIDWORKS allows us to have multiple closed contours in a sketch for one operation as long as they don't intersect or touch each other in one point. To add intelligence to our model (Design intent) we'll use an end condition for the cut called **Up to Surface**; with this end condition we can define the stopping face for the cut instead of giving it a depth. Select the "**Extruded Cut**" icon and select "Up To Surface" from the "Direction 1" options drop down selection box. A new selection box is displayed and activated; this is where we'll select the face where we want the cut to stop. Select the face indicated as the end condition, click OK to finish the feature and rename it "*Corner Cuts.*"

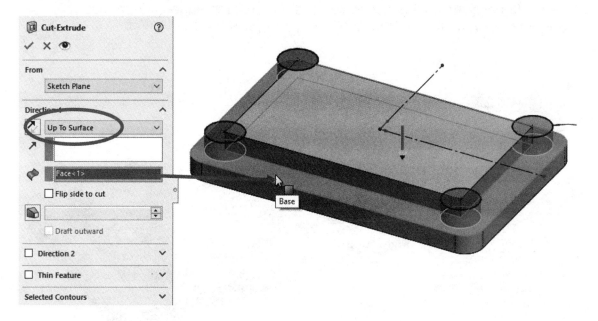

 The reason for selecting a face as an end condition is if the height of the "Top Boss" changes, the cut will still go up to the intended depth. This is how design intent is maintained and intelligence added to our model.

Our part now looks like this:

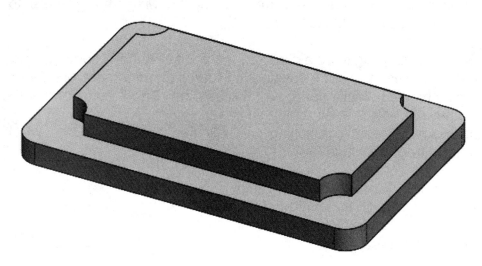

4.14. - Using the "**Fillet**" command add a 0.25″ radius fillet to the edges indicated in the corner cuts. There are eight (8) edges to be rounded. To make the selection easier, SOLIDWORKS has a built-in tool to help us select multiple edges. Make sure the "Show selection Toolbar" is activated in the "**Fillet**" options, after selecting the first edge, a pop-up toolbar gives the user options to select different groups of edges. Moving the mouse over the different selection options highlights the edges that would be selected. In this example the "Connected to end loop" option selects exactly the edges we are interested in. In subsequent exercises the user will be able to explore other selection options.

After clicking on the icon to select the rest of the edges, the mouse changes to give us the OK/Finish in the right mouse button.

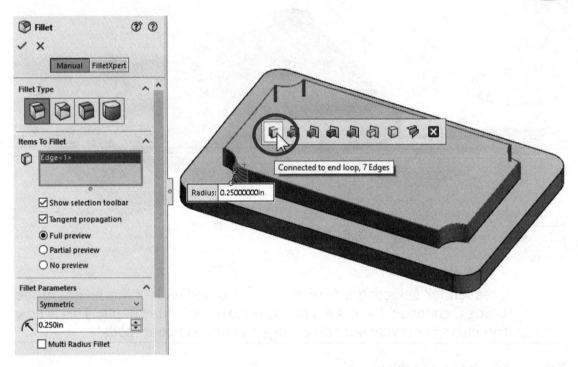

 If selecting small edges is difficult, try using the "**Magnifying Glass**" to make selection easier (default shortcut "G").

4.15. - Since this is going to be a cast part, we want to remove some material from the inside, and make its walls a constant thickness (common practice for castings and injection molded plastic parts). In this case the "**Shell**" command is the best tool for the job. The Shell creates a constant thickness part by removing one or more faces from the model.

Select the "**Shell**" icon from the Features tab in the CommandManager. This is an applied feature, which means it does not require a sketch.

In the Shell's PropertyManager under "Parameters," set the wall thickness to 0.125″, rotate the part and select the bottom face; this is the face that will be removed making the remaining faces in the part 0.125″ thick. Turn on the "Show Preview" option to see the result. Click OK to finish the command. Since we only have one shell feature in this part, there is no need to rename it.

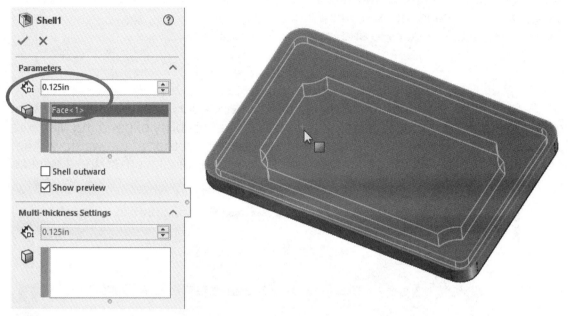

 If no faces are selected the "**Shell**" command creates a hollow part.

 The "Shell outward" option adds material outside, essentially growing the part by the shell thickness value, making the "outside" faces, "inside" faces.

After the shell feature the part looks like this, where every face in the part is 0.125″ thick.

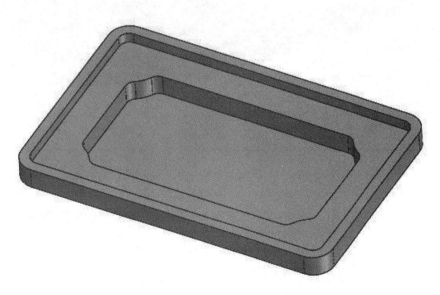

4.16. - How can we tell if the walls are really 0.125″ thick? Using the "**Measure**" tool. It is located under the Evaluate tab in the CommandManager. This tool is like a digital measuring tape, where we can select faces, edges, vertices, axes, planes, coordinate systems, or sketch entities to measure to and from.

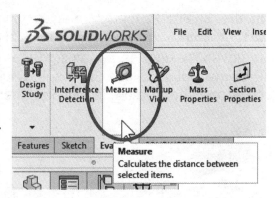

Select the Evaluate tab in the CommandManager, activate the "**Measure**" tool and select the indicated face and edge; for better visibility expand the Measure box clicking in the double arrow icon in the upper right corner of the window. After the entities are selected, the measurements will be displayed in the Measure window and the tag in the graphics area.

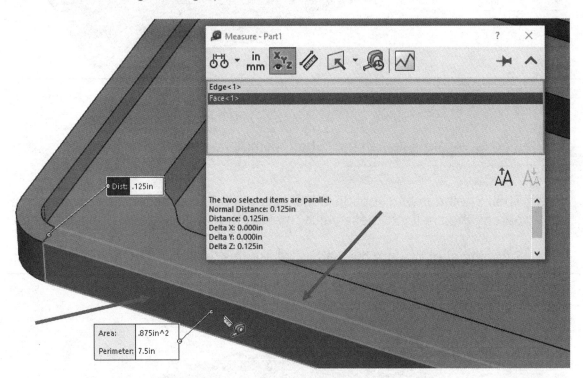

 If the Point-to-Point option is turned ON, we will be shown the distance between the selected points in all three axes (Delta X, Y and Z).

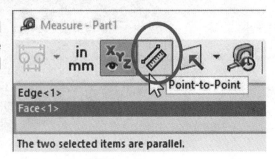

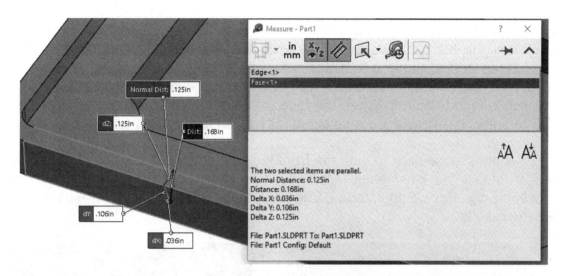

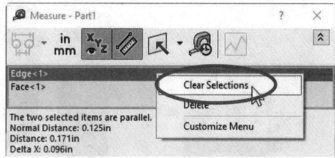

To measure a different set of entities, click on an empty area of the graphics window, or make a right-mouse-click inside the selection box, and select "**Clear Selections**." This option works with any selection box.

4.17. - To select a hidden face without having to rotate the component, make a right-mouse-click near the face that we want to select, and use the "**Select Other**" command from the pop-up menu. SOLIDWORKS automatically hides this face allowing us to see other faces behind it.

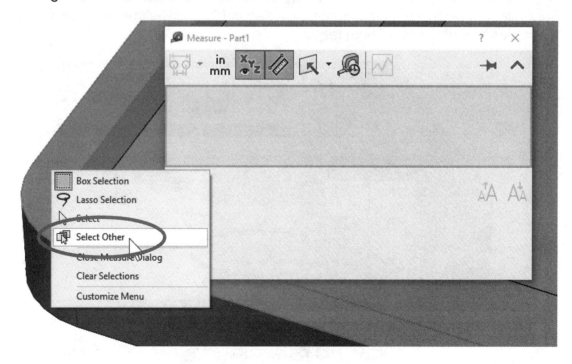

After this face is hidden, when we touch any of the remaining model faces they are highlighted; this way we know which face we are selecting. We also get a list of faces behind the one that was removed where we can select the face we need. If we still cannot see the face we need to select, we can right-mouse-click in any visible face to hide it. When the face we want to select is visible, left-mouse-click to select it. After a face is selected, all hidden faces are made visible again.

Right-mouse-click to hide the front face as indicated, and then left-mouse-click to select the inner shell face. Feel free to make different measurements between Edge-Face, Edge-Vertex, Edge-Edge, Face-Vertex, and Vertex-Vertex to see the results.

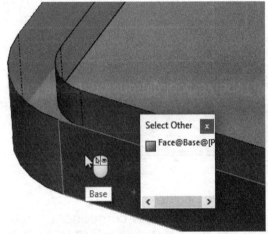

Right mouse button to hide face   **Left mouse button** to select hidden face

When we select a face first we are immediately presented with the area and perimeter of the selected face. Selecting a second face gives the distance between them plus the total area of both faces.

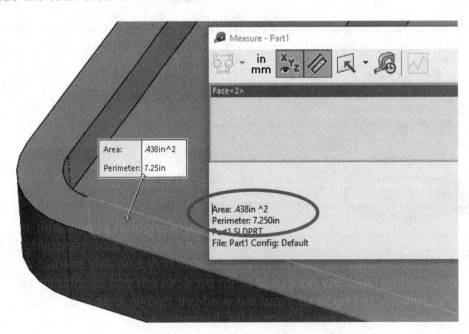

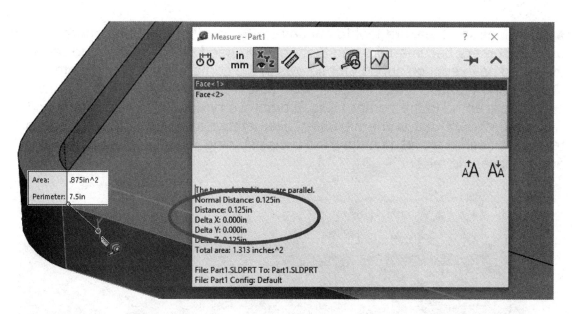

 In the "**Measure**" tool options, we can change the units of measure and precision if needed.

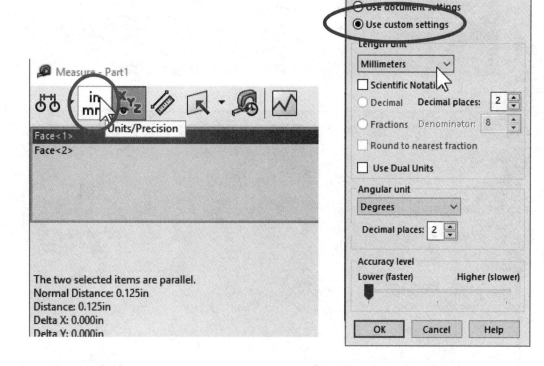

4.18. - Close the Measure tool to continue. After adding the Shell operation, the outside corners become thin, and therefore we need to add material to reinforce them and have enough support for screws in the corners. Switch to a Bottom view (Ctrl+6) and create a sketch on the bottommost face as indicated and draw four circles concentric to the corner fillets. Do not worry about the size of the circles; we'll take care of that in the next step. Remember to touch the round edges to reveal their centers, and then draw the circles starting in the center to automatically add a concentric relation.

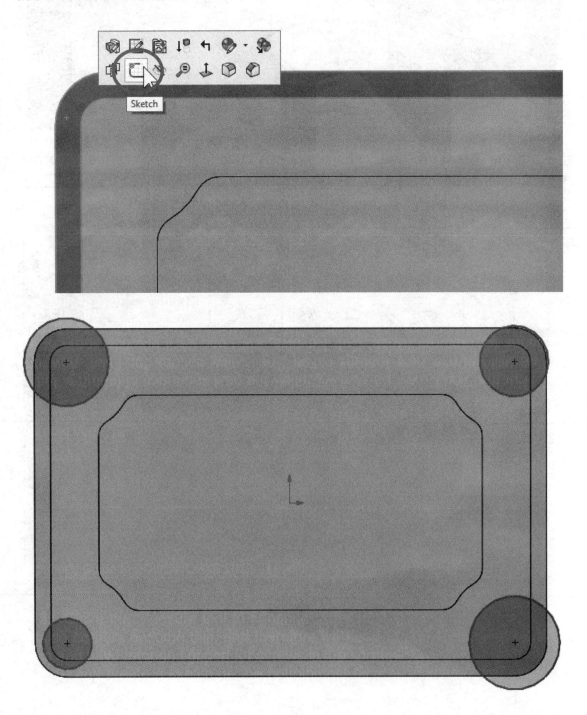

4.19. - To maintain the design intent we are going to make the circles the same size as the corner fillets using an Equal geometric relation. Select the "**Add Relation**" icon from the Sketch Tab in the CommandManager. Select all four circles, and under "Add Relations" select **Equal** to make all circles the same size.

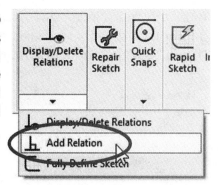

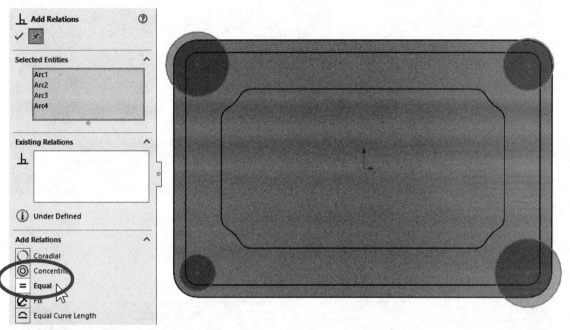

Optionally, window select or Ctrl + Select all four circles and add the relation from the pop-up menu or the PropertyManager.

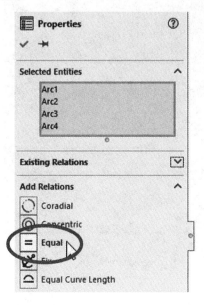

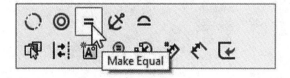

After the four circles are made the same size, select one of them and another geometric relation to make it either "Coradial" (same size and concentric) or "Equal" to any rounded corner to fully define the sketch.

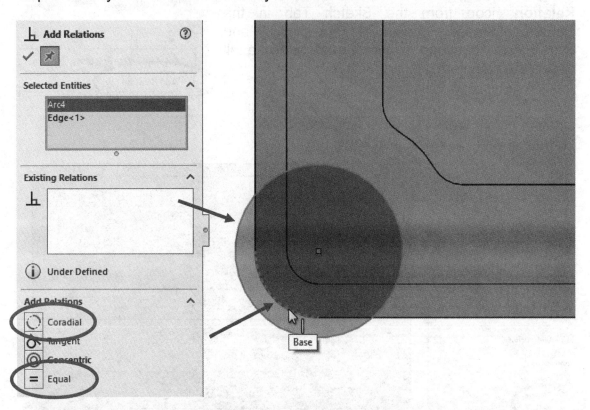

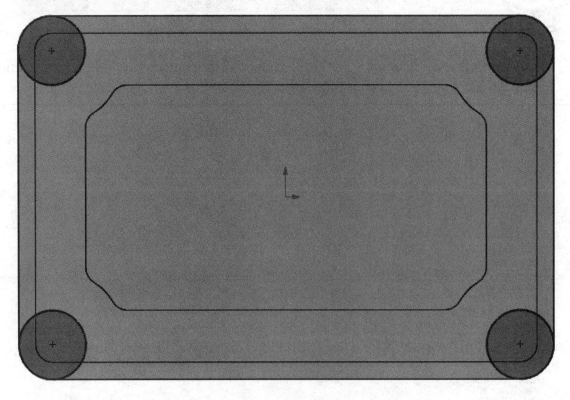

4.20. - We will now make the extrusion using the "**Up to Surface**" end condition as we did in the last cut. Select the "**Extrude Boss/Base**" command from the Features tab and use the option "**Up to Surface**" end condition. Select the face indicated as the end condition and click OK. By doing the extrusion this way we can be sure that our design will update as expected if any of the previous features changes.

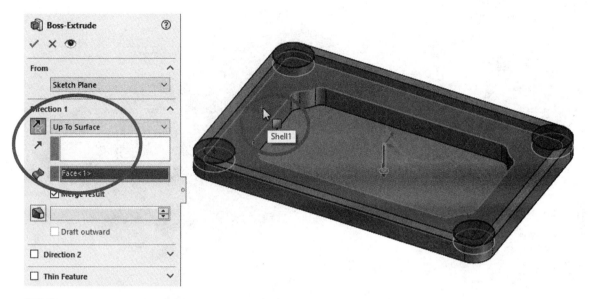

 Shortcut: Double-clicking a face automatically sets the end condition to "Up to Surface" using the selected face.

Our part should now look like this:

4.21. - Now we need to add a 0.031″ radius fillet using the "**Fillet**" command from the Features tab. Select the two faces on top of the cover as indicated. Notice all the edges on the top side of the part are rounded with only two selections, maintaining our design intent and making our job easier at the same time.

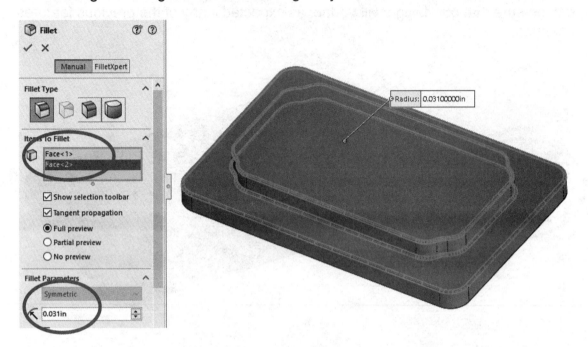

4.22. - To add fillets to all the inside edges of the part we will use a slightly different approach. Instead of individually selecting the inside edges or faces, we will only select the "Shell1" feature from the fly-out FeatureManager and add a 0.031″ radius. Adding the fillet using this technique will round every edge of the "Shell1" feature, making it faster and convenient maintaining our design intent better.

Selecting a feature to fillet as in this example can only be done using the fly-out FeatureManager or pre-selecting it in the FeatureManager before selecting the "**Fillet**" command.

Now every edge inside is rounded in one operation with a single selection.

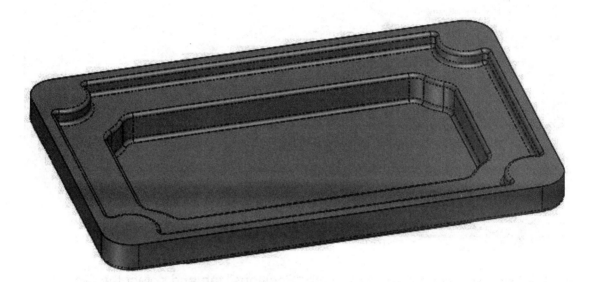

4.23. - Now we need to add four clearance holes for #6-32 screws in the corners. We'll use the "**Hole Wizard**" feature from the Features tab. Switch to a Top view for visibility and select the "**Hole Wizard**" icon.

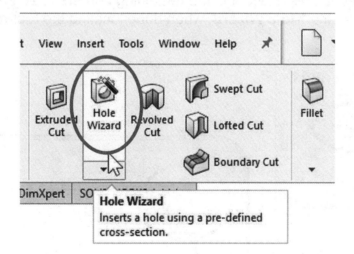

In the first step select the "Hole" specification icon. From the "Type" drop down list select "**Screw Clearances**." From the "Size" selection list pick "# 6" and from "End Condition" select "Through All." This will create a hole big enough for a #6 size screw to pass freely through it.

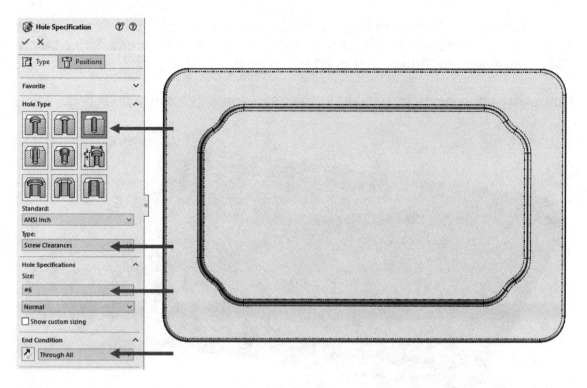

4.24. - For the second step, click in the "Positions" tab and select the face indicated to add the hole's location sketch. After selecting the face, the **"Point"** tool is automatically selected, and we are ready to add four points for the holes' centers. Touch the round corner edges to reveal their centers and click in one center to locate the hole. Repeat in the remaining corners to add all four holes. Click OK to finish the "**Hole Wizard**" when done locating the four holes.

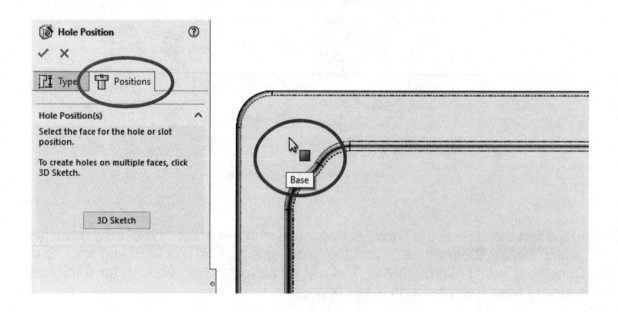

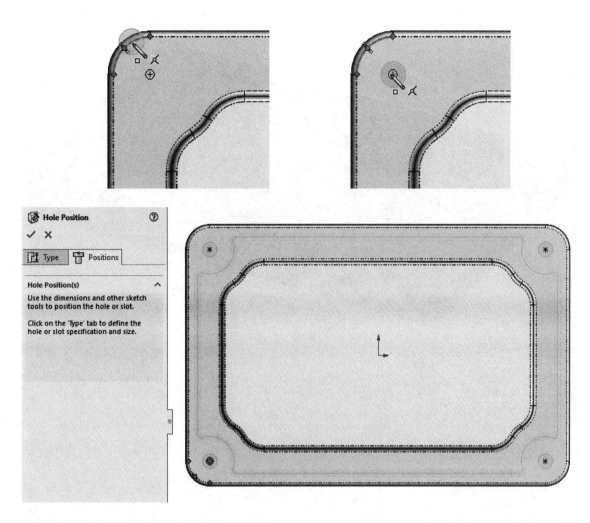

4.25. - Save the part as *'Top Cover'* and close the file. The finished part should look like this.

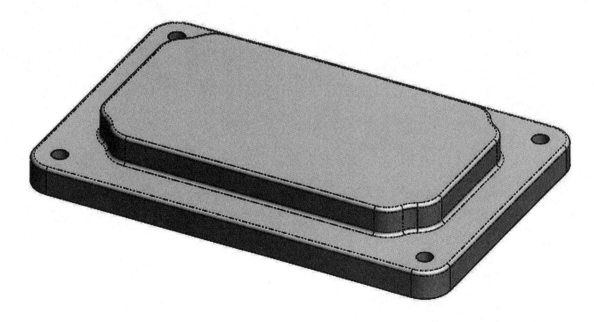

Exercises: Build the following parts using the knowledge acquired in this lesson. Try to use the most efficient method to complete the model.

Exercise 7
DIMENSIONS: INCHES
MATERIAL: AISI 1020

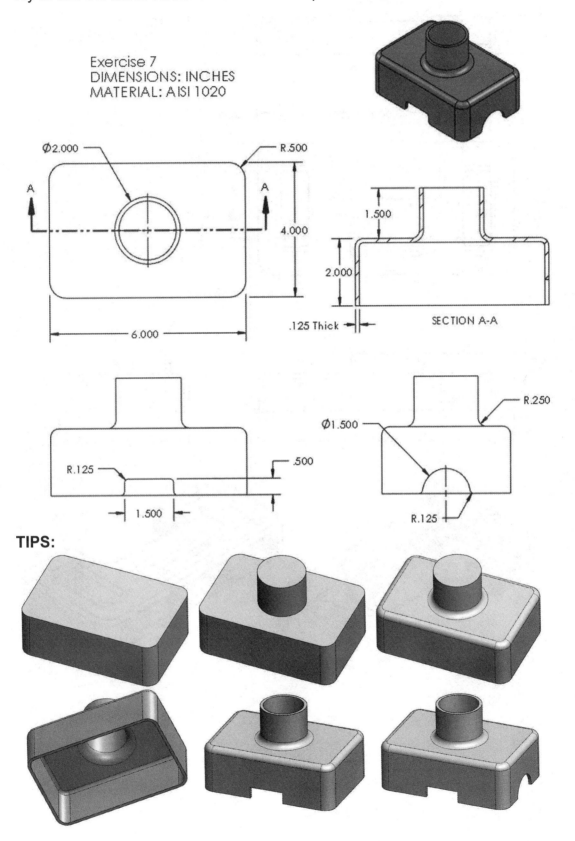

Ø2.000

R.500

A

A

4.000

6.000

1.500

2.000

.125 Thick

SECTION A-A

R.125

.500

1.500

Ø1.500

R.250

R.125

TIPS:

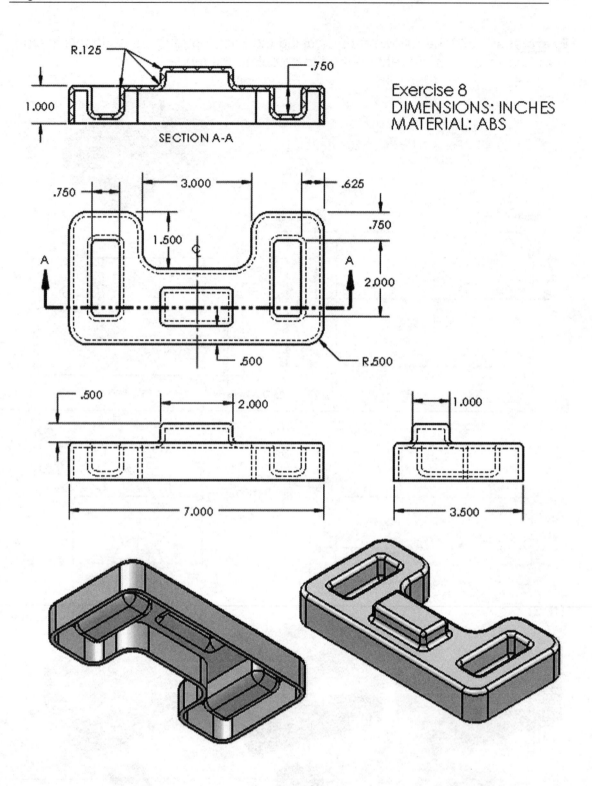

R.125

.750

1.000

SECTION A-A

Exercise 8
DIMENSIONS: INCHES
MATERIAL: ABS

.750

3.000

.625

1.500

.750

2.000

A

A

.500

R.500

.500

2.000

1.000

7.000

3.500

TIPS:

- Add the top boss and the cuts before adding the fillets.
- When adding the fillets select the features, not the faces or edges. All fillets can be done with three selections.
- Add the shell feature at the end.

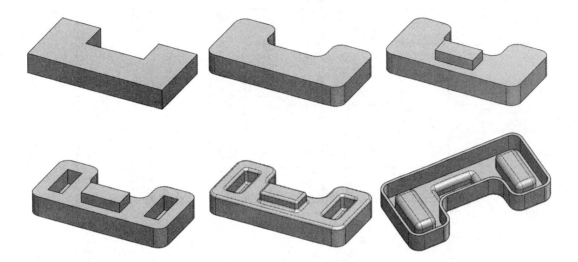

Engine Project Parts: Make the following components to build the engine. Save the parts using the name provided. High resolution images are included on the accompanying files.

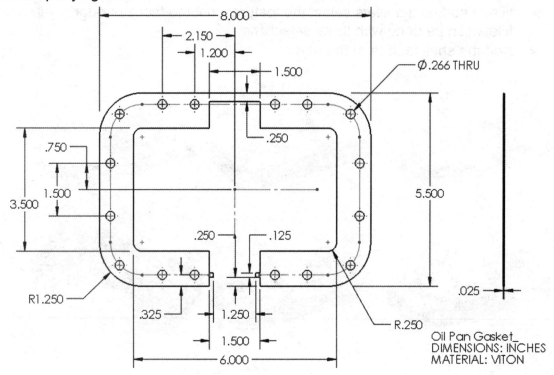

Oil Pan Gasket_
DIMENSIONS: INCHES
MATERIAL: VITON

TIPS:

- The gasket's profile can be made in the first sketch, then add the holes in the second step.

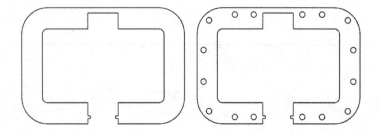

- Alternatively, the gasket can be drawn using a closed contour first, then add the cuts as secondary operations using separate features.

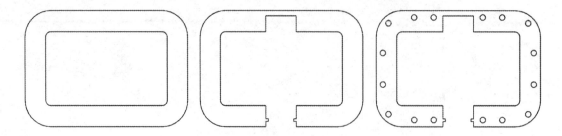

- To add the sketch for the gasket holes, pre-select the outside edges using a right-mouse-click and select "Select Tangency" before using the "Offset Sketch" tool.

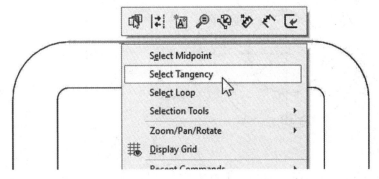

- Make the sketch offset to locate the gasket holes using the "Construction Geometry: Offset geometry" option; this way the newly created offset geometry will be automatically converted to construction geometry

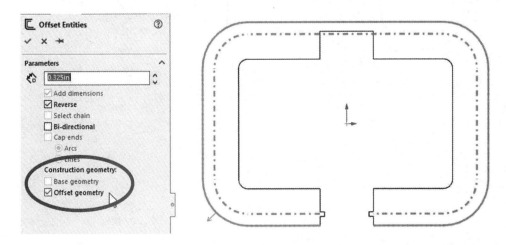

- Add one quarter of the holes and mirror vertically and horizontally either in the sketch or as a feature.

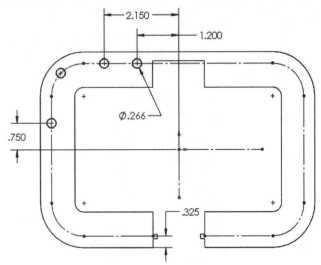

159

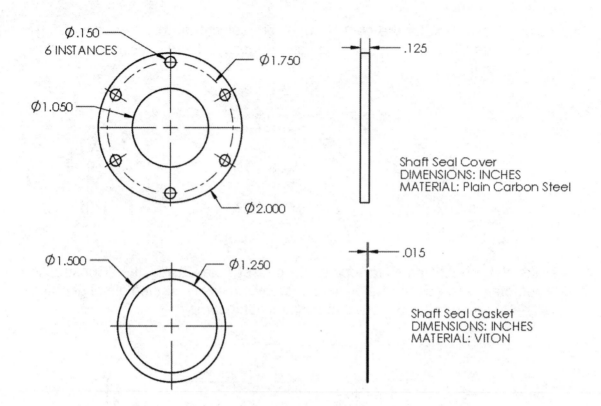

Ø.150
6 INSTANCES

Ø1.750

Ø1.050

Ø2.000

.125

Shaft Seal Cover
DIMENSIONS: INCHES
MATERIAL: Plain Carbon Steel

Ø1.500

Ø1.250

.015

Shaft Seal Gasket
DIMENSIONS: INCHES
MATERIAL: VITON

The Offset Shaft

Notes:

For the *'Offset Shaft'* we'll follow the next sequence of operations. In this part we'll learn how to make polygons in the sketch, a new option for the **"Cut Extrude"** feature, auxiliary planes and a Revolved cut.

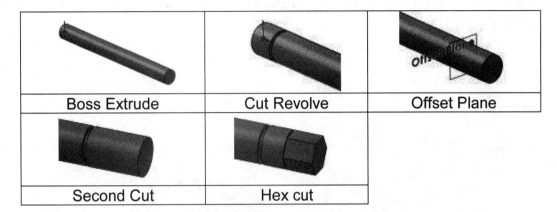

Boss Extrude	Cut Revolve	Offset Plane
Second Cut	Hex cut	

5.1. - Start by making a new document using a part template in Inches and set the material to "Chrome Stainless Steel." For the first feature in this part create a sketch in the *"Right Plane."* A different way to create a sketch to how we've been doing it so far is to select the *"Right Plane"* in the FeatureManager, and from the pop-up toolbar select **"Sketch."**

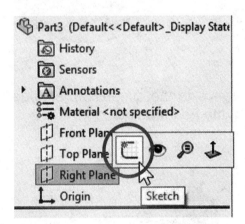

The view will be oriented to the Right view automatically. Draw a circle starting at the origin and dimension it 0.600" using the **"Smart Dimension"** tool. Since this shaft will need to meet certain tolerances for assembly, we will give a tolerance to the diameter.

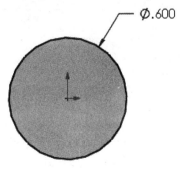

5.2. - To add (or change) the tolerance of the shaft's diameter, select the dimension in the graphics area. Notice the dimension's properties are displayed in the PropertyManager. This is where we can change the tolerance type. For this shaft select "Bilateral" from the **"Tolerance/Precision"** options box and add +0.000"/-0.005".

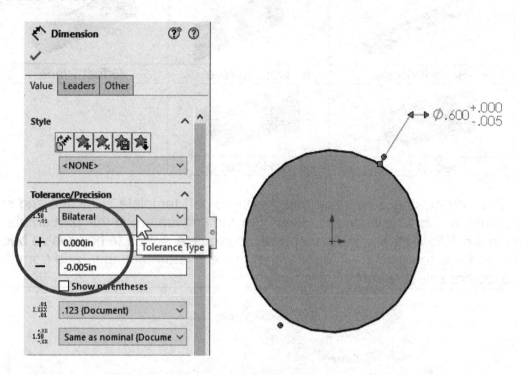

5.3. - After adding the tolerance to the dimension, extrude the shaft 6.5" using **"Extruded Boss/Base"** from the Features tab.

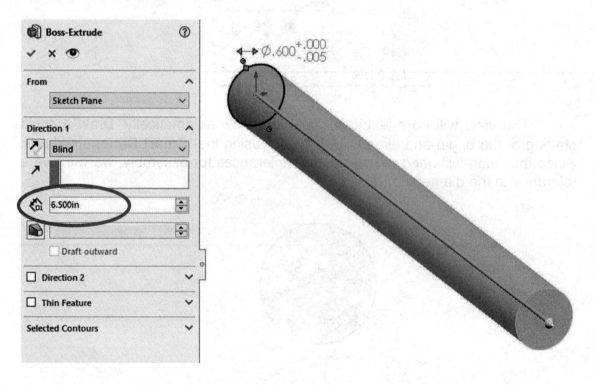

5.4. - For the second feature we'll make a "**Revolved Cut**." As its name implies, we'll remove material from the part similar to a turning (Lathe) operation. Switch to a Front view and select the "*Front Plane*" from the FeatureManager. From the pop-up menu select the "**Sketch**" icon as before to create a new sketch on it. Draw the following sketch and be sure to add the centerline. This is the profile that will be used as a "cutting tool." The centerline will be automatically selected as the axis of revolution for the cut.

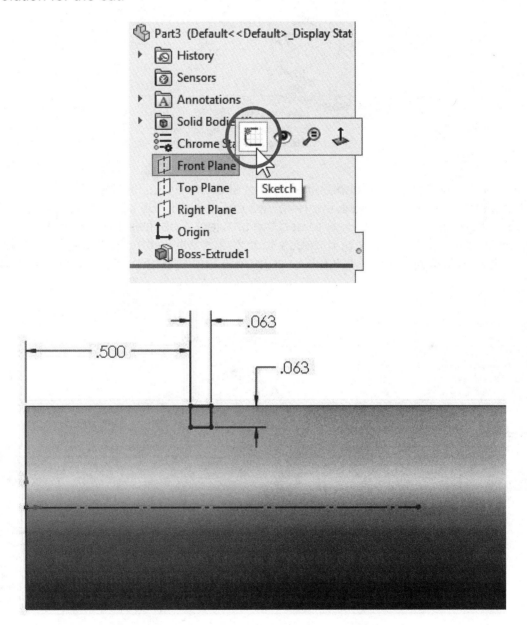

 The reason for selecting the "*Front Plane*" for this sketch is because there are no flat model faces that could be used to create the sketch in this orientation.

5.5. - Now that the sketch is finished, select the "**Revolved Cut**" icon from the Features toolbar.

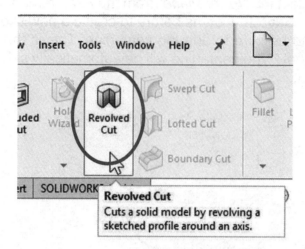

Since we only have one centerline in the sketch, SOLIDWORKS automatically selects it as the axis to make the revolved cut about it. By default, a revolved cut is 360 degrees. Rotate the view to see the preview. Feel free to use the value spin box to change the number of degrees to cut; this will better illustrate the effect of a revolve cut very clearly. Click OK to complete the feature and rename it *'Groove'*.

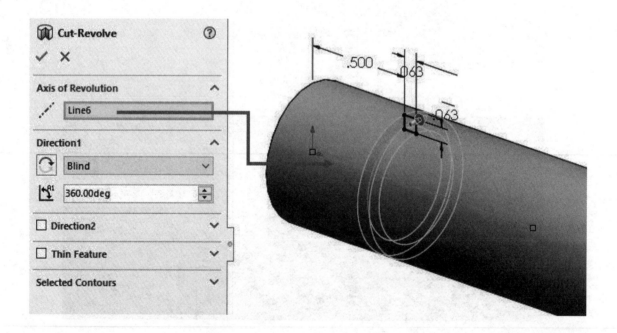

 If there are two or more centerlines in the sketch or none, we will be asked to select a line or model edge to be used for the axis of revolution.

Our part now looks like this.

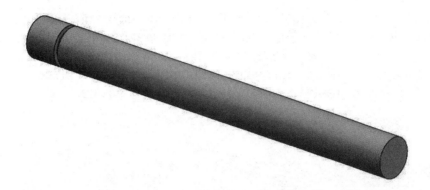

5.6. - For the next feature, we need to make a cut exactly like the one we just did, but in the right side of the shaft. We could have done it at the same time with the previously made Revolved Cut feature by adding an extra profile, but we'll show a different way to make it and learn additional functionality at the same time.

For this feature, we'll create an auxiliary plane for a new sketch. Auxiliary Planes help us locate new features or sketches where we don't have any flat faces or planes to use. To create an auxiliary plane, select "**Reference Geometry, Plane**" from the Features tab in the CommandManager or from the menu "**Insert, Reference Geometry, Plane**." Planes can be defined using many different options using model vertices, edges, faces, sketch geometry, existing planes, and/or axes.

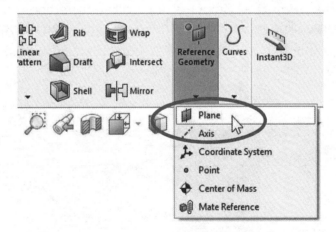

As we start selecting references to define a new plane, the possible options to define it are dynamically shown in the PropertyManager, along with a preview of the plane in the graphics area. Some plane definitions require 1, 2, or 3 references depending on each case, but SOLIDWORKS helps us by letting us know when the necessary options have been selected with a "**Fully Defined**" message at the top of the PropertyManager.

 Notice that references are color coded in the PropertyManager and matched in the graphics area to easily identify them.

Here are some of the most common ways to define an auxiliary plane:

Offset Distance – Select a Plane or flat face and define the distance from it to create the new Plane. "Flip offset" can be used to change the side on which the new plane is created. In this case the face on the left was selected and the new plane created to the right. We can optionally create multiple parallel planes changing the number of planes to create (default is 1). One reference is needed.

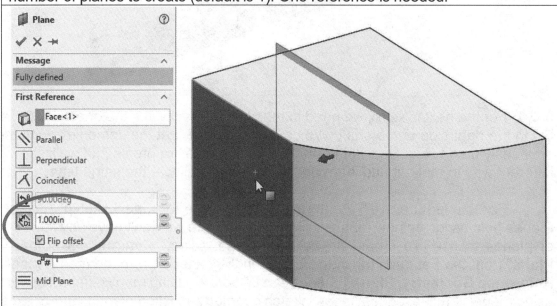

Through Lines/Points – Select 3 non-collinear vertices, or one straight edge and one non-collinear vertex. A line (or edge) and a point or three points (vertices) are needed.

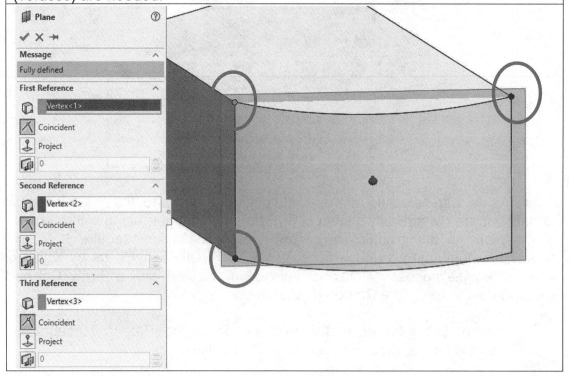

Parallel Plane at Point – Select an existing Plane or a flat face and a point/vertex. Note that when we select a Plane or flat face we get additional options like Parallel, Perpendicular, Coincident, at angle or distance. Two references are needed.

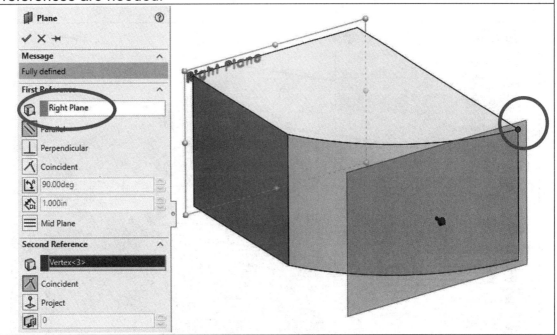

At an angle – Select a Plane or flat face and an Edge and enter the angle. The Plane's direction can be reversed using the "Flip" checkbox. In this case the face in the back was selected and the indicated Edge. The Edge acts as a "hinge" to the new plane. Change the angle using the value spin box to see the effect. The "Flip" checkbox will change the direction of the plane. A plane/flat face and an edge/linear sketch entity are needed.

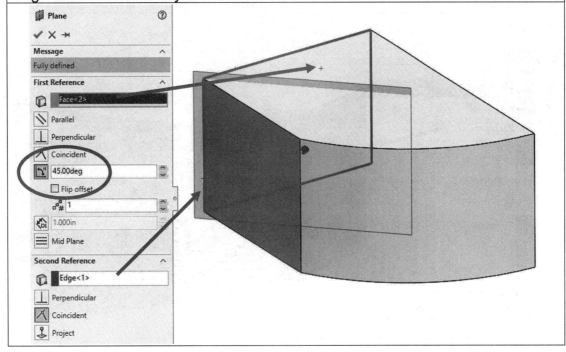

Normal to Curve – The plane is created by selecting an edge and a vertex (or a point) in the edge. The plane created is perpendicular to the curve at the vertex (point). In this case we selected the curved edge and the vertex indicated. Two references are needed.

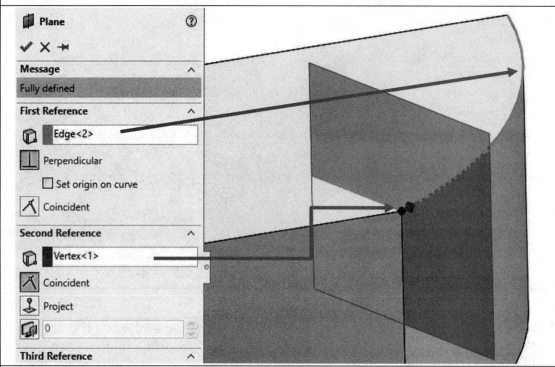

On Surface – The plane is created selecting a surface (any surface) and a vertex (or sketch point/endpoint) on the surface. In this case the face is curved and the resulting plane is tangent to the face at the selected vertex. Two references are needed.

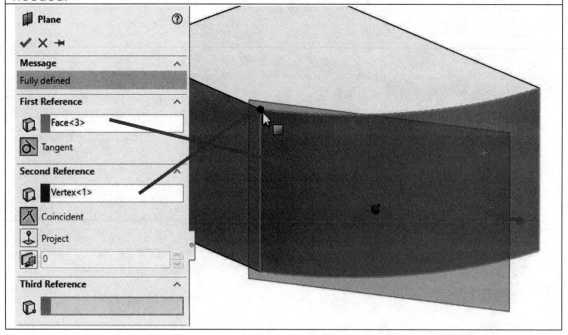

Mid Plane – The plane is created between two selected planes or faces. If the planes and/or faces are not parallel the new plane will be created at an angle half the angle between the reference planes/faces.

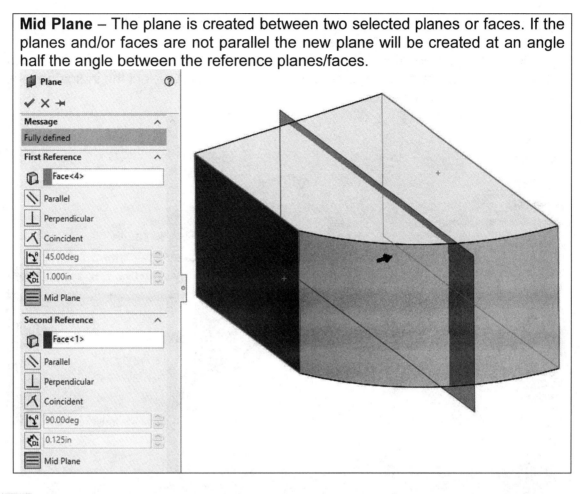

When creating auxiliary planes, we can use as references existing Planes, Faces, Vertices, sketch elements, Axes and Temporary Axes, the Origin, etc. To make sketch elements visible, expand a feature in the FeatureManager, select the sketch that we want to make visible, and from the pop-up toolbar select the **Show/Hide** icon. If the sketch is hidden, it will be made visible and vice versa.

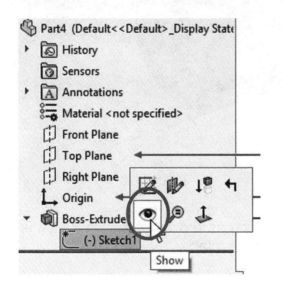

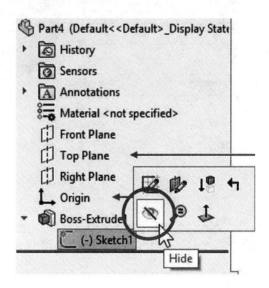

5.7. - Back to our part, we'll make an Auxiliary Plane parallel to a face of the groove. Select "**Auxiliary Geometry, Plane**" from the Features tab. We'll make an auxiliary plane a set distance from the right face of the groove. Using the "**Select Other**" function (right mouse button menu or pop-up icon), select the indicated face below (hidden in this view).

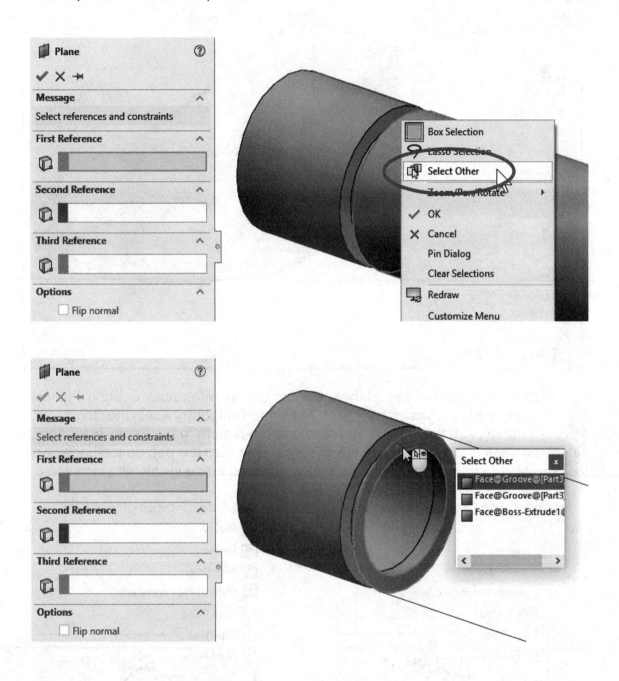

 Notice the mouse pointer changes, letting us know that the left button is to select a face, and the right button will hide a face.

Set the distance to 5″ and select the "Flip offset" checkbox if necessary. Notice the preview. After creating the plane, rename it to *"Offset Plane."*

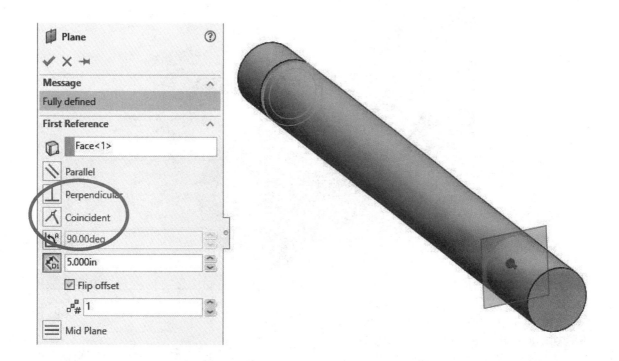

 If the new plane is not visible, use the "**Hide/Show Items**" command. "*Front Plane*," "*Top Plane*" and "*Right Plane*" are hidden by default and can be hidden/shown using the "Hide/Show Primary Planes" command; you can hide and show any plane the same way we can hide a sketch. Notice we added the keyboard shortcut "P" to hide/show auxiliary planes.

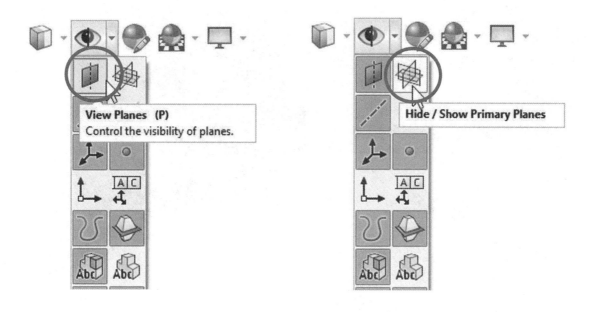

5.8. - Switch to an Isometric view (for visibility) and select the *"Offset Plane"* in the graphics area; from the pop-up toolbar click on "**Sketch**" to create a new sketch on it.

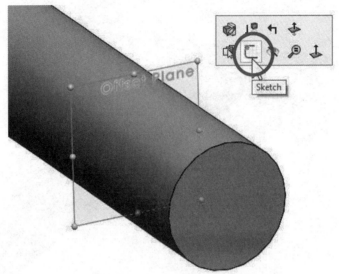

The "**Convert Entities**" command is used to project existing geometric entities onto a sketch, such as model edges or other sketch entities and convert them to new sketch entities at the same time. Click on "**Convert Entities**" from the Sketch tab in the CommandManager; select the two edges indicated from the "*Groove*" feature and click OK to finish.

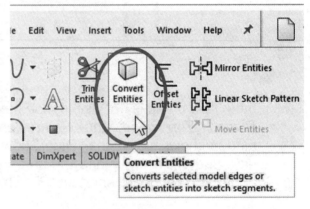

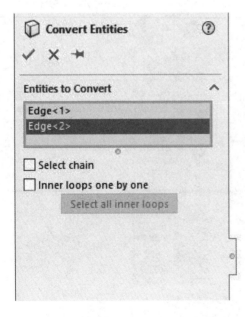

5.9. - The two edges are projected onto the sketch plane and are automatically fully defined since they are a projected copy of the edges they came from, and at the same time adding an "On Edge" geometric relation. One advantage of using "**Convert Entities**" is if the original geometry changes, so will the converted entities.

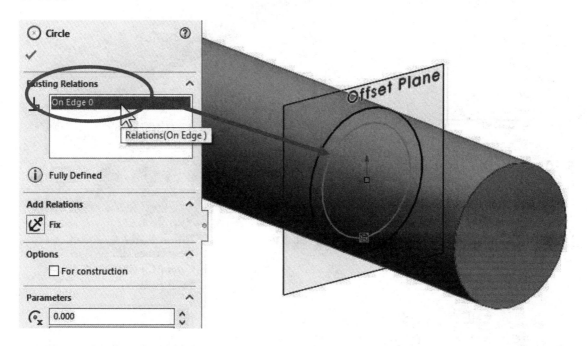

Making an "**Extruded Cut**" feature using nested closed loops (one inside the other) will result in removing the material between both entities. Use "**Extruded Cut**" using the "Blind" option 0.063" deep, going to the right. Rename the new feature *"Offset Groove."*

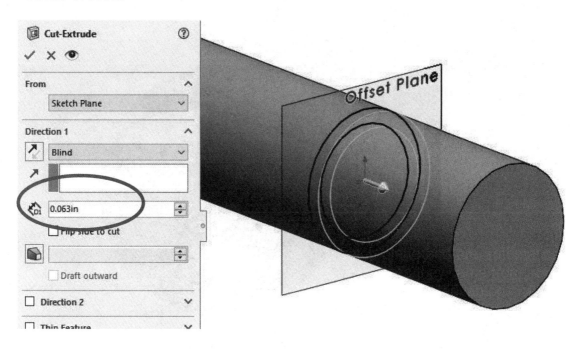

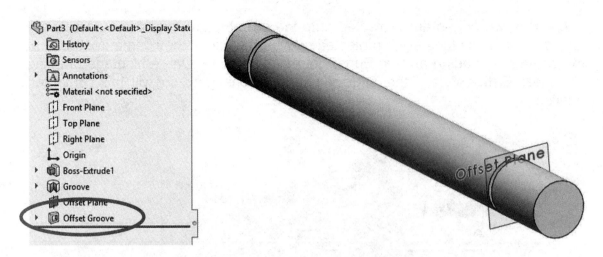

5.10. - For the last feature, we'll add a hexagonal cut at the right end of the shaft. Hide the "*Offset Plane*" (or turn off all planes) for easier visibility and switch to a Right view. Insert a sketch in the rightmost face of the shaft as indicated.

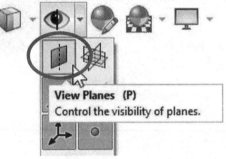

View Planes (P)
Control the visibility of planes.

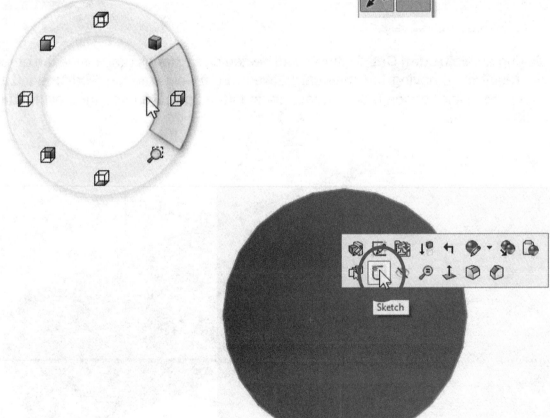

Sketch

5.11. - We can make a polygon using lines, dimensions and geometric relations, but we really want to make it easy, so we'll use the "**Polygon**" tool. Go to the menu, "**Tools, Sketch Entities, Polygon**" or select the "**Polygon**" tool from the Sketch tab in the CommandManager.

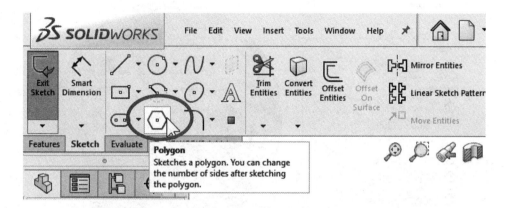

 SOLIDWORKS toolbars can be customized to add or remove icons as needed. Right-mouse-click on any toolbar, select "**Customize**" and from the "Commands" tab drag command icons to and from toolbars.

5.12. - After we select the "**Polygon**" tool, we are presented with the options in the PropertyManager. This tool helps us to create a polygon by making it either inscribed or circumscribed to a circle. For this exercise we'll select the "Circumscribed circle" with 6 sides in the "Parameters" options. Don't worry too much about the rest of the options, as we'll fully define the hexagon using two additional geometric relations.

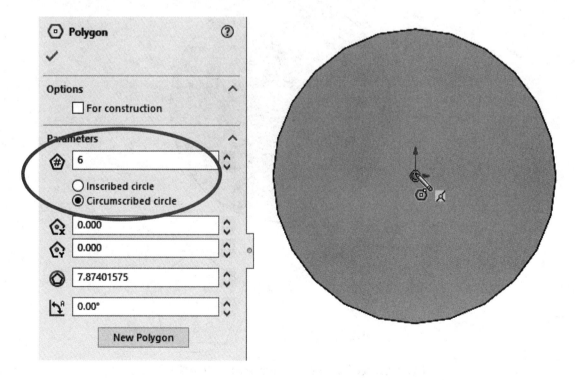

Since the polygon is defined by a circle (Inscribed or circumscribed), it is drawn like a circle. Start at the center of the shaft as shown and notice that we immediately get a preview of the hexagon, its radius and angle of rotation.

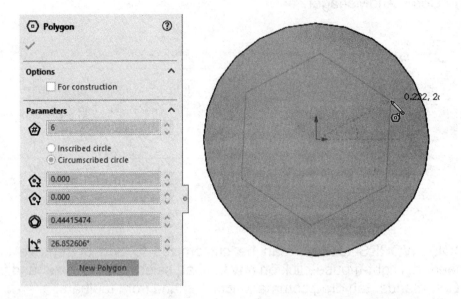

Draw the circle a little smaller (or larger) than the shaft. The idea is to make the construction circle the same size as the shaft using a geometric relation. Hit "Esc" or OK to close the polygon tool when done. Finishing the polygon at the edge of the face would capture a coincident relation making the circle the same size as the shaft's face, but we are going to practice manually adding a geometric relation.

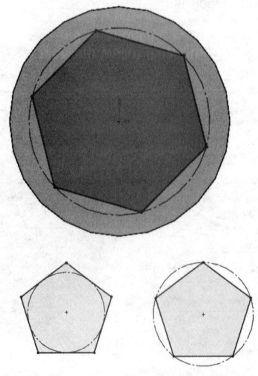

Inscribed Circle   Circumscribed Circle

5.13. - Now select the "**Add Relation**" tool from the Sketch tab and select the polygon's construction circle and the edge of the shaft; add an "Equal" geometric relation to make them the same size.

 Remember that the "**Add Relation**" tool can also be found in the right mouse button menu or by pre-selecting geometric elements.

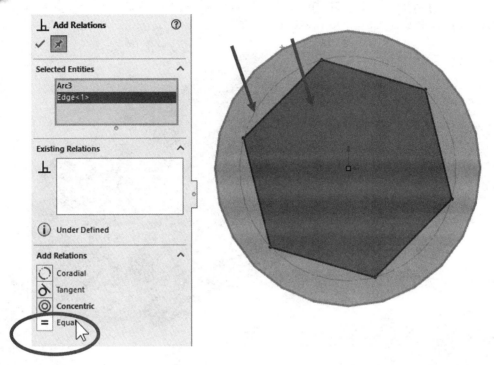

5.14. - Now select one of the hexagon's lines (anyone will do) and add a "**Horizontal**" relation to fully define the sketch.

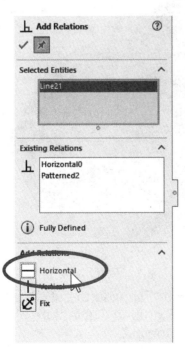

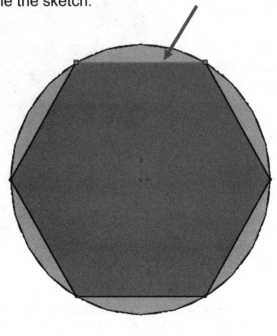

5.15. - Now we are ready to make the extruded cut. For this operation we'll use a seldom used but very powerful option in SOLIDWORKS. Select the "**Extruded Cut**" icon from the Features tab as before, but now activate the checkbox "**Flip side to cut**."

This option will make the cut *outside* of the sketch, not inside. Notice the arrow indicating which side of the sketch will be used to cut. Make the cut 0.5″ deep and rename the resulting feature *'Hex Cut'*.

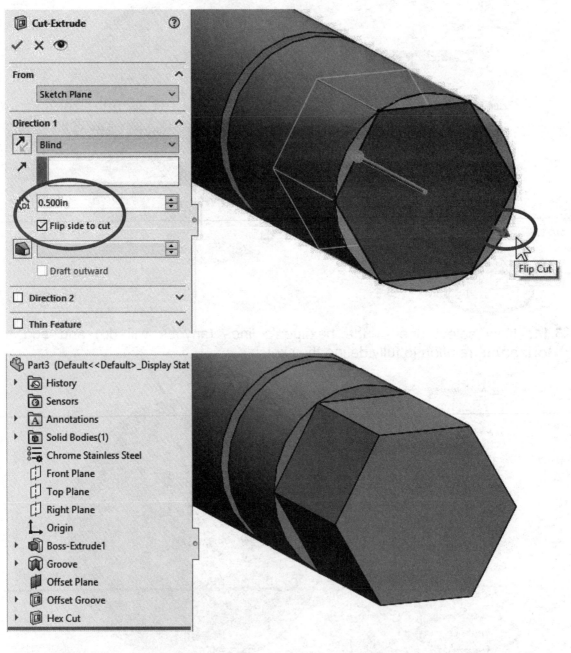

 The same result can be obtained without the "Flip side to cut" by adding a profile larger than the polygon; this way the Extruded Cut feature would use the region between the two profiles to make the cut.

5.16. - If not already set, change or set the material to "Chrome Stainless Steel" from the materials library (or from the Favorites list if available).

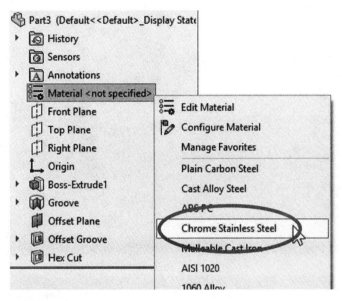

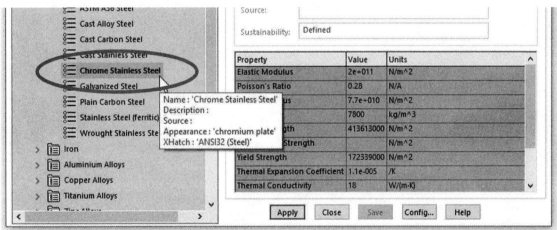

Save the part as *'Offset Shaft'* and close the file.

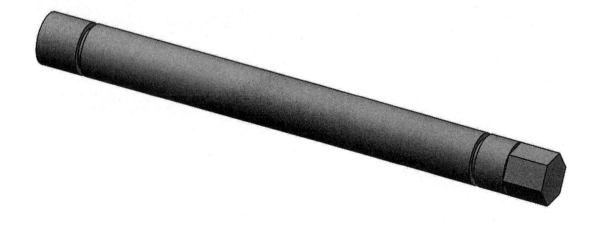

Notes:

Auxiliary Geometry:

Auxiliary geometry includes **Planes**, **Axes**, **Coordinate Systems**, **Points**, **Mate References and Center of Mass**. All of these can be created from the "**Reference Geometry**" icon in the Features tab in the CommandManager. Reference geometry can be used for many reasons, including locating features and components, as reference, or to use as part of a feature. An axis can be used to define the direction for a circular pattern and planes to add sketch geometry.

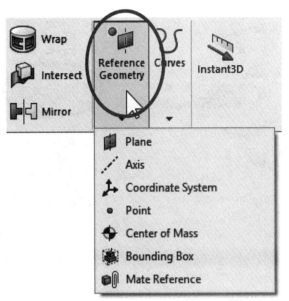

The main options to create **Planes** were covered previously; here are other types of Reference Geometry. Open the '*Knob*' part from the included files to practice.

An **Axis** can be made using the following options:

One Line/Edge/Axis – Any linear edge, sketch line, or axis. Every cylindrical and conical face has an axis (Temporary Axis) running through it. To reveal it use the menu "**View, Temporary Axis.**"

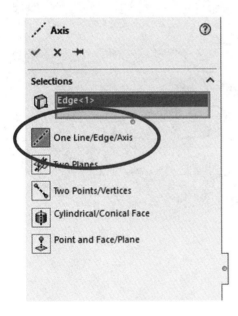

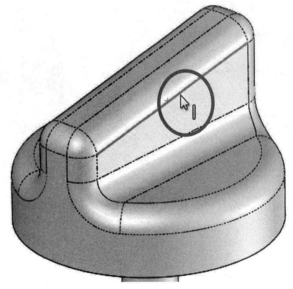

Two Planes – An axis can be created at the intersection of any two planes and/or faces.

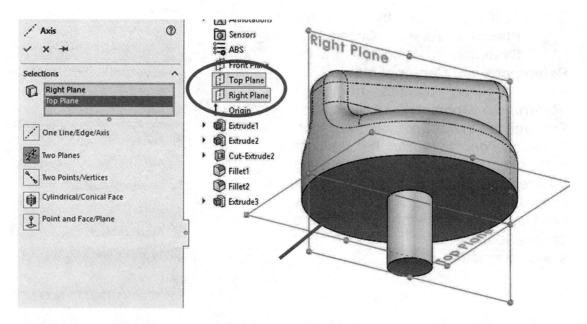

Two points/Vertices – Using any two vertices and/or sketch points/endpoints.

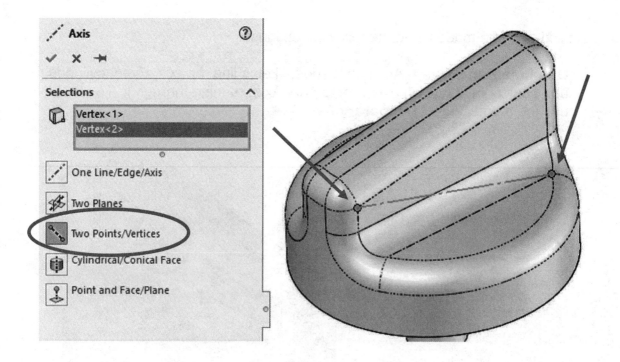

Cylindrical/Conical Face – Selecting any cylindrical or conical face will make an axis using the face's temporary axis.

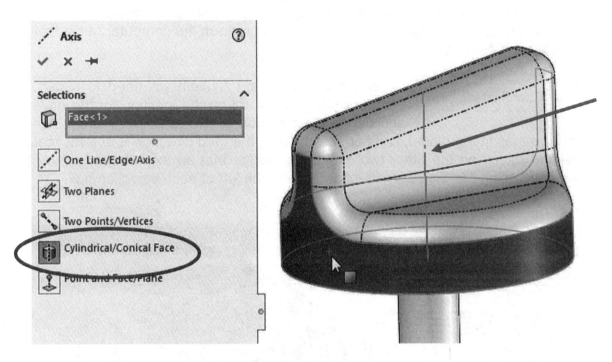

Point and Face/Plane– Selecting a point, sketch point, endpoint, or vertex and a flat face or plane will make an axis perpendicular to the face/plane that passes through the point/vertex.

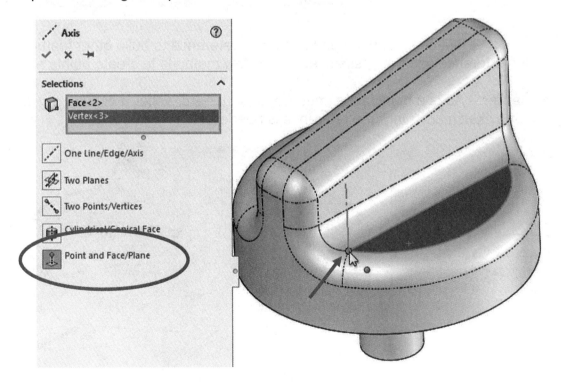

Coordinate Systems can be used for a number of things including to calculate a component's mass properties referenced to a specific location, to export geometry for manufacturing using Computer Assisted Manufacturing (CAM), when, more often than not, the origin the designer used for the component is not the best location for the CAM operator to program the computer numerically controlled (CNC) equipment.

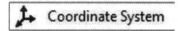 To create a Coordinate System we need to select a vertex or point which will be the location for the origin, and define the direction of two axes (X, Y or Z) using linear edges in the model, axes or sketch lines; the third direction is automatically defined based on the other two directions. Notice that an axis direction can be reversed using the "Reverse direction" icon to the left of each selection box.

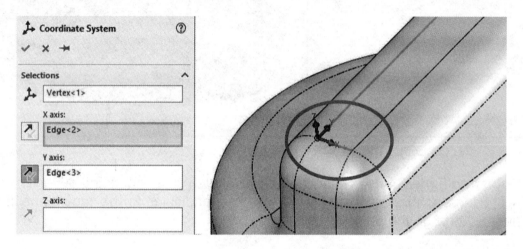

A Point can be added and used as a reference to build other features, including reference geometry, sketches, etc. A few methods to create points are:

Arc Center – Adds a point in the center of a planar arc, either a circular model edge or a sketch arc. Only one reference is needed.

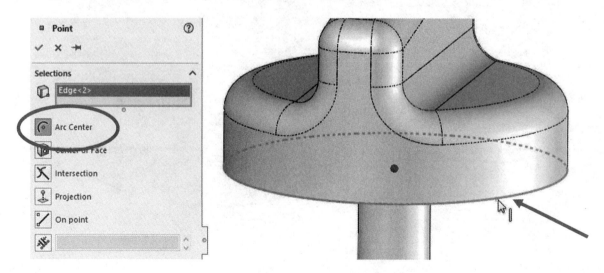

Center of Face – Creates a point at the centroid of the selected face; it can be either a planar or non-planar face. Only one reference is needed.

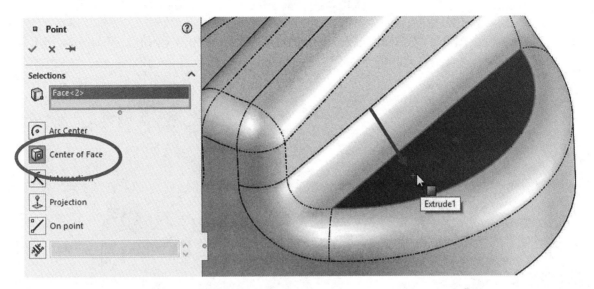

Intersection – Creates a point at the intersection of edges, curves or sketch segments. Two references are needed.

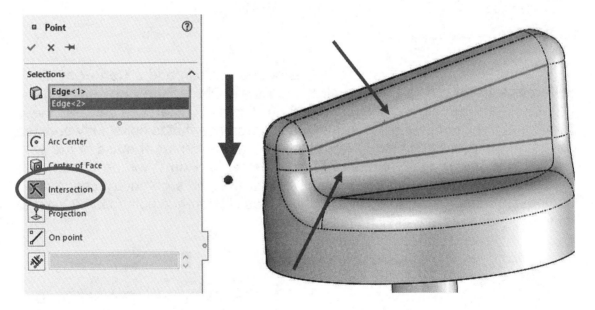

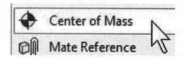

The **Center of Mass** (COM) adds a visual marker at the center of mass of the model.

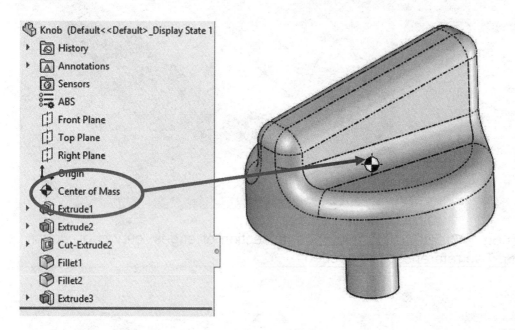

After the Center of Mass (COM) is added, we can add a **Center of Mass Reference Point** by right mouse clicking in the COM feature and selecting the COM Reference Point from the pop-up menu. The main difference between these two is that the **Center of Mass** is a dynamic reference that is constantly updated based on the part's geometry. It will always be at the center of mass of the part. On the other hand, the "**Center of Mass Reference Point**" will be added as a feature and will remain at this location even if the part's geometry changes by adding more features after it. The Center of Mass Reference point can be used to dimension features and sketch entities from it.

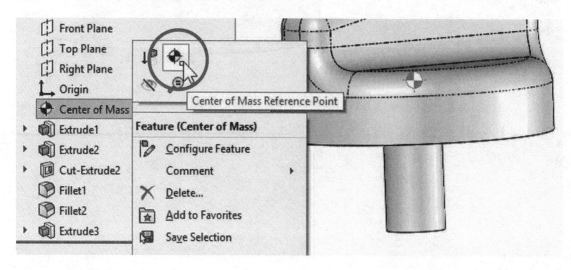

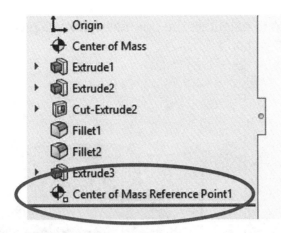

 Important: The "COM Reference Point" is not dynamic; after it is added it will be fixed. What this means is that if the model is modified by adding more features, the **COM** will change, but the **COM Reference Point** will not.

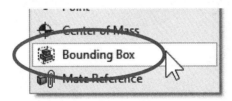

A **Bounding Box** adds a virtual box around a body represented with a 3D sketch showing the smallest box in which the body fits. When a bounding box is added to a part, by default, the box is aligned using the "Best Fit" option but can also be aligned to a model face or plane.

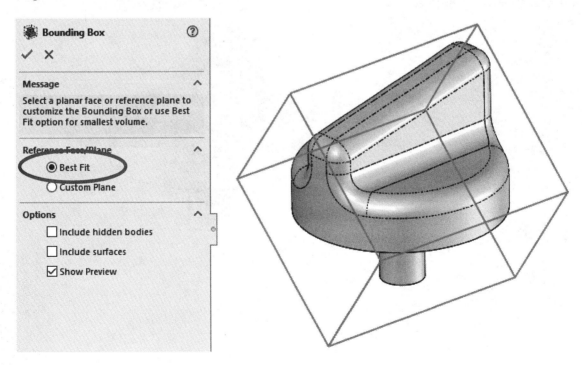

Selecting a plane or face to calculate the bounding box will probably make it bigger than using the "Best Fit" option.

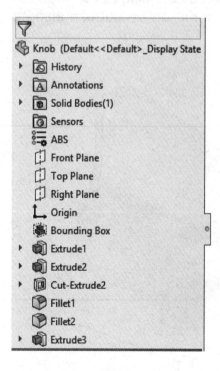
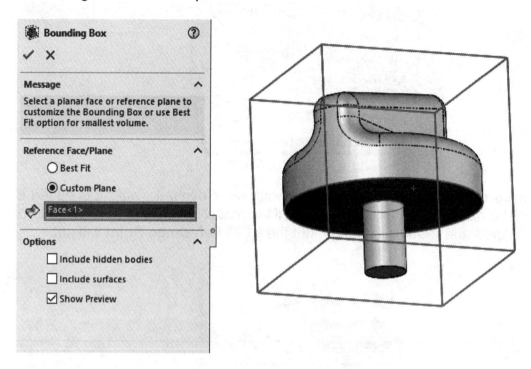

After creating the bounding box, it is added to the FeatureManager and can be shown, hidden, suppressed or unsuppressed as needed.

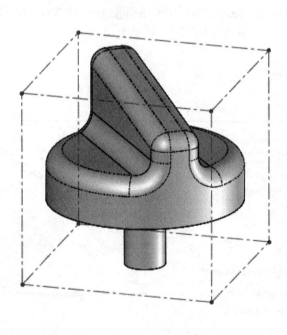

The bounding box is particularly useful for manufacturing, it helps us to find the smallest material stock size from which the part can be made. The bounding box's volume, width, depth and height information is added to the document's properties for reference. Use the menu "**File, Properties**" and select the "Configuration Specific" tab. The box dimensions can be used later in the detail drawing.

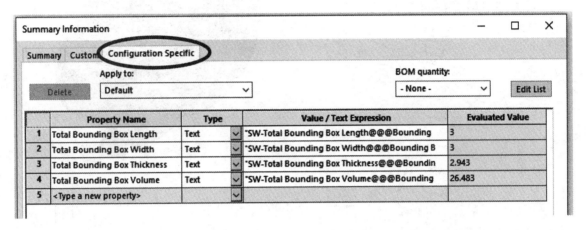

Remember that we can use sketch geometry from any sketch to create reference geometry (Planes, Axes and Coordinate Systems), but before we can use it, we need to make the sketch visible using the "**Show/Hide**" button in the pop-up toolbar. If you cannot see a sketch after showing it, the "View Sketches" button in the "**Hide/Show**" drop-down command may be inactive.

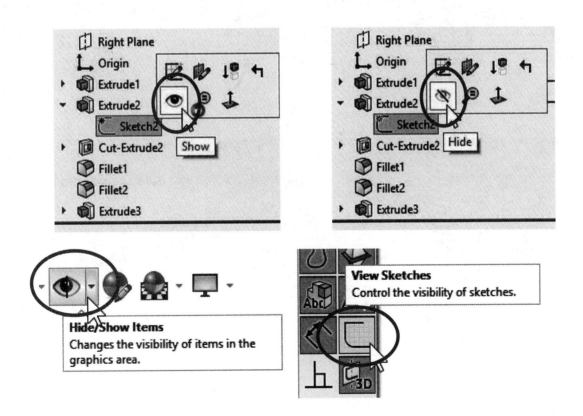

Exercises: Build the following parts using the knowledge acquired in this lesson. Try to use the most efficient method to complete the model.

Exercise 9
DIMENSIONS: INCHES
MATERIAL: 6061-T4 (SS)

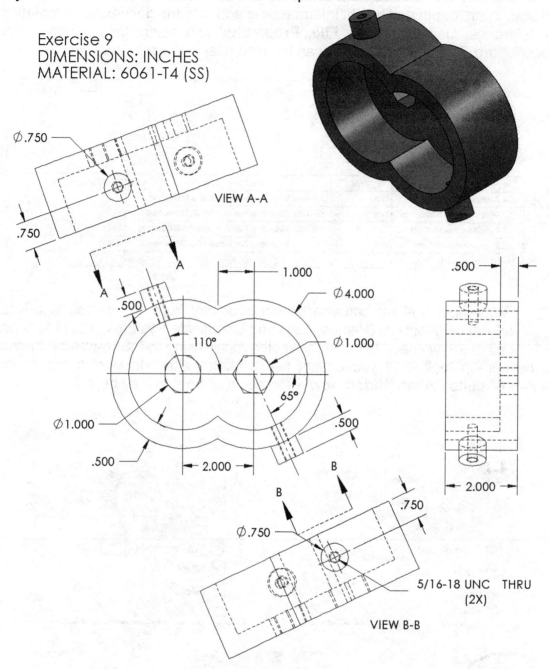

Ø.750

VIEW A-A

.750

A

A

.500

1.000

Ø4.000

110°

Ø1.000

65°

Ø1.000

.500

.500

2.000

B

B

.500

.750

.500

2.000

Ø.750

.750

5/16-18 UNC THRU
(2X)

VIEW B-B

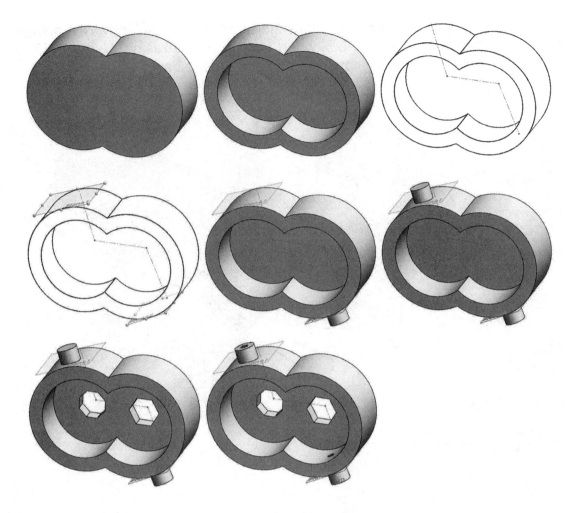

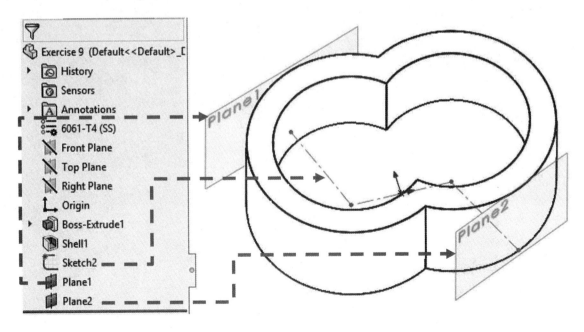

HINT: Draw a sketch with centerlines ("Sketch2"), exit the sketch, and use it as a layout to make the auxiliary planes needed ("Plane1" and "Plane2"). Hide the layout sketch when finished.

To make the cylindrical extrusions indicated in the auxiliary views, activate the "**Direction 2**" option box when making the extrusion. Use the "Up to Next" end condition to match the curvature of the base.

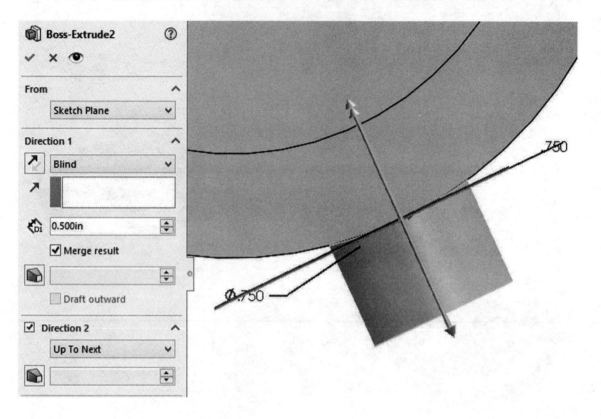

Exercise 10
DIMENSIONS: INCHES
MATERIAL: 2024-O

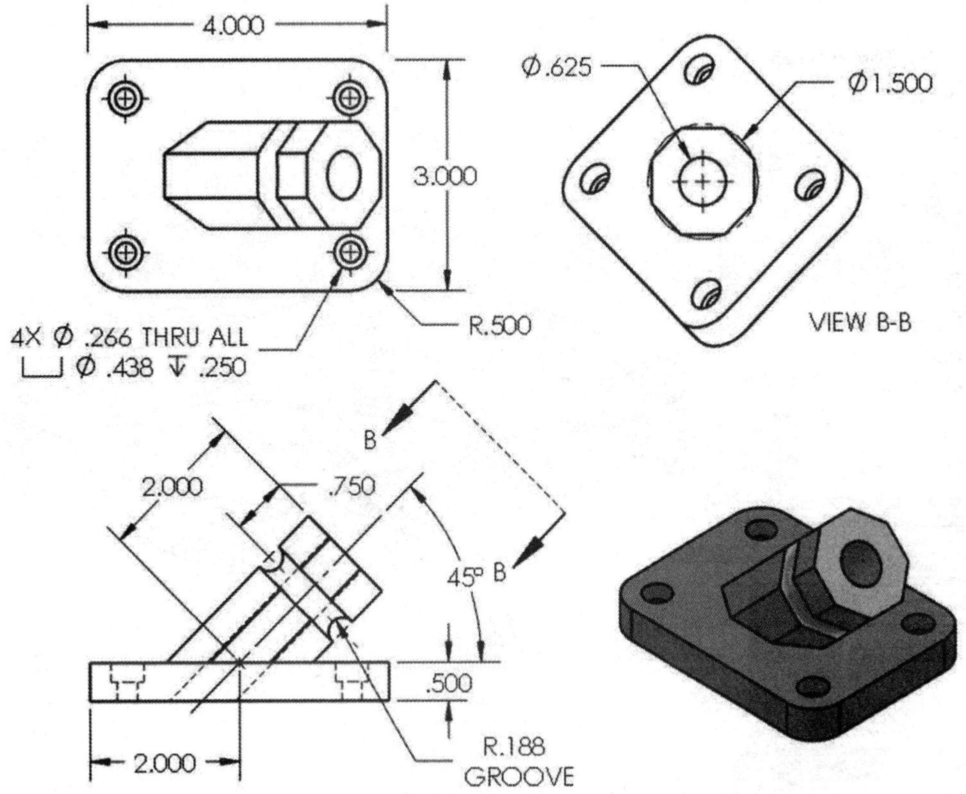
4.000
3.000
4X Ø .266 THRU ALL
⌴ Ø .438 ↧ .250
R.500
Ø.625
Ø1.500
VIEW B-B
B
B
2.000
.750
45°
.500
2.000
R.188
GROOVE

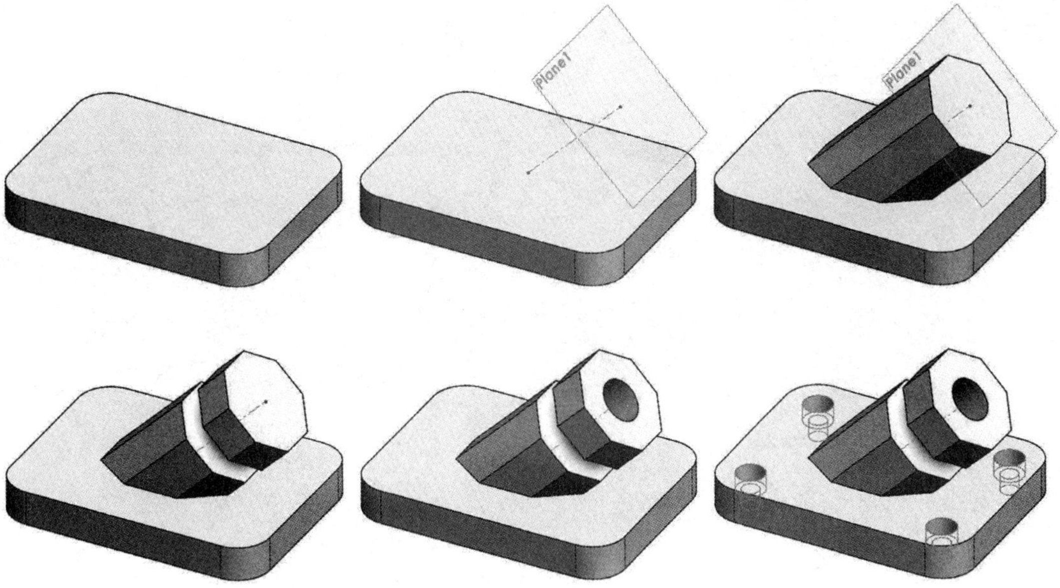
Plane1
Plane1

HINT: Draw a sketch ("Sketch2") with a centerline, exit the sketch, and use it to add an auxiliary plane ("Plane1") perpendicular to it. Add the hexagonal sketch ("Sketch3") in this plane and extrude it using the "Up to Surface" end condition. Hide the layout sketch when finished.

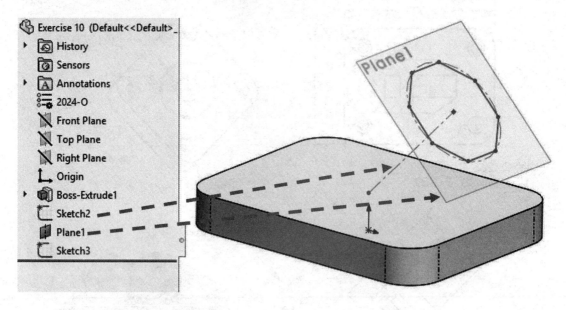

Engine Project Parts: Make the following components to build the engine. Save the parts using the name provided. High resolution images are included on the accompanying exercise files.

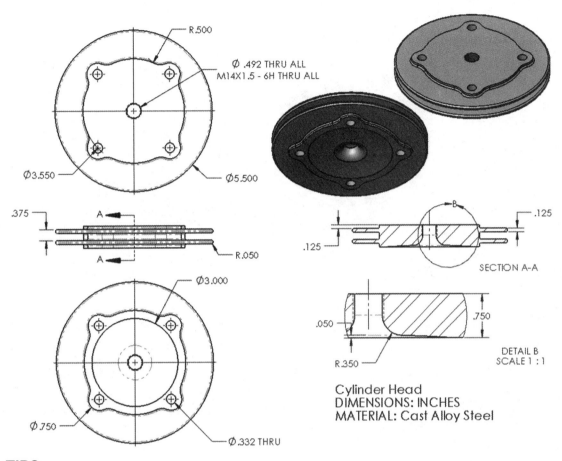

R.500

Ø .492 THRU ALL
M14X1.5 - 6H THRU ALL

Ø3.550

Ø5.500

.375

A

A

R.050

.125

.125

SECTION A-A

Ø3.000

.050

.750

R.350

DETAIL B
SCALE 1 : 1

Ø.750

Ø.332 THRU

Cylinder Head
DIMENSIONS: INCHES
MATERIAL: Cast Alloy Steel

TIPS:

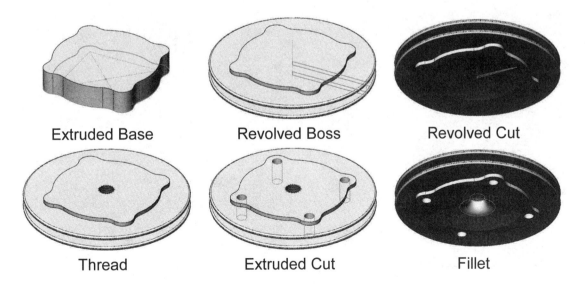

Extruded Base

Revolved Boss

Revolved Cut

Thread

Extruded Cut

Fillet

Notes:

The Worm Gear

Notes:

For the *'Worm Gear'* we will make a simplified version of a gear without teeth, and after covering the Sweep feature it will be modified to add the gear. The intent of this book is not to go into gear design, but rather to help the user understand and learn how to use basic SOLIDWORKS functionality. With this part we'll learn a new extrusion (or cut) end condition called "Mid Plane," how to chamfer the edges of a model, a special dimensioning technique to add a diameter dimension to a sketch used for a revolved feature, and how to dimension to a circle's perimeter. We'll also practice previously learned commands. For the *'Worm Gear'* we will follow the next sequence of features.

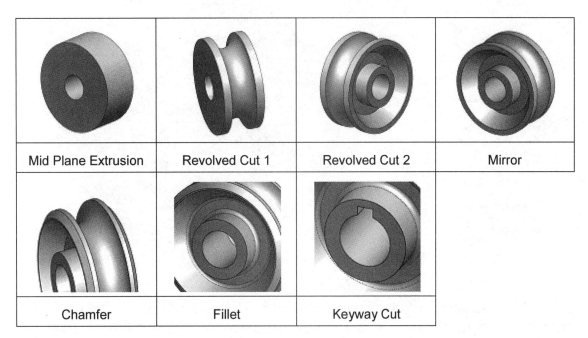

Mid Plane Extrusion	Revolved Cut 1	Revolved Cut 2	Mirror
Chamfer	Fillet	Keyway Cut	

6.1. - Start by making a new part using with a template with English units, set the material to "AISI 1020" Steel and create the following sketch in the *"Front Plane."*

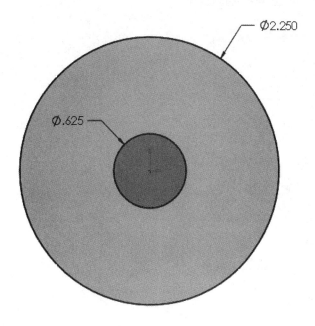

6.2. - In this case we want the part to be symmetrical about the "*Front Plane*." To achieve this, we'll make an "**Extruded Boss/Base**" using the "**Mid Plane**" end condition; this condition extrudes half of the distance in one direction and half in the second direction giving us a symmetric extrusion about the sketch's plane. Change the end condition to "**Mid Plane**" and extrude it 1". (The result will be 0.5" going to the front and 0.5" going to the back.) Rename the feature "*Base*."

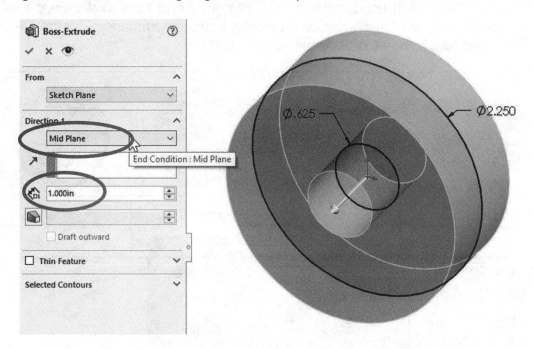

6.3. - For the second feature we will use a "**Revolved Cut**" to make the slot around the part. Change to a Right view, select the "*Right Plane*" from the FeatureManager, and click on the "**Sketch**" icon from the pop-up toolbar as shown. Make the center of the circle coincident to the Midpoint of the cylinder's top edge and <u>don't forget to add the centerline</u>.

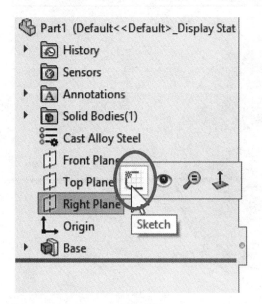

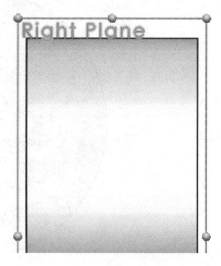

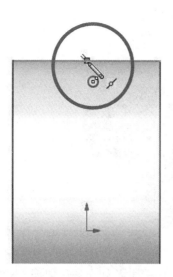

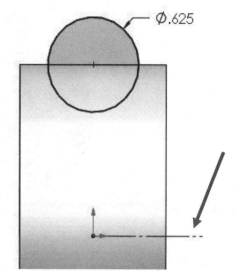

6.4. - Select the "**Revolved Cut**" command from the Features tab to complete the feature using the default settings. Rename the feature *'Groove'*.

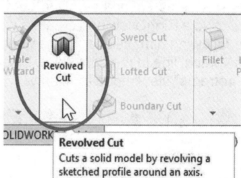

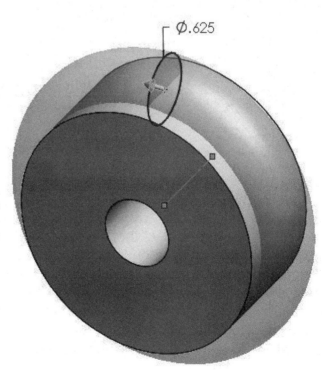

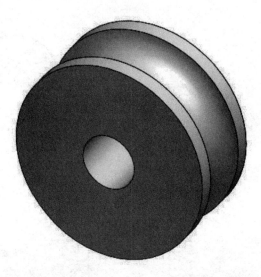

6.5. - Make a second revolved cut to remove material from one side of the part. Switch to a Right view and create a new sketch in the "*Right Plane*" as before using the following dimensions. Remember, it is a closed Sketch (four lines, don't forget the vertical line in the right side). If needed, turn off the Sketch Grid and change to "Wireframe" view mode for clarity.

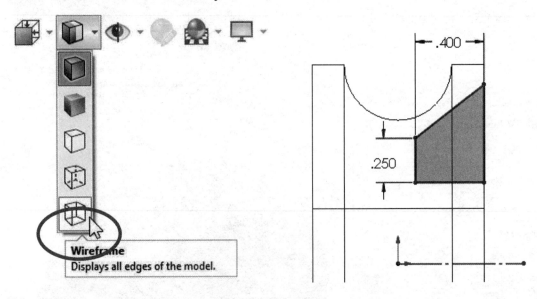

6.6. - We'll now use a new dimensioning technique to add diameter dimensions for Revolved Features; this way we will have a diameter dimension in the revolved feature when finished. Select the "**Smart Dimension**" tool and add a dimension from the Centerline (not an endpoint!) to the top endpoint of the sketch. Before locating the dimension, cross the centerline and notice how the dimension's value doubles. Locate the dimension and change it to 2″. Immediately after adding the first dimension the Smart Dimension tool remains in Doubled dimension mode. Select the bottom horizontal line to add a 1″ dimension from the centerline.

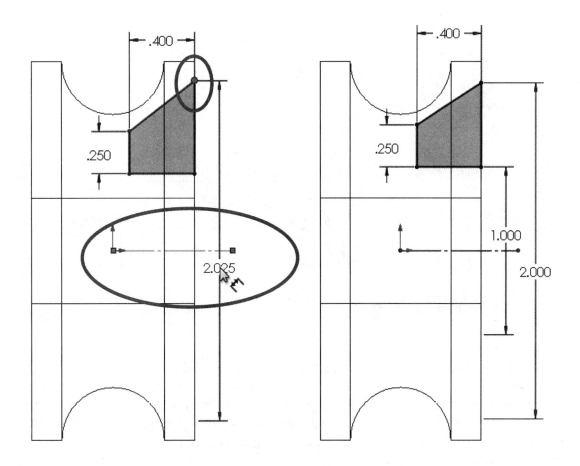

6.7. - In the next step we will make a "**Sketch Mirror**" to make the same profile in the other side of the part. Draw a vertical centerline at the origin as shown and select the "Sketch Mirror" command from the Sketch tab. Select the profile lines in the "Entities to mirror" selection box and the vertical centerline in the "Mirror about:" selection box. Click OK to finish the "**Sketch Mirror**."

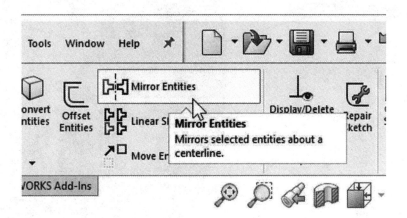

 To select the four lines that make the profile, we can click inside the closed profile to automatically select them.

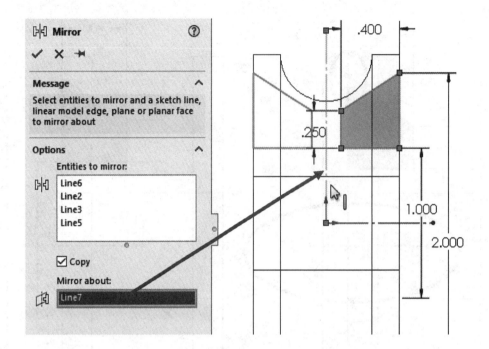

6.8. - After completing the sketch, select the "**Revolved Cut**" command from the Features tab in the CommandManager. In previous operations we only had one centerline and it had been selected automatically; in this case, since we have more than one centerline in the sketch, we have to select the horizontal centerline to make the revolved cut about it. Notice the doubled diameter dimensions; their purpose is more obvious in this image.

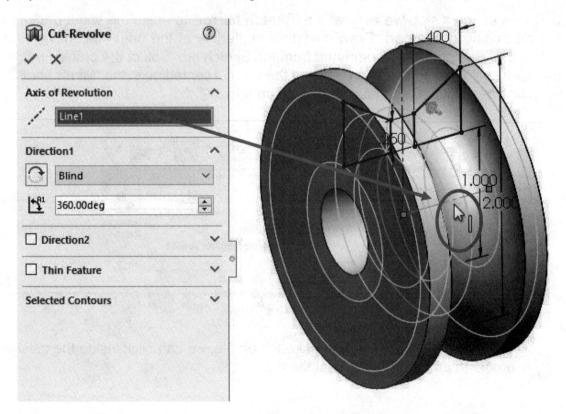

6.9. - To eliminate the sharp edges on the outside perimeter we'll add a 0.1″ x 45° "**Chamfer**." The Chamfer is an applied feature similar to the Fillet, but instead of rounding an edge, it adds a bevel to it. Select it from the drop-down menu under "**Fillet**" in the Features tab or in the menu "**Insert, Features, Chamfer**."

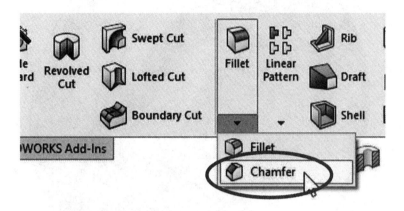

Set the chamfer type to "Angle Distance," make it 0.1″ x 45° and select the two edges indicated. This way the chamfer added will be .01″ in the direction of the arrow and at 45°. Click OK to finish.

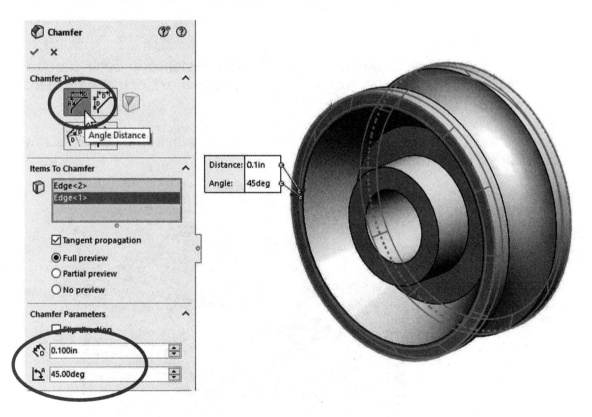

 A Chamfer can also be added by defining a distance in each direction (Distance-Distance), a direction along 3 edges (Vertex), or the width of the chamfer's face (Face-Face).

- *Angle-Distance Chamfer*. Defined by entering a distance and an angle, the chamfer's direction can be flipped.

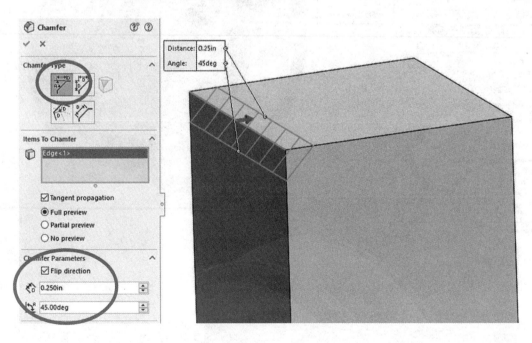

- *Asymmetric Distance-Distance Chamfer.* Defined by entering a different distance to each side of the edge.

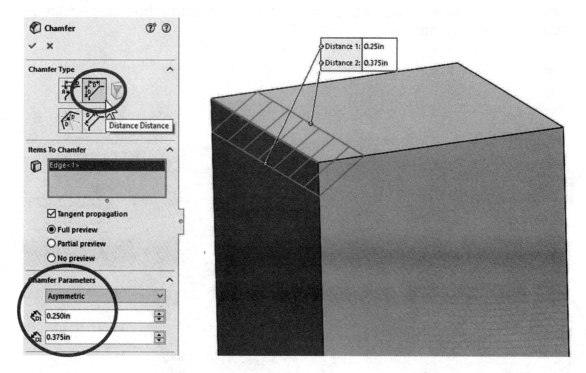

- *Symmetric Distance-Distance Chamfer.* Defined by entering the distance of the chamfer measured from the edges.

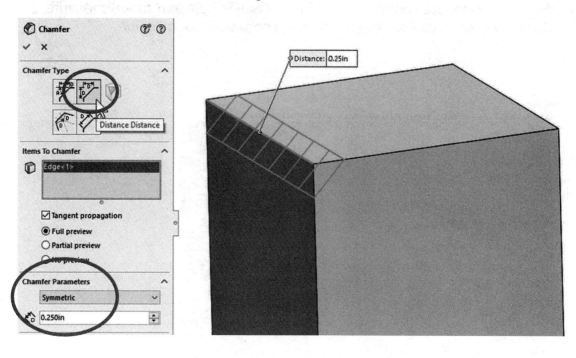

- *Offset Face*. Is defined by offsetting the adjacent faces by the defined distance and measuring perpendicular from the intersection. Can be symmetric or asymmetric. Review the image to better understand how the chamfer dimensions are measured.

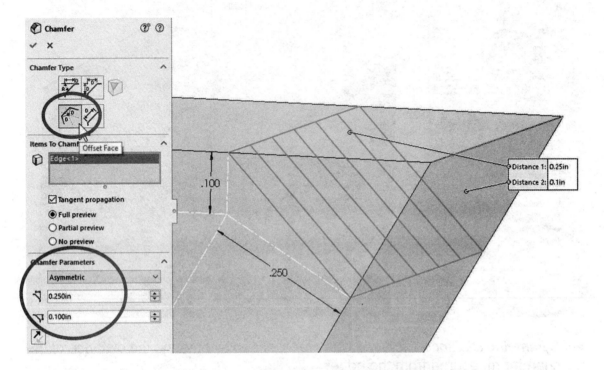

- *Face-Face*. Select two faces to add a chamfer between them, measure the distance in each direction, symmetrical, chord width or hold line.

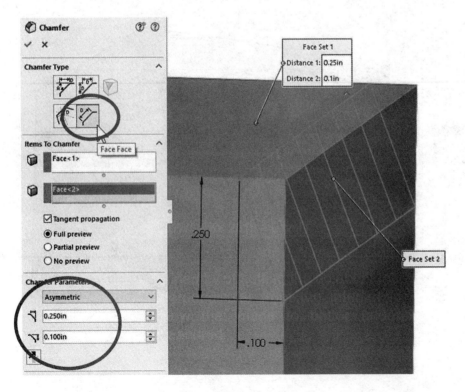

- *Chord Width* option defines the width of the chamfer's face.

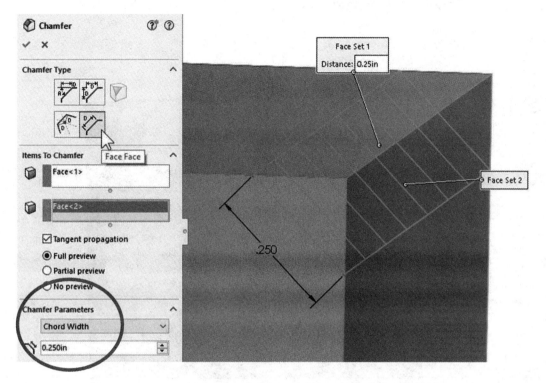

- *Hold Line option.* The hold line is defined by a model edge, which can be made using a split line to divide a face (covered later).

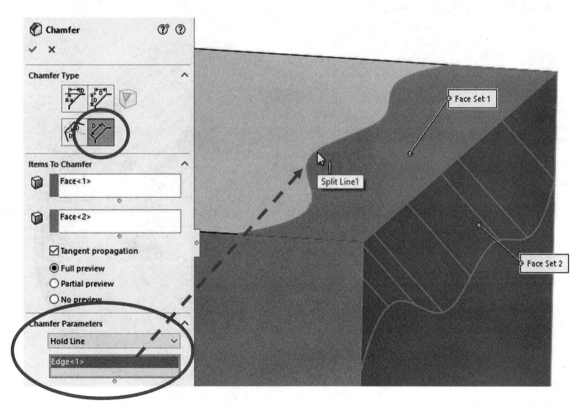

- *Vertex*. Defined by selecting a vertex and entering the distance along each edge. All 3 edges can be made equal.

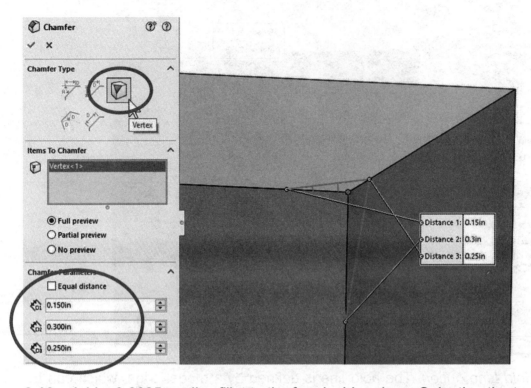

6.10. - Add a 0.0625″ radius fillet to the four inside edges. Selecting the two inside faces will make selection easier.

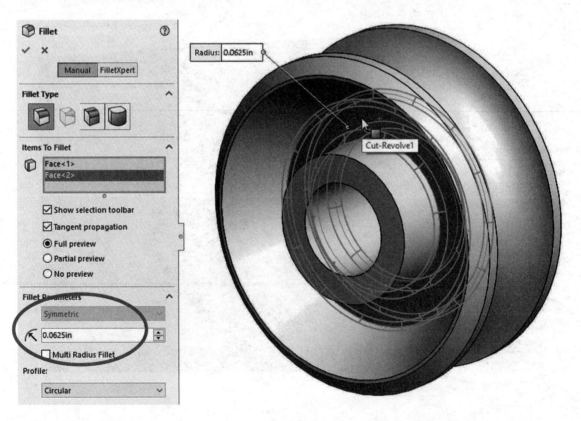

6.11. - For the last step, we'll make a keyway. Switch to a Front view and add a sketch in the "*Front Plane*." Draw a rectangle as indicated without capturing any geometric relations to other geometry. In order to temporarily disable automatic relations while adding the rectangle, hold down the "Ctrl" key while drawing it. Notice the absence of coincidence feedback icons while moving the mouse pointer. Release the

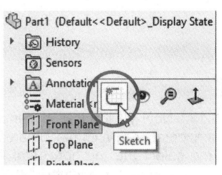

"Ctrl" key when finished and add a Midpoint relation between the rectangle's lower horizontal line and the part's Origin to center it about the origin.

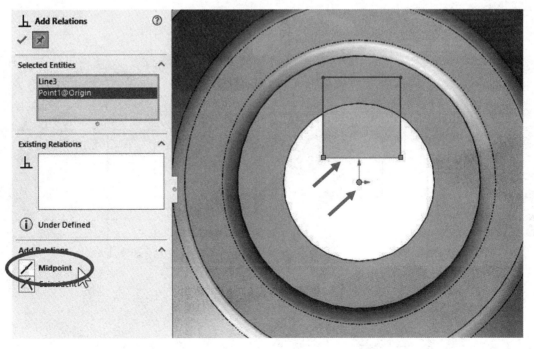

 We could have added the sketch in the front face, but we placed it in the "*Front Plane"* to show additional functionality.

6.12. - By adding the Midpoint relation, the rectangle will be coincident to, and centered about the origin. Add a 0.188″ width dimension as indicated. If the top horizontal line (blue) is below the center hole's edge, click-and-drag it until it's above the circle as shown. This will allow us to add a dimension to the circular edge in the next step.

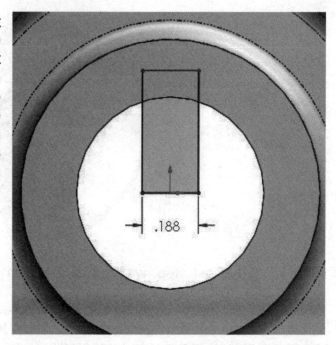

6.13. - Now we'll add a dimension from <u>the top of the circular edge</u> to the top horizontal line. With the Smart Dimension tool active, <u>press and hold</u> the "**Shift**" key, select the top horizontal line and the top of the circular edge, locate the dimension and release the "**Shift**" key. Change the dimension's value to 0.094″.

 If we don't hold the "**Shift**" key the dimension will be referenced to the center of the circular edge.

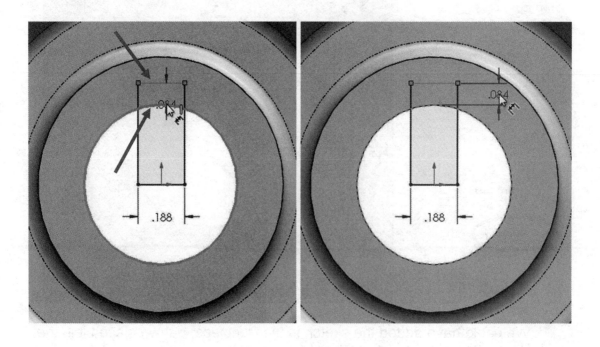

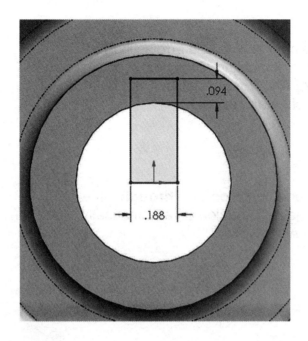

6.14. - To finish the part, we need to make a "**Cut Extrude**." Since the sketch is in the "*Front Plane,*" which is in the middle of the part, making a cut "**Through All**" will only go from the middle through one side. In this case we need to activate the "Direction 2" checkbox and select the end condition to "**Through All**" also for "Direction 2." Being able to select different end conditions in each direction can be very useful. Rename the feature "Keyway."

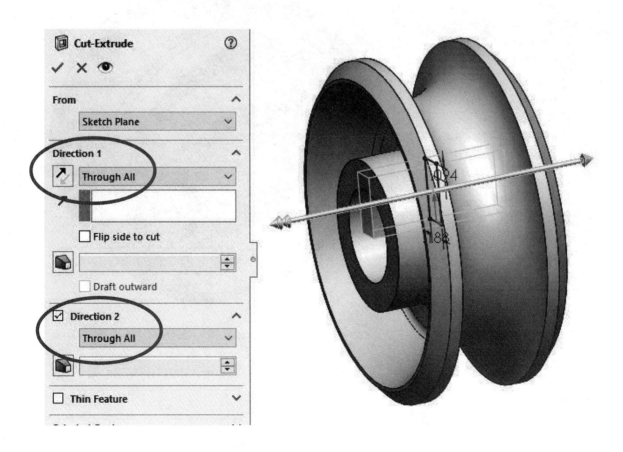

Notice the single and double arrows in the graphics area, indicating "Direction 1" and "Direction 2." Any "End Condition" can be used for either direction as needed.

 Setting the end condition to "**Through All – Both**" will make "Direction 2" automatically to "Through All."

6.15. - Save the part as *'Worm Gear'* and close the file.

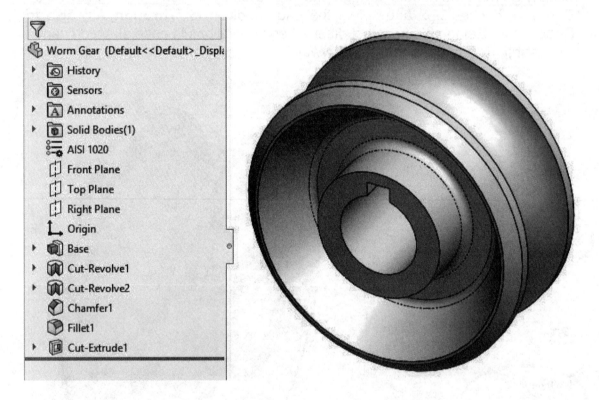

Exercises: Build the following part using the knowledge acquired in this lesson. Try to use the most efficient method to complete the model.

Exercise 11
DIMENSIONS: INCHES
MATERIAL:
Chrome Stainless Steel

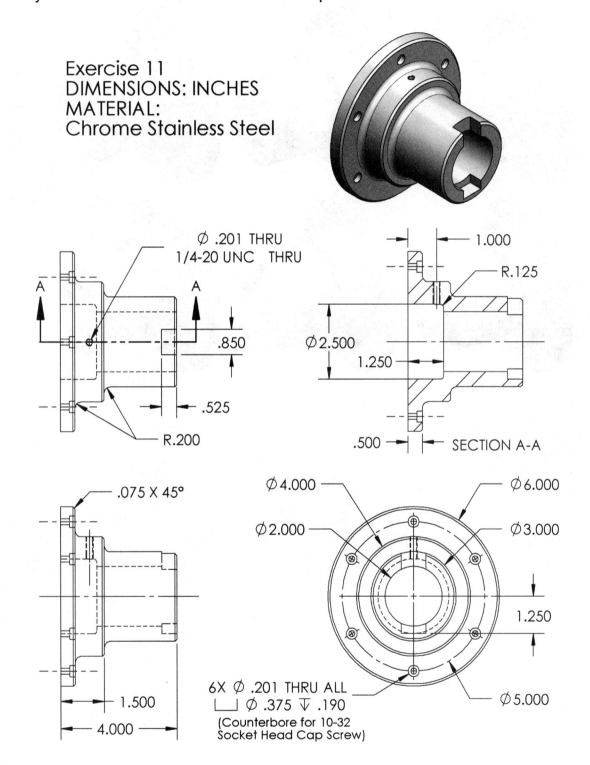

Ø .201 THRU
1/4-20 UNC THRU

.850

.525

R.200

1.000

R.125

Ø 2.500

1.250

.500

SECTION A-A

.075 X 45°

1.500

4.000

Ø 4.000

Ø 2.000

6X Ø .201 THRU ALL
⌴ Ø .375 ▼ .190
(Counterbore for 10-32
Socket Head Cap Screw)

Ø 6.000

Ø 3.000

1.250

Ø 5.000

217

TIPS:

Engine Project Parts: Make the following components to build the engine. Save the parts using the name provided. High resolution images at mechanicad.com.

Bushing Top Con Rod
DIMENSIONS: INCHES
MATERIAL: Brass

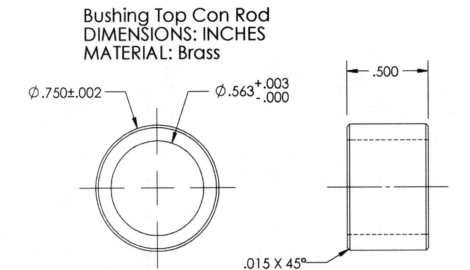

Con Rod Crankshaft Half Bushing
DIMENSIONS: INCHES
MATERIAL: Brass

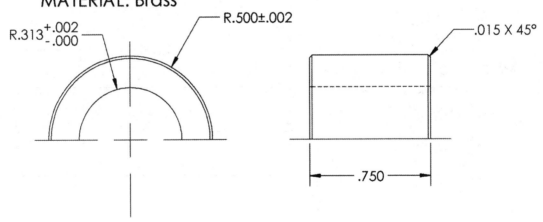

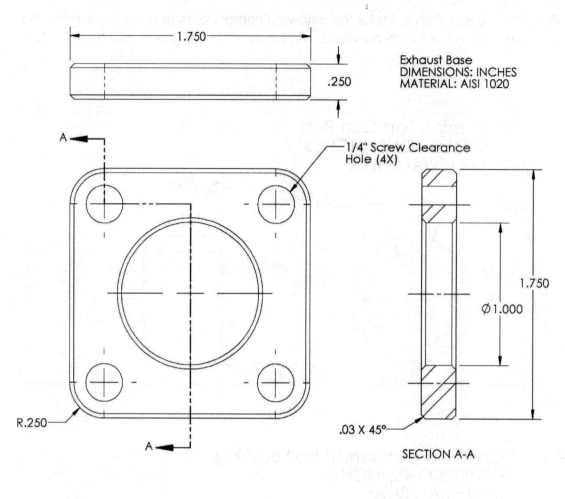

1.750

.250

Exhaust Base
DIMENSIONS: INCHES
MATERIAL: AISI 1020

A

1/4" Screw Clearance
Hole (4X)

R.250

A

1.750

Ø1.000

.03 X 45°

SECTION A-A

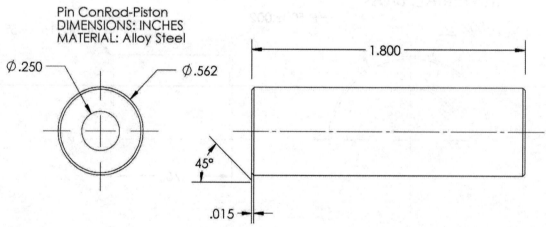

Pin ConRod-Piston
DIMENSIONS: INCHES
MATERIAL: Alloy Steel

Ø.250

Ø.562

1.800

45°

.015

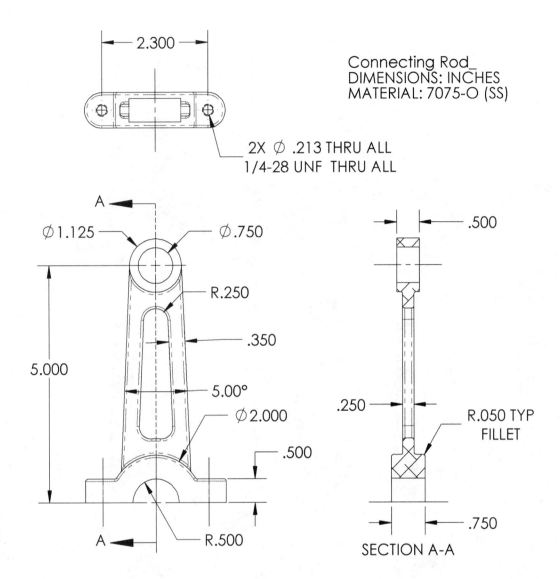

2.300

Connecting Rod_
DIMENSIONS: INCHES
MATERIAL: 7075-O (SS)

2X ⌀ .213 THRU ALL
1/4-28 UNF THRU ALL

A

⌀1.125   ⌀.750

R.250

.350

5.000   5.00°

⌀2.000

.500

A   R.500

.500

.250

R.050 TYP
FILLET

.750

SECTION A-A

TIPS:

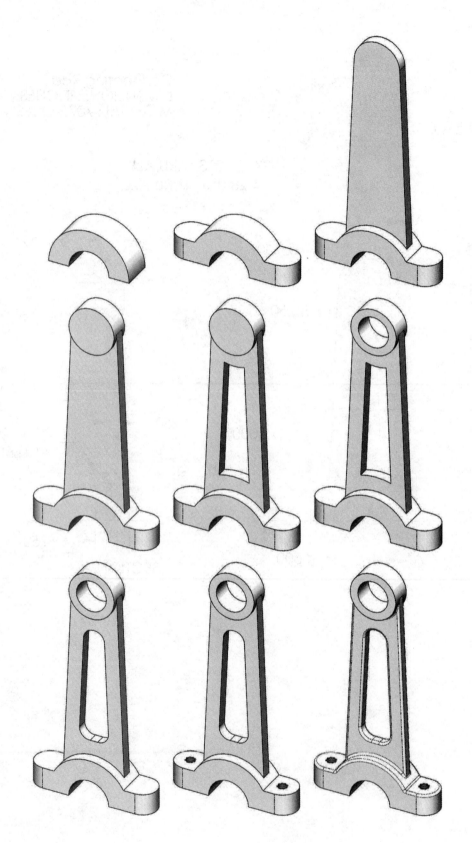

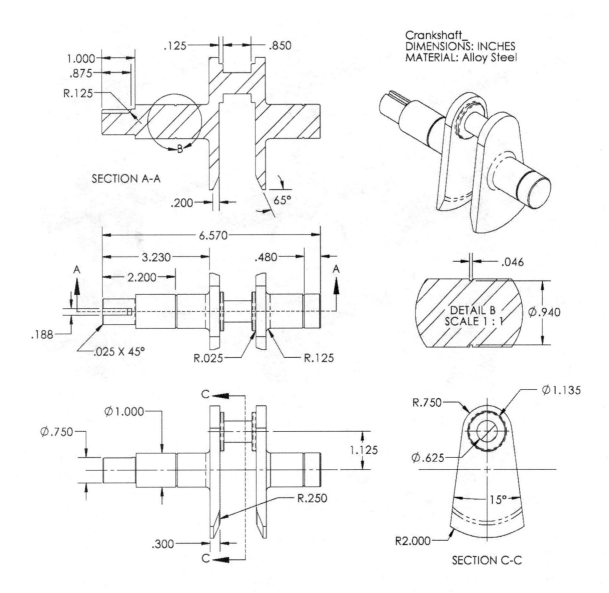

SECTION A-A

.125

.850

1.000

.875

R.125

B

.200

65°

Crankshaft
DIMENSIONS: INCHES
MATERIAL: Alloy Steel

6.570

3.230

.480

A

2.200

A

.188

.025 X 45°

R.025

R.125

.046

DETAIL B
SCALE 1 : 1

Ø.940

Ø1.000

Ø.750

1.125

R.250

.300

C

C

Ø1.135

R.750

Ø.625

15°

R2.000

SECTION C-C

223

TIPS:

The Worm Gear Shaft

Notes:

For the *'Worm Gear Shaft'* we will review the Revolved Feature previously learned, the sketch Polygon tool, and a Mid Plane cut.

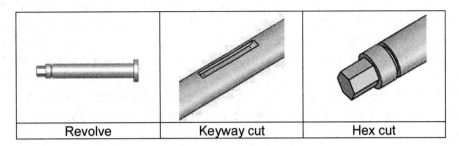

Revolve	Keyway cut	Hex cut

7.1. - In this part we only need to make three features. The first feature will be a revolved extrusion. Make a new part file using a template in inches and insert a sketch in the *"Front Plane"* as shown. It's very important to add the centerline, as we'll need it to add the diameter dimensions as we did in the previous part. Select the **"Revolved Boss"** command to complete the first feature. Rename the feature *'Base'*. Notice the three doubled diameter dimensions using the centerline.

 In a sketch like this it is better to add the smaller dimensions before the larger ones. The reason is if the large dimensions are added first, when the geometry updates, the small features may behave unexpectedly.

Add a "Collinear" relation between the two lines indicated before dimensioning.

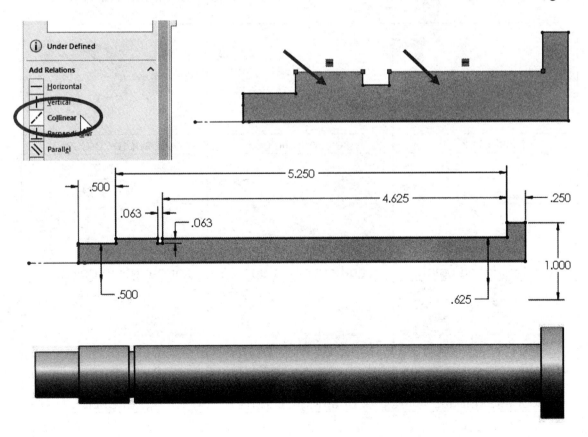

7.2. - The second feature is the keyway. Select the *"Front Plane"* from the FeatureManager and click in the "**Sketch**" icon from the pop-up toolbar. We'll make a cut using the "**Mid Plane**" end condition for the keyway. Making the top line beyond the top of the part allows us to "cut air" beyond the end of the part; this way we are sure we will cut the part. This is a common practice and is OK as long as we make the sketch big enough to accommodate possible future changes. Adding a dimension to ensure the sketch extends past the part is a good idea too. The problem that *may* occur depending on the geometry and the feature being made if we don't extend the sketch upward is that the top line could generate a "zero thickness" face error and would not let us continue.

 Remember that we can use Mouse Gestures (right mouse click-and-drag) for the most commonly used commands or configure it for the tools we use most to help us speed up the design process.

A useful option when drawing lines is to transition from drawing a line to a tangent arc. To do this we can use the default shortcut key "A" or move the mouse back to the last endpoint. After moving the mouse back, a tangent arc will be started. Depending on the direction that we move out, that's the direction the tangent arc will be created. Select the "**Line**" tool and draw a line, click in one location to start, click in the second point to complete the first line, and instead of clicking a third time to add a second line, move back to the second endpoint and move out to add a tangent arc. Press the "A" key to toggle between arc and line.

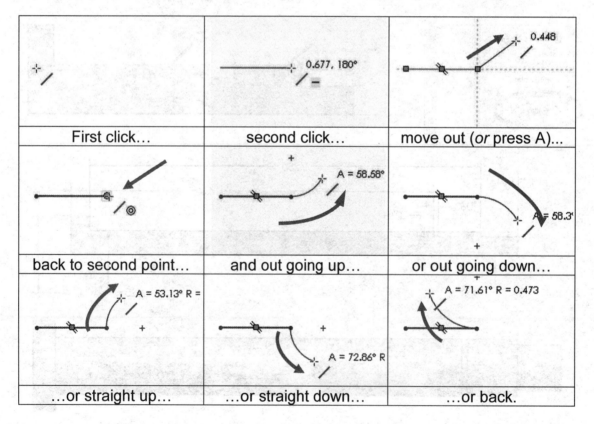

First click…	second click…	move out (*or* press A)…
back to second point…	and out going up…	or out going down…
…or straight up…	…or straight down…	…or back.

Note the dynamic feedback as we draw the profile using this technique. After finishing the sketch add an equal relation between the two arcs—this way we only need to dimension one of them.

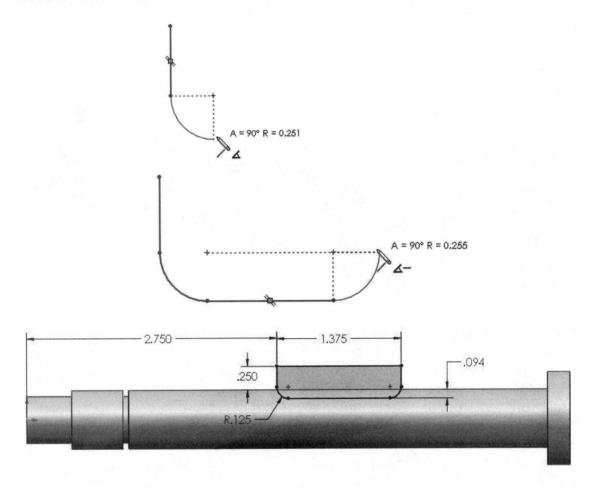

The 0.250" dimension's only purpose is to make sure the sketch cuts through the part by extending the sketch beyond the part. Since this dimension will not be required in the detail drawing, we will turn off the dimension's "Mark for Drawing" option. This option controls whether a dimension is imported to the detail drawing or not. To change it, right mouse click in the 0.25" dimension and uncheck it. The default setting is checked. Notice that dimensions not marked for drawing are displayed using a different color.

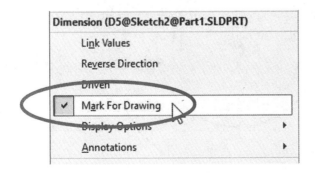
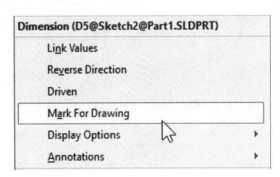

7.3. - With the sketch finished, select the "**Cut Extrude**" command from the Features tab and use the "**Mid Plane**" end condition with a dimension of 0.1875"; this way the keyway will be exactly at the center of the shaft. Notice that the "Direction 2" option box is unavailable when we select the Mid Plane end condition. Rename this feature "*Keyway*."

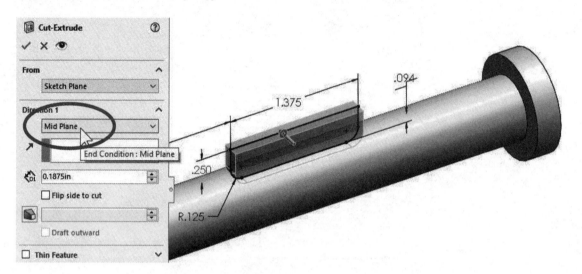

7.4. - For the last feature we'll make a hexagonal cut just as we did in the '*Offset Shaft*'. Switch to a Left view, select the round face, and add a sketch. Using the "**Polygon**" tool from the Sketch tab draw a hexagon using the "Circumscribed circle" option and make the construction circle coincident to the edge at the shaft's end as seen. We changed the view to "Hidden Lines Removed" mode for clarity.

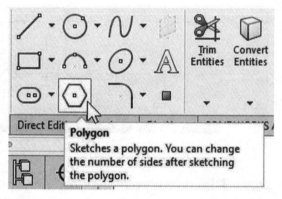

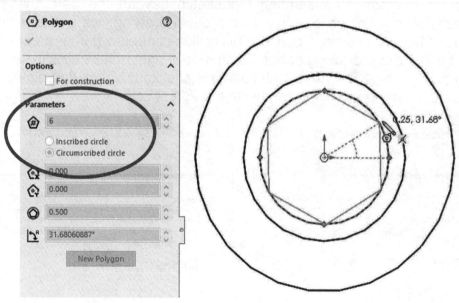

Finally select one line from the hexagon and add a horizontal relation to fully define the sketch.

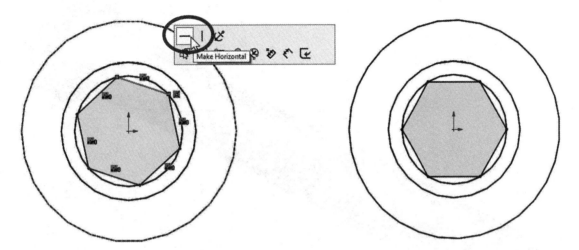

7.5. - Make a "**Cut Extrude**" using the "Up to Surface" end condition as indicated and activate the option "Flip side to cut" to cut *outside* the hexagon.

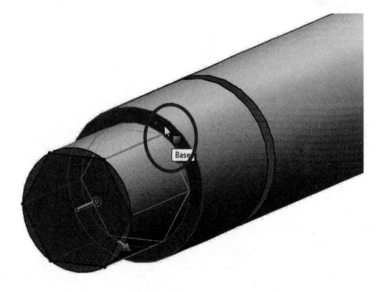

7.6. - Set the part's material to "Chrome Stainless Steel." Feel free to add the most commonly used materials using the "**Manage Favorites**" option in the "**Edit Material**" menu.

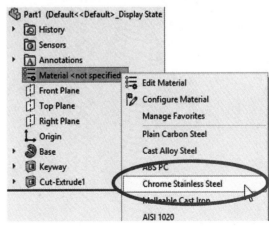

7.7. - Save the part as *'Worm Gear Shaft'* and close the file.

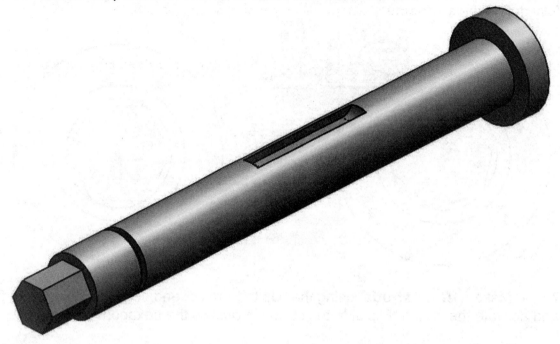

Engine Project Parts: Make the following components to build the engine. Save the parts using the name provided. High resolution images are included in the exercise files.

The following drawing shows the part in three stages of modeling for clarity.

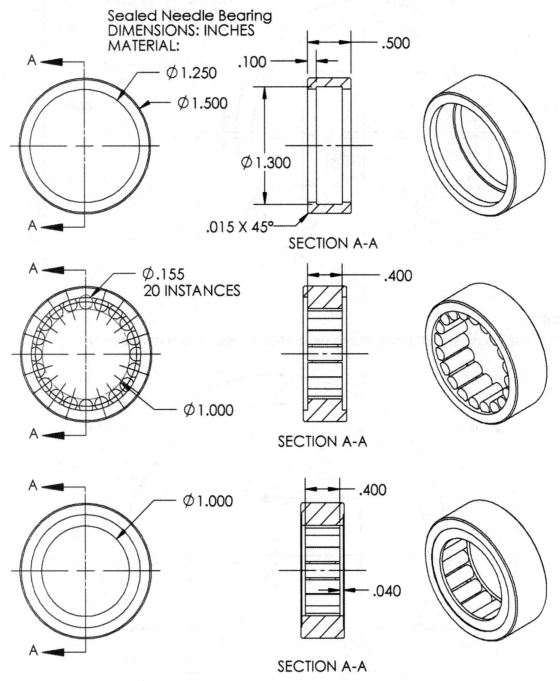

Sealed Needle Bearing
DIMENSIONS: INCHES
MATERIAL:

Ø1.250
Ø1.500
.100
.500
Ø1.300
.015 X 45°
SECTION A-A

Ø.155
20 INSTANCES
Ø1.000
.400
SECTION A-A

Ø1.000
.400
.040
SECTION A-A

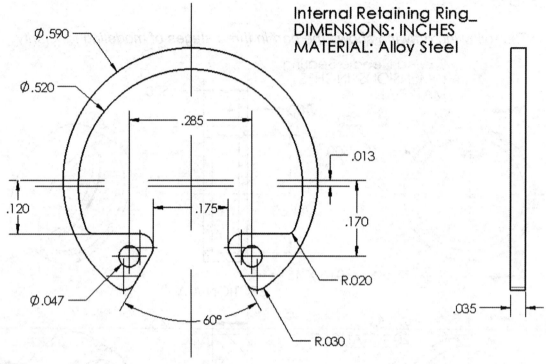

Internal Retaining Ring_
DIMENSIONS: INCHES
MATERIAL: Alloy Steel

TIPS:
Make a single sketch using these dimensions; use sketch mirror to save time.

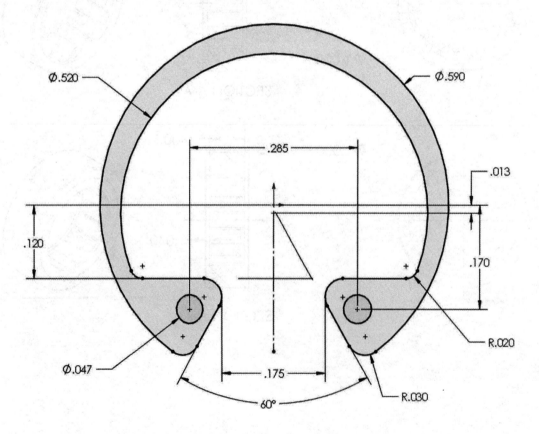

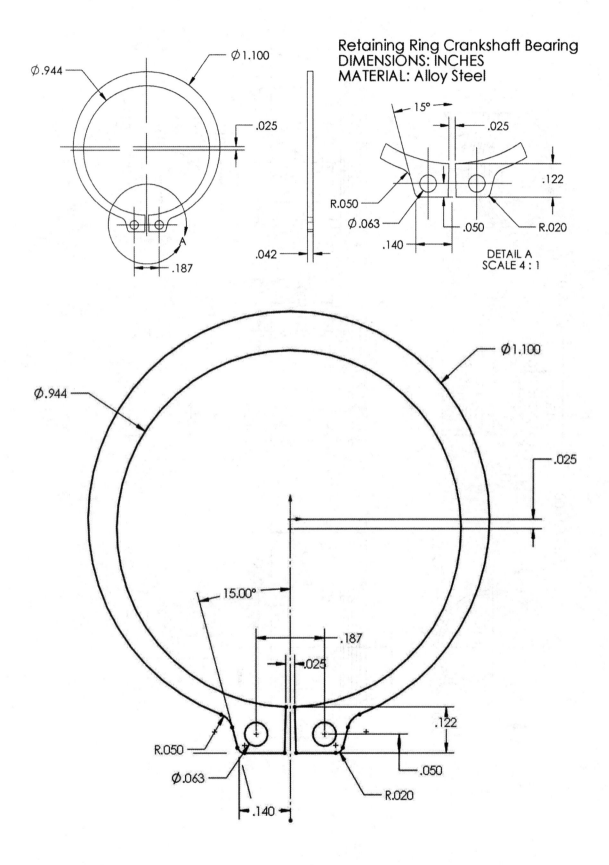

Retaining Ring Crankshaft Bearing
DIMENSIONS: INCHES
MATERIAL: Alloy Steel

Ø1.100

Ø.944

.025

15°

R.050

Ø.063

.025

.122

.050

R.020

.140

DETAIL A
SCALE 4 : 1

.187

.042

Ø1.100

Ø.944

.025

15.00°

.187

.025

.122

R.050

Ø.063

.050

R.020

.140

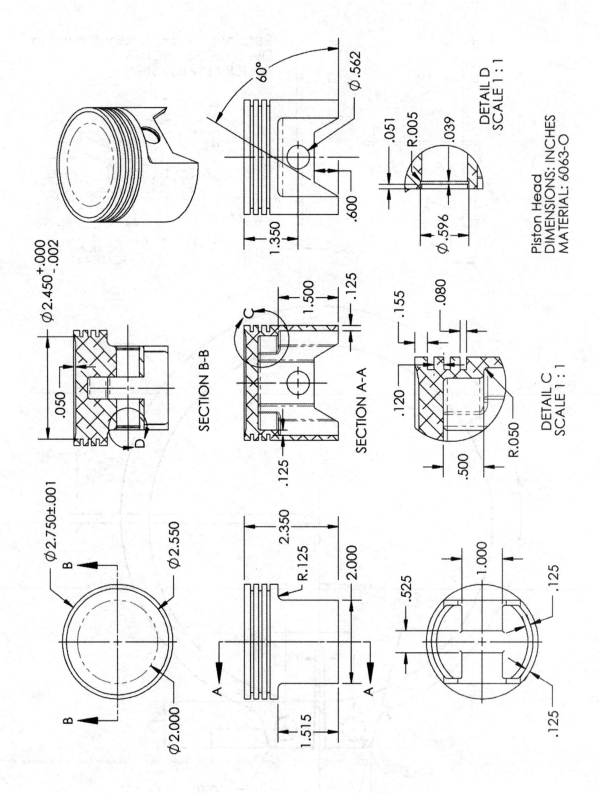

60°

⌀.562

.051

R.005

.039

DETAIL D
SCALE 1 : 1

⌀.596

Piston Head
DIMENSIONS: INCHES
MATERIAL: 6063-O

1.350

.600

⌀2.450 +.000 -.002

.050

SECTION B-B

1.500

.125

.155

.080

.120

R.050

.500

DETAIL C
SCALE 1 : 1

SECTION A-A

.125

⌀2.750±.001

⌀2.550

B

B

⌀2.000

2.350

R.125

2.000

.525

1.000

.125

1.515

.125

Here is a suggested sequence of features to build the Piston Head for reference.

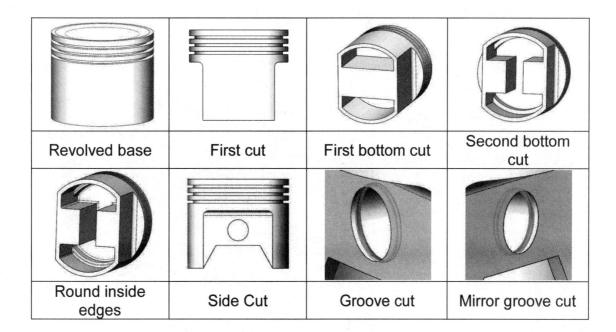

Revolved base	First cut	First bottom cut	Second bottom cut
Round inside edges	Side Cut	Groove cut	Mirror groove cut

Notes:

Chapter 3: Special Features: Sweep, Loft and Wrap

There are times when we need to design components that cannot be easily defined by prismatic shapes. For those features that have 'curvy' shapes we can use **Sweep** and **Loft** features, which let us create almost any shape we can think of. These are the features that allow us to design consumer products, which, more often than not, have to be visually attractive and pleasant to touch, making extensive use of curvature and organic shapes. These products include things like your remote control, a computer mouse, a coffee maker, perfume bottles, appliances, etc., and very often, their commercial success can be directly attributed to their appearance. They have to look nice, 'feel' right, and of course perform the task that they were intended for. Sweeps and Lofts (also used to create "organic" shapes like those found in nature) are widely used in the automotive, consumer products and aerospace industry, where cosmetics, aerodynamics, and ergonomics are a very important aspect of the design.

Sweeps and Lofts have many different options that allow us to create anything from relatively simple to extremely complex shapes. Considering the vast number of variations and possibilities for these features, we'll keep these examples as simple as possible without sacrificing functionality, to give the reader a good idea as to what can be achieved.

Sweeps and Lofts are usually referred to as advanced features, since they usually require more work to complete, and a better understanding of the basic concepts of solid modeling. Having said that, these exercises will assume that commands that we have done more than a couple of times up to this point, like creating a sketch, are already understood and we'll simply direct the reader to create it providing the necessary details. This way we'll be able to focus more on the specifics and options of the new features.

The Wrap feature is a special tool that helps us, as the name implies, to 'wrap' a sketch around a cylindrical surface. This tool helps us create features like cylindrical cams, slots on cylinders or cylindrical surfaces, projected profiles on irregular surfaces, etc.

These are some examples of designs made using advanced modeling techniques; some of them are covered in the Level II book.

Sweep: Cup and Springs

Notes:

The Cup

For this exercise we are going to make a simple cup. In this exercise we will learn a new option when creating features called "**Thin Feature**," the basics of the Sweep command, a new Fillet option to create a Full Round fillet and a review of auxiliary Planes. This is the sequence of features to complete the cup:

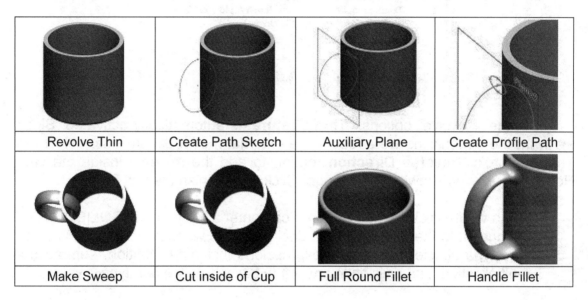

Revolve Thin	Create Path Sketch	Auxiliary Plane	Create Profile Path
Make Sweep	Cut inside of Cup	Full Round Fillet	Handle Fillet

8.1. - For the first feature we will create a "**Revolved Feature**" using the "Thin Feature" option. This option makes a feature with a specified thickness based on the sketch that was drawn. Make a new part with a template based in inches. Select the "*Front Plane*" and create the following sketch. Notice the sketch is an open profile with two lines, an arc and a centerline. (Remember to make the diameter dimension about the centerline.)

Thin Features can be made using either an open or closed sketch but using an open sketch will always make a thin feature.

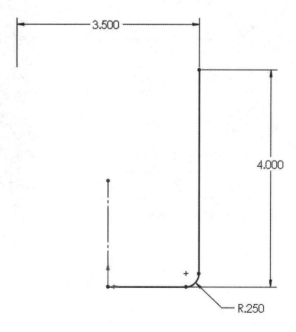

243

8.2. - After selecting the **"Revolve Boss/Base"** command we get a warning telling us about the sketch being open. Since we want a thin revolved feature, select "No."

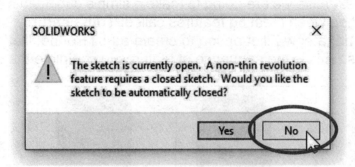

In the "Revolve" options, **"Thin Feature"** is automatically activated. Since we want the dimensions we added to be external model dimensions, select the Thin Feature's **"Reverse Direction"** option to add the material inside the cup. Notice the preview showing the change. Click OK to finish the first feature.

 In the value box we typed 3/16; we can enter a fraction and SOLIDWORKS changes it to the corresponding decimal value when we click OK. We can also type simple mathematic expressions including addition, subtraction, multiplication, and division in any value box where we can type a value.

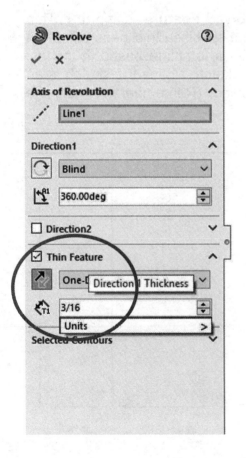

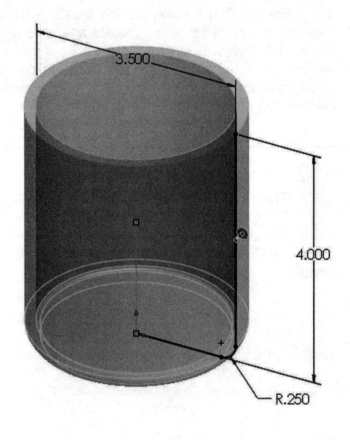

Our part looks like this and *"Revolve-Thin1"* is added to the Feature Manager.

8.3. - Select the *"Front Plane"* and create the following sketch using an ellipse. Switch to a Front view and select the "**Ellipse**" command from the Sketch tab in the CommandManager or from the menu "**Tools, Sketch Entities, Ellipse**." To draw an ellipse, first click to locate its center, click again to locate one axis and again to locate the other axis. Add the corresponding dimensions between the ellipse's points at the major and minor axes. Be sure to add a "Vertical" geometric relation between the top and bottom points of the ellipse (or "Horizontal" between left and right) to make them vertical (or horizontal) to each other to fully define the sketch.

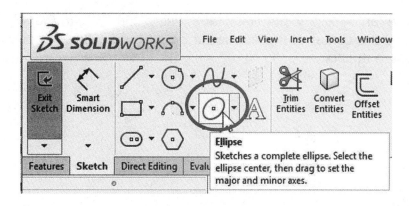

 When looking at a round surface's profile from an orthogonal view we can add a dimension to its silhouette. This option will be needed to add the 1.375" dimension.

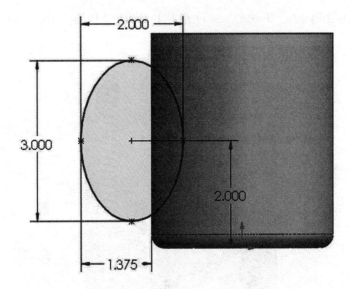

Exit the sketch and rename it *"Path Sketch."* We will not use the sketch for a feature just yet.

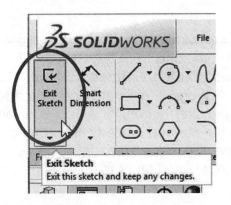

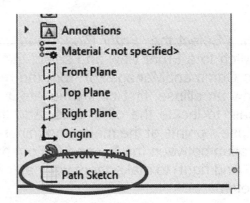

8.4. - Create an Auxiliary parallel plane using the *"Right Plane"* as the first reference and the center of the ellipse as the second reference as shown. Click OK to complete the plane.

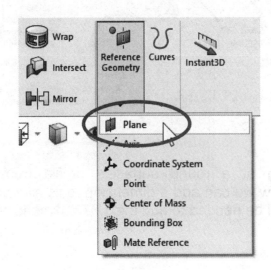

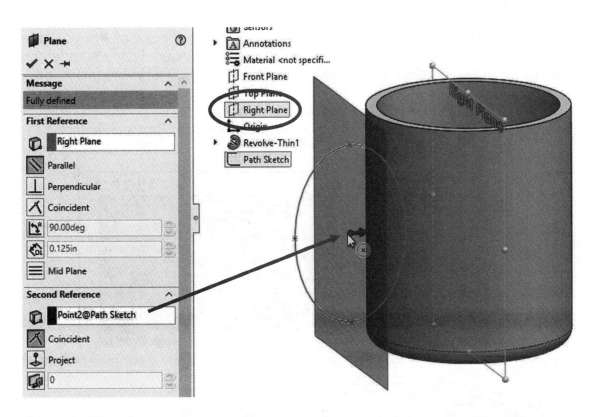

8.5. - Select the plane just created and draw the next sketch in it. Looking at the plane perpendicular to it helps us better visualize the sketch. Before adding the sketch, select the plane and press the "Normal To" command (shortcut Ctrl+8). Selecting "Normal To" again will rotate the view 180°.

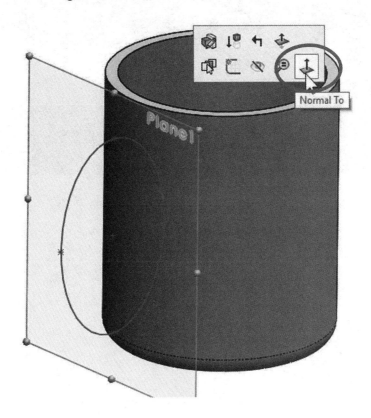

247

Add a new ellipse and start drawing it at the top point of the previous sketch's ellipse. Add a coincident relation to locate it if needed. Remember to add a horizontal relation between the points of the major (or vertical to the minor) axis.

If necessary, rotate the view to have a better view to select the top point of the previous sketch's ellipse. Exit the sketch and rename it "*Profile Sketch.*"

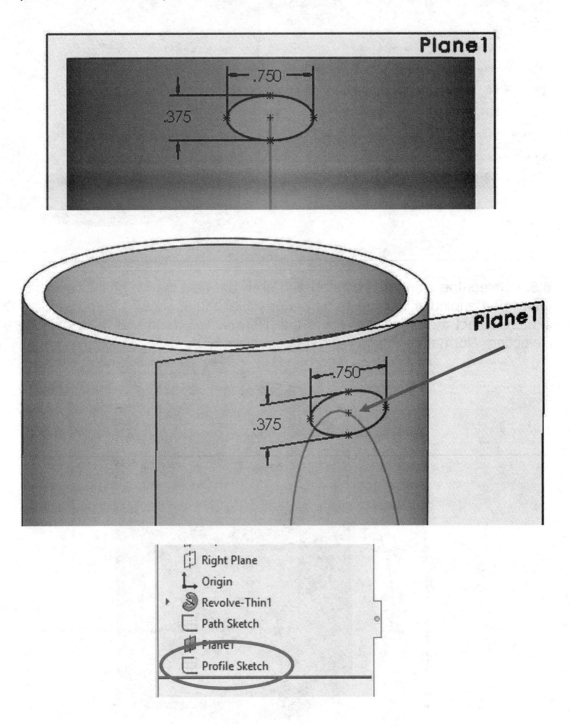

8.6. - Select the "**Sweep**" icon from the Features tab in the CommandManager or from the menu "**Insert, Boss/Base, Sweep.**" The sweep is a feature that requires a minimum of two sketches: one for the sweep's profile and one for the path. (For the path of the sweep instead of a sketch we can also use a model edge or a user defined curve that cross the profile.)

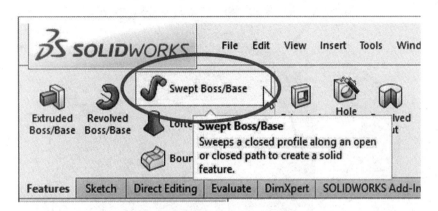

In the "**Sweep**" properties, select the *"Profile Sketch"* in the Profile selection box, and the "*Path Sketch*" in the Path selection box. Optionally, a Sweep can have guide curves and other parameters to better control the resulting shape; in this case we are making a simple sweep feature. Notice the preview and click OK when done to finish the sweep. Hide the auxiliary plane using the "**Hide/Show Items**" toolbar if so desired.

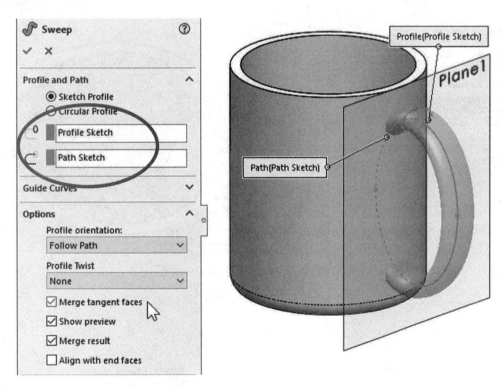

 Turn on the "**Merge Tangent Faces**" checkbox under "Options" to have a single, continuous sweep surface by merging tangent faces.

8.7. - Notice the sweep also goes inside the cup. To fix this we'll make a cut. Create a sketch in the flat face at the top of the cup.

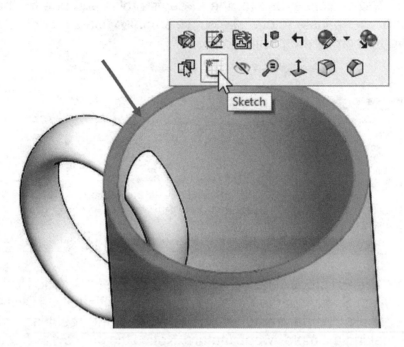

Select the <u>inside</u> edge at the top and use "**Convert Entities**" from the Sketch tab to convert the edge to sketch geometry. Click OK to continue.

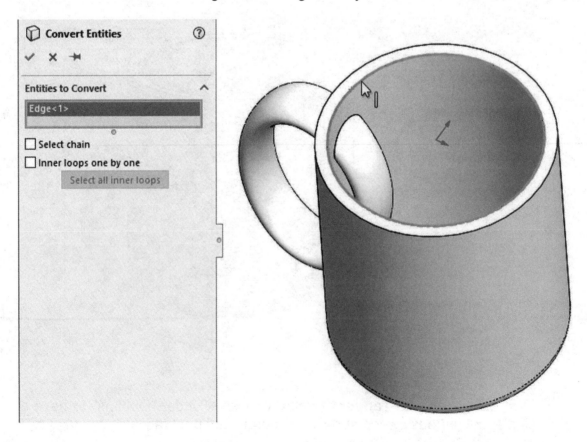

8.8. - Select the "**Extruded Cut**" icon and use the "Up to next" end condition from the drop down selection list. This end condition will make the cut until it finds the next face (the bottom) effectively cutting the part of the handle inside the cup.

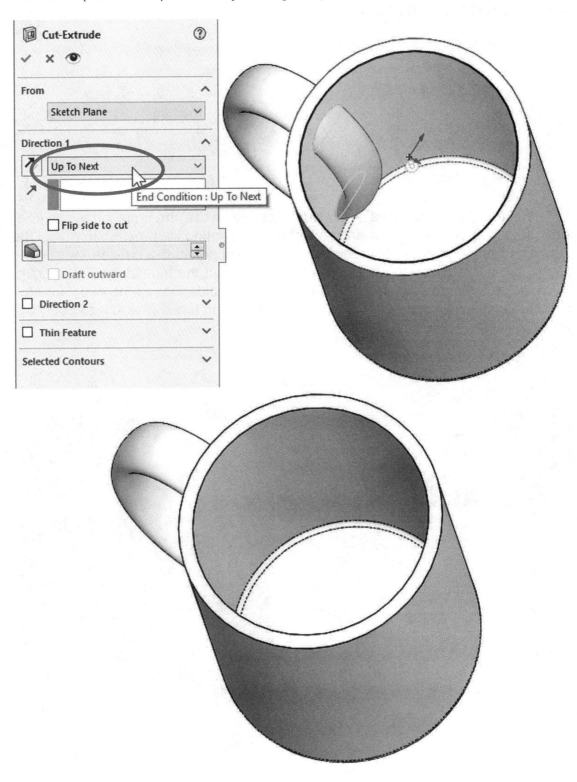

8.9. - Now we need to round the flat face at the top lip of the cup. To do this we'll add a "**Fillet**" using the "**Full Round Fillet**" option. The full round will essentially remove the flat face at the top (the middle face) and replace it with a rounded face tangent to both the start and end faces.

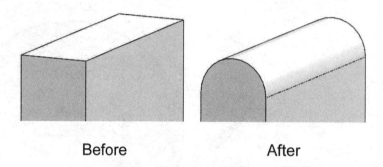

Before   After

We need to select three faces; the middle face will be the one replaced by the fillet. After selecting the Fillet command, select the "Full round fillet" option to reveal the selection boxes. With the first selection box active select the outside face of the cup. Click inside the second selection box and select the top flat face of the cup. Click inside the third selection box and select the inside face of the cup. Note the selected faces are color coded. When done selecting faces click OK to apply the fillet.

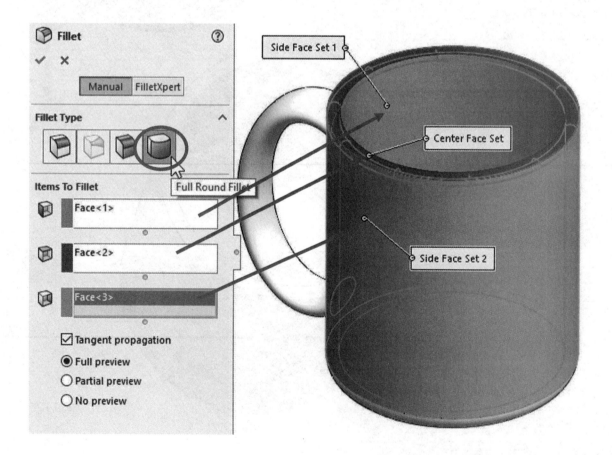

8.10. - To finish the cup, add a fillet to round the edges where the handle meets the cup. Select the fillet command with the "Constant Radius" option, select the handle's surface, and change the radius to 0.25″. Click OK to finish.

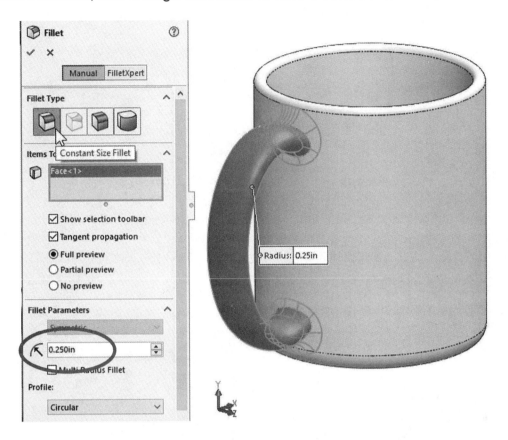

8.11. – Set the material as "Ceramic Porcelain" and save the finished part as *'Cup'*.

Notes:

Simple Spring

In the next exercise we are going to show how to make a simple and a variable pitch spring. To make these springs we'll have to learn how to make a simple and a variable pitch helix to be used as a sweep path. This is the sequence of features to complete the springs:

Simple Spring

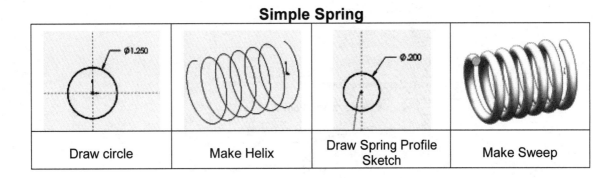

Draw circle	Make Helix	Draw Spring Profile Sketch	Make Sweep

Variable Pitch Spring

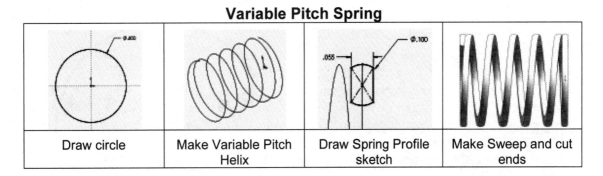

Draw circle	Make Variable Pitch Helix	Draw Spring Profile sketch	Make Sweep and cut ends

9.1.- First Make a new part in inches. To make a helix, we need to make a sketch with a circle. This circle is going to define the helix's diameter. Select the "*Front Plane*" and make a sketch using the following dimensions. Exit the sketch when done.

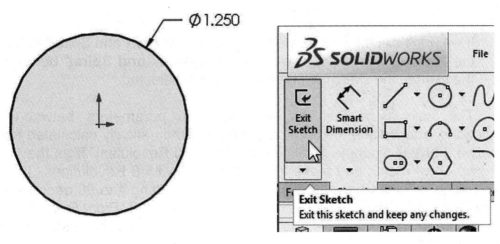

9.2. - In the Features tab select "**Curves**, **Helix and Spiral**" from the drop-down icon or the menu "**Insert, Curve, Helix/Spiral**." If asked to select a plane or a sketch, select the sketch we just drew.

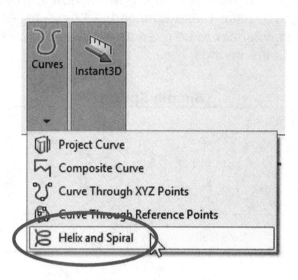

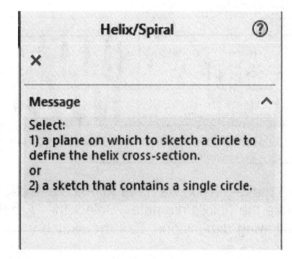
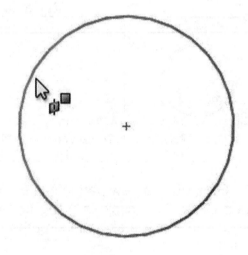

 If we don't exit the sketch, we'll only have the "Helix and Spiral" option from the "Curves" command. If we select the "**Helix and Spiral**" before exiting the sketch, the sketch will be automatically selected.

The helix is defined by selecting two parameters between Pitch, Revolutions, and Height; the third parameter is automatically calculated from the other two. For this example, we'll select "Pitch and Revolution" from the "Defined By:" drop down menu, and make the pitch 0.325" with 6 Revolutions. The "Start Angle" value defines where the helix will start. By setting it to 90 degrees we are defining the start of the helix at the top, coincident with the "*Front Plane*." If we had made it 0 degrees, it would be coincident to the "*Top Plane*" instead. (Feel free to explore the options.) Click OK to complete the helix.

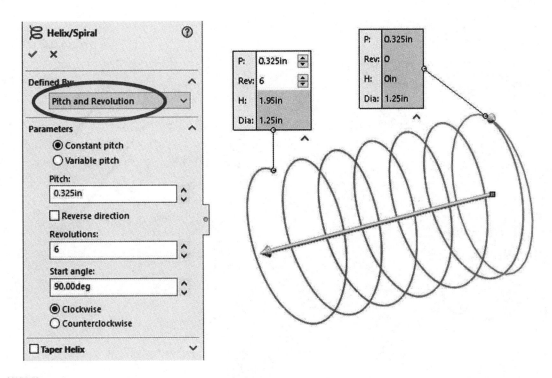

 Note the Helix command has options to make it Counterclockwise, Clockwise, Tapered, Variable pitch and reversed (going right or left).

9.3. - Once the Helix is done, we need to make the *profile* sketch for the sweep. Switch to a Right view and add a new sketch in the "*Right Plane*" as shown. Make the circle close to the Helix...

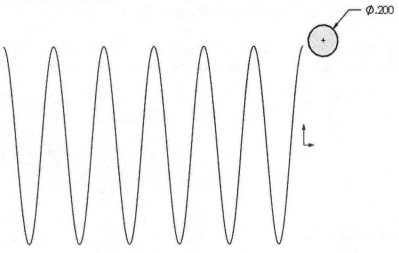

... and add a "**Pierce**" geometric relation between the center of the circle and the helix. This way the profile will be located at the beginning of the helix. This relation will make the sketch fully defined. Exit the profile sketch when done.

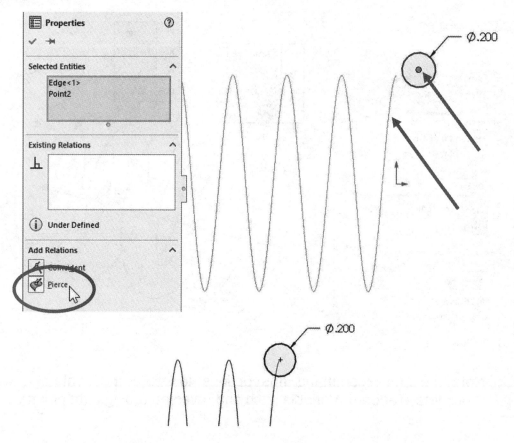

A pierce relation is done between an element that is oblique or perpendicular to the sketch plane and a point (or endpoint) in the sketch. Think of it as a needle piercing through a fabric, the sketch being the fabric and the helix (or curve, model edge or another sketch element) the needle.

9.4. - Select the Sweep command and make the sweep using the last sketch as a profile and the Helix as a Path. Note the Preview and click OK to finish.

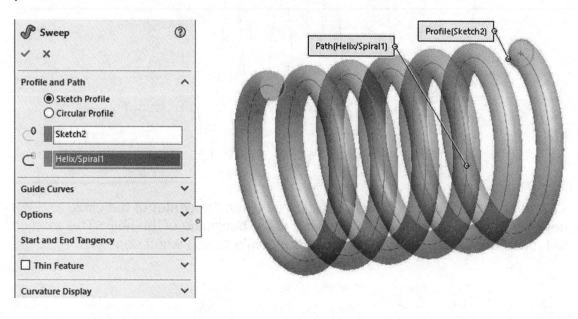

If needed/desired, hide the "*Helix/Spiral1*" in the FeatureManager, save the part as '*Spring*' and close the file.

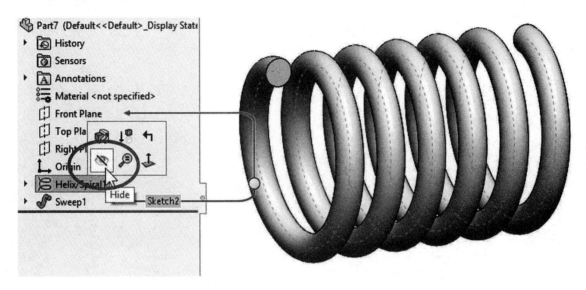

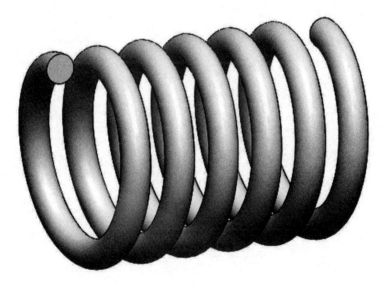

Notes:

Variable Pitch Spring

10.1. - For the variable pitch spring we'll start the same way and make the following sketch in the "*Front Plane*." This will be the spring's outside diameter.

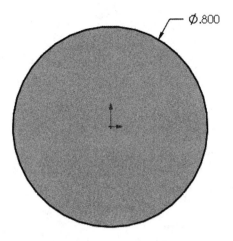

Ø.800

10.2. - While still editing the sketch, in the Features tab from the "**Curves**" drop down icon select "**Helix and Spiral**"; notice it is the only option available.

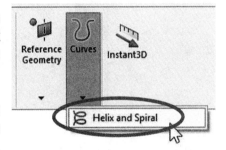

Helix and Spiral

In the **Helix/Spiral** command select "Pitch and Revolution" from the "Defined by:" selection box and "Variable Pitch" in the Parameters box. After selecting it we are presented with a table; fill in the values for the helix using the next table. Click OK when finished to build the helix.

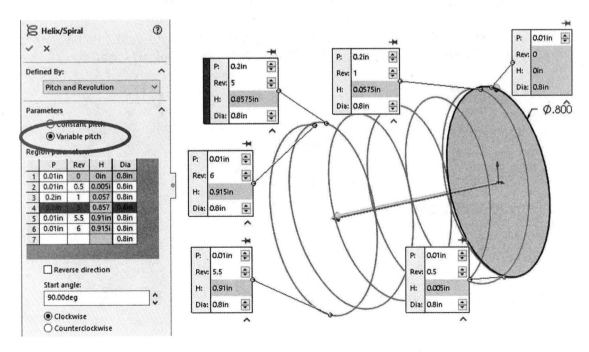

Helix/Spiral

Defined By:

Pitch and Revolution

Parameters

○ Constant pitch
● Variable pitch

Region parameters

	P	Rev	H	Dia
1	0.01in	0	0in	0.8in
2	0.01in	0.5	0.005i	0.8in
3	0.2in	1	0.057	0.8in
4	0.2in	5	0.857	0.8in
5	0.01in	5.5	0.91in	0.8in
6	0.01in	6	0.915i	0.8in
7				0.8in

☐ Reverse direction

Start angle:

90.00deg

● Clockwise
○ Counterclockwise

P: 0.2in
Rev: 5
H: 0.8575in
Dia: 0.8in

P: 0.2in
Rev: 1
H: 0.0575in
Dia: 0.8in

P: 0.01in
Rev: 0
H: 0in
Dia: 0.8in

P: 0.01in
Rev: 6
H: 0.915in
Dia: 0.8in

Ø.800

P: 0.01in
Rev: 5.5
H: 0.91in
Dia: 0.8in

P: 0.01in
Rev: 0.5
H: 0.005in
Dia: 0.8in

10.3. - After the helix is complete add a new sketch in the *"Right Plane."* Draw the circle first, then add a center rectangle, trim and dimension as needed. Locate the sketch to either side of the helix.

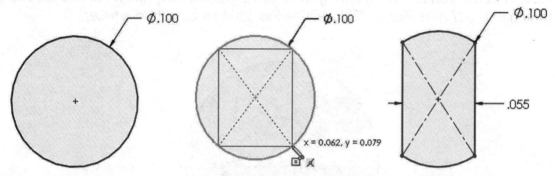

Add a centerline from the center to the top line (make sure it is coincident). Add a **"Pierce"** geometric relation between the top endpoint of the centerline and the helix to fully define the sketch. Exit the sketch and optionally rename it *'Profile'*.

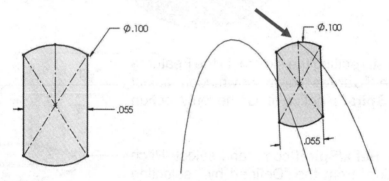

 The Pierce relation allows us to fix the profile to the path and can be added to any part of the sketch. We chose to add it to the top because the original sketch used for the helix is the spring's outside diameter.

10.4. - Just as we did before, select the **"Sweep"** command, add the path and profile as shown and click OK to finish and hide the *"Helix/Spiral1"* feature.

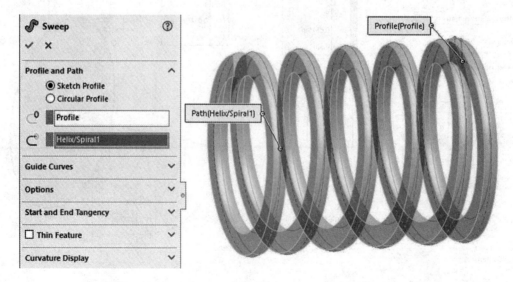

10.5. - As a finishing touch we'll make a cut to flatten the sides of the spring. Change to a Right view and add a sketch in the *"Right Plane."* Draw a single line starting at the midpoint of the indicated edge and long enough to cross the part.

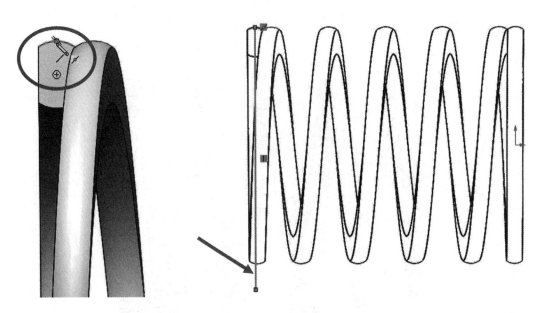

10.6. - Select the **"Extruded Cut"** command; if we use an open sketch to make a cut, selecting either the "Through All" or the "Through All - Both" option, it automatically activates the "Direction 2" checkbox with the "Through All" end condition and one side of the model is cut using the open sketch. The small arrow located at the center of the line indicates which side of the model will be cut. Use the "Flip side to cut" option if needed to cut the left side. Click OK to complete the cut and repeat the same sketch to cut the right side. Note the cutting plane is shown in the graphics area.

When using an open sketch to cut a model, we can only use a single open profile. If we have multiple open profiles we cannot make the Cut Extrude using these options, we would get multiple thin cuts with a different behavior.

10.7. - Save the model as *'Variable Pitch Spring'*.

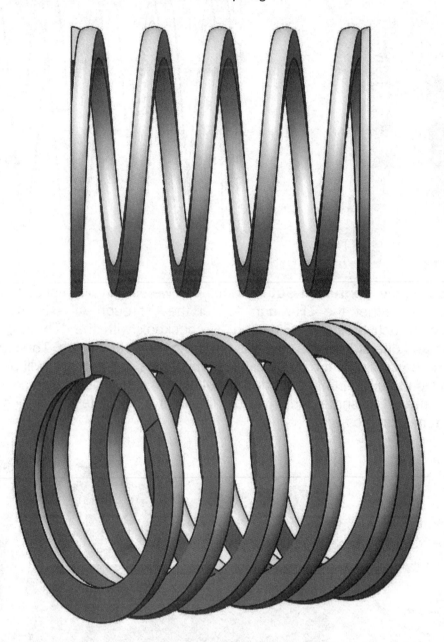

Thread Feature

Threads can be added to a model using a Sweep Boss or a Sweep Cut with a helix, and defining every thread parameter, or simplify the process using the "**Thread**" tool. When modeling screws and fasteners in general, it is *almost always* unnecessary to add a helical thread, as it consumes a large amount of computing resources, and a simple representation using a Revolved Boss/Cut or a Cosmetic Thread is usually enough. It is strongly advised to only add threads when required by the model, as in the bottle exercise later in this lesson.

11.1. - To learn how to use the "**Thread**" command open the part '*Threads*' from the accompanying files. This part has two cylinders; the first one (Blue) is the size of a ¼-20 screw's <u>minor</u> diameter, and the second (Green) is the size of a ¼-20 screw's <u>major</u> diameter. The hole (Pink) is the size needed to make a ¼-20 tapped hole.

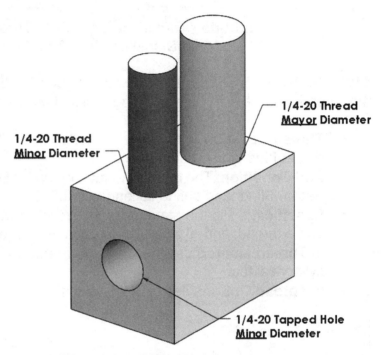

1/4-20 Thread <u>**Mayor**</u> Diameter

1/4-20 Thread <u>**Minor**</u> Diameter

1/4-20 Tapped Hole <u>**Minor**</u> Diameter

The "**Thread**" command is located in the menu "**Insert, Features, Thread.**" This command allows us to select a standard thread size and apply it either as an external thread (male) to a cylindrical boss or an internal thread (female) to a hole.

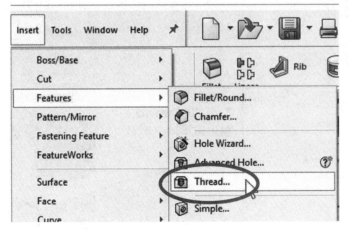

There are two ways to add an external (male) thread: The first way is to add (**extrude**) the thread in a cylinder the size of the thread's *minor* diameter. The second way is to start with a cylinder the same size as the screw's *major* diameter and **cut** away the thread.

 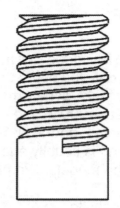

Extrude - Thread added to Minor Diameter   **Cut** -Thread removed from Major Diameter

11.2. - Select the Blue cylinder's top edge to add the thread. In this case we'll add the thread using the following settings; click OK when done to finish the thread feature:

- **Thread Location** is to define the edge to start the thread. Select the top edge of the blue cylinder's face.
- **End Condition**: The thread's length, number of revolutions or up to surface. Set to "Blind" and define its length to 0.375".
- **Specification** is the standard and size of the thread. Select "Inch Tap" to add a thread, and "0.2500-20" to define the thread size as ¼-20.
- In **Thread Method** use the "Extrude thread" to add the thread on top of the cylindrical face.
- In **Thread Options** select Right-hand thread.

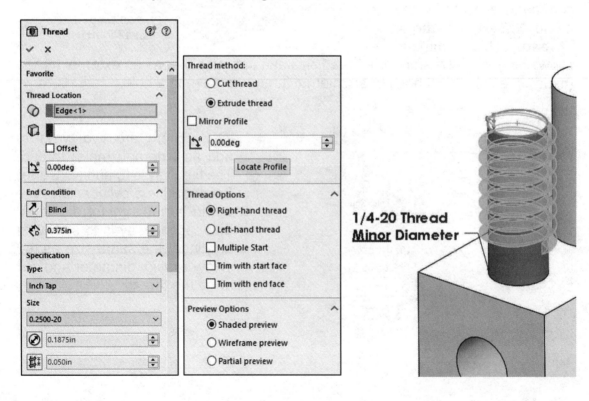

Our first finished thread looks like this:

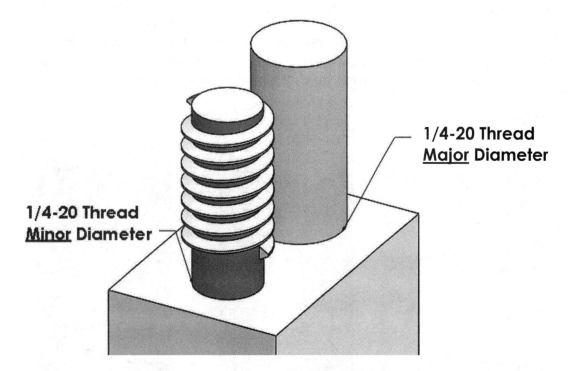

1/4-20 Thread
Major Diameter

1/4-20 Thread
Minor Diameter

11.3. - To illustrate the second way to add an external thread, in this step the thread will be added to the green cylinder using the same settings as the previous one, except under "Specification" we'll use the "Inch Die" and "Cut thread" options to cut into the cylinder instead of adding material.

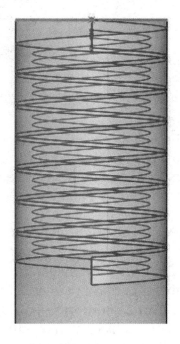

Another option that will be changed in the second thread is to turn on the "Offset" option under "Thread Location." The reason to turn this option ON is to have a thread that starts to cut the cylinder outside, making a more true-to-life thread. Change the direction of the offset and make it 0.040" as indicated; this way the thread will be cut starting away from the face. Optionally we can also control the angle where the thread starts by changing the "Start Angle" value.

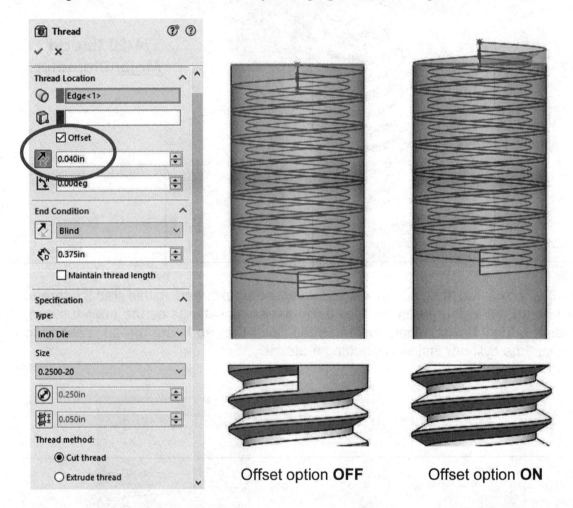

Offset option **OFF**   Offset option **ON**

11.4. – The last thread will be the internal (tapped) thread. In this case the hole in the part has a 0.201" diameter (#7 drill bit). This is the drill size we'd use in real life to make a ¼-20 tapped hole. Select the "**Thread**" command, and this time select the pink hole's edge. These are the options we'll use:

- **Thread Location**, select the edge and turn on the "Offset" option to have the same result as in the previous step to start the thread outside the hole.
- In **Specification** select "Inch tap," size "0.2500-20."
- **Thread Method** use the "Cut thread" option to cut into the part.
- For **End Condition** select "Up To Selection," and select the face at the other end of the block to make the thread go through the entire block, also using the "Offset" option to make the thread go past the end face.
- Under **Thread Method** select "Cut thread."

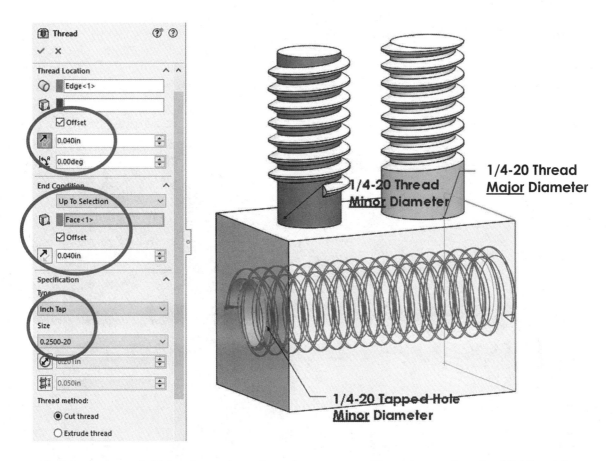

After finishing all three threads, orient the part to a Top view using the Hidden Lines Visible option. Here we can see the external threads made at the top fit perfectly inside the tapped hole.

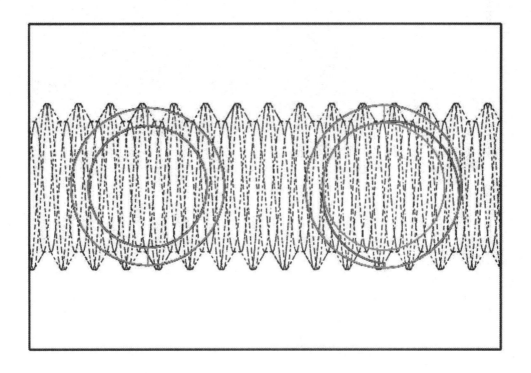

Notes:

Sweep: Bottle

For the next exercise we'll build the following bottle; notice the complete absence of straight linear edges and *almost* no flat faces in the model. This is a good example of a model that would be very difficult, if not impossible to complete using simple prismatic features.

*Image made using RealView graphics

In this model we'll use two sweep features, one for the body and one for the thread. For the body we'll make a sweep with two guide curves, therefore we need to make four sketches for the first sweep feature: Path, Guide Curve 1, Guide Curve 2 and Profile, in that order. The reason to make the Profile last is that it has to pierce the Path and both Guide Curves for the sweep feature to work as expected.

271

12.1. – Make a new part in inches with three decimal places, set the material to a clear plastic, "ABS PC." Add the Path sketch as indicated in the "*Front Plane*" and Exit the sketch. Rename *'Path'*.

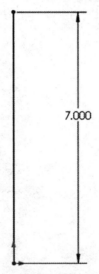

7.000

12.2. - Add a second sketch also in the "*Front Plane*." Make arcs and lines tangent to each other; use centerlines as reference for tangency. Add geometric relations to make all arcs equal. Exit the sketch and rename it *'Guide 1'*.

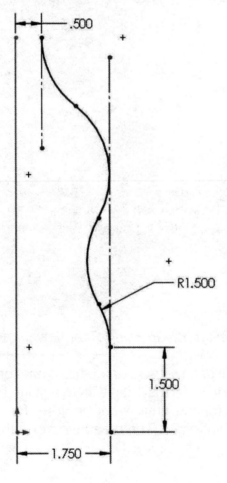

.500

R1.500

1.500

1.750

12.3. – Switch to a Right view and add a third sketch in the *"Right Plane."* Add geometric relations to previous sketches to maintain design intent. Make the topmost endpoint Horizontal to the topmost endpoint of the '*Path*' sketch. Arcs are equal size and tangent. Add a horizontal geometric relation to the endpoints indicated. Only geometric relations and two dimensions are needed to fully define the sketch. Exit the sketch and rename it '*Guide 2*.'

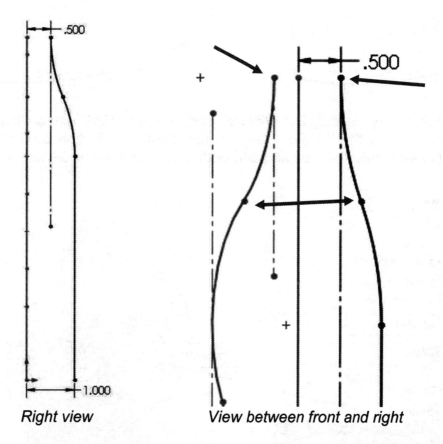

Right view   *View between front and right*

12.4. - Add a new sketch in the *"Top Plane."* Make an ellipse starting at the origin adding Pierce relations between the ellipse's major and minor axes to '*Guide 1*' and '*Guide 2*'. View is in Isometric for clarity. Exit the sketch and rename it '*Profile*'.

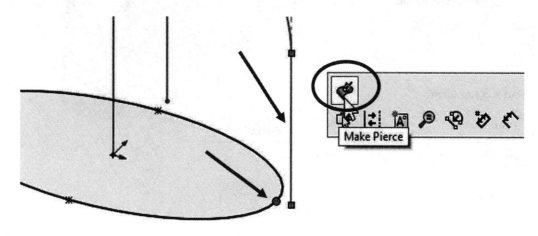

273

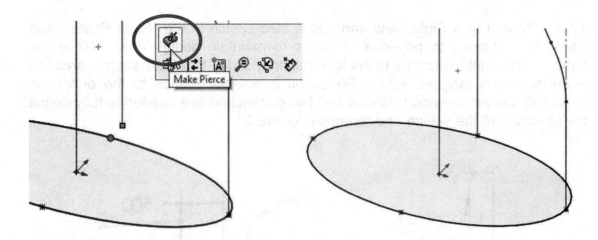

12.5. - Select the "**Sweep Boss/Base**" icon, add the *'Path'* sketch to the "Path" selection box and the *'Profile'* sketch to the "Profile" selection box. Expand the "Guide Curves" selection box and add both guides to it. Click OK to finish.

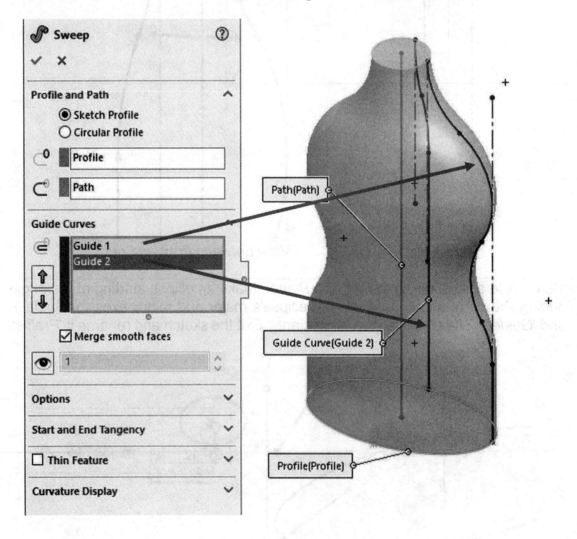

12.6. – To make the neck of the bottle, add a sketch on the top face, use "**Convert Entities**" to project the edge of the top circular face…

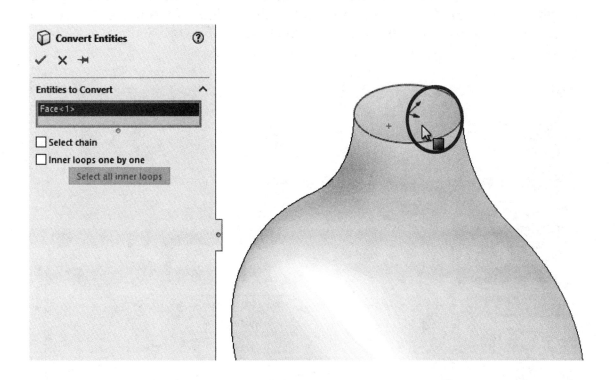

…and extrude it 1".

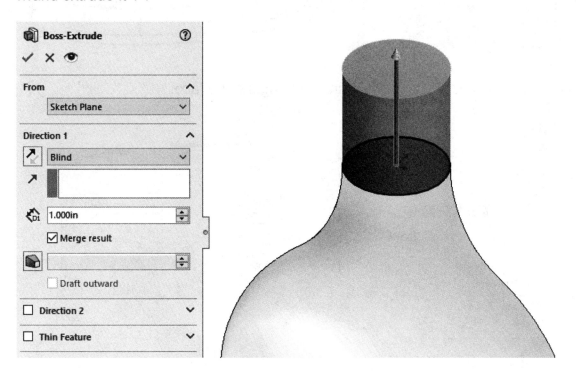

12.7. – Next, we need to add a fillet to the bottom of the bottle. Add a fillet to the bottom edge using the "Asymmetric" option. An asymmetric fillet allows us to define the distance the fillet goes into each of the faces. Activate the option and set the first distance to 0.375" going up, and 0.25" at the bottom of the bottle. Be aware that Distance 1 and 2 may be reversed in your case.

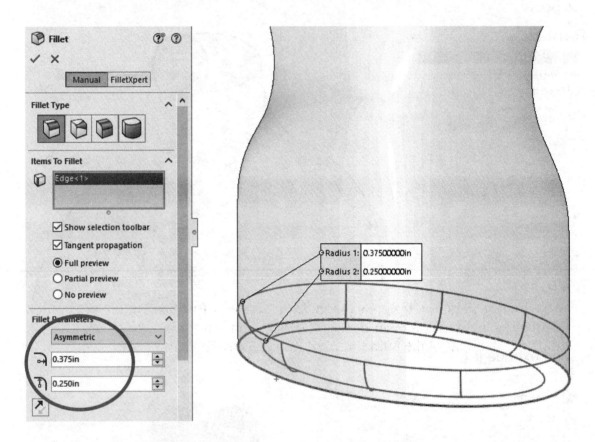

12.8. - Add a 0.050" shell to the part removing the top face.

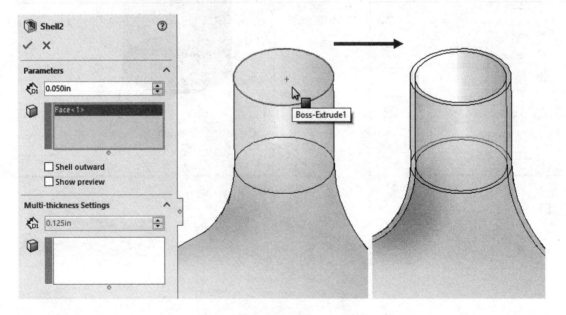

12.9. - Add a parallel auxiliary plane 0.125" below the topmost face.

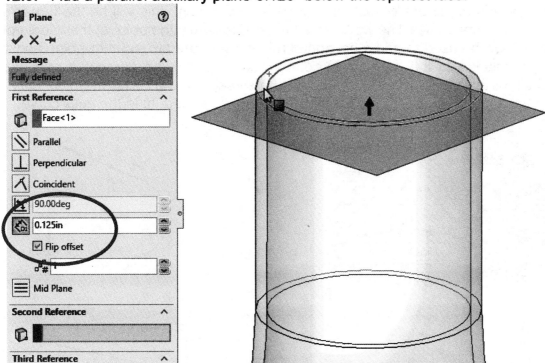

12.10. - Add a sketch in the new plane. Use "**Convert Entities**" to project the top outside edge and make a Helix defined by Pitch and Revolution using 0.2" for pitch and 2.5 revolutions. Make sure the Start Angle is set at 0 degrees.

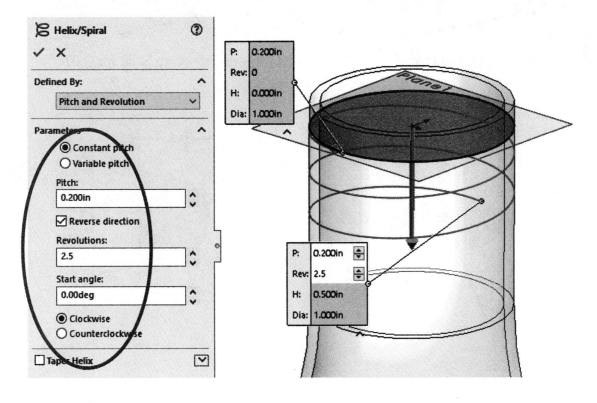

12.11. - Rotate to a Right View and add the following sketch in the "*Right Plane*" near the top of the bottle, close to the helix's start point; this will be the thread's Profile. For the vertical line we'll use a "**Midpoint Line**"; the point at the middle of the line will be used to add a horizontal relation with the arc's center point, and a Pierce relation to the helix. Draw, dimension and add the geometric relations indicated, then Exit the sketch and make a sweep feature using the helix curve and this thread profile. After creating the sweep, if needed, hide the helix curve.

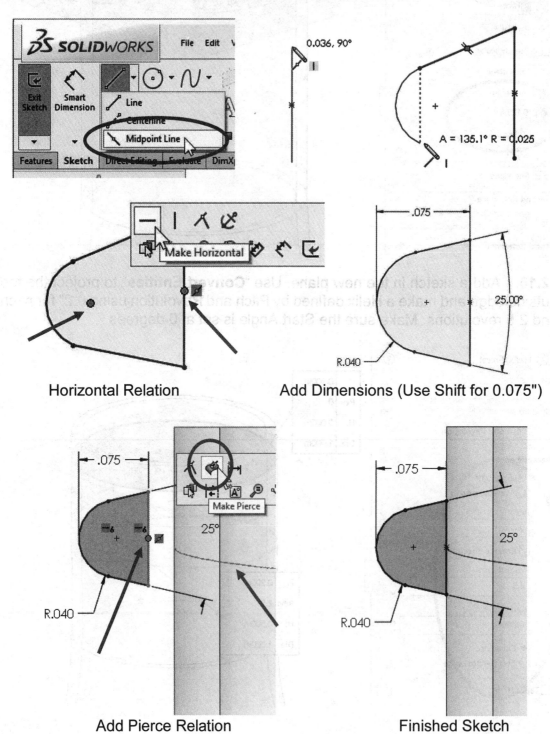

Horizontal Relation

Add Dimensions (Use Shift for 0.075")

Add Pierce Relation

Finished Sketch

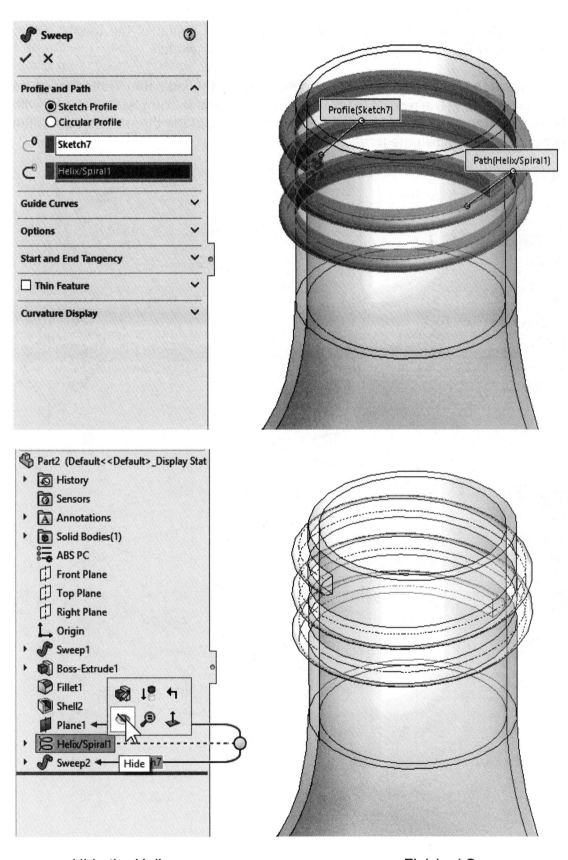

Hide the Helix curve   Finished Sweep

12.12. - Add a new sketch to the flat face at the beginning of the thread, using "**Convert Entities**" to project the face's edges into the sketch.

Make a "Revolved Boss" 120 degrees using the indicated vertical edge as axis of revolution. The reason to make it 120 degrees is to completely merge it with the body of the bottle. Add another sketch at the end of the thread and repeat the same process to finish the thread.

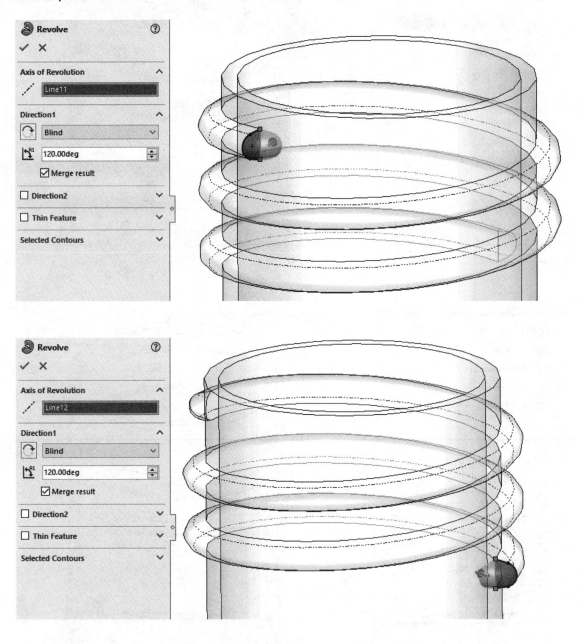

12.13. - Add a 0.02" fillet to the thread as a finishing touch. You may need to select multiple edges. Save the part as '*Bottle Sweep*' and close it.

Engine Project Parts: Make the following components to build the engine. Save the parts using the name provided. High resolution images are included on the exercise files.

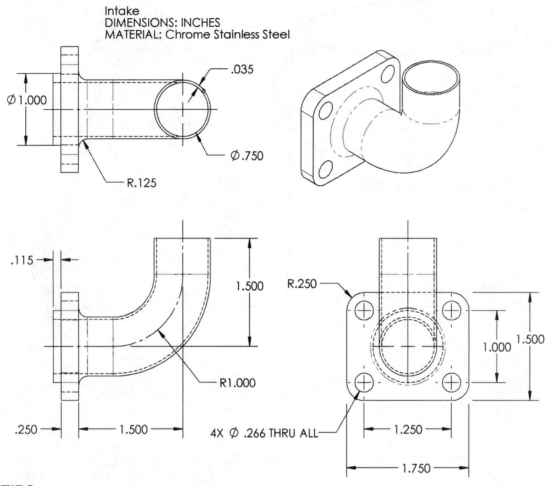

Intake
DIMENSIONS: INCHES
MATERIAL: Chrome Stainless Steel

Ø 1.000

.035

Ø .750

R.125

.115

1.500

R.250

R1.000

1.000 1.500

.250

1.500

4X Ø .266 THRU ALL

1.250

1.750

TIPS:

Extrude base

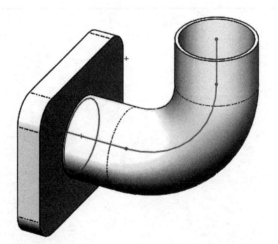

Add Path & Profile sketches. Create a sweep using the "Thin Feature" option

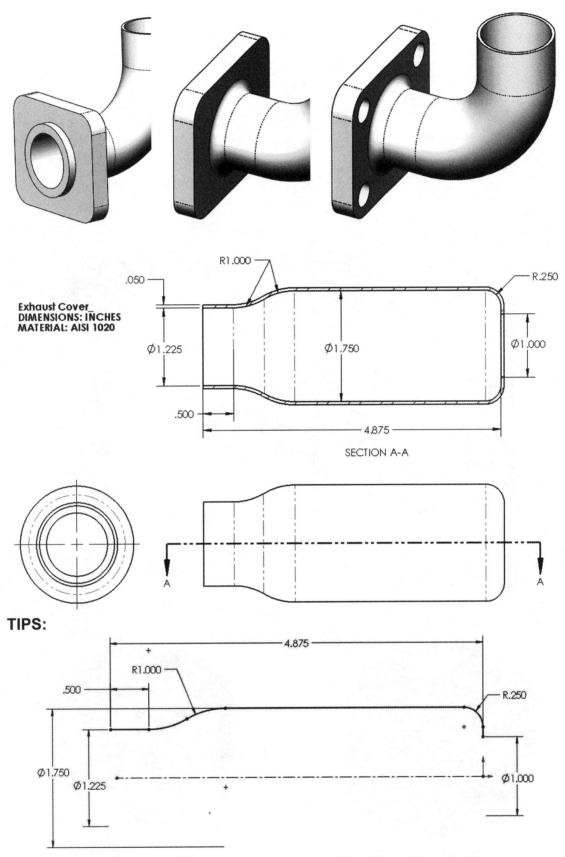

Exhaust Cover_
DIMENSIONS: INCHES
MATERIAL: AISI 1020

R1.000

.050

R.250

Ø1.225

Ø1.750

Ø1.000

.500

4.875

SECTION A-A

A   A

TIPS:

4.875

R1.000

.500

R.250

Ø1.750

Ø1.225

Ø1.000

Make a "Revolved Boss" using the "Thin Feature" option, adding material outside

283

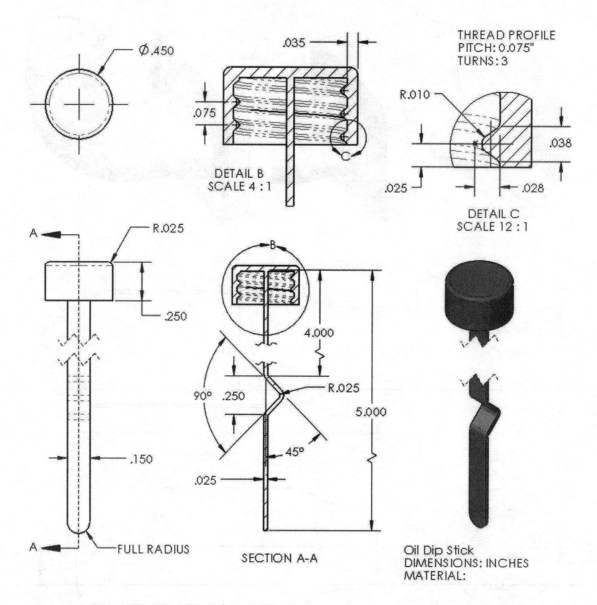

Ø.450

.035

DETAIL B
SCALE 4 : 1

THREAD PROFILE
PITCH: 0.075"
TURNS: 3

R.010

.038

.025   .028

DETAIL C
SCALE 12 : 1

A

R.025

.250

.150

A   FULL RADIUS

B

4.000

R.025

90°   .250

5.000

45°

.025

SECTION A-A

Oil Dip Stick
DIMENSIONS: INCHES
MATERIAL:

TIPS:

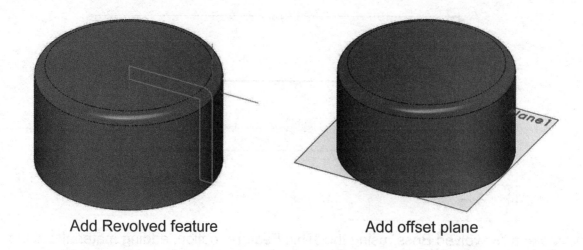

Add Revolved feature   Add offset plane

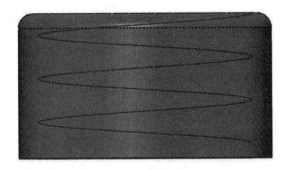

Add Helix

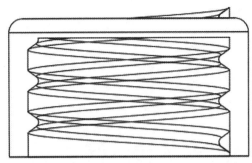

Add thread profile and create sweep. Thread will go through the top; it will be trimmed later

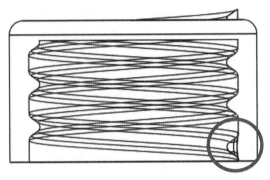

Add Revolved boss and fillet to finish thread

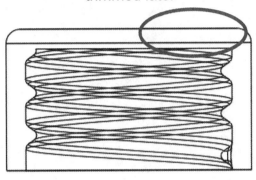

Trim the excess thread from the top

Add extrusion for dip stick using the "Thin Feature" option, adding material on both sides

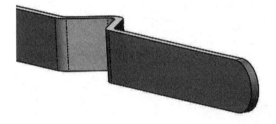

Add full round fillet

Notes:

Loft: Bottle

Notes:

The "**Loft**" is a feature that helps us design complex shapes by defining the different cross sections of the model, accurately controlling the final shape. It requires at least two different sketches and/or faces, and optionally guide curves. In this exercise we will make a bottle using a loft with four different sketches.

13.1. - Make a new part using a template based in Inches and create three auxiliary planes using the "**Plane**" command. Select the *"Top Plane"* as reference, set the number of planes to 3 and space them 2.5″ as shown. Click OK when done.

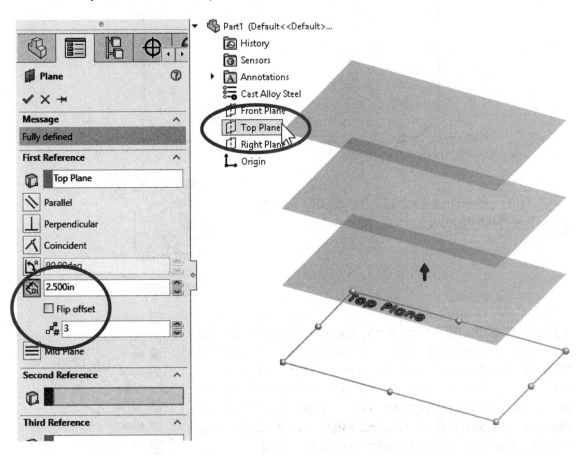

13.2. - For clarity, select the *"Top Plane"* in the FeatureManager and show it using the "Hide/Show" command from the pop-up toolbar. Turn on Plane visibility if needed in the "Hide/Show Items" command.

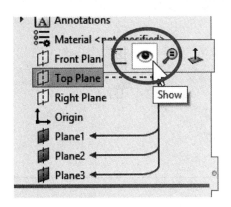

13.3. - Switch to a Top view, select the *"Top Plane"* and draw the following sketch using the "**Center Rectangle**" and "**Sketch Fillet**" tools. Exit the sketch when finished. Plane visibility and "**Shaded Sketch Contours**" were turned OFF for clarity.

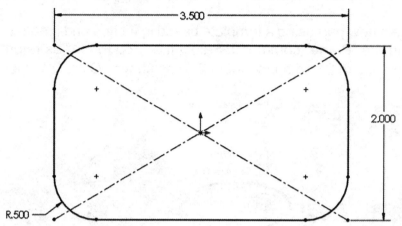

 Turn off Plane visibility for clarity using the "Hide/Show Items" menu.

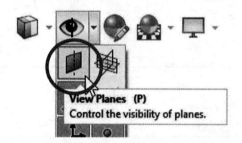

13.4. - Still in the Top view, select *"Plane1"* in the FeatureManager and create the second sketch. Use the "**Center Rectangle**" tool; be sure to start in the origin and make the rectangle's corner coincident to the previous sketch's diagonal line. Add a 2.5″ width dimension and the 0.5″ "**Sketch Fillet**" to fully define the sketch. Exit the second sketch when finished.

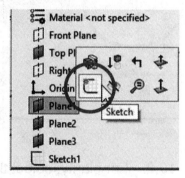

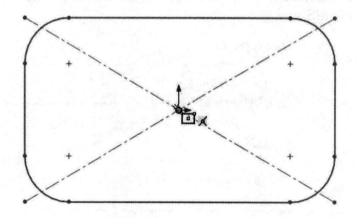

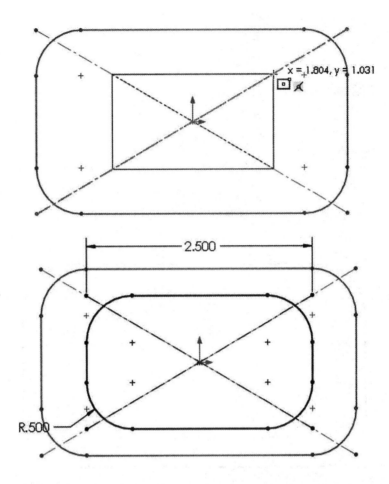

13.5. - For the third profile, select *"Plane2"* from the FeatureManager and create a new sketch in it. This sketch will be exactly the same as the first one. To help us save time and maintain design intent, we'll use the "**Convert Entities**" tool. In the "**Convert Entities**" selection box, select *"Sketch1"* from the fly-out FeatureManager and click OK to project *"Sketch1"* in the new sketch. Exit the sketch when done.

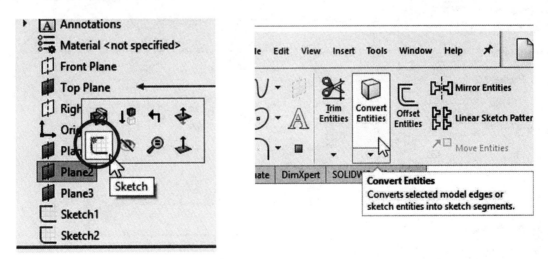

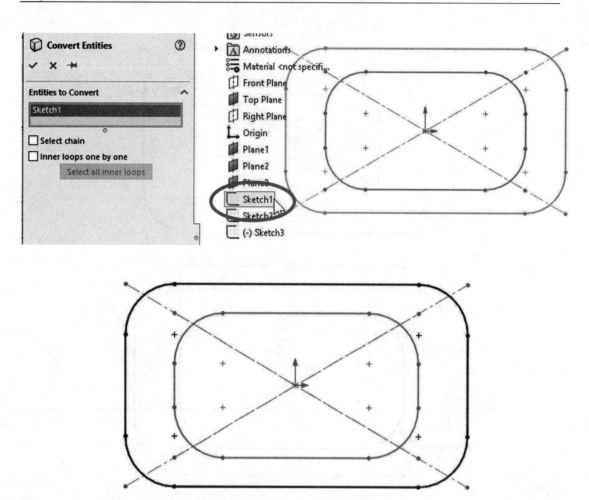

13.6. - For the last profile select *"Plane3"* from the FeatureManager and create a new sketch. Draw a circle and add a geometric relation to make it "Tangent" to the horizontal line in *"Sketch2"* as indicated. This relation will fully define the sketch. Exit the sketch to finish.

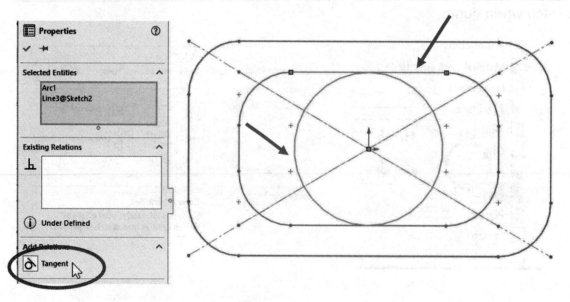

The finished sketches will look like this with the planes visible:

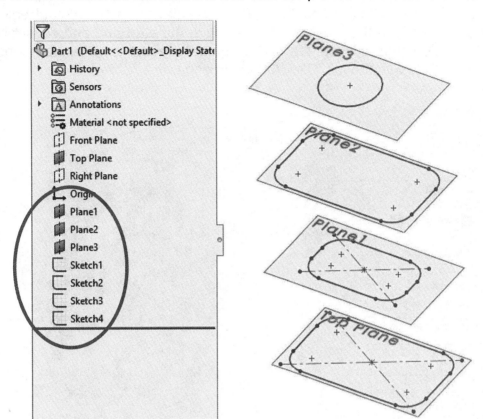

13.7. – After completing the sketches, each of which will be a different cross section of the bottle, we are ready to make the "**Loft**" feature. The loft feature requires two or more sketches and/or faces, and we'll use the four sketches we just made to build the bottle.

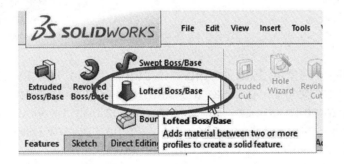

We will select the profiles in order starting with the first one we made at the bottom and finishing with the last one at the top (or from top to bottom, but in order). It is important to select the sketches thinking that the point where we select the profile will affect the result. Click to select each sketch in the graphics area near the green 'dots'. This line indicates the segment of one profile that will be connected to the next profile. If we select points randomly in the profiles, the loft would twist and produce undesirable results. Optionally, guide curves can be added to improve control of the resulting shape as we did with the sweep bottle.

After selecting the profiles, from the "Start/End Constraints" options select "Normal to Profile" for both the "Start" and "End" constraint using the default values. Notice the difference in the preview after selecting the start and end constraints. In the "Options" section activate the "Merge tangent faces" to generate a single, continuous surface with the loft. Click OK when done.

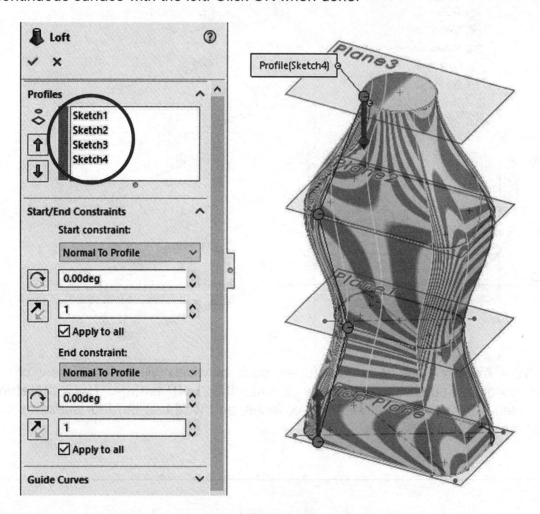

Optionally, we can turn on the curvature display for the preview; this will allow us to better visualize the final shape using a mesh, zebra stripes and/or curvature combs. The zebra stripes help us visualize the continuity of the curvature in the model's surfaces. This is particularly useful when designing consumer products with a high aesthetic requirement.

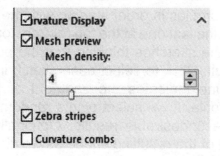

13.8. - Add a 0.25″ radius fillet in the bottom edge of the bottle to round it off.

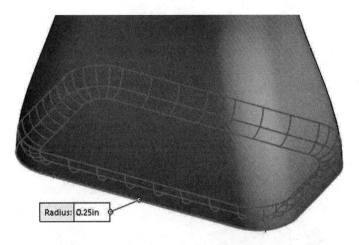

13.9. - Add a 0.75″ extrusion at the top of the bottle and finish using the "**Shell**" command removing the top face making the wall thickness 0.125.″

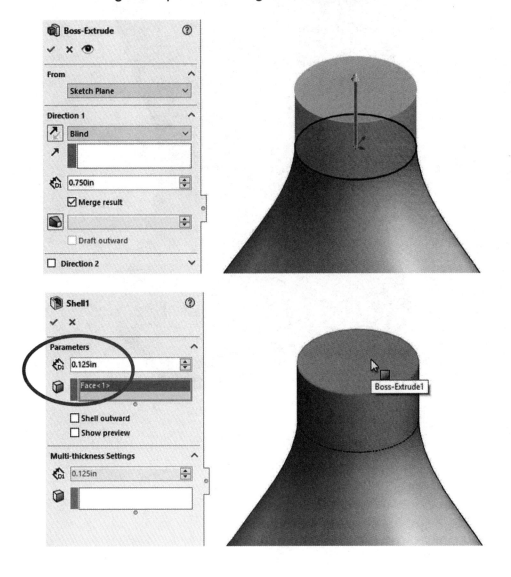

Image using Real View

13.10. - To view the inside of the part, select the "**Section View**" icon from the View toolbar or the menu "**View, Display, Section View**." We can define which plane to cut the model with, the depth of the cut and optionally add a second or third section plane. If we click OK in the Section View, the model will be displayed as cut, but this is only for display purposes; the part is not actually cut. To turn off the Section View, select its icon or menu command again. This section view can be used along with the "**Measure**" tool to inspect the part.

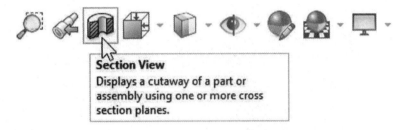

Section View
Displays a cutaway of a part or
assembly using one or more cross
section planes.

The depth of the section can be changed by changing the distance in the PropertyManager or by dragging the arrow in the center of the plane. The section view plane's rotation can also be changed by dragging the rotation handles or entering a value in the rotation value boxes.

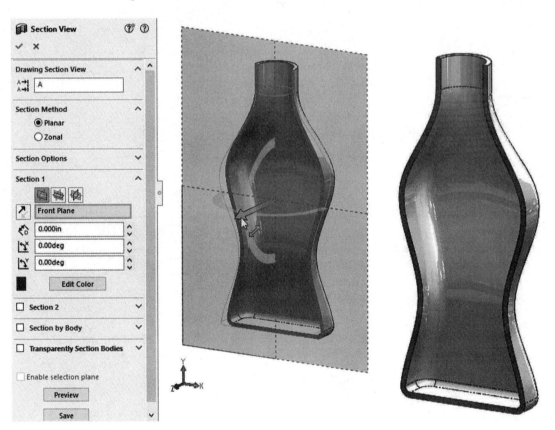

Save the part as '*Bottle Loft*' and close.

Notes:

Wrap Feature

 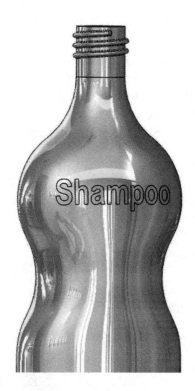

As its name implies, this feature allows us to wrap a sketch around any face and add material, remove material, or split (divide) the face. In this lesson we'll learn how to create a cylindrical cam, and how to wrap a text sketch onto an irregular surface. To wrap a sketch on a cylindrical (or conical) surface we need to add a sketch on a plane tangent to the surface we want to make the wrap on.

14.1. – The first step is to create a cylinder to add the wrap feature to. Make a new part using a template with Inches and three decimal places, add this sketch on the *"Front Plane"* and extrude it 3". Be sure to add a Sketch Point at the top of the circle; this point will be used later as a reference to create an auxiliary plane.

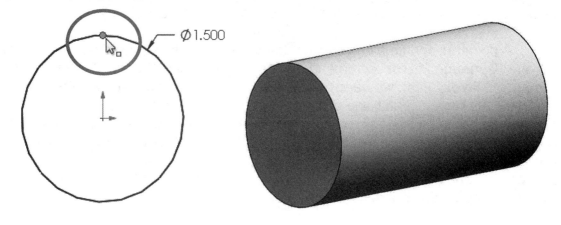

14.2. - The next step is to make an auxiliary plane to add the sketch that will be used for the wrap feature. We'll define the new auxiliary plane parallel to the *"Top Plane"* and tangent to the cylindrical surface, coincident to the sketch point added in the previous step. Make the sketch of the first feature visible and add a new plane's first reference parallel to the *"Top Plane"* and the second reference coincident to the Point in the first sketch. Click OK to finish the plane and hide the first sketch when done.

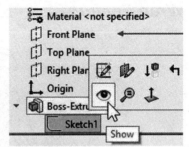

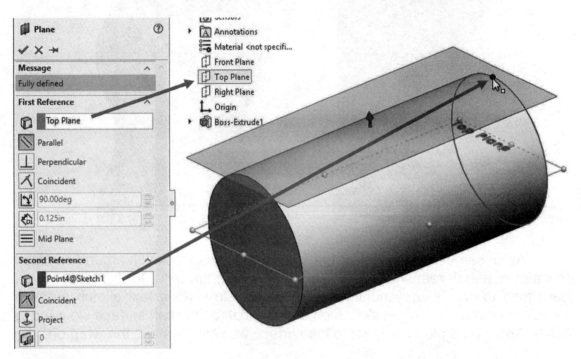

14.3. - Add a sketch in the new plane and draw the following profile. Make sure it's symmetrical about the centerline. All arcs are equal radii. The top dimension's value is entered as *1.5 * pi*, since we want the sketch to wrap completely around the cylinder. After entering the equation in the "Modify" box the resulting value will be automatically calculated. Switch to a Top view for clarity.

A powerful and time efficient tool to make a sketch symmetrical as we draw it is the "**Dynamic Mirror**" command, located in the menu "**Tools, Sketch Tools, Dynamic Mirror**." First, we need to draw and optionally pre-select a centerline (or pre-select any linear sketch line or linear model edge), and then turn the "**Dynamic Mirror**" command On. To know if the Dynamic Mirror command is active, SOLIDWORKS will add an equal (=) sign at both ends of the mirror centerline. While the dynamic mirror is active, everything we draw on either side of the mirror line will be automatically duplicated on the other side, and at the same time Symmetric geometric relations will be added.

 If a line/edge is not pre-selected, we'll be asked to select one.

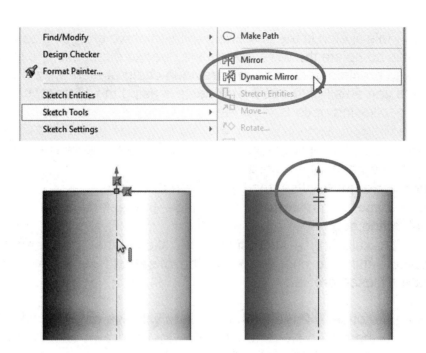

 For the Dynamic Mirror command, we can pre-select a centerline, line or a model's linear edge

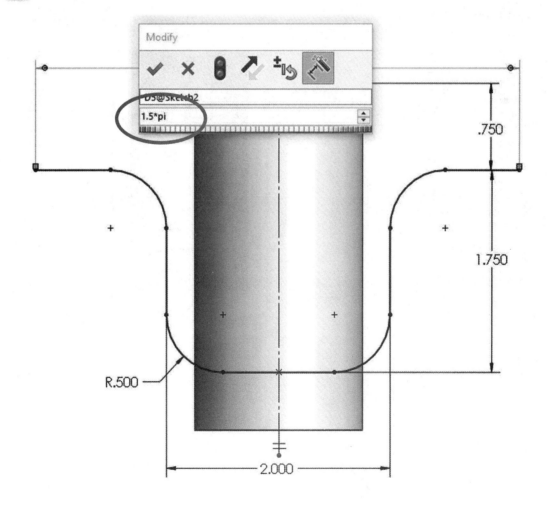

14.4. – So far, this sketch is the centerline of the slot we are going to wrap around the cylinder. To complete the wrap sketch, we need to offset this line to both sides, create a closed profile and change the previous chain of elements to construction geometry. We can do all of this at the same time using the "**Offset**" command and turning on the following options:

- "**Select chain**" to automatically select the entire centerline when we pick any one segment of the chain
- "**Bi-directional**" to offset the line in both directions
- "**Make base construction**" to change the selected line (or chain) to construction geometry
- And after selecting a segment of the sketch the "**Cap ends**" option is enabled, letting us choose to cap the offset with either Arcs or Lines to create a closed profile.

Making the offset 0.25" in both directions will give us a 0.5" wide slot. Click OK to complete the offset. Note how the previously drawn sketch lines are now changed to construction geometry and now we have a closed (capped) profile. Exit the sketch when done.

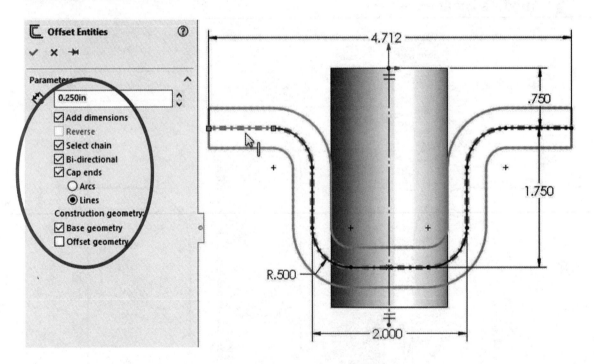

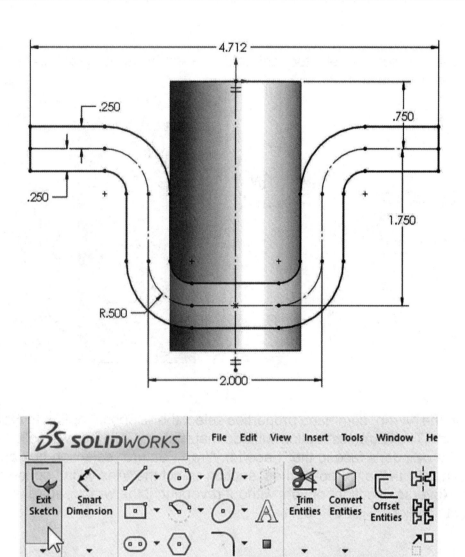

14.5. - Select the "**Wrap**" feature from the Features tab or the menu "**Insert, Features, Wrap**."

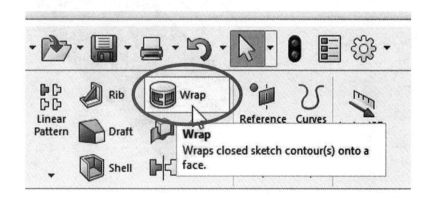

If the sketch is not pre-selected, select the sketch from the graphics area or the fly-out FeatureManager.

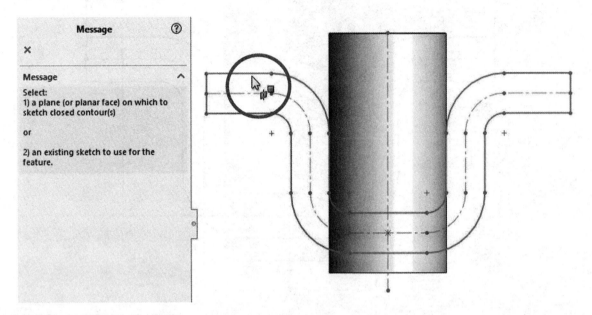

In the "Wrap" command properties select the "Deboss" option to make a cut in the part. (The "Emboss" option will add material and "Scribe" will split the face.) Select the cylinder's face in the "Face for Wrap Feature" selection box and make the depth 0.1". In the "Wrap Method" section select the "Analytical" option and turn on the "Reverse direction" option; without reversing it the wrap feature will give an incorrect solution. Click OK to finish.

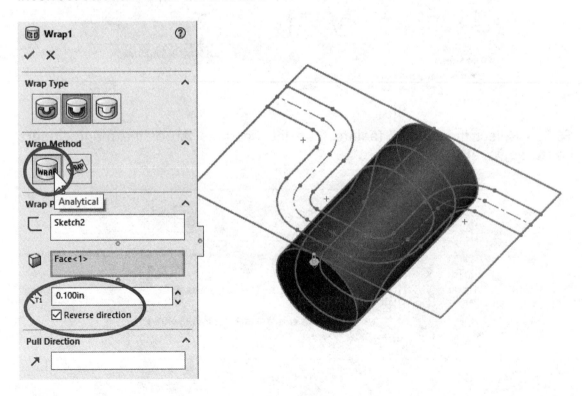

14.6. - Add a 0.02" fillet to the wrap feature to finish the part and save it as *'Cylindrical Cam'*.

Finished part image made using Real View and material set to AISI 1020 Steel

14.7. – To show how to apply the wrap feature to an irregular face (non-analytical), open the '*Bottle_Wrap*' part from the provided files. If asked to proceed with feature recognition click "No" to continue.

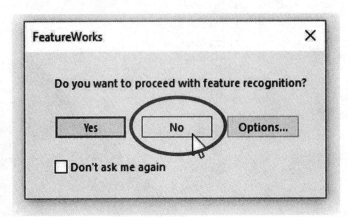

14.8. – Add a new sketch in the Front plane. This is the sketch that will be wrapped onto the bottle's surface. Switch to a Front view and add the following two construction lines. The horizontal line's midpoint is located at the top of the vertical line.

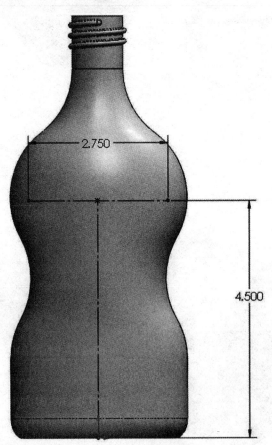

 For the horizontal line we can use the "Midpoint line" command, and then convert it to construction geometry.

14.9. – To add text to a sketch and use it to create new features like an extrude or cut, select the menu "**Tools, Sketch Entities, Text**." The sketch text can be located freely by adding relations and dimensions to its location point (located in the lower left corner of the text), or it can be made to follow a continuous sketch line or curve, or a model edge. For our example we'll add the text along the horizontal line.

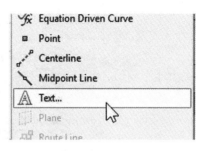

Add the horizontal line to the "Curves" selection box, and in the "Text" box type the word "Shampoo." Depending on the default font settings the word may not completely fill the line.

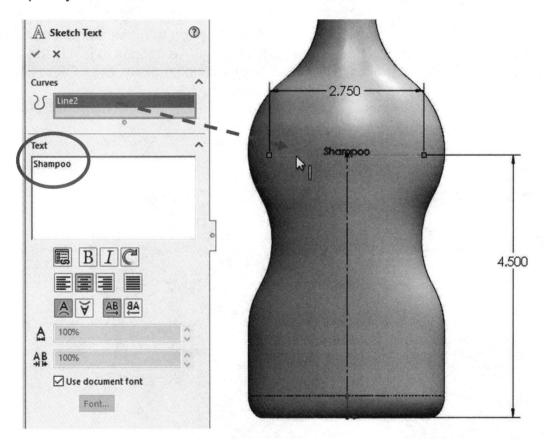

14.10. – To make the text fit the line, we must turn Off the "Use document font" checkbox and click in the "Font…" button. Select the "Arial Rounded MT" font; make it Bold and 0.4" units high. Press OK to continue.

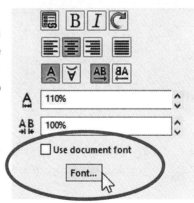

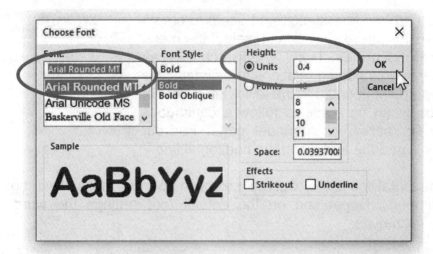

In the "Text" options select the "Full justify" option to make the text use the full length of the construction line. Note that we also have options to justify the text left, center and right, as well as to flip it vertically and horizontally. Click OK to finish adding the sketch text and exit the sketch to continue.

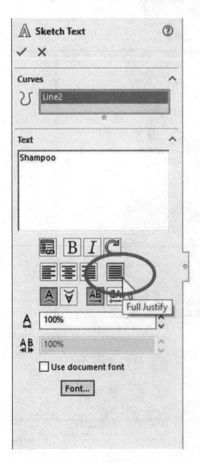

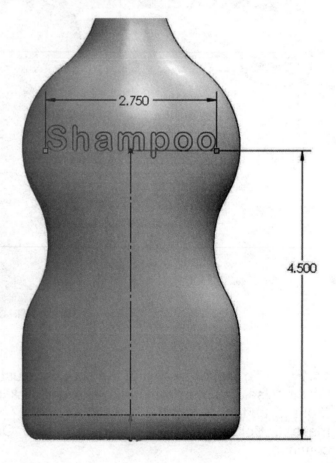

 To edit the sketch text after adding it to a sketch, double click the text to display and modify its properties.

14.11. – From the "Features" tab select the "Wrap" command, or the menu "**Insert, Features, Wrap.**" If the sketch is not pre-selected, we'll be asked to select it either in the graphics area or the fly-out Feature-Manager.

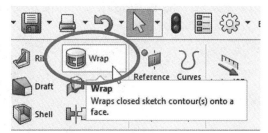

In the "Wrap Type" section select the "Emboss" option to add material; in the "Wrap Method" we'll use the "Spline Surface" option because the bottle's surface is irregular. In the "Wrap Parameters" select the bottle's face to wrap the sketch onto it and enter a 0.010" height for the emboss. If needed, we can reverse the direction to wrap the sketch in the other side of the surface. Click OK to finish.

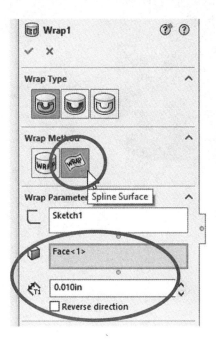

If we try to use the "Analytical" wrap method in this case, we will not be able to select the bottle's face because it is an irregular surface.

14.12. – This is what the final bottle looks like. Save and close the file.

Exercises: Build the following parts using the knowledge acquired in this lesson. Try to use the most efficient method to complete the model.

Eccentric Coupler

Notes:

- Both circles are centered horizontally (Right view).
- Add a guide sketch at the bottom.
- Set the Start and End Conditions for the loft to "Normal to Profile."
- Make as Thin Feature *or* Shell it after making the loft feature.

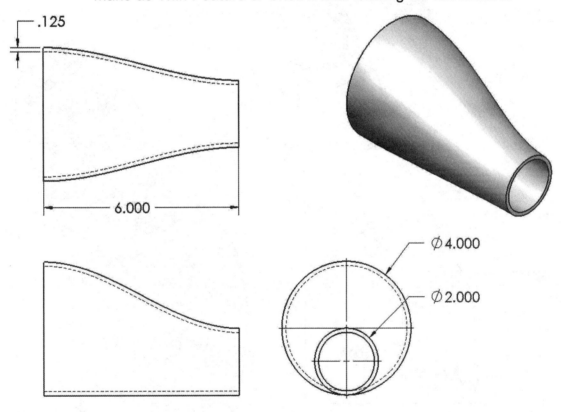

TIPS:

- Draw the 4" diameter circle in the *"Right Plane."*
- Make an Auxiliary Plane 6" parallel to the *"Right Plane."*
- Draw a 2" circle in the Auxiliary Plane.
- Draw a sketch in the *"Front Plane"* to make the guide curve.
- Select the guide sketch in the loft's "Guide Curves" selection box.

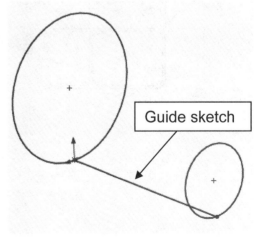

Guide sketch

Bent Coupler

Build the following part using a Loft feature and a shell. The part is 0.15″ thick.

Notes:

- Add a guide sketch along the right side of the part.
- Start and End Conditions "Normal to Profile."
- Make as a Thin Feature or add a Shell feature after the loft.

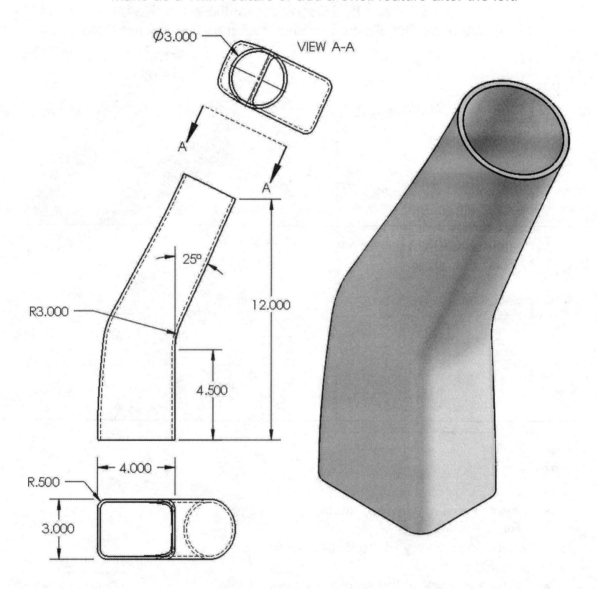

TIPS:

- Make the rectangular sketch in the Top Plane first.

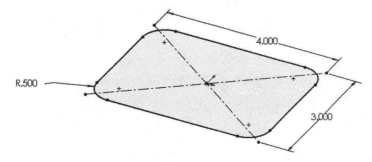

- Add the Guide Sketch in the Front Plane.

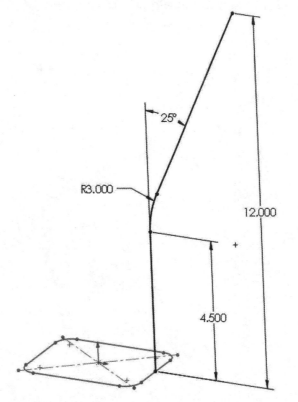

- Make an Auxiliary Plane perpendicular to the endpoint of the Guide sketch.

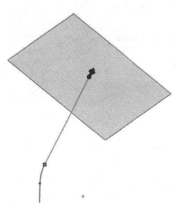

- Add the circular sketch in the Auxiliary Plane.

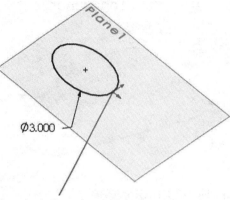

- Make Loft using Guide sketch as a guide curve.

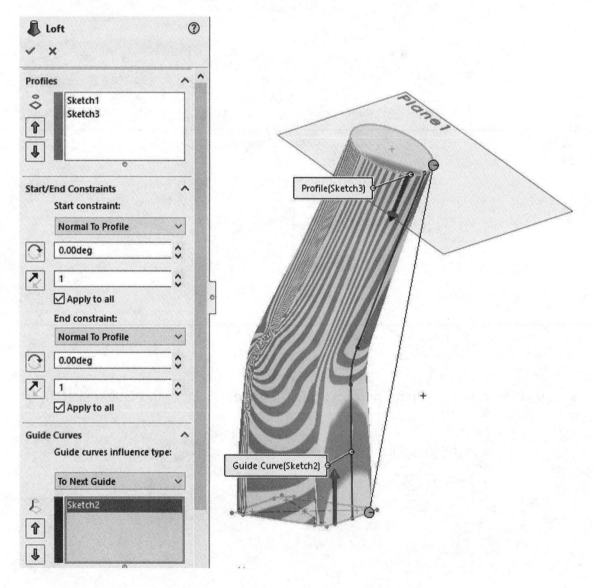

- Shell the part 0.15″ thick.

Challenge Exercises: Build the *'Worm Gear'* and *'Offset Shaft'* complete gears using the knowledge learned so far with the information given. High resolution images at www.sdcpublications.com.

Worm Gear Shaft Complete

DISCLOSURE: The gears modeled in this tutorial are not intended for manufacturing, nor is this tutorial meant to be a gear design guide. Its sole purpose is to show the reader how to apply the learned knowledge using a simplified version of the gears.

15.1. - Offset Shaft: Open the previously made *'Offset Shaft'* and add the following sketch in the "*Front Plane*"; exit the sketch and rename it "*Gear Width*." This will be the length of the full-size helix.

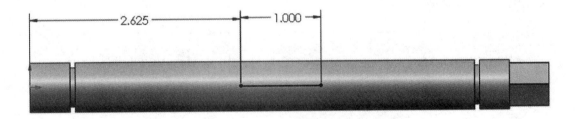

15.2. - Add two reference planes parallel to the "*Right Plane*," one at each end of the "*Gear Width*" sketch.

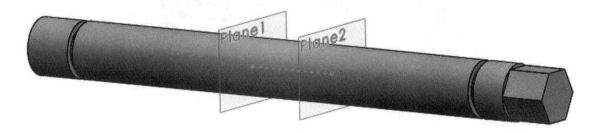

15.3. - Add the gear's profile sketch in the "*Front Plane*" at the left side of the "*Gear Width*" sketch. Use the "Dynamic Mirror" command to maintain symmetry. When finished, exit the sketch and rename it "*Thread Profile*."

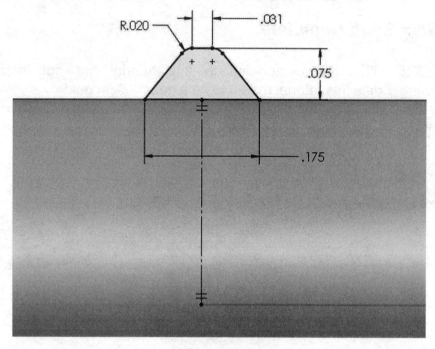

15.4. - In the plane located at the left of the "*Gear Width*" sketch, add a new sketch; use the "Convert Entities" drop down icon to select "**Intersection Curve**." This command will create sketch entities at the intersection of the selected surface(s) and the current sketch plane. This tool is particularly useful when we have irregular surfaces intersecting the sketch plane. In this case it works the same as "Convert Entities," but this is the best option when we need to obtain a surface's profile that intersects the sketch plane.

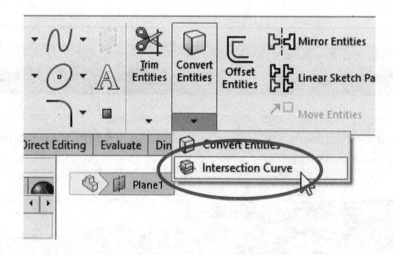

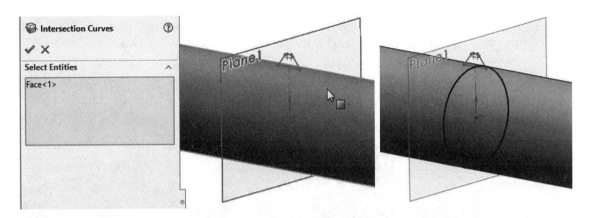

15.5. – Using the "Helix and Spiral" command, add a Helix using the "Height and Pitch" option, and make the Helix 1" High with a 0.25" Pitch. Be sure to start the helix at the location of the "Thread Profile" sketch, to accomplish this make the helix's start angle 90 degrees.

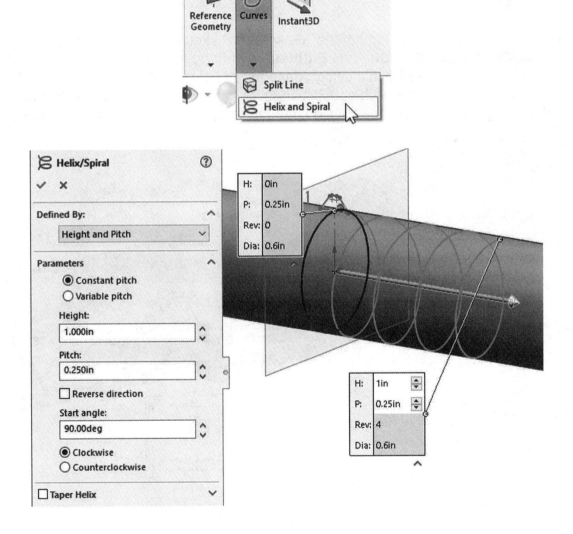

15.6. – After the helix is complete, use the "**Sweep**" command to make the first part of the gear using the profile and the first helix. Hide the helix when finished.

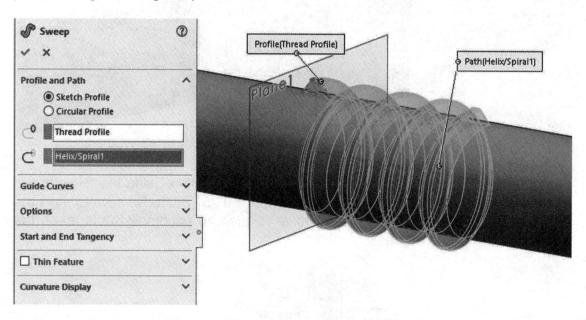

15.7. - In the plane to the right of the "*Gear Width*" sketch, add a new sketch for a second Helix. Using the "**Convert Entities**" command, project the round edge of the shaft to make the sketch's circle. Make the helix using the "height and Pitch" option, 0.625" high and a pitch of 0.25". Be sure to turn on the "Taper Helix" option with a 7deg taper to make the helix go into the part, continuing the previous helix.

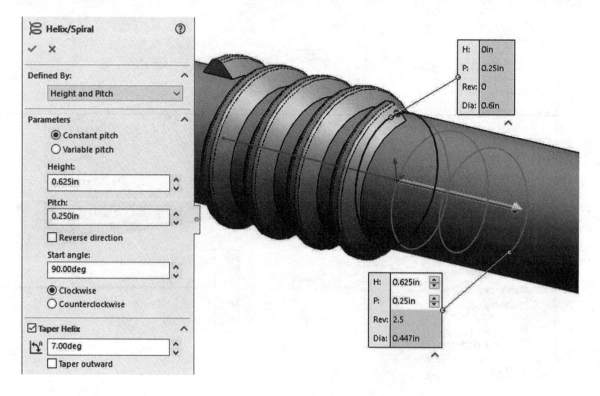

15.8. – For the second sweep we'll use the face at the end of the first sweep as the profile and the tapered helix to continue the gear. Notice we are looking at the back of the part. Hide the second helix when the sweep is complete.

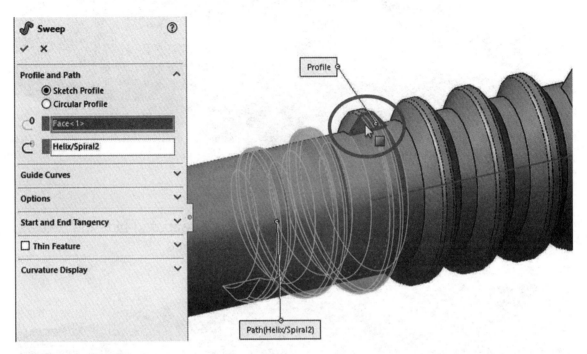

 The sweep profile could have been done by adding a sketch in the flat end of the thread, and then projecting the profile using "**Convert Entities**."

15.9. – Using the same process, add new tapered helix at the other end of the part going to the left. You will need to reverse the helix's direction and make it counterclockwise. Make the third sweep after completing the helix.

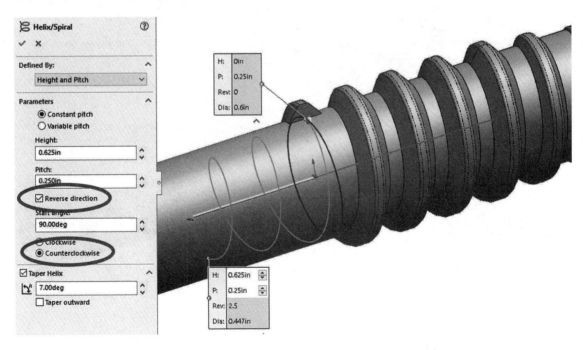

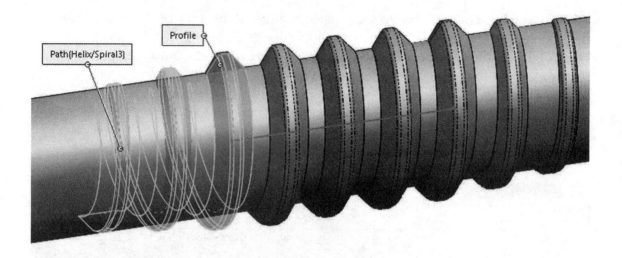

15.10. - Add a 0.015" fillet to round the edges of the thread and hide the helixes, sketches and planes. Save as '*Offset Shaft Gear*' and close.

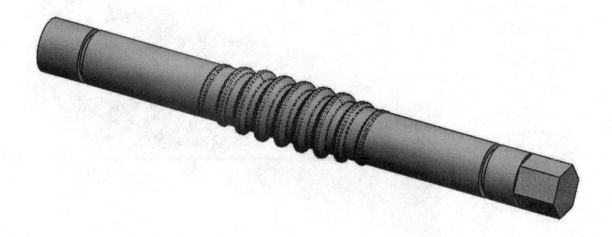

Worm Gear Complete

16.1. - Before adding the actual gear, we must make a few changes to the original part. Open the '*Worm Gear*' part and change "*Fillet1*" to a 0.031" radius.

16.2. – Edit the "Cut-Revolve1" sketch, change the following dimensions and rebuild the model to update and continue.

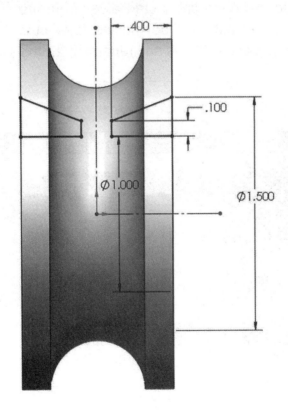

16.3. - Add a new plane 0.95" parallel above the "Top Plane."

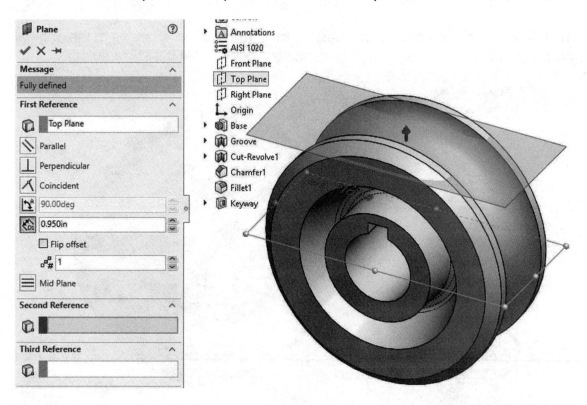

16.4. - Add the following sketch in the plane previously made; use the "**Midpoint Line**" command starting at the origin. The reason to make the line's start away from the part is to make sure the cut we'll be making later cuts through the entire part. Exit the sketch and rename it "Path."

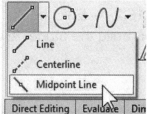

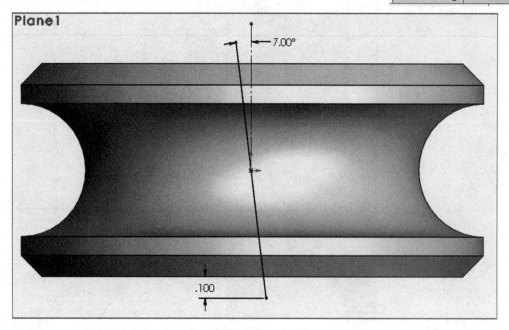

16.5. - Add a new auxiliary plane perpendicular to the "*Path*" sketch and rename it "*Profile Plane*."

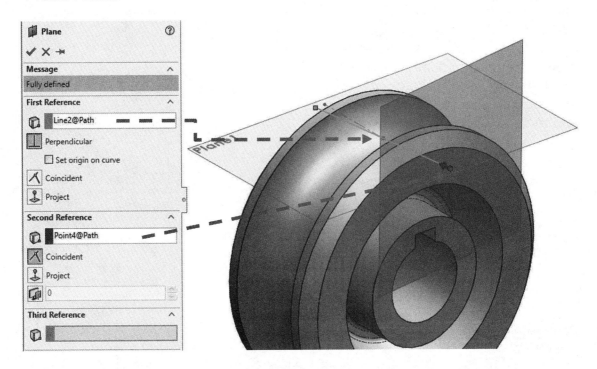

16.6. – The original part is a simplified version of a gear; in this step we'll remove some material to make a better representation of the gear teeth. Add this sketch in the "*Right Plane*" and make a revolved cut. Notice the doubled diameter dimension about the centerline. (Auxiliary planes have been turned off for clarity.)

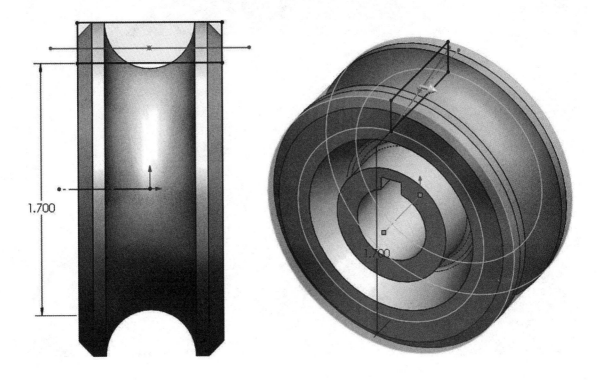

16.7. - In the "*Profile Plane*" add the following sketch. This will be the profile to make the gear cut. Press Ctrl+8 to view the plane normal to the screen. Locate the profile sketch by adding a pierce relation to the "Path" sketch at the indicated point. Exit the sketch and rename "*Gear Profile.*"

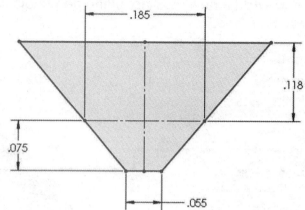

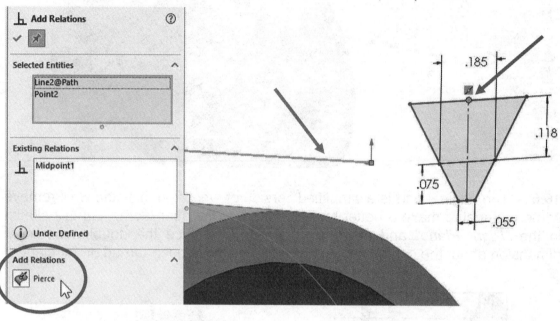

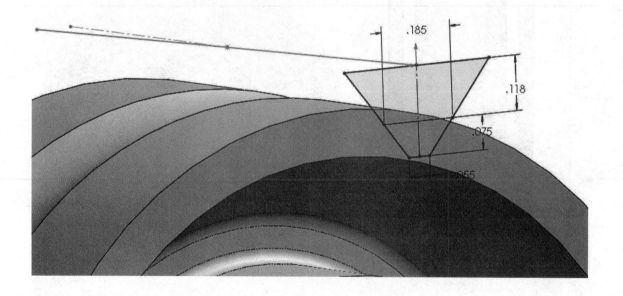

16.8. - Make a "**Swept Cut**" using the "*Path*" and "*Gear Profile*" sketches. In this case a "Cut Extrude" would give us the same result; however, we chose to teach the reader a different approach in this exercise. If the "*Path*" sketch had not been a straight line, the "**Swept Cut**" command would be the only option to make the cut.

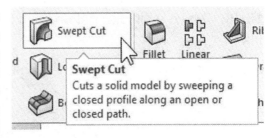

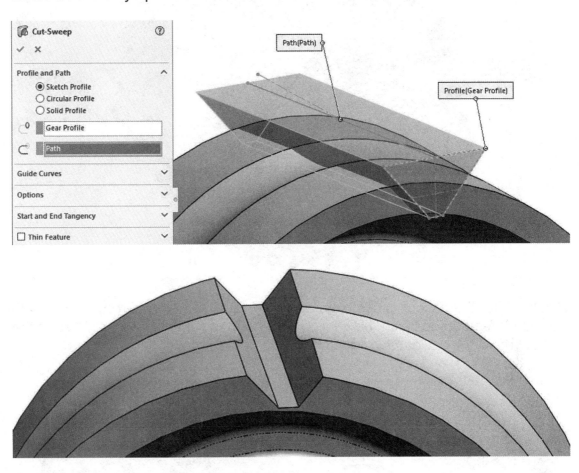

16.9. - Add a 0.015" fillet at the bottom of the swept cut.

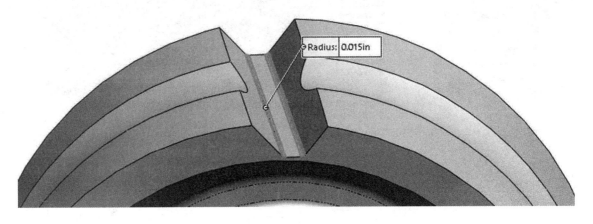

16.10. - Make a circular pattern with 22 copies of the Cut-Sweep and the fillet.

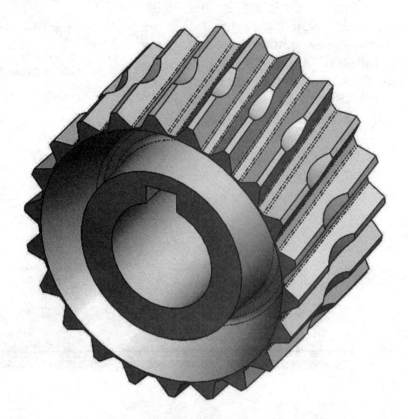

16.11. - After the pattern is complete add a 0.015" fillet to the "*Groove*" and "*Cut-Revolve2*" features to finish the gear. Using the menu "**File, Save as…**" save the part as '*Worm Gear Complete*' and close.

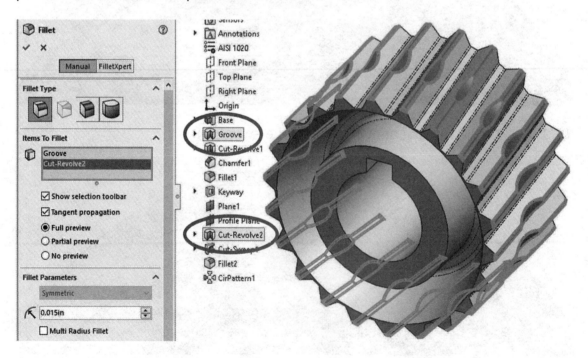

Engine Project Parts: Make the following components to build the engine. Save the parts using the name provided. High resolution images are included with the exercise files.

Hint: Make a revolved boss with the option "Thin Feature." Note the dimensions provided are external.

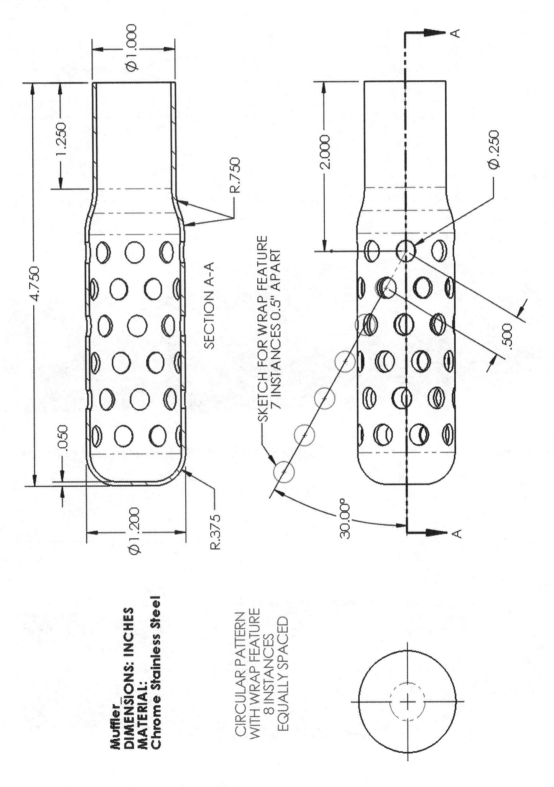

SECTION A-A

Muffler
DIMENSIONS: INCHES
MATERIAL:
Chrome Stainless Steel

SKETCH FOR WRAP FEATURE
7 INSTANCES 0.5" APART

CIRCULAR PATTERN
WITH WRAP FEATURE
8 INSTANCES
EQUALLY SPACED

Tips:

- Draw the sketch and make a Thin Revolved Boss feature.

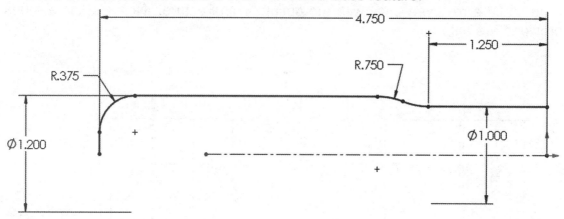

- Add an auxiliary plane at the top of the part.

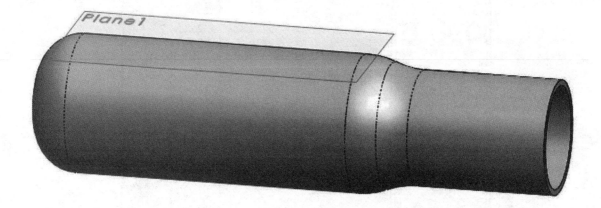

- Draw the sketch for the "**Wrap**" feature. Use the "**Sketch Pattern**" command.

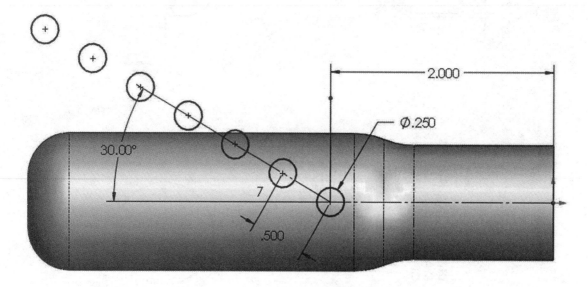

- Make the "**Wrap**" feature.

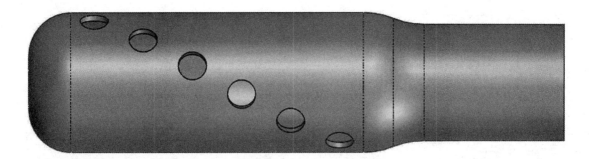

- Add the Circular Pattern to copy the "Wrap" feature.

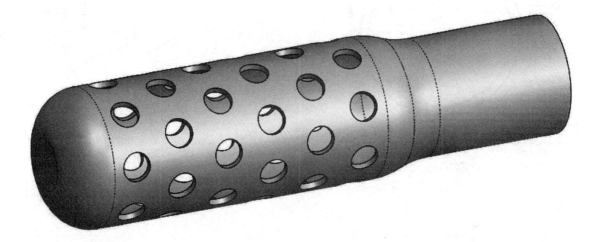

The 3 compression rings for the piston head.

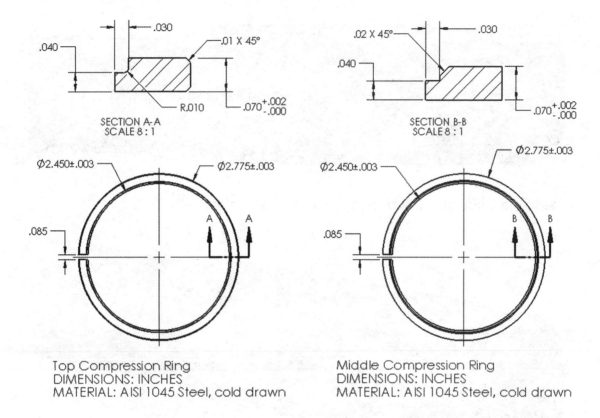

SECTION A-A
SCALE 8 : 1

SECTION B-B
SCALE 8 : 1

Top Compression Ring
DIMENSIONS: INCHES
MATERIAL: AISI 1045 Steel, cold drawn

Middle Compression Ring
DIMENSIONS: INCHES
MATERIAL: AISI 1045 Steel, cold drawn

TIPS:

- Sketches for revolved feature

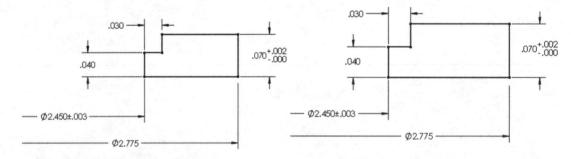

- Add Chamfers and Fillets

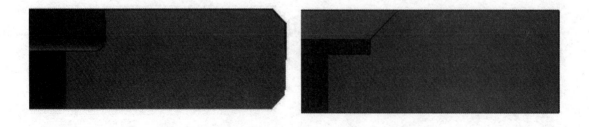

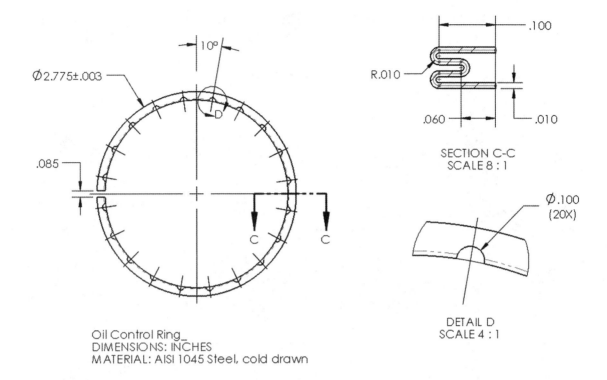

SECTION C-C
SCALE 8 : 1

DETAIL D
SCALE 4 : 1

Oil Control Ring_
DIMENSIONS: INCHES
MATERIAL: AISI 1045 Steel, cold drawn

TIPS:

• Sketch for Thin Revolved Feature (Thin Feature option: Mid-Plane)

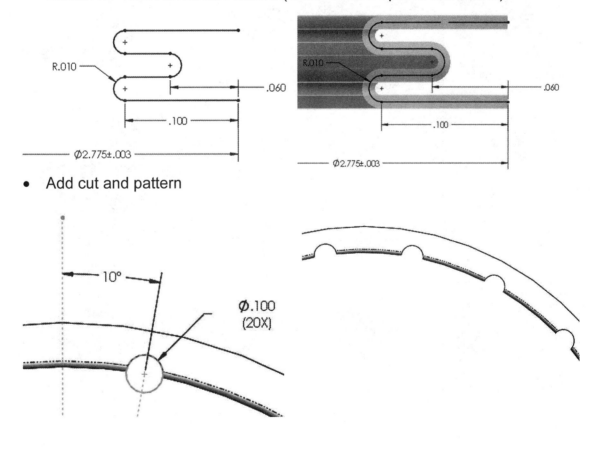

• Add cut and pattern

Final Parts for the Engine Project: The following components are the last required to complete the engine. Save the parts using the name provided.

Suggested sequence of features for the **Crank case top**

Main body	Corner Fillets	Flange	Bearing mount boss
Mirror bearing mount	Shell part	Top cut	add 0.375" offset plane inside
Add inside bosses. Use Intersecting curve command	Add front reinforcement	Add rear reinforcement	Add bearing mount
Add plane for oil dipstick	Add reference sketch for dipstick	Add plane at top of reference sketch	Add dipstick extrusion
Make hole for dipstick	Add sweep cut for threaded cap	Add fillets	Add holes and threads

Crank Case Top Drawing (2 Pages)

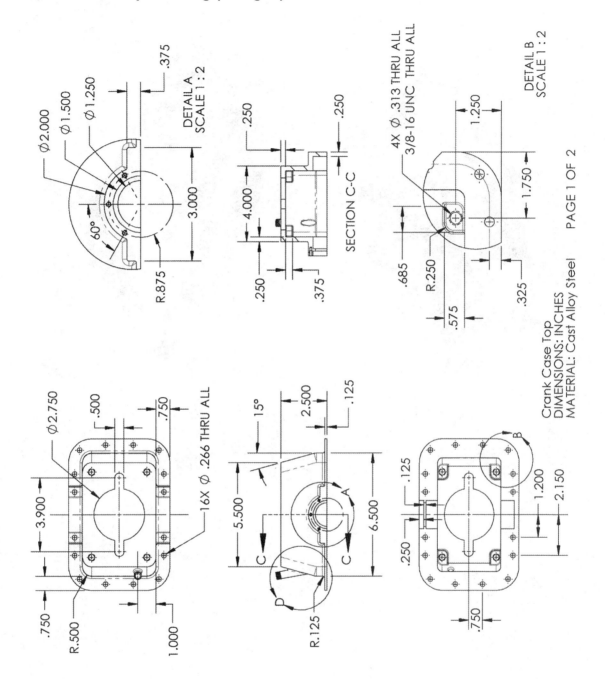

Crank Case Top
DIMENSIONS: INCHES
MATERIAL: Cast Alloy Steel

PAGE 1 OF 2

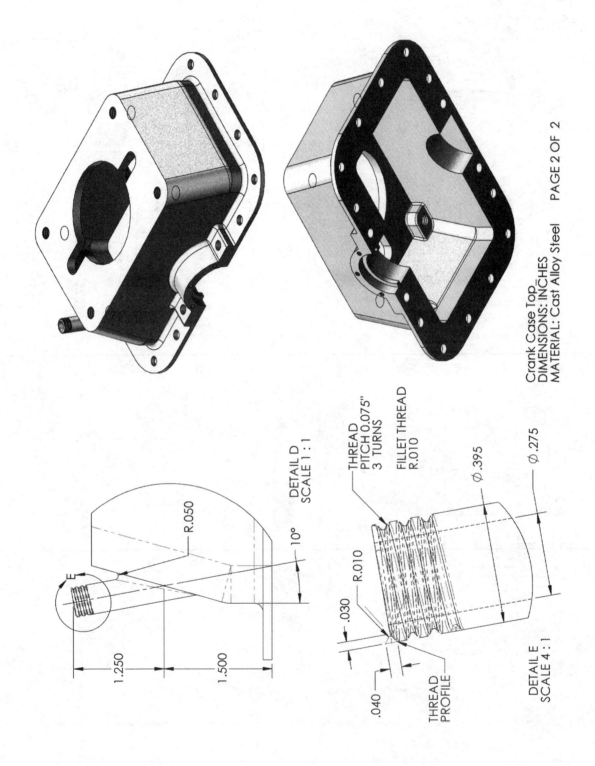

DETAIL D
SCALE 1 : 1

R.050

10°

E

1.250

1.500

THREAD
PITCH 0.075"
3 TURNS

FILLET THREAD
R.010

∅.395

∅.275

R.010

.030

.040

THREAD
PROFILE

DETAIL E
SCALE 4 : 1

Crank Case Top
DIMENSIONS: INCHES
MATERIAL: Cast Alloy Steel   PAGE 2 OF 2

Oil Pan

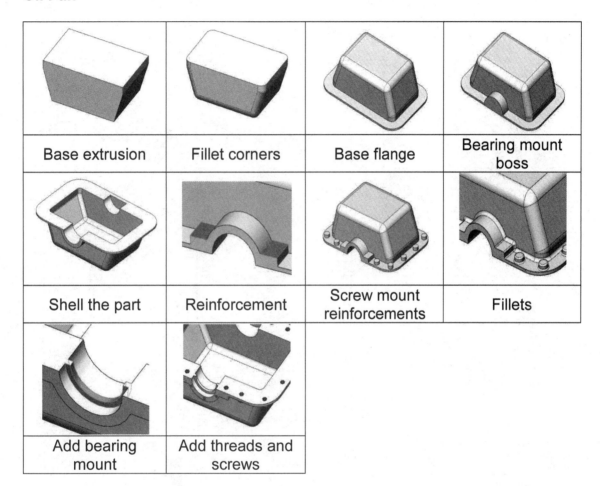

Base extrusion	Fillet corners	Base flange	Bearing mount boss
Shell the part	Reinforcement	Screw mount reinforcements	Fillets
Add bearing mount	Add threads and screws		

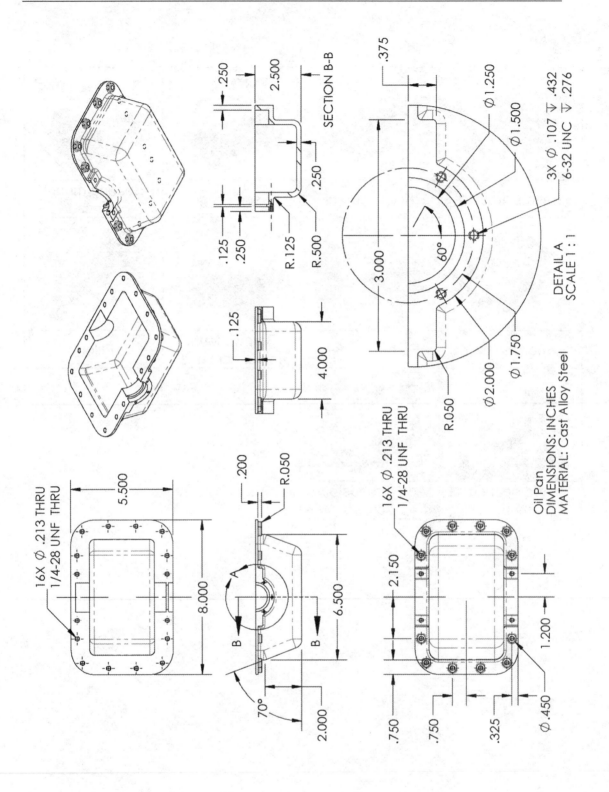

Oil Pan
DIMENSIONS: INCHES
MATERIAL: Cast Alloy Steel

Engine Block

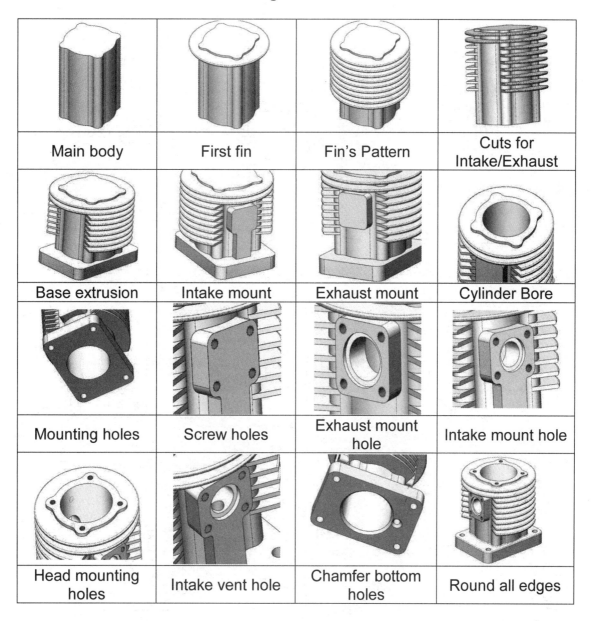

Main body	First fin	Fin's Pattern	Cuts for Intake/Exhaust
Base extrusion	Intake mount	Exhaust mount	Cylinder Bore
Mounting holes	Screw holes	Exhaust mount hole	Intake mount hole
Head mounting holes	Intake vent hole	Chamfer bottom holes	Round all edges

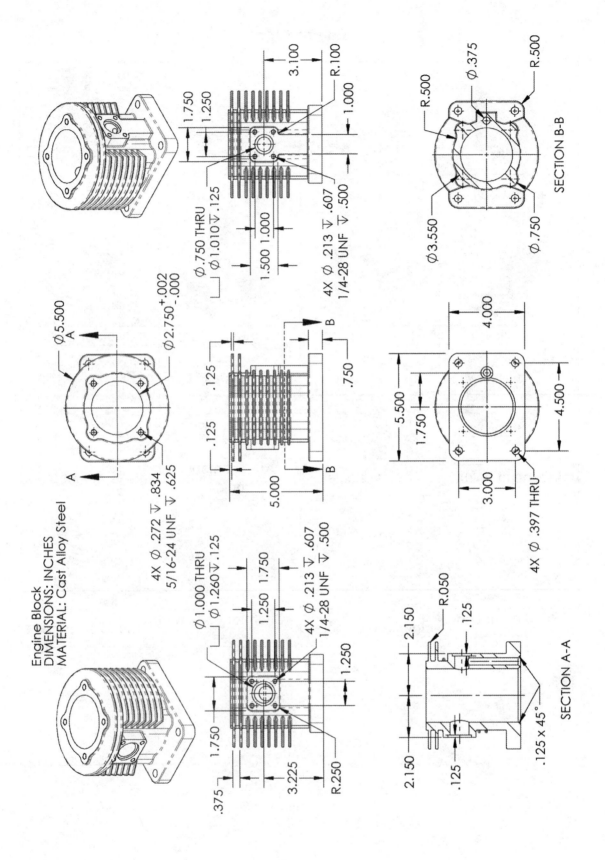

Engine Block
DIMENSIONS: INCHES
MATERIAL: Cast Alloy Steel

Extra Credit: Visit our web site to download details for the *Gas Grill* project as well as the finished parts for this book, higher resolution images of the exercises and some extra topics not covered in the book. Build these components and after the Assembly lesson make the Gas Grill assembly.

http://www.mechanicad.com/download.html

Notes:

Chapter 4: Detail Drawing

Now that we have completed modeling the components, the next logical step would be to assemble them, and this is what is usually done at this point in the design. However, we are going to make their 2D detail drawings instead.

The reason we are doing this is to show additional functionality before jumping into assembly modeling, including understanding component configurations, and later in the assembly lesson the effect of modifying a component in the assembly and propagating those changes back to the part and its corresponding drawing.

In most modern design software packages, including SOLIDWORKS, the 3D models are created first, and from them a 2D manufacturing drawing is generated. By deriving the drawing from the solid model, the 2D drawing is linked and associated to the 3D model. This means that if the part is changed, the drawing will be updated and vice versa. In SOLIDWORKS, 2D detail drawing files have the extension *.slddrw and each drawing can contain multiple sheets; each corresponding to a different printed page of the same or different 3D models (parts and/or assemblies).

SOLIDWORKS offers an easy to use environment where we can quickly create 2D drawings of parts and assemblies. In this section we'll cover Part drawings only. Assembly drawings will be covered after the assembly section. We will add all the different model views, annotations, dimensions, section and detail views necessary for a manufacturing department to fabricate the component without missing any details. In this section we will also introduce a new concept called **Configurations**, which allow us to have similar but different versions of the same part using the same 3D model, for example, a version of the part as it comes out of a foundry and a version for the machine shop with all the details to machine the finished part.

IMPORTANT: 2D detail drawings can be made using any of the many dimensioning standards available in the industry. In this book we will use the ANSI standard. After, or while creating a detail drawing, the dimensioning standard can be easily changed to a different one by going to the menu "**Tools, Options**" and selecting the "Document Properties" tab. In the Drafting Standard section, a different standard can then be selected. It is important to note that after changing to a different standard, SOLIDWORKS updates the dimension styles, arrowhead type, etc. accordingly. For more information about changing units and dimensioning standards see the Appendix.

The detail drawing environment in SOLIDWORKS is a true *What-You-See-Is-What-You-Get* interface. When we make a new drawing, we are asked to select a template with a predefined sheet size, which corresponds to the printed sheet size, or define the size of the drawing sheet we want to use. Do not be too concerned about selecting the correct sheet size at first; it can be changed to a larger or smaller size at any time during the detailing process if we find out our drawing would fit better on a different sheet size. Also, any drawing sheet or drawing view can be scaled up or down to fit the page and/or printer size.

Just like part templates can be made using different units, materials, standards and settings to fit our needs, the same can be done with drawing templates. In this section we'll start by using the default document templates and later, after we learn more about the different options and settings available for detail drawings, we'll learn how to customize drawing templates.

Now, let's create manufacturing detail drawings for the parts previously designed.

Drawing One: The Housing

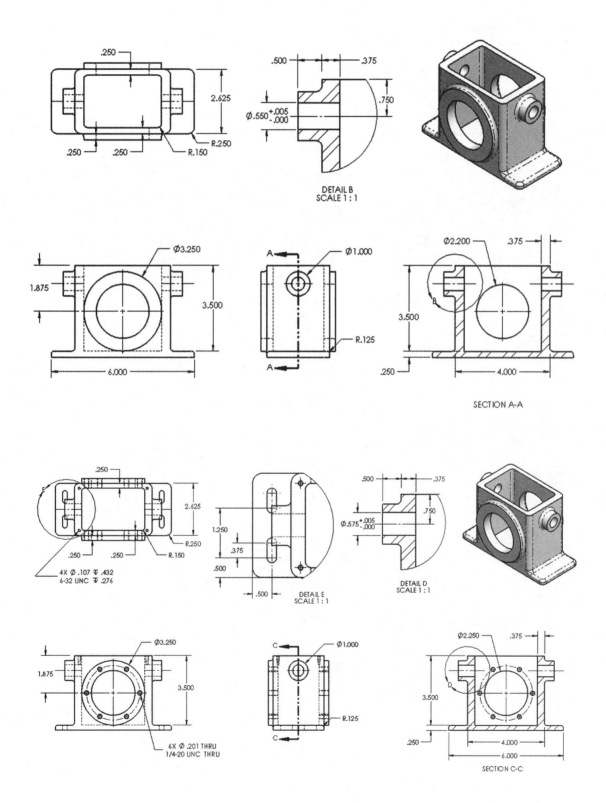

.250

2.625

R.250

.250   .250   R.150

.500   .375

.750

Ø.550 +.005 / -.000

DETAIL B
SCALE 1 : 1

Ø3.250

1.875   3.500

6.000

A

Ø1.000

A

R.125

SECTION A-A

Ø2.200   .375

3.500

.250   4.000

.250

2.625

R.250

.250   .250   R.150

4X Ø .107 ∇ .432
6-32 UNC ∇ .276

1.250

.375

.500

.500

DETAIL E
SCALE 1 : 1

.500   .375

.750

Ø.575 +.005 / -.000

DETAIL D
SCALE 1 : 1

Ø3.250

1.875   3.500

6X Ø .201 THRU
1/4-20 UNC THRU

C

Ø1.000

C

R.125

Ø2.250   .375

3.500

.250   4.000
6.000

SECTION C-C

343

Notes:

In this lesson we'll learn how to make multiple configurations of a part, how to make a multi-sheet drawing showing these different configurations, add main views, sections and details, change a drawing's view display style, move views within the drawing sheet, import model dimensions and manipulate them.

The detail drawing of the *'Housing'* part will follow the next sequence.

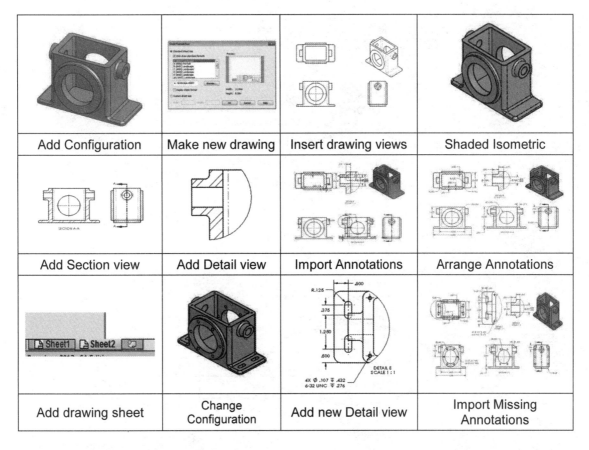

Add Configuration	Make new drawing	Insert drawing views	Shaded Isometric
Add Section view	Add Detail view	Import Annotations	Arrange Annotations
Add drawing sheet	Change Configuration	Add new Detail view	Import Missing Annotations

17.1. - The first thing we are going to do is to make a new **Configuration** of the *'Housing'* part. SOLIDWORKS' Configurations allow us to make slightly or considerably different versions of a part without having to make a new part, in order to show or document different states of a part or a different but similar component. In our example, we'll have a configuration of the *'Housing'* as it comes out of the foundry, and another of the finished part for the machine shop including all the details that need to be machined in the cast part. Using configurations, we can control different aspects of a component including dimensions, tolerance, whether features are present or not, etc.

Open the *'Housing'* part and select the **"ConfigurationManager"** tab. It is in the Configuration Manager where we can add, delete, edit and switch between the different configurations of a part. We can find it at the top of the FeatureManager as indicated.

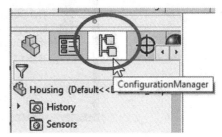

In every part there will always be one configuration called *'Default'*. Once the Configuration-Manager is selected, we can add a new configuration by right-mouse-clicking at the top of the ConfigurationManager in the part's name, in this case *'Housing'*, and select the "**Add Configuration**..." option. Name the new configuration *"Forge"*, set the description to *"Model for Foundry"* and click OK. Adding comments and/or descriptions is optional but often useful.

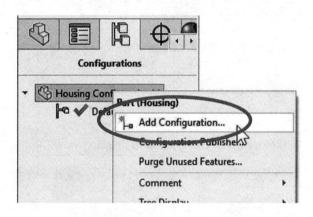

After adding the configuration, you may be asked if you'd like to link the display state to show the material's appearance in the configuration. Press "Yes" to continue.

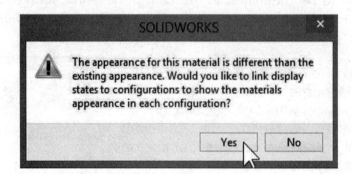

17.2. - After adding the new configuration, we'll rename the "Default" configuration. Right-mouse-click in it, select "Properties" and change its name to *"Machined."* As before, adding a description is optional. Click OK when finished.

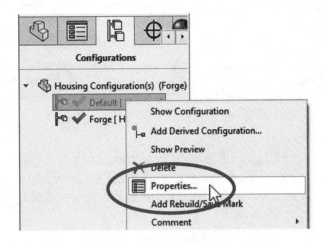

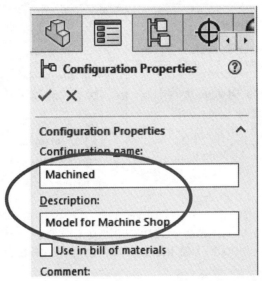

 Configurations can also be renamed using the slow double click method in the ConfigurationManager.

17.3. - In the ConfigurationManager the currently active configuration is the one with the green checkmark; all other configurations are grayed out. To switch to a different configuration, we can double-click an inactive configuration's name, or right-mouse-click and select "**Show Configuration**" from the pop-up menu.

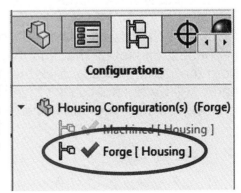

347

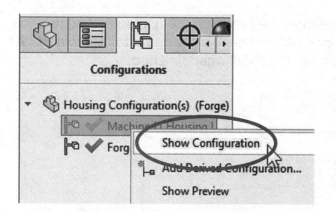

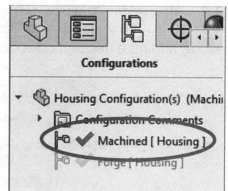

17.4. - Right-mouse-click or double click to activate the *"Forge"* configuration (it will have the green checkmark) and change to the FeatureManager tab. Notice that next to the name of the part at the top of the FeatureManager, the currently active configuration's name is shown.

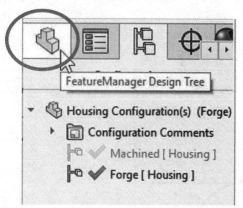

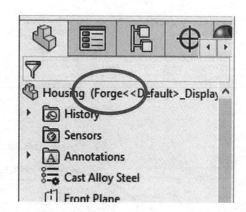

With the *"Forge"* configuration active, we will "**Suppress**" from the model all the features that cannot be made when casting the part and make some of the holes smaller than they need to be; the reason to do this is to be able to machine them to size later.

When we suppress a feature, we are essentially telling SOLIDWORKS to *NOT* make it. The feature will still be listed in the FeatureManager, but it will be grayed out, and for all practical purposes it doesn't exist in the model, and, since it is not created, the model's mass properties are affected.

17.5. - The first feature that will be suppressed is the *'1/4-20 Tapped Hole1'*, because it cannot be made in the foundry. Select it in the FeatureManager, and from the pop-up toolbar click the "**Suppress**" command. After suppressing it its name will be automatically grayed out. Remember that **we are not deleting the feature**; it's still there, but it is suppressed.

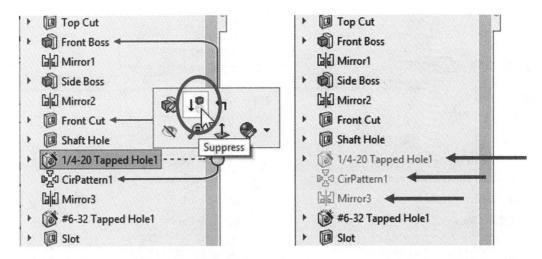

Notice that the two features after "¼-20 Tapped Hole1," "CirPattern1" and "Mirror3" were also suppressed. These features were also suppressed because they are 'children' of the "1/4-20 Tapped Hole1". Look at it this way: if there is no tapped hole, we cannot pattern it, and if there is no pattern, we cannot mirror it either. The "Parent/Child" relations are created when we add features that use or reference existing features in the model. For example, when a sketch is created in a face, the face becomes a parent to the sketch. If the face is deleted, so will the sketch and so on. If we dimension a sketch entity or a feature referencing another feature, we are creating a Parent/Child relation automatically.

 Parent/Child relations can be found two ways: By right-mouse-clicking a feature and selecting the "**Parent/Child**" command from the pop-up menu.

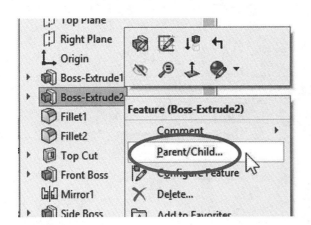
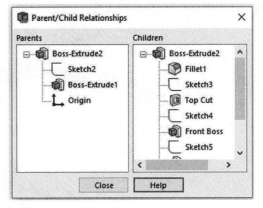

Or using the "Dynamic Reference Visualization (Parent/Child)" option, which is active by default and graphically indicates with arrows the Parent and Children features of any feature selected in the FeatureManager tree. This option can be turned On or Off by right-mouse-clicking at the top of the FeatureManager and switching either one ON or OFF. When the mouse hovers over a feature, the parent feature(s) above are indicated with blue arrows, and the child feature(s) below are indicated with purple arrows.

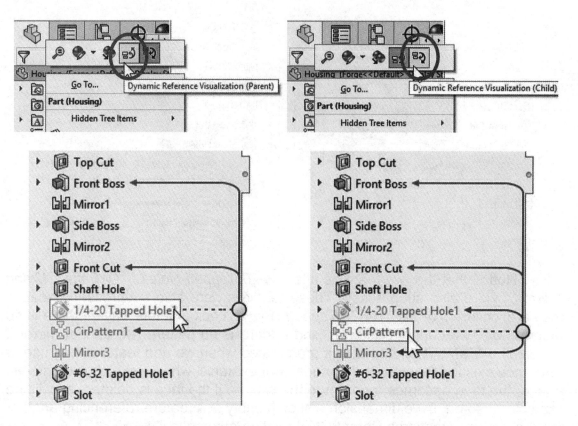

17.6. - The next feature that will be suppressed from the "Forge" configuration will be the *"#6-32 Tapped Hole1"* and the *"Slot"* features. Select and suppress them individually. Notice the children features of *"Slot"* are also suppressed.

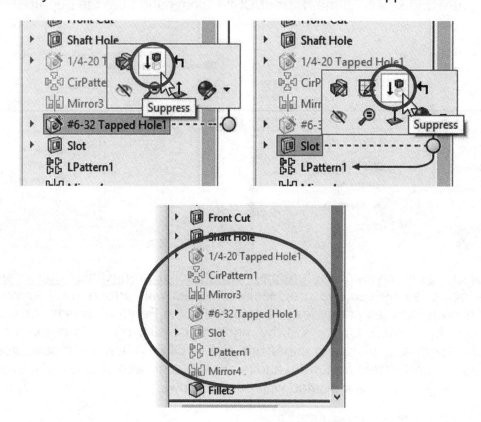

The reverse process to Suppressing a feature is "**Unsuppress**." Select a suppressed feature and click the "Unsuppress" icon (the icon with the arrow pointing up). Do not Unsuppress any features at this time; this is just an explanation of how it's done—we'll get to it later.

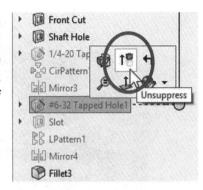

Our "*Forge*" configuration now looks like this:

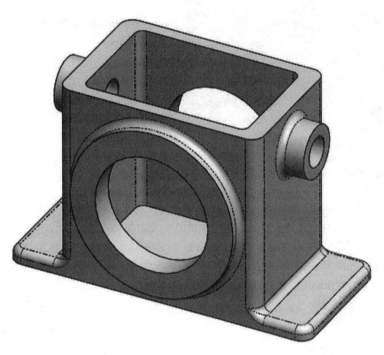

When a feature is Unsuppressed in the FeatureManager, its *child* features can be either Unsuppressed individually, or Unsuppressed at the same time as the parent feature using the "**Unsuppress with Dependents**" command. Select a parent feature (one with children) and use the menu "**Edit, Unsuppress with Dependents**." Also, when a feature is unsuppressed, all the features needed for it to exist (its *parent* features) are automatically unsuppressed.

17.7. - The next step is to change the size of the two circular cuts in the *"Forge"* configuration; the reason is that those holes are made smaller in the foundry, and later machined to size within the indicated tolerance. Make sure the *"Forge"* configuration is active. If the "**Instant 3D**" command is active, select a face of the *"Front Cut"* in the screen, OR double click in the feature in the FeatureManager, or one of its faces in the graphics area to reveal the feature's dimensions. Right-mouse-click the diameter dimension and select the option "**Configure Dimension**" from the pop-up menu.

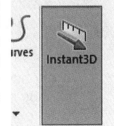

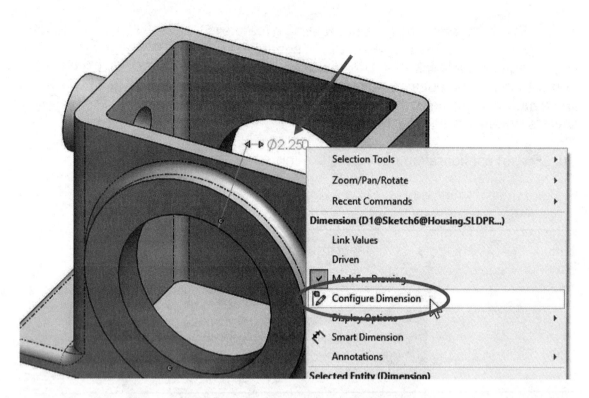

In the "**Modify Configurations**" table, we can change the value of the diameter's dimension for each configuration. Change the *"Forge"* configuration's dimension to 2.200″. The idea is to have a smaller hole made in the forge, and machine it to size later (*Machined* configuration). Leave the "*Machined"* configuration dimension's value as 2.250″ and click "OK" to finish. Notice the change in the hole size.

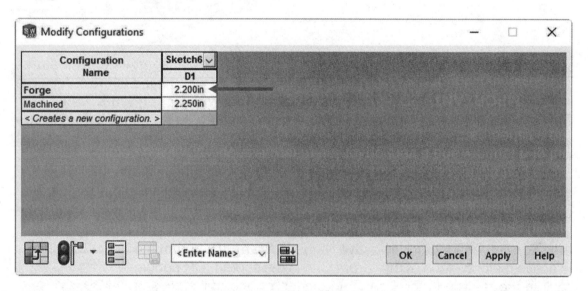

 A big advantage of the "**Configure Dimension**" command is that we can change multiple dimension's values for multiple configurations at the same time regardless of the active configuration in a convenient tabulated format.

17.8. - A different way to configure a dimension is by double-clicking it, and when the "Modify" dialog comes up we can select which configuration(s) to modify; if a component has configurations, the "Modify" dialog adds a configuration icon at the end of the value box. This icon helps us define if we want to change the dimension's value for "This Configuration," "All Configurations" or "Specify Configurations." Keep in mind that this icon will only be visible *IF* we have more than one configuration in the part.

Double-click the *"Shaft Hole"* feature to display its dimensions, double-click the diameter dimension, and change the value to 0.55". Make sure we are only changing the value in this configuration by selecting the "Configuration" button and selecting "This Configuration." Click OK to finish and rebuild the model. Using this option, the dimension's change only affects the active configuration *"Forge."*

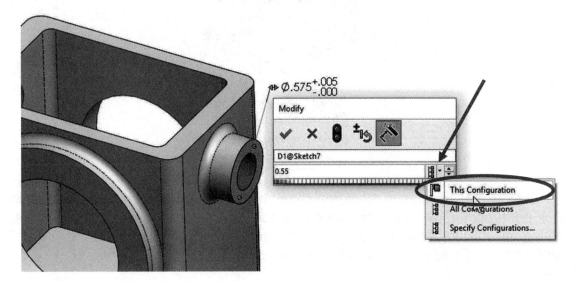

17.9. - An easy way to work with configurations is by viewing both the **FeatureManager** and the **ConfigurationManager** at the same time by splitting the FeatureManager pane. To do this, place the mouse pointer at the top of the FeatureManager and look for the "split" feedback icon, then click-and-drag down. Select the FeatureManager in the top pane and the ConfigurationManager in the bottom pane.

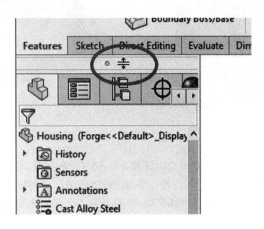

FeatureManager

ConfigurationManager

Now that we have made some changes to the *"Forge"* configuration, double click each configuration to activate it and see what each one of them looks like. Notice the missing features (suppressed) and different hole sizes in the *"Forge"* configuration. Save the changes to the *'Housing'* part.

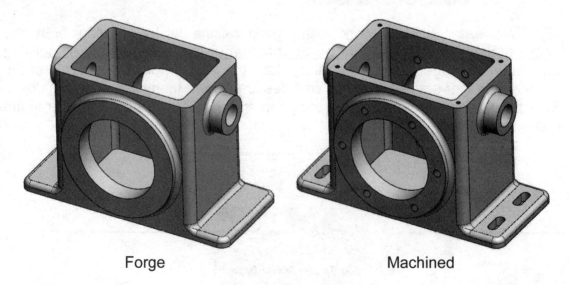

Forge   Machined

 We can add as many configurations as needed, but when we have more than 2 or 3 configurations it is usually easier to manage them with a table. Design Tables are very powerful and will be covered later in the book.

17.10. - After adding the configurations to the '*Housing*' we'll make a detail drawing of each one.

For the first drawing, make sure the *"Forge"* configuration is the currently active configuration and select the menu **"File, Make Drawing from Part"** or from the **"New"** document drop-down command, go to the "Templates" tab which has the default templates and select the template named "Drawing."

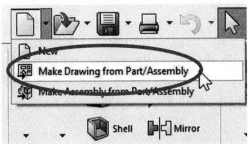

By starting a drawing with the *"Forge"* configuration selected, the drawing views created will have the forge version of the '*Housing*'. The drawing views for the *"Machined"* configuration will be added later in the lesson. Just like we can change units and detailing standards, we can also change the configuration referenced in a drawing view as needed.

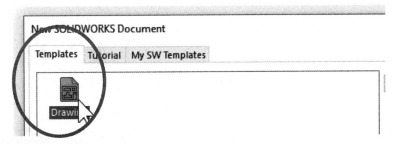

17.11. - After selecting the drawing template, we are asked to pick a sheet size to use for this drawing. Select the "B-Landscape" Sheet Size and turn off the **"Display Sheet Format"** option. Click OK to continue.

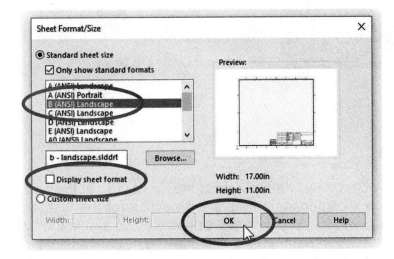

The "Sheet Format" is the part of the drawing that contains the title block and related annotations. It will be covered in detail later; we are not including it at this time to focus on basic drawing functionality instead.

17.12. - After selecting the sheet size, a new detail drawing file is opened and SOLIDWORKS is ready for us to choose the views that we want in the drawing.

The **"View Palette"** is automatically displayed at the right side of the screen; make sure the "Auto-start projected view" option is active; this option will save us time locating additional views by automatically showing the projected views.

From the **"View Palette"** drag-and-drop the "Front" view onto the sheet, in the lower left quadrant of the sheet.

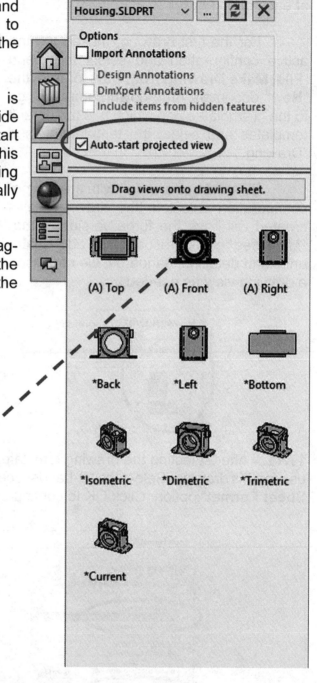

The selected view corresponds to the Front view in the part. When we start modeling a component, the plane selection for the first feature affects the orientation of the default views, but as we can see in the "View Palette" we can drag any model view onto the detail drawing, regardless of its name, use it as the main view, and project additional views from it regardless of which named view it is, or its orientation.

17.13. - Adding the first view in the lower left of the sheet will give us enough room to add the rest of the projected views. After locating the "Front" view in the sheet, SOLIDWORKS automatically starts the "**Projected View**" command; to add the "Top," "Right" and "Isometric" views, move the mouse in the direction of the missing view, and when the preview is visible left-mouse-click to locate it on the sheet.

As a guide, after adding the "Front" view, move the mouse pointer up to locate the "Top" view, move to the top-right for the "Isometric" view and finally move towards the right and click to locate the "Right" view under the "Isometric" view. After the four views are added click "OK" or press the "Esc" key to finish the "**Projected View**" command. Your drawing should now look approximately like this:

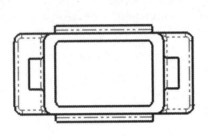

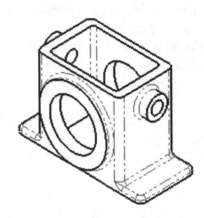

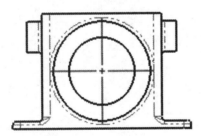

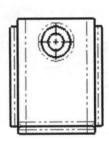

 Depending on the system options, your drawing may look slightly different; these options and differences will be explained in the next few steps.

Adding standard and projected drawing views

*When we make a new drawing from a part using the menu "**Make Drawing from Part**," SOLIDWORKS automatically displays the "**View Palette**." However, if we make a new drawing using the "**New Document**" icon, we get a different behavior.*

*To test it, select the "**New Document**" icon and create a Drawing by selecting the "**Drawing**" template using the same settings as before. In this case, we are presented with an empty drawing. (If the "**Model View**" command is loaded, cancel it.)*

To activate the View Palette, click in the View Palette icon from the Task Pane on the right side of the screen to display it.

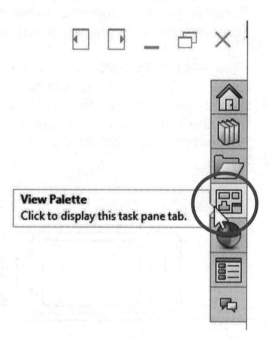

Once the "View Palette" is displayed, we can browse for a part or assembly or select one of the open documents from the drop-down list. After selecting a part or assembly the view's list in the lower pane is automatically populated, and we are ready to start adding model views to the drawing sheet as we did in the previous step.

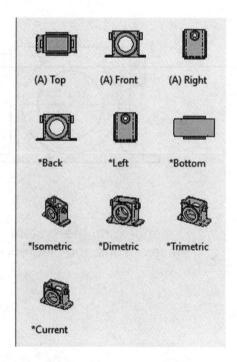

17.14. - Click to Select the "Front" view in the graphics area and, if different from the next image, change its Display Style to "**Hidden Lines Visible**" from the view's PropertyManager at the left. Since the rest of the views were projected from the "Front" view, this change will update the rest of them because by default a projected view uses the parent view's display style.

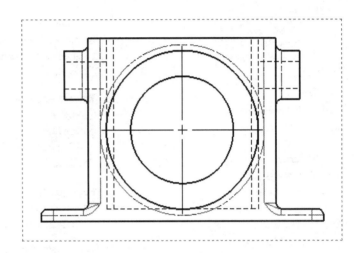

 Another way to change a view's display style is by selecting the drawing view and using the "**Display Style**" icon from the "View" toolbar at the top of the screen, just like we did when working with the solid models.

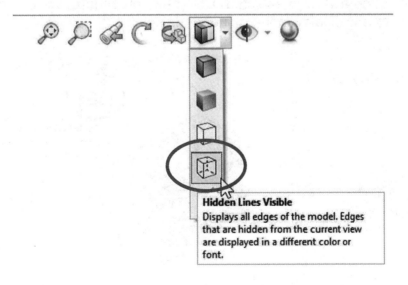

Hidden Lines Visible
Displays all edges of the model. Edges that are hidden from the current view are displayed in a different color or font.

The option to set the default display and tangent edge style for new views can be changed in the menu "**Tools, Options, System Options**" under "**Display Style**."

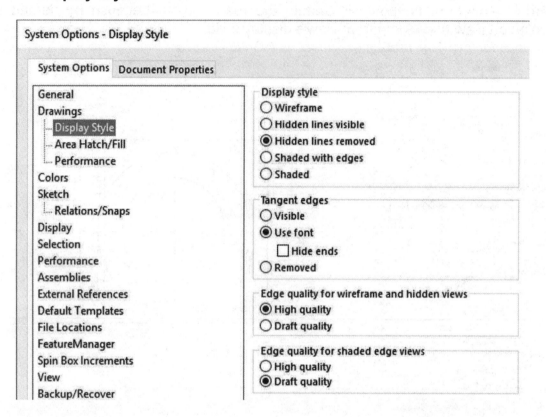

17.15. - Now that we have the views in place, we want to show the Isometric view in "**Shaded with Edges**" mode. Click to select the "Isometric" view on the screen and click the "**Shaded with Edges**" icon either in the "**Display Style**" toolbar or in the PropertyManager as we did in the previous step. Notice the dotted line around the view indicating a view is selected. Using this procedure, we can change any drawing view to any display mode as needed.

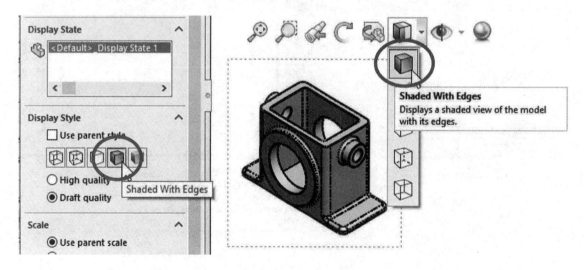

17.16. - In the drawing views we may or may not want to see the tangent edges of a model. A tangent edge is where a curved face and a flat face have a tangent transition, as in fillets. SOLIDWORKS has three different ways to show them: Visible, With Font, or Removed. Right-mouse-click inside the Front view, and from the pop-up menu select the option **"Tangent Edge, Tangent Edges Removed**." Repeat for the Top and Right views. Optionally we can hold down the "Ctrl" key and select different views to change their display state at the same time.

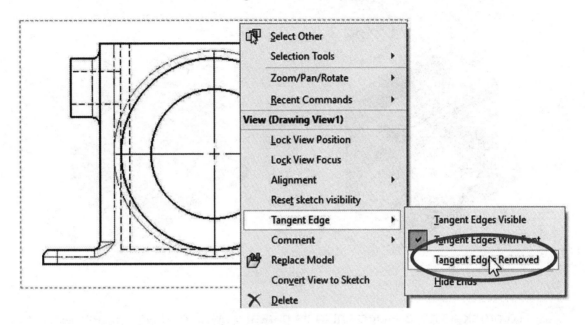

 The differences between tangent edge display types are visible next. **"Tangent Edges Visible"** shows all model edges with a solid line, **"Tangent Edges with Font"** shows tangent edges with a dashed line, and **"Tangent Edges Removed"** eliminates the tangent edges from the view.

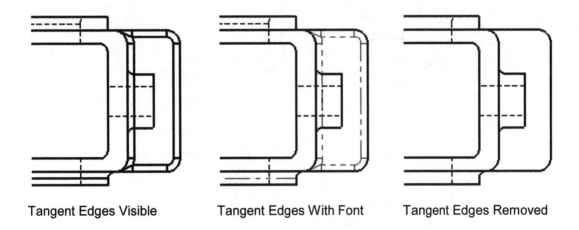

Tangent Edges Visible   Tangent Edges With Font   Tangent Edges Removed

The default Tangent Edge display settings can also be changed in the menu **"Tools, Options, System Options, Drawings, Display Style**."

17.17. - The next thing we want to do in the drawing is to move the views to arrange them in the sheet. To move a view, click-and-drag it, either from the border or any model edge in the view. Projected views are automatically aligned to their parent view; therefore, the Top view can only be moved up and down, and the Right view from side to side, and the Isometric view is free to move around since it is not aligned to any other view.

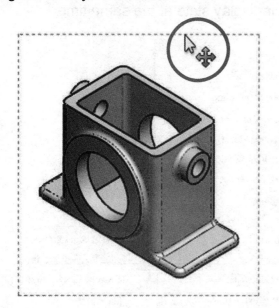
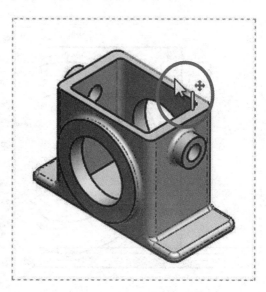

 To break a view's alignment to its parent view and move it freely within the drawing sheet, right-mouse-click it and select "Alignment, Break Alignment."

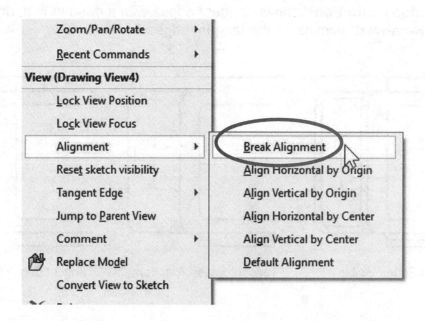

17.18. - Click-and-drag the drawing views in the sheet and arrange them as shown, with "Tangent Edges Removed" for all views, and the Isometric with "Shaded with Edges" display. This layout will allow enough space to import the dimensions into the drawing. Make sure the drawing's units are set to inches and three decimal places.

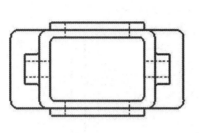

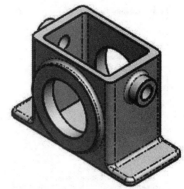

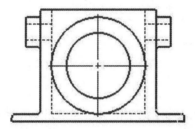

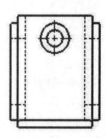

 Notice the toolbars available in the CommandManager were automatically changed to match the detail drawing environment in which we are now. The Features tab was replaced by View Layout and Annotation, detailing and annotation tools.

17.19. - To make the drawing easier to read and add more information to the detail drawing, we are adding a section of the Right view. A section view essentially cuts the model along a defined line to let us "see" inside to understand internal model details that would not be visible otherwise. To make the section view, select the "**Section View**" command from the View Layout tab in the CommandManager.

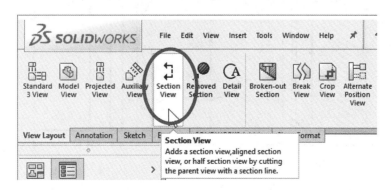

In the "**Section View**" properties, select the "Vertical" option from the "Cutting Line" section. In the graphics area move the mouse pointer close to the center of the "Right" view, touching a round edge. To set the section line, click when the line's preview snaps in the middle of the '*Housing*'.

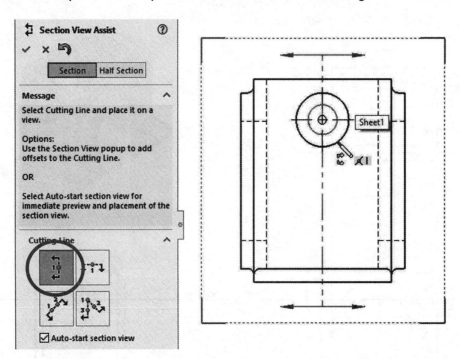

 We can zoom in or out in the Drawing using the mouse scroll wheel while adding a section view. SOLIDWORKS will zoom in or out at the location of the mouse pointer.

17.20. - Immediately after locating the section line, moving the mouse will reveal a dynamic preview showing the section of the '*Housing*' overlapping the Right view.

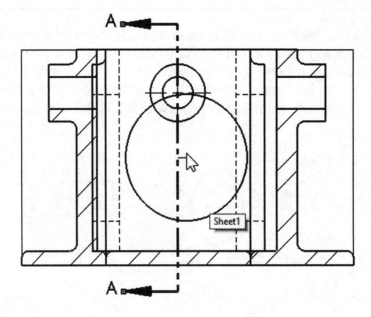

The last thing to do is to locate the section view in the sheet. Move the mouse to the right side of the drawing and click to the right of the Right view to locate it. If needed, after locating the Section view move the other views to arrange and space them. Notice the section view's alignment is locked to the Right view.

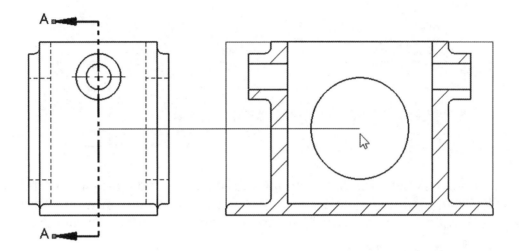

 Using the mouse wheel, we can manipulate the drawing sheet's view. Use the wheel to zoom in and out and the middle mouse button (Press the wheel down) to move the drawing sheet (Pan). We can do this while positioning the section view or any other view.

17.21. – By selecting the Section View we can see its Properties where we can change different options such as the Section Label, reverse the section direction ("Flip Direction"), Display Style, Scale, etc.

By default, projected views inherit the display style of the view from which it was made, but section views are automatically set to "**Hidden Lines Removed**" for clarity.

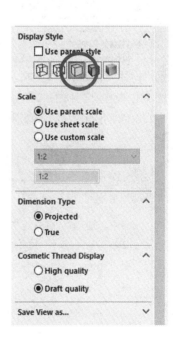

17.22. - The next step is to add a "**Detail View"** to zoom in an area of the drawing, to make smaller model details easier to visualize. From the View Layout tab in the CommandManager, select the "**Detail View**" command. After selecting it we are asked to draw a circle in the area that we want to make a detail of.

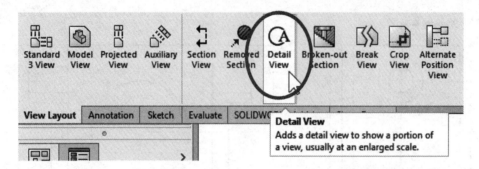

To draw sketch elements without automatically adding geometric relations to other geometry, hold down the "Ctrl" key while drawing them. This technique works in the part, assembly, and detailing environments.

Draw a circle in the upper left area of the "Section" view we just made. When drawing the circle be sure not to capture automatic relations to existing geometry by holding the Ctrl key while drawing it, and just like the "Section" view...

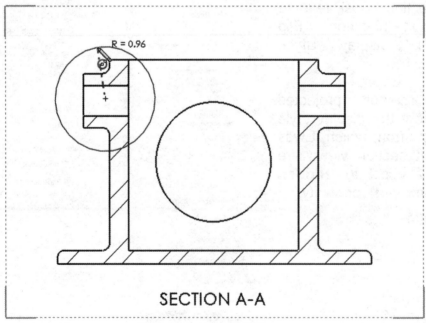

SECTION A-A

...move the mouse to locate the Detail view next to the Top view using the dynamic preview. By default, detail views are two times bigger than the view they came from, and if needed this option can be changed in the menu "**Tools, Options, System Options, Drawings**" and change the "**Detail view scaling**" factor to multiply the scale of the view the detail came from.

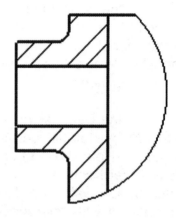

 If the detail area is not as big as needed, click-and-drag the detail's circle to resize it. If the location of the detail needs to be redefined, drag its center to move the detail area; the Detail view will update dynamically. That is the reason why we don't want to add any geometric relations when drawing the detail's circle, to be able to move and resize it if needed.

DETAIL B
SCALE 1 : 1

 Alternatively, we can draw a circle or any closed contour like an ellipse, a polygon, or a spline and then select the "**Detail View**" icon and use that profile for the detail view.

17.23. - Now that we have added all the necessary views to our drawing, the next step is to import the '*Housing's*' dimensions from the part (the 3D model) into the drawing sheet (the 2D drawing). If you remember, when we modeled the part we added all the necessary dimensions to define it, and now we can import those dimensions into the detail drawing reducing the amount of work needed to complete this task.

Select "**Model Items**" from the Annotation tab in the CommandManager or go to the menu "**Insert, Model Items**."

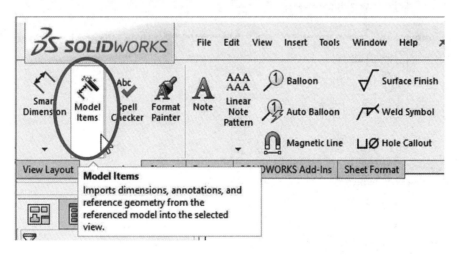

In the "Model Items" options select the type of dimensions and annotations we want to import into the drawing. For this exercise select the options "Source: Entire Model" to import the dimensions of the entire *'Housing'* and activate the checkbox "**Import items into all views**" to import annotations and dimensions to all drawing views.

In the Dimensions section, the option "Dimensions Marked for Drawing" is selected by default; activate the following options to import other annotations including "Hole Wizard Locations" to import the location of holes made using the Hole Wizard command, "Hole Callout" to add the machining annotations to the holes made with the Hole Wizard, and Tolerenced Dimensions to import dimensions with tolerances. 3D model items available for import into a detail drawing include notes, datum, welding annotations, Geometric Tolerances (GD&T), reference geometry (axes, planes, coordinate axes, etc.), surfaces, center of mass and others. Click OK to import the selected annotations.

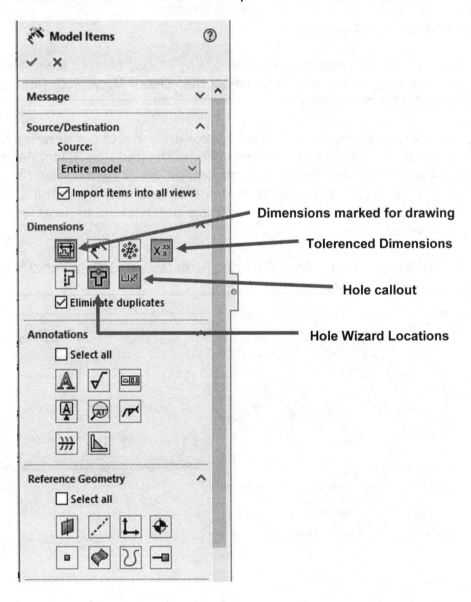

17.24. - SOLIDWORKS will import the dimensions and annotations to all views, attempting to arrange them automatically in the view that displays it better.

Dimensions are first added to Detail views, then to Section views, and finally main views. The reason for this order is because Detail Views show smaller model details, Section Views often show otherwise hidden features, and finally the main views show the major features. While SOLIDWORKS does a good job adding dimensions to views, they are not always added to the view that best displays it; here is where we must do some work. Move the views to make room for the dimensions as shown. Notice the dimensions attached to a view move with it.

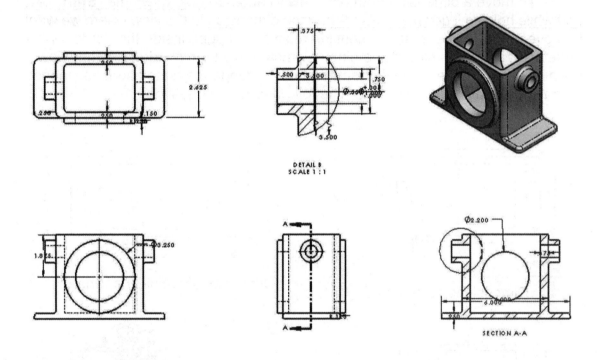

 If the drawing's dimensions are not the desired units and/or decimal places, they can be changed in the menu "**Tools, Options, Document Properties, Units**" or from the Unit System button in the status bar.

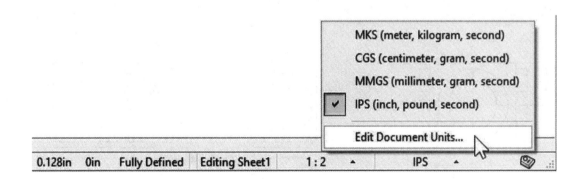

369

17.25. - All the dimensions that were added to the 3D model when we made it, including sketch and feature dimensions, have been imported into the drawing. Now we need to arrange them to make our drawing easier to read. To arrange annotations in the drawing click-and-drag to locate them as needed.

Some of the imported dimensions may have been incorrectly added to a drawing view, and we'll need to move them to a different view. One way to tell which dimensions belong to a view is to move the view and the annotations attached to it will move at the same time.

To move a dimension from one view to another, <u>hold down the "**Shift**" key, and while holding it down, click-and-drag the dimension</u> to the view where we want to move it. Remember that a dimension must be dragged inside the border of the target view. Keep in mind that a dimension can only be moved to another view if the dimension can be correctly displayed in it. Arrange the annotations as needed to make the drawing easy to read. The following is a suggestion.

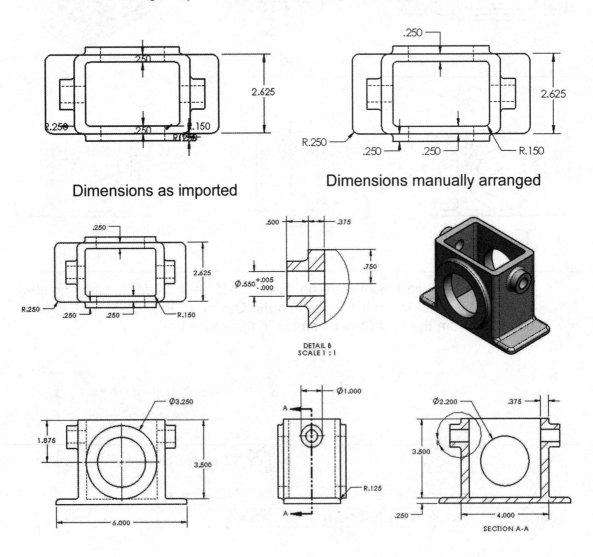

Dimensions as imported

Dimensions manually arranged

17.26. – In a detail drawing, Center Marks are used to locate the center of circular features such as round holes or an arc's center. If the center marks are not automatically added or are missing, they can be manually added using the "**Center Mark**" command from the Annotation toolbar. Select the "**Center Mark**" command and click on the edges of the circular features missing a center mark as needed. When done, click OK to finish the command.

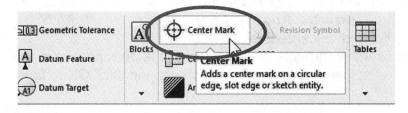

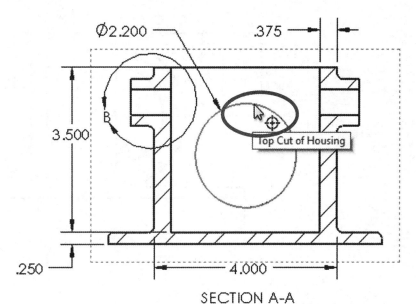

SECTION A-A

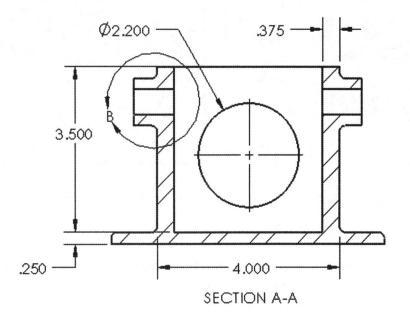

SECTION A-A

371

17.27. - Another annotation that may be missing from our drawing is the centerlines, which are used to indicate the axis of a cylindrical surface. To add missing centerlines to our drawing select the "**Centerline**" command from the Annotation tab in the CommandManager, or the menu "**Insert, Annotations, Centerline**." To add a centerline, select

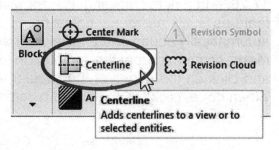

two edges or sketch segments, a cylindrical (as is our case) or a conical surface. Add the missing centerlines to all drawing views and click OK to finish.

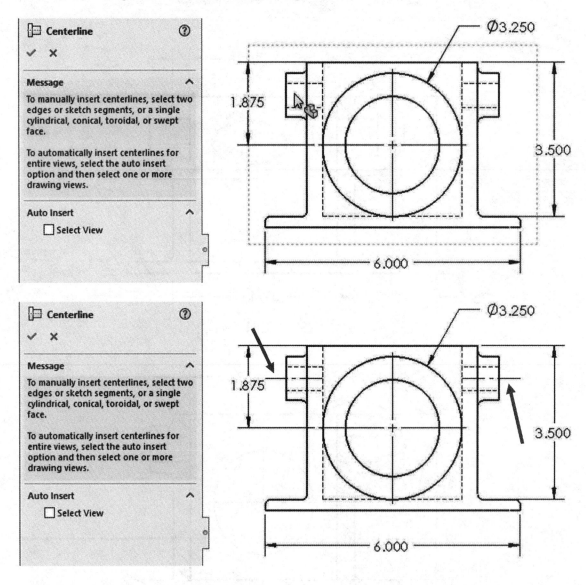

 In the Auto Insert section, using the "Select View" option will add a centerline to every cylindrical face in the selected view.

17.28. - When arranging the dimensions in a drawing we may want (or need) to reverse a dimension annotation's arrows. To do this, select a dimension and click the dots in the arrowheads to reverse them or the option in the Dimension's properties.

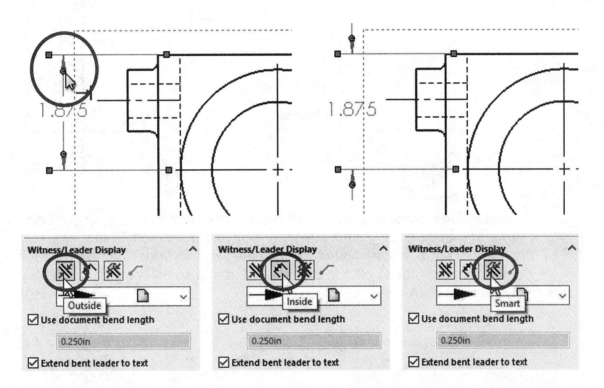

 Dimension's extension lines can also be dragged to change its location to a more convenient position where it doesn't overlap model geometry and/or annotations.

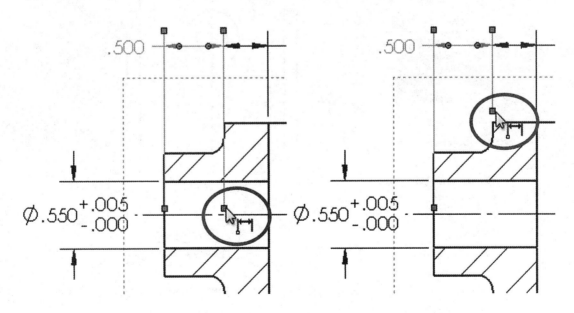

A dimension's leader can be reversed by selecting the leader's corner.

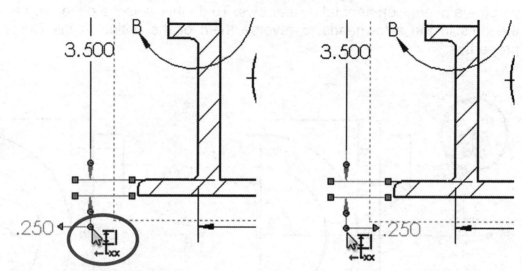

 Diameter dimensions displayed with linear witness lines can be changed to Diameter witness lines. Select the dimension, in the PropertyManager select the "Leaders" tab, and activate the "Diameter" option under Witness/Leader Display, or right-mouse-click in the dimension and select the option "Display Options, Display as Diameter."

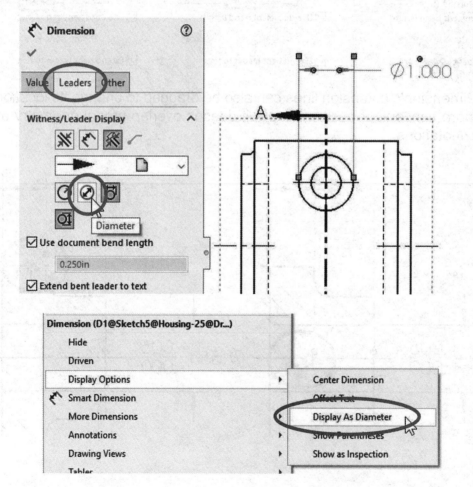

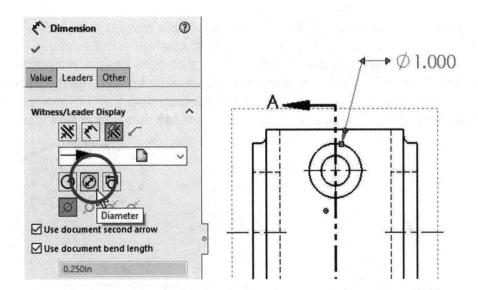

 If a fillet's radial dimension is not added in the best location or if we need to move it, it can be re-attached to an equal radii arc, as in the next image by clicking and dragging the dimension arrow's endpoint onto a different arc.

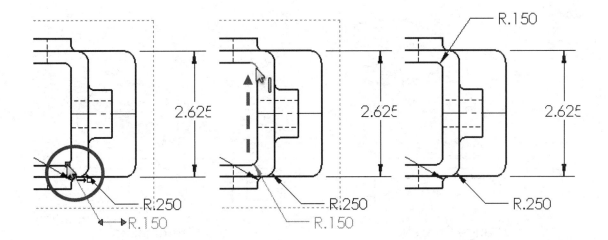

 To delete duplicate or unwanted dimensions select them and hit the "Delete" key. Be aware that when a dimension is deleted in the detail drawing, we are only deleting the annotation and not the dimension from the 3D model.

If annotations and dimensions are erroneously deleted from the drawing, using the "**Model Items**" command again will re-import the deleted and/or missing annotations; keep in mind that SOLIDWORKS will only add the missing/deleted dimensions and will not duplicate existing annotations.

17.29. - Notice the tolerance for the *"Shaft Hole"* was carried over to the drawing from the 3D model. If we need to add tolerance, change a dimension's precision or other parameters we can change them using either the PropertyManager *or* the "**Dimension Palette**." In the Dimension Palette we can change the precision, tolerance, add notes, parentheses, etc. Select a dimension and move the mouse to the Dimension Palette icon to expand it and access dimension options.

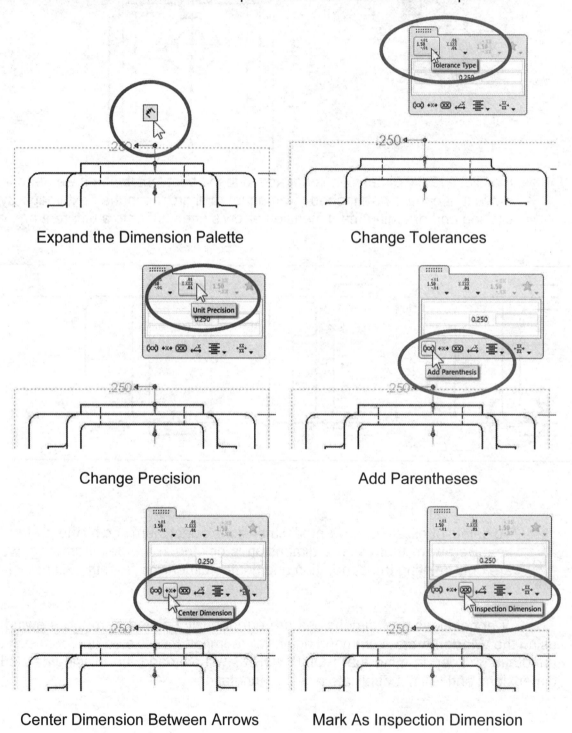

Expand the Dimension Palette   Change Tolerances

Change Precision   Add Parentheses

Center Dimension Between Arrows   Mark As Inspection Dimension

The finished drawing for the *"Forge"* configuration should look like the next image.

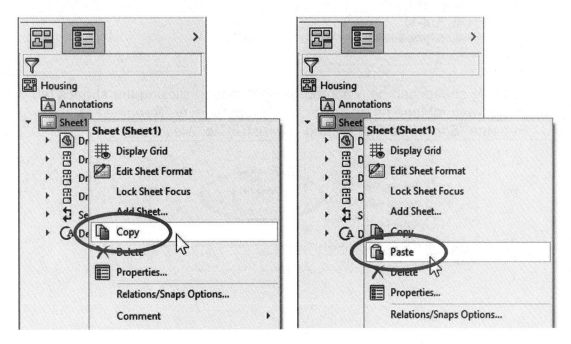

17.30. - Now that the Forge configuration drawing is finished, we need to add a second sheet to the drawing for the Machined configuration details and complete the '*Housing's*' detail drawing. We can add a second sheet and add views and dimensions just as we did in the first sheet, but since the machined drawing is very similar to the forge drawing, a faster way is to copy the drawing sheet we just finished, paste it into a new sheet and modify it to add the missing details. In the FeatureManager, right-mouse-click in *"Sheet1"* and select "**Copy,**" then repeat and select "**Paste**" *or* select the menu "**Edit, Paste**" (shortcut Ctrl+V).

377

After selecting **"Paste"** from the menu, the **"Insert Paste"** dialog box is presented. Select the "After selected sheet" option to add the new sheet after the first one. When asked about renaming the drawing views, click 'Yes'; this option will rename the section and detail views to use the next available view labels. In our case the new section will become "**C**" and the new detail will be "**D**."

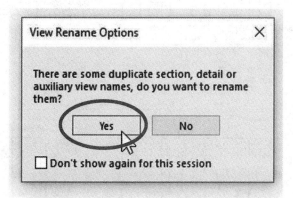

17.31. - Now we have a second sheet in the drawing exactly as the first one, except for the view labels. If needed, rebuild the drawing to update the views using the "Rebuild" command (Ctrl-B shortcut). A force-rebuild updates all geometry in the 3D model and the detail drawing, not only what needs to be updated using the shortcut Ctrl-Q.

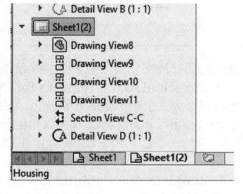

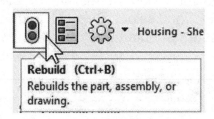

 Drawing sheets can be renamed by right-mouse-clicking the sheet's tab or in the FeatureManager and selecting "Properties" or "Rename." Rename *"Sheet1"* to *"Forge"* and *"Sheet1(2)"* to *"Machined."*

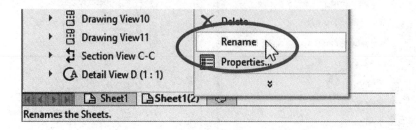

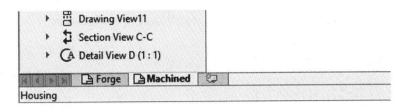

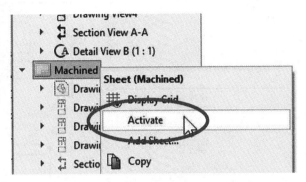

To change from one drawing sheet to the next select the sheet's tab at the bottom of the screen, or right-mouse-click in the feature manager and select "Activate."

17.32. - After adding the second sheet we need to tell SOLIDWORKS to show the *"Machined"* configuration in this sheet. If it's not already active, select the *"Machined"* sheet's tab on the lower left corner of the screen to activate it.

Since this is a copy of the *"Forge"* drawing, we'll need to change the drawing views to show the *"Machined"* configuration. To do this, select the Front view in the screen, from the drawing's properties select the "Reference Configuration" drop-down menu and select the *"Machined"* configuration. After selecting the new configuration, the drawing view is automatically updated. The other views will have an extra option to select a new configuration or link it to their parent view, which in our case is the Front view. The Section and Detail views don't have this option because they are linked to their parent view regardless. As soon as their parent view is updated they will be updated too.

Front View *"Forge"* configuration

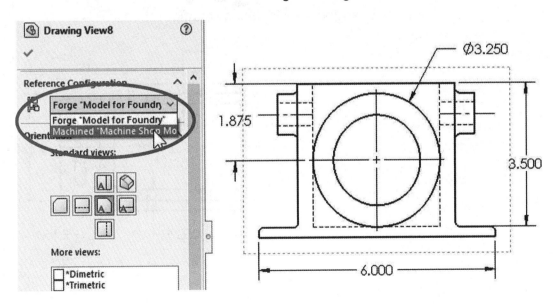

Front view changed to "*Machined*" configuration

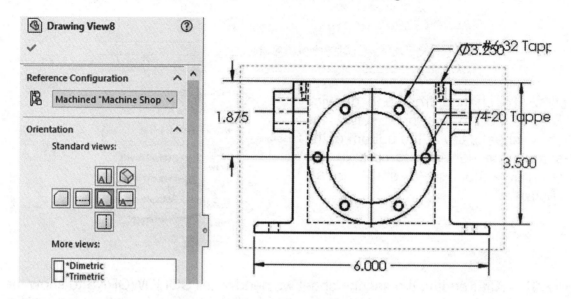

Top View "*Forge*" configuration

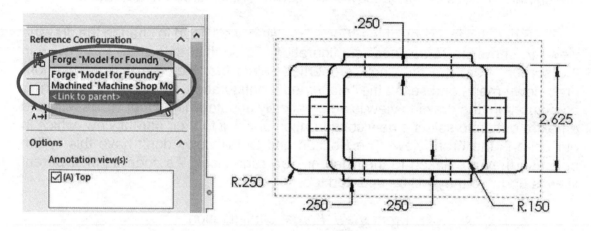

Top View changed to "*Machined*" configuration

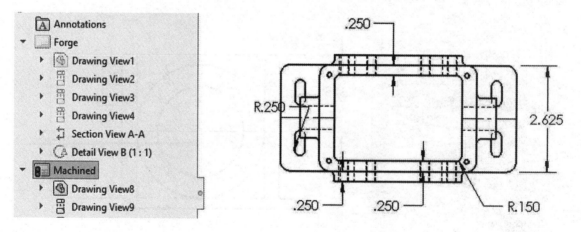

17.33. - Change all views in this sheet to the "*Machined*" configuration or linked to the parent view; repeat the "**Model Items**" command to import the missing dimensions and annotations for this configuration and be sure to include the "Hole Wizard Locations" and "Hole Callout" options.

The holes made with the Hole Wizard include annotations with machining information for manufacturing. Turning ON the "Eliminate duplicates" option will prevent the dimensions already in the drawing from being imported again. After the missing dimensions are added to the drawing, arrange them as needed in the different drawing views as we did before to make the detail drawing easier to read.

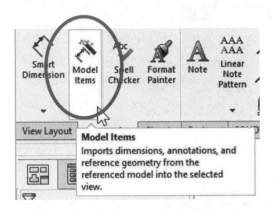 

When hole wizard annotations are imported, if a hole wizard feature made multiple holes at the same time, the annotation is added to one of them, but it may not be the best location to display it. To move the annotation, select it and then click and drag the tip of the arrow to another instance of the hole. Be aware that if the arrow is added to a different hole wizard feature, the annotation will change to show the hole's specification.

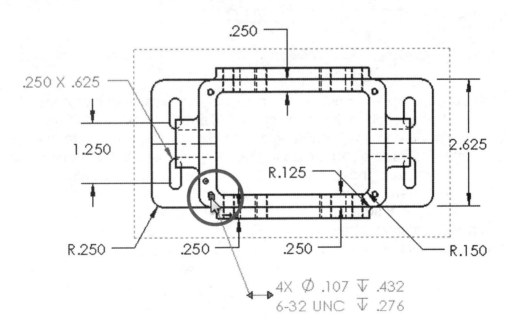

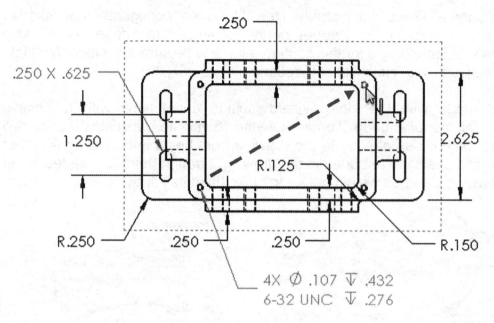

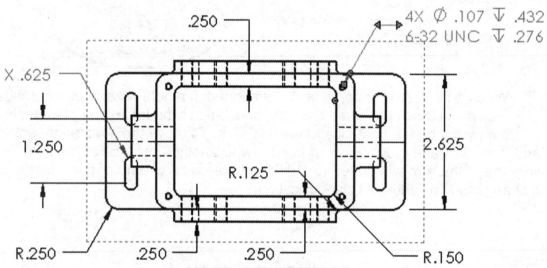

17.34. – After updating the drawing, we need to add the missing center marks to the circular array of tapped holes. Select the "**Center Mark**" icon from the Annotation tab, click on one hole's edge, and then click on the Propagate pop-up icon to add Center Marks to all the holes in the array of holes at the same time. SOLIDWORKS recognizes the pattern and adds the Circular Center Marks automatically. In the Property Manager we can see the additional options to manually add center marks and change their style if needed. Centermarks are added per view and cannot be added to multiple views at the same time.

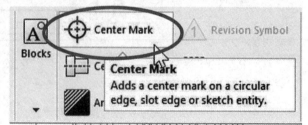

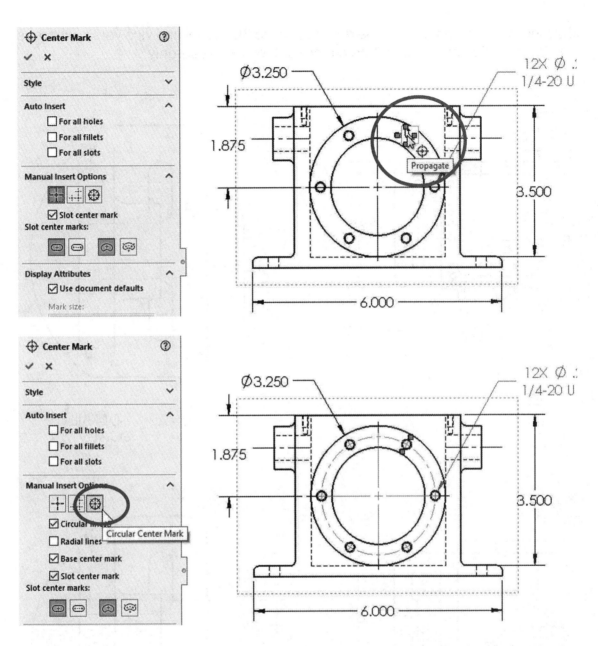

17.35. - To make the slot dimensions easier to see, add a new detail view in the slot area. Move the Isometric and Detail views to the right and locate the new detail next to the Top view.

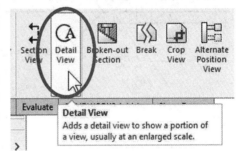

After the detail view is added, delete the imported slot annotation and move the imported dimensions from the Top view to the new detail view using the "Shift-Drag" method.

Select the "**Smart Dimension**" command and add the missing dimensions to the slots. These are now reference and not parametric dimensions. The difference is that a parametric dimension can be changed by double clicking in the drawing, and the changes will be reflected in the 3D model and vice versa, whereas a reference dimension cannot be changed; they are "read-only".

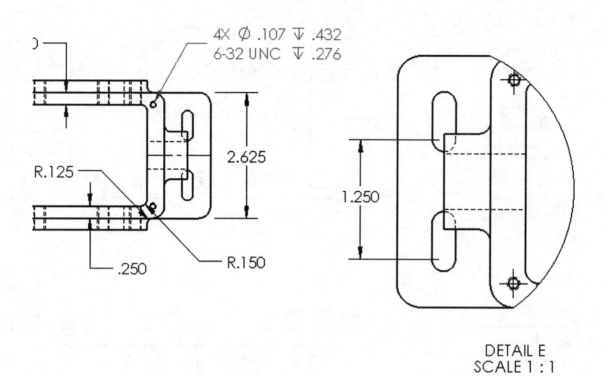

DETAIL E
SCALE 1 : 1

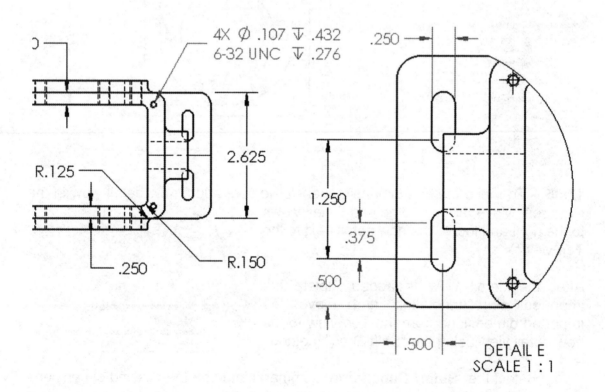

DETAIL E
SCALE 1 : 1

17.36. – Now we need to add the missing "**Center marks**" to the slots, using the option "Slot Ends."

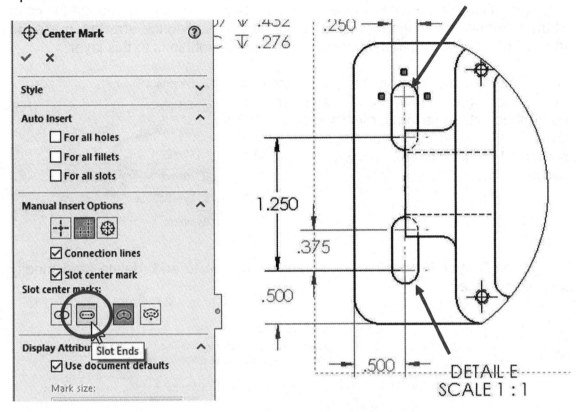

The Top and Detail views should now look similar to this.

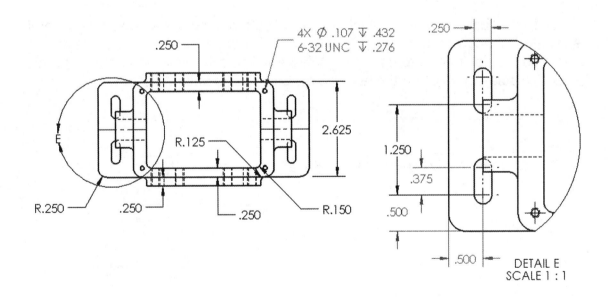

17.37. - Depending on system and template settings, manually added dimensions (reference) and annotations are grey by default, whereas parametric dimensions and annotations (imported from the model) are black. One way to change the color of the reference annotations is by adding a new "Layer" to the drawing, make the layer's color black, and then move the reference annotations to this layer.

If not visible, turn on the "Layer" toolbar. Go to the menu "**View, Toolbars, Layer**" or right-mouse-click in a toolbar and select "Layer." By default, it is in the lower left corner of the screen, but it can be moved around for convenience.

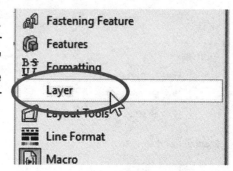

The "**Layer Properties**" command allows us to add, delete, and modify layers as needed.

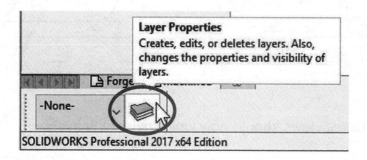

In the "Layers" dialog select the "New" button to create a new layer. After it is created change its name to "Reference." The default color for new layers is black, so we don't need to change it. We'll add the reference annotations to this layer to display them using the layer color. Click OK to finish.

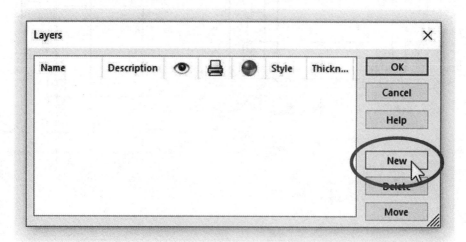

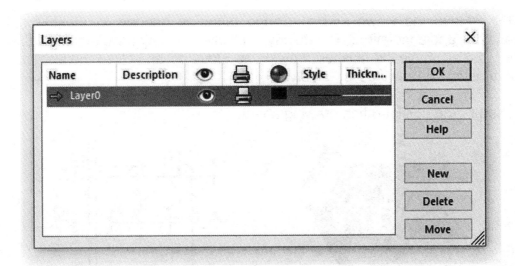

 In the "Layers" properties we can change their name, optionally add a description, hide or show them on the screen, select them to print or not, change their color, the line style, and its thickness.

17.38. - To move annotations or models to a different layer select them in the graphics area and pick the new layer from the Layer toolbar. Select all reference annotations (individually or multi-selected using Ctrl + Select) and assign them to the newly created layer. Notice the annotation's color updating after the change.

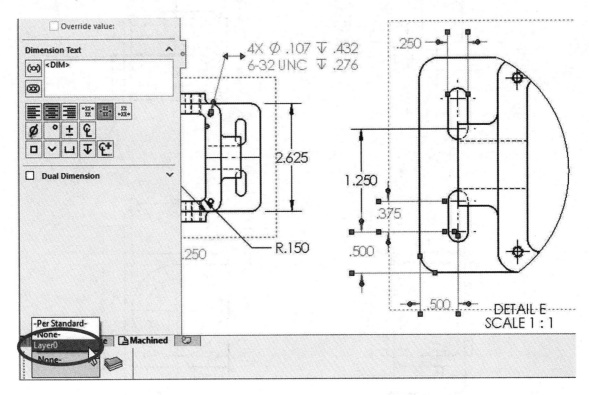

17.39. - Move the views and rearrange them as needed to make room for dimensions and annotations. Delete any duplicate dimensions to make the drawing easier to read. Save the finished drawing as *'Housing'* and close the file. Note the file extension for drawings is *.slddrw.

Finished Forge configuration drawing sheet.

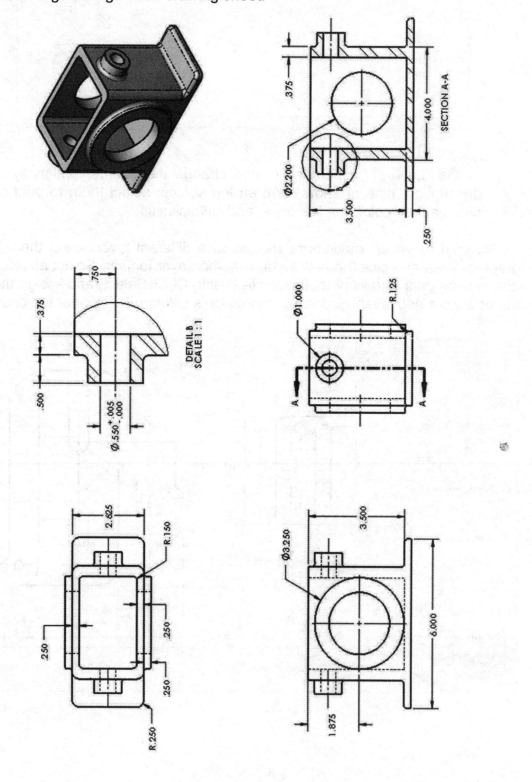

Finished Machined configuration drawing sheet.

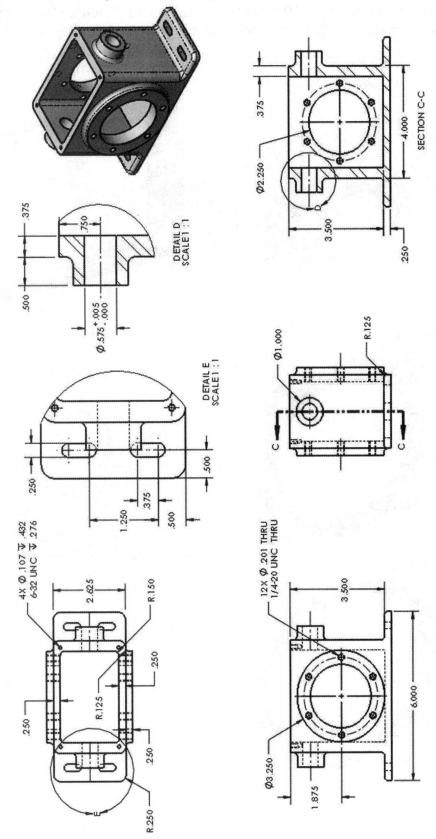

Exercises: Make the detail drawings of the following Engine Project parts that were previously made in the Part Modeling section, to match the drawing supplied to make each part. High resolution images are included with the book files.

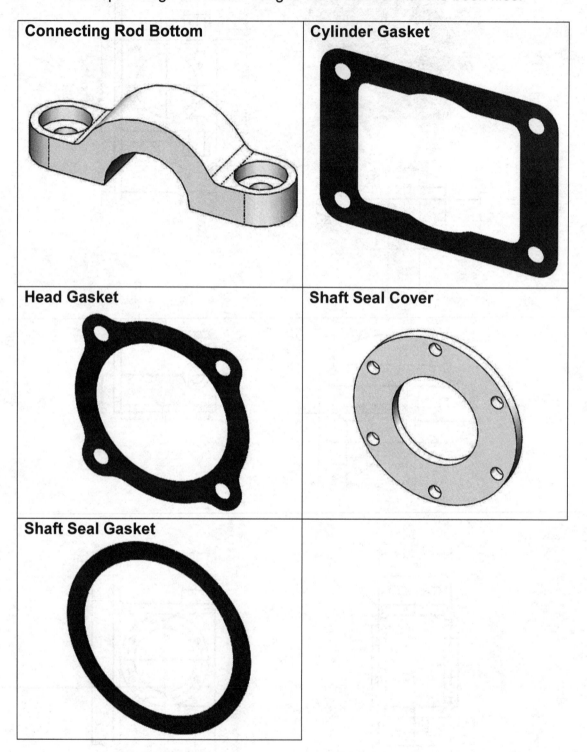

Connecting Rod Bottom	Cylinder Gasket
Head Gasket	Shaft Seal Cover
Shaft Seal Gasket	

Drawing Two: The Side Cover

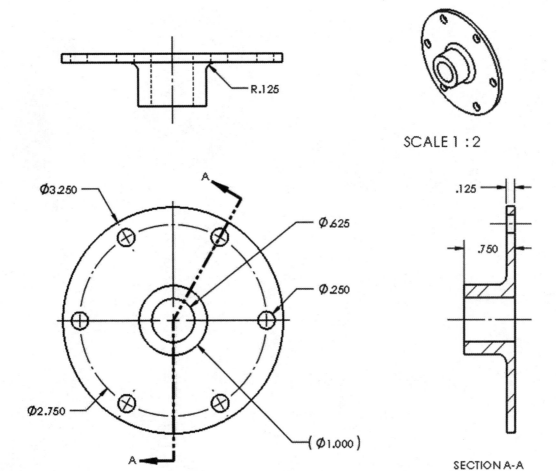

R.125

SCALE 1 : 2

Ø3.250

Ø.625

Ø.250

Ø2.750

(Ø1.000)

.125

.750

SECTION A-A

391

Notes:

In this lesson we'll review material previously covered, including how to add configurations to a part, add standard views and importing annotations from the model to the drawing. We'll also learn how to add missing dimensions, formatting dimensions, add an aligned section view, and change a sheet and view's scale.

The drawing of the *'Side Cover'* part will follow the next sequence.

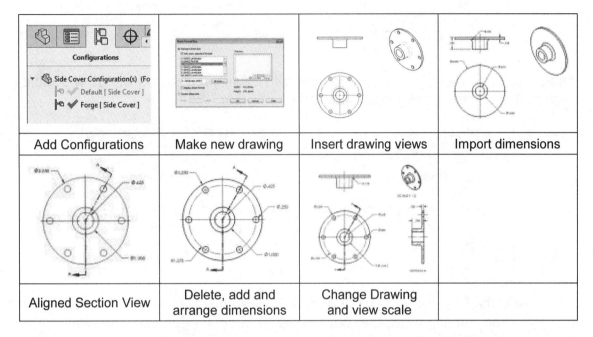

Add Configurations	Make new drawing	Insert drawing views	Import dimensions
Aligned Section View	Delete, add and arrange dimensions	Change Drawing and view scale	

18.1. - For this exercise we will repeat the process that we used with the *'Housing'* by adding a configuration for a forged and a machined configuration to reinforce the material just covered. Open the *'Side Cover'* part and go to the Configuration Manager tab. Right-mouse-click on the part's name and select "Add Configuration." Enter the name "*Forge*" and click OK to continue.

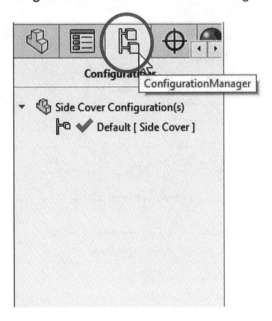

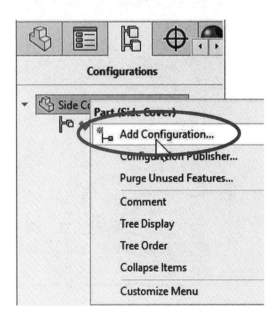

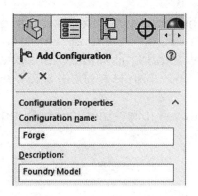 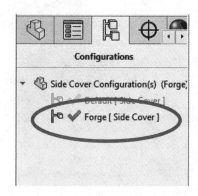

18.2. - Rename the "*Default*" configuration to "*Machined*" using the slow-double-click method or the configuration properties.

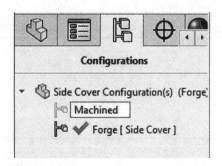 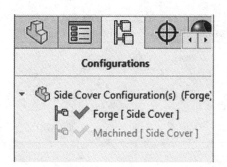

18.3. - Split the FeatureManager and show the Feature Manager at the top and the Configuration Manager at the bottom. With the "*Forge*" configuration active suppress the "*CirPattern1*" feature to suppress the pattern of holes. After suppressing the pattern, the original hole is left. Since the pattern was made using the face of the hole and not the feature, suppressing the "*Cut-Extrude1*" feature would also suppress the center hole, which we don't want to suppress.

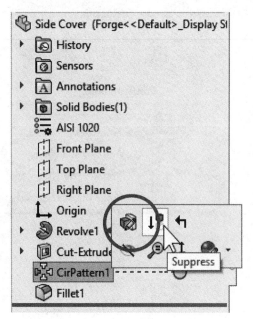 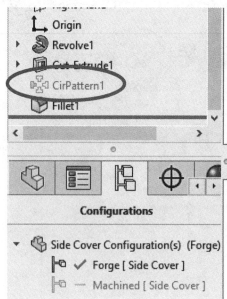

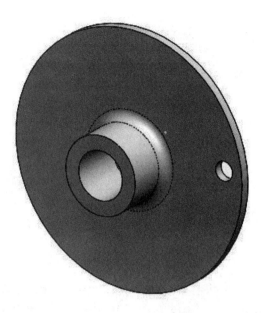

18.4. - The next step is to eliminate the small hole from the "Forge" configuration. Since suppressing "*Cut-Extrude1*" would also eliminate the center hole because both holes were made using the same sketch and feature, we'll use a different option.

If not already visible, right-mouse-click in a CommandManager's tab to activate the "**Direct Editing**" tab and select the "**Delete Face**" command, or from the menu "**Insert, Face, Delete.**" Deleting a face from a solid model allows us to directly manipulate the solid body without modifying existing features, and the "**Delete Face**" command is itself a feature which can be suppressed.

Select the inside face of the small hole as indicated with the "Delete and Patch" option. The "**Delete Face**" command will remove the hole's cylindrical face from the model, and the "Delete and Patch" option will close the openings left behind by extending the flat faces on both sides of the part to close the round holes and form a continuous surface to maintain a solid body. Click OK to finish.

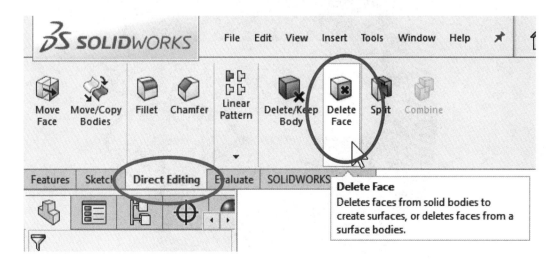

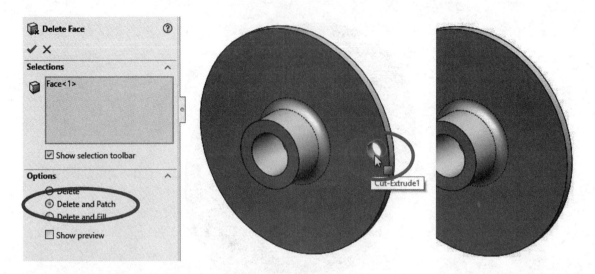

Using the "Delete" option would not close the openings causing our model to become an open surface body. In the following image it *looks* like the hole's surface is still there, but all we have is a surface with a hole in them and open inside. Surfaces are covered in more depth in *Beginner's Guide to SOLIDWORKS Level II*. The "Delete and Fill" option is usually used to delete multiple contiguous faces and replace them with a single surface. In this particular case the end result is the same using "Delete and Patch" or "Delete and Fill".

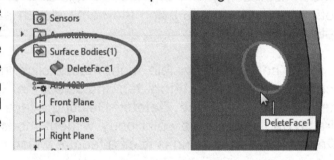

18.5. - Now we need to configure the center hole's diameter dimension. Double click on the center hole to reveal its dimension and then double-click on it. Change its value to 0.6" for the Forge configuration using the "This Configuration" button in the "Modify" value box and rebuild the model.

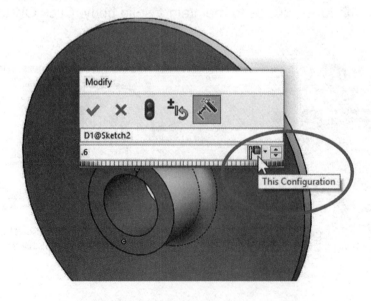

18.6. - Activate the "Machined" configuration and make sure the "DeleteFace1" feature is suppressed.

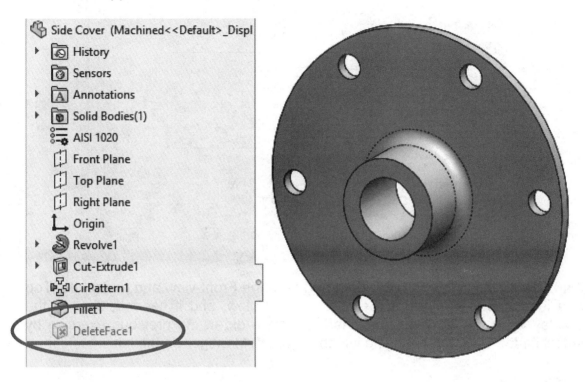

18.7. - After the configurations are complete, save the part, and with the "*Forge*" configuration active, select the menu "**File, Make Drawing from Par**t" or the icon from the "**New Document**" drop down menu to start a new drawing.

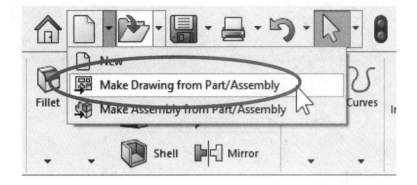

For the *'Side Cover'* part we'll use the "A-Landscape" drawing template without sheet format. After selecting the sheet size add the Front, Top and Isometric views. Drag the Front view from the View Palette and project the Top and Isometric views from it. Make sure our drawing is set to Inches and three decimal places.

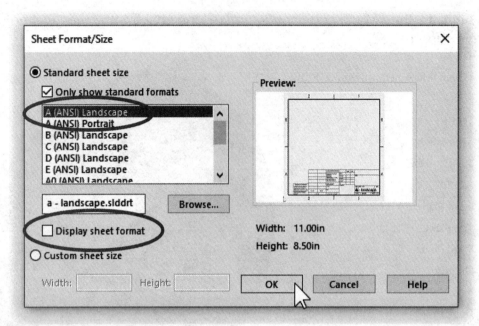

18.8. - From the "**View Palette**" drag and drop the Front view and project the Top and Isometric views from it. Select the Front view, and if needed, change the Display Style to "**Hidden Lines Visible**" as we did in the previous exercise by selecting the view and changing it in the PropertyManager on the left.

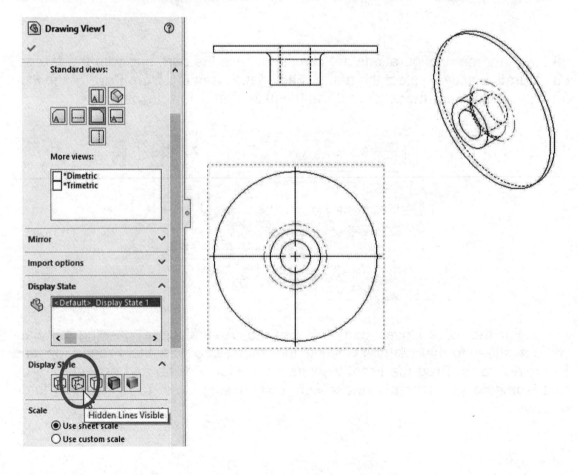

Change the Tangent Edge display for the Front and Top views to "**Tangent Edges Removed**" and the Isometric view to "**Hidden Lines Removed**" mode and "**Tangent Edges Visible**" to make the drawing easier to understand.

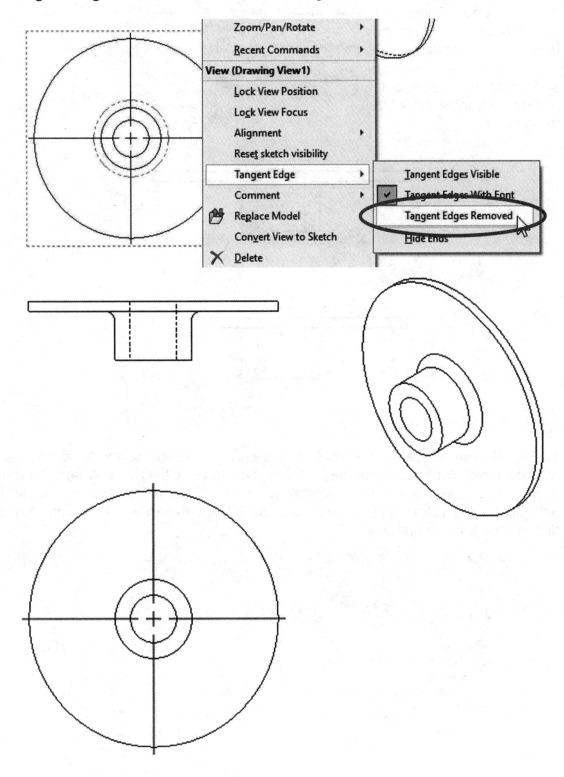

18.9. - After changing the view's display mode, we need to add a centerline to the Top view. Select the **"Centerline"** command from the Annotations tab and click in the cylindrical surface of the Top view to add the centerline. Click OK to continue.

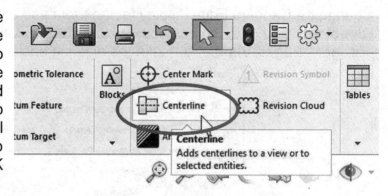

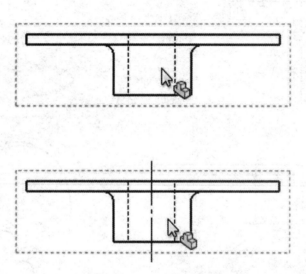

18.10. - Now we are ready to import the dimensions from the 3D model. Select the **"Model Items"** command from the Annotation tab or go to the menu "**Insert, Model Items**" and select the "Entire Model" from the "Source/Destination" section, activate the checkbox "Import items into all views" and click OK to import the dimensions and annotations.

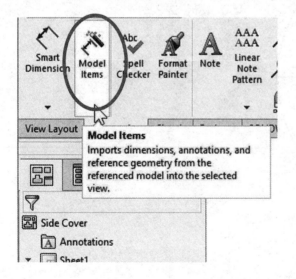

18.11. - After importing the model's dimensions, if the drawing's units are not in inches with three decimal places, use the "Units" command in the lower right corner of the screen to change them. Delete and arrange the dimensions as shown. The diameter dimensions will be added in the next step.

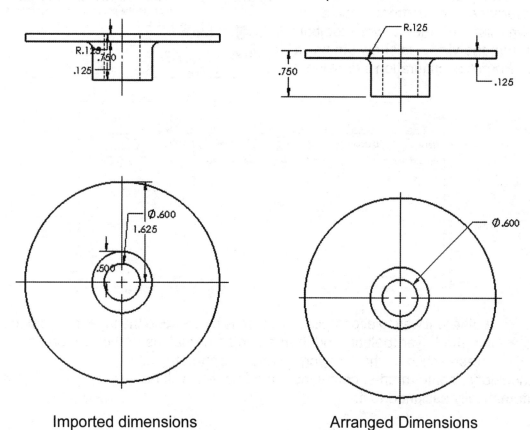

Imported dimensions   Arranged Dimensions

18.12. - Select the "**Smart Dimension**" tool and manually add the missing diameter dimensions. Remember we can also select the Smart Dimension command using mouse gestures.

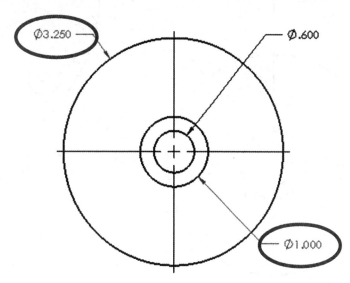

18.13. – After adding the diameter reference dimensions, add a new layer with the **"Layer"** command, make its color black and rename it "Reference." To change the reference dimensions to this layer, pre-select them and then select the "Reference" layer from the drop-down list in the Layer toolbar. Arrange the views as shown to finish the "*Forge*" configuration's drawing.

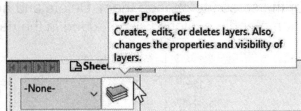

Layer Properties
Creates, edits, or deletes layers. Also, changes the properties and visibility of layers.

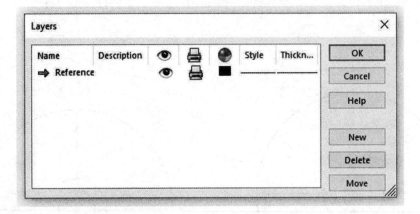

A different way to accomplish this step is by creating the layer first, selecting it in the Layer toolbar, and then adding dimensions and/or annotations; by pre-selecting the layer, any annotations, dimensions, center marks, centerlines, models, etc. will be automatically assigned to it.

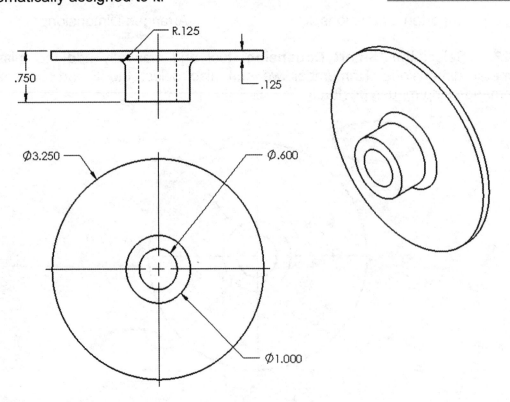

18.14. - Now we need to add a second sheet to our drawing to add the "*Machined*" configuration drawing. Instead of copying and pasting a complete sheet as we did with the '*Housing*', we'll only copy the Front view.

First, select the "**Add Sheet**" command in the lower left corner, and rename the sheets "*Forge*" and "*Machined*" respectively to match the configurations.

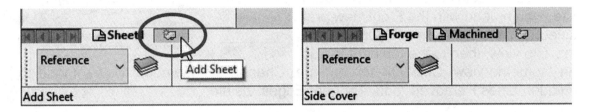

18.15. - Activate the "*Forge*" sheet, select the Front view and either press **Ctrl-C** or the menu "**Edit, Copy**" to copy the view. Select the "*Machined*" sheet's tab to activate it and press **Ctrl-V** or the menu "**Edit, Paste**" to paste the view. Select the newly added Front view and from the PropertyManager select the "Machined" configuration from the "Reference Configuration" drop down list.

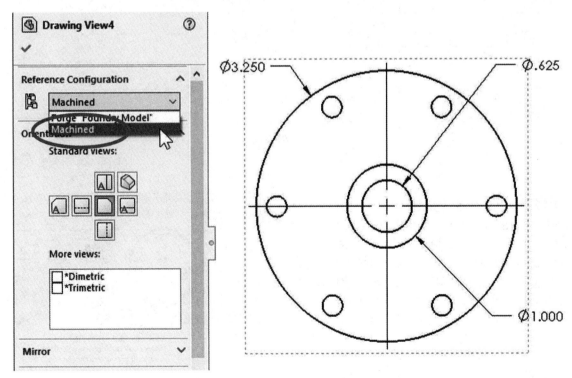

18.16. - To add the Top and Isometric views we'll use the "**Projected View**" command. Select the Front view, go to the View Layout tab, and click in the "**Projected View**" command.

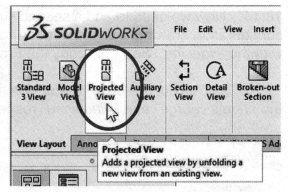

After activating the "**Projected View**" command with the Front view pre-selected, move the mouse "up" to add the Top view, then to the top right to add an Isometric view. Click OK to continue. Change the Isometric view's mode to "Hidden Lines Removed" and "Tangent Edges Visible."

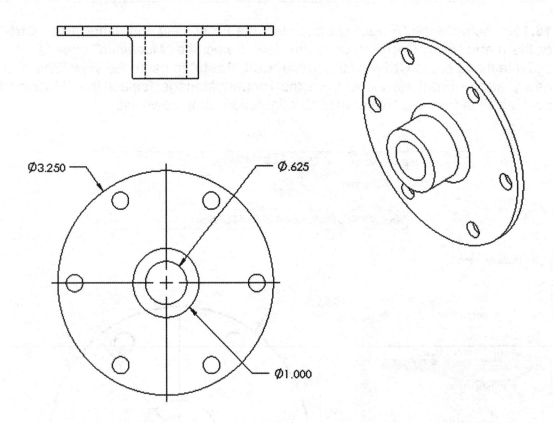

18.17. – In the '*Side Cover*', a vertical section view would not cut through the bolt holes; in this case we can use an aligned section view using an angled cutting plane to include features that would not be visible using a regular section view. In an Aligned Section view the section plane is revolved and then projected to the side, in order to show the section flat to the side.

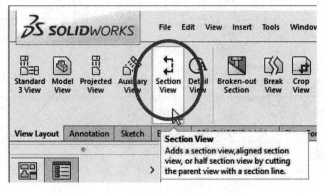

After selecting the "**Section View**" command, activate the "Aligned" option under the "Cutting Line" section.

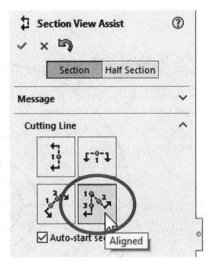

 Notice the icon shows the sequence needed to create it, identifying the order of the selections with the numbers 1, 2 and 3.

Follow the next sequence of selections to define the aligned section view needed.

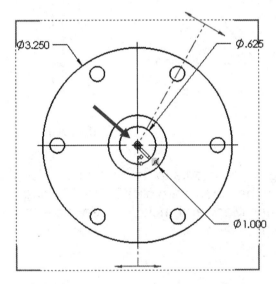

1st selection: center of the Side Cover

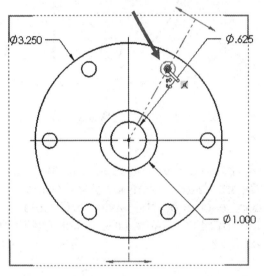

2nd selection: center of the top right hole

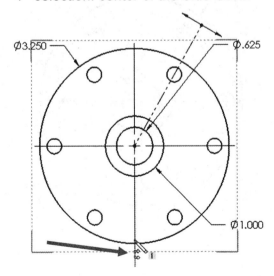

3rd selection: make the second line vertical.

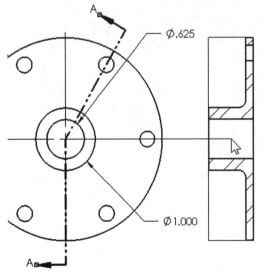

Locate the section view to the right.

405

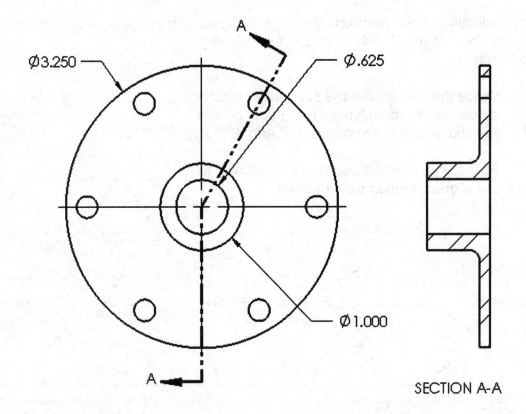

SECTION A-A

18.18. – Now we need to add the missing center marks to the bolt holes. Use the "**Center Mark**" command and select one of the holes, then click in the "Propagate" icon to change the style to Circular Center Mark and include every hole in the pattern at the same time. Click OK to finish.

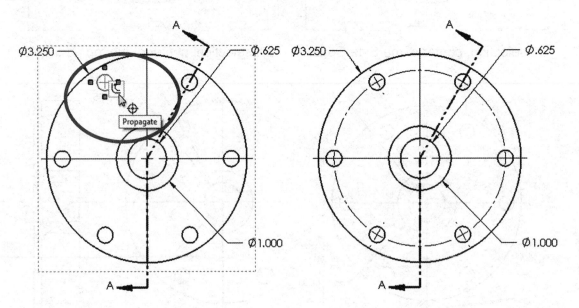

If missing, add the centerlines to the Top and Section views using the "**Centerline**" command.

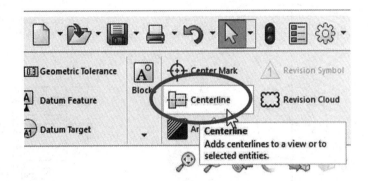

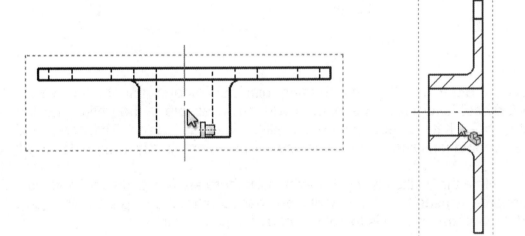

SECTION A-A

18.19. - To import the missing dimensions and annotations for the "Machined" configuration drawing, select the "**Model Items**" command from the Annotation tab using the options "Entire model" and "Import items into all views." In the "Dimensions" section include "Hole Wizard Locations," "Hole Callout" and "Tolerenced Dimensions" and click OK to import the missing annotations.

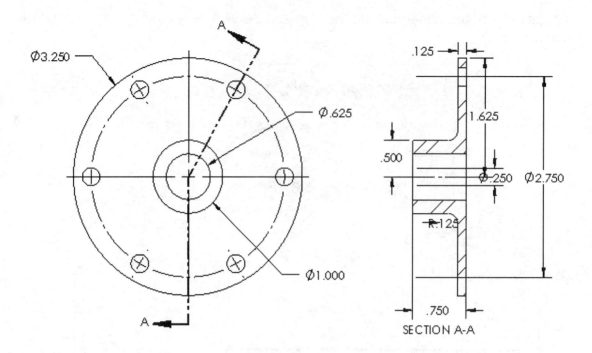

18.20. - Before adding the missing dimensions and annotations select the "*Reference*" layer created earlier. Layers are available in the entire drawing, not just a sheet. If it was not renamed, it will be listed as "*Layer0.*" By pre-selecting a layer, annotations added to the drawing will be automatically assigned to this layer.

Note that in the layer's drop-down list there are two additional options; "-Per Standard-" is used to define annotations based in the drawing's drafting standard, and the "-None-" option is to not select or assign a layer.

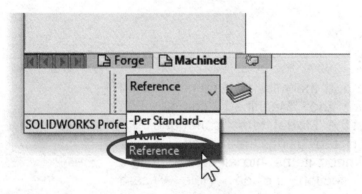

After selecting the layer use the "**Smart Dimension**" tool to add the missing dimensions, delete the extra dimensions and move them around to match the next image. Note that by selecting the new layer, new reference dimensions are added to it and shown in the selected layer's color, in this case black.

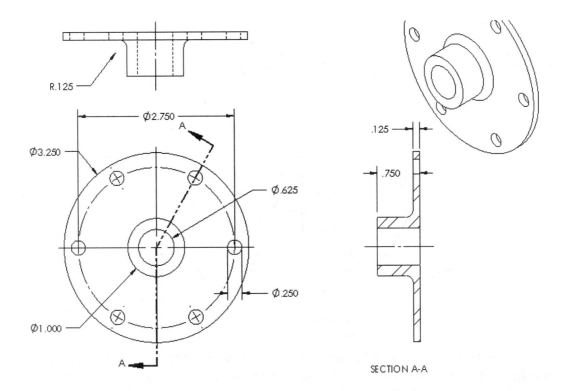

SECTION A-A

18.21. – If a dimension is displayed as a radius instead of a diameter, it can be changed as needed. Right-mouse-click on the dimension and from the pop-up menu select "Display Options, Display as Diameter" or from the "Leaders" section in the dimension's properties.

 If the "Smart Dimension" command is selected, the "Display As Diameter" option will be displayed immediately, otherwise the option will be listed under the "Display Options" sub-menu.

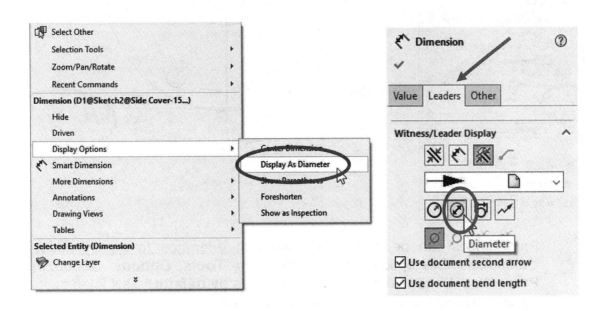

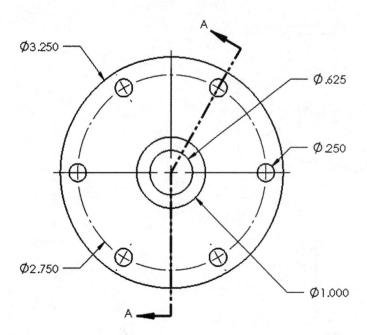

18.22. - Dimensions manually added to a drawing can be displayed with a parenthesis to indicate they are **Reference Dimensions**. Reference dimensions are used to provide additional information to the drawing, don't have a tolerance and oftentimes their value depends on other dimensions.

Reference dimensions are usually not critical to the design and are not used for manufacturing or inspection. In this component, the 1" diameter dimension is not critical, and it will be marked as reference. To add the parenthesis, select the dimension and in the Property Manager turn on the "Parenthesis" option.

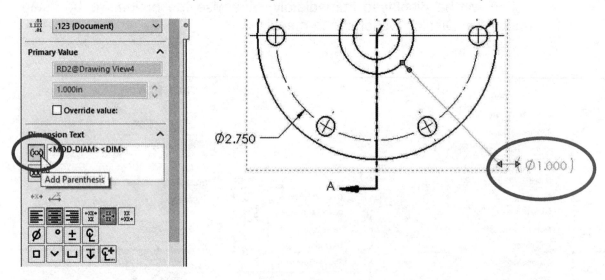

 Parentheses can be added by default to reference (manually added) dimensions. This option is set in the menu "**Tools, Options, Document Properties, Dimensions, Add parentheses by default**." This is an option set by document or template and is not a system option.

18.23. - Another way to add or remove parentheses and other annotation options is to use the "**Dimension Palette**." When we select a dimension the dimension palette appears near it; move the mouse over it to expand the window and modify the dimension's appearance. Here we can change a dimension's parameters including parenthesis, tolerance, precision, etc.

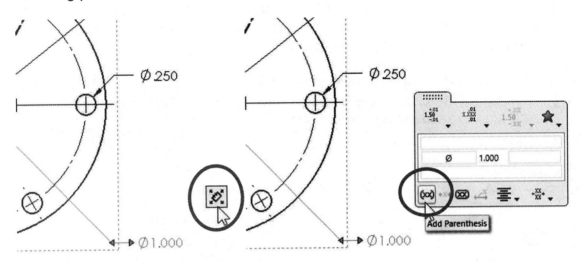

18.24. – In certain instances, radial dimensions Leaders may need to be modified to make the dimension easier to read, as is the case with the fillet dimension. To change it, select the dimension and from the "Leaders" tab in the dimension's properties, turn on the "Arc extension line or opposite side" option, and reverse the direction of the arrow to point towards the radius by selecting the "Inside" option for the Leader display.

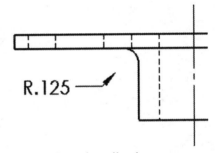

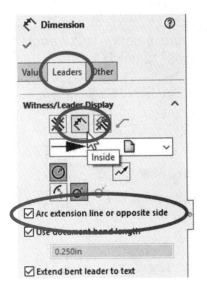

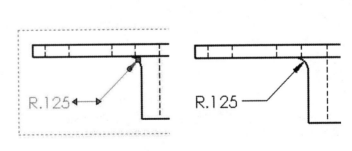

18.25. – When we start a new detail drawing, oftentimes we don't know for sure what scale to use in order to comfortably fit the necessary views in the sheet. In these instances, when the drawing sheet is too big or too small, we can change the sheet's and/or an individual view's scale. In this case, the best scale to fit the views in this sheet needs to be 1:1, which means the drawing is the exact size as the real part. To change or confirm the scale is correct, right-mouse-click *in the sheet* (not a drawing view), or in the sheet's name in the FeatureManager, and from the pop-up menu select "Properties."

 If the option you want is not visible in the pop-up menu, click at the bottom of the menu to expand the hidden options.

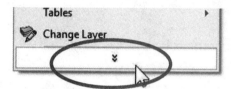

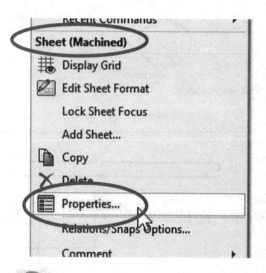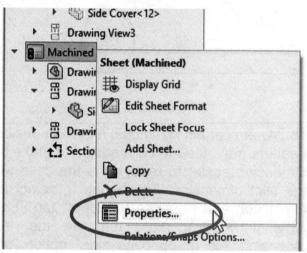

 In pop-up menus notice the commands are grouped under a heading; in this case the options for "*Sheet (Machined)*" are listed.

In the "Sheet Properties" dialog, set the scale to 1:1 and select "Third angle" projection to match the book images. Click "Apply Changes" to finish.

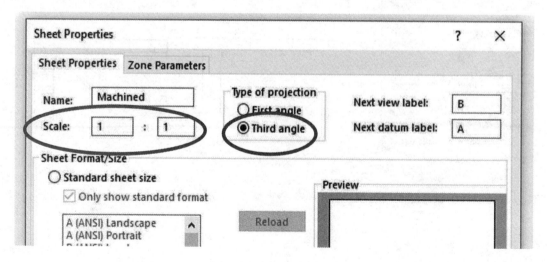

The "Type of projection" refers to how the views are projected from the 3D model. The "**First angle**" projection is widely used in Europe and the "**Third angle**" projection is more commonly used in America. This is a document property saved in the template. See the Appendix for more information in creating, modifying, and using templates.

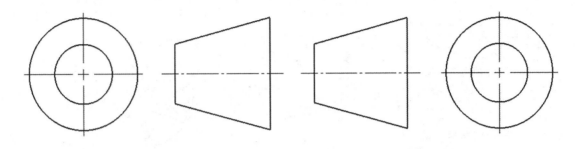

Third Angle Projection   First Angle Projection

18.26. - After arranging the views and dimensions, we notice the Isometric view is too big to fit in our sheet, so we need to change the isometric view's scale.

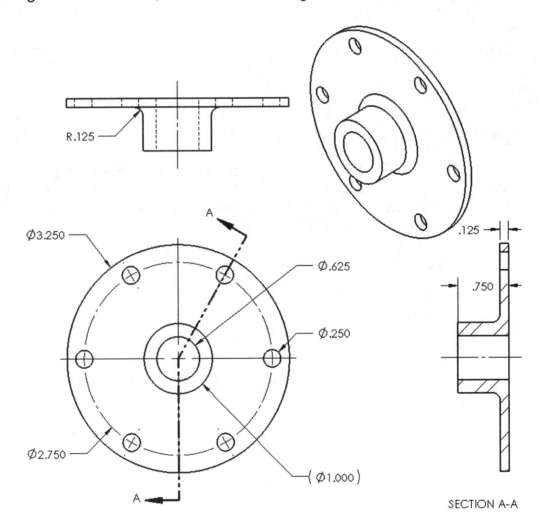

To change the Isometric's view scale, select it in the graphics area, and from its PropertyManager settings, change the view's scale to 1:2 from the "Use custom scale" drop down menu in the "Scale" options box. This means that the drawing view will be ½ the size of the real part.

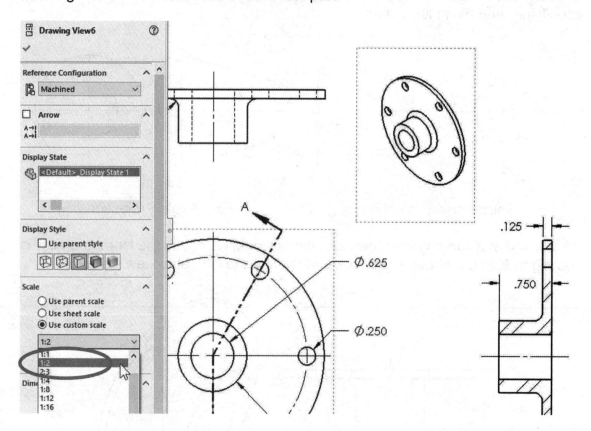

 If the option "**Tools, Options, Document Properties, Views, Other, Add view label on view creation**" is ON when the views are first added to the drawing, changing a view to a scale different than the sheet's scale, a note will be automatically added to the view indicating its scale.

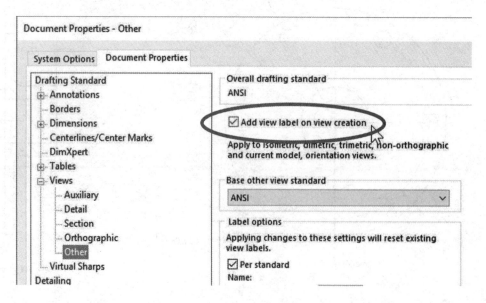

18.27. - The drawing is finished. Save it as '*Side Cover*' and close the file.

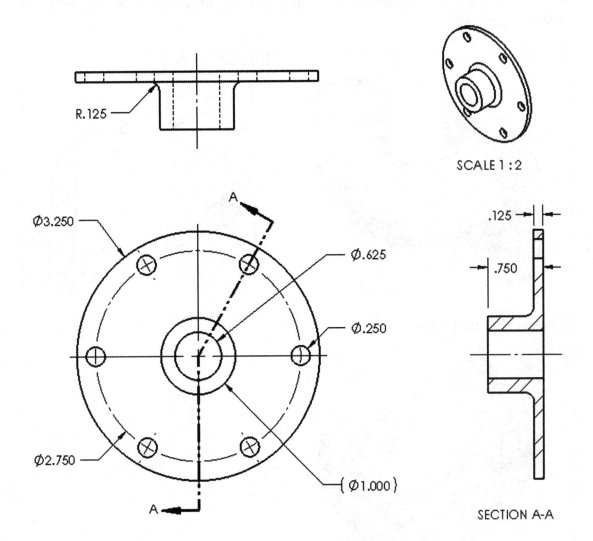

SCALE 1 : 2

SECTION A-A

Exercises: Make the detail drawings of the following Engine Project parts that were done in the Part Modeling section to match the drawing previously supplied to make each part.

Intake	Oil Pan Gasket
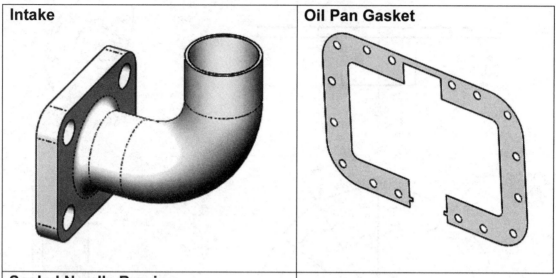	

Sealed Needle Bearing

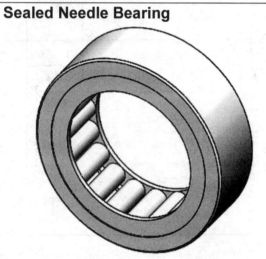

Add three configurations to show the different steps to model the part and match the original drawing suppressing features as needed in each configuration.

Drawing Three: The Top Cover

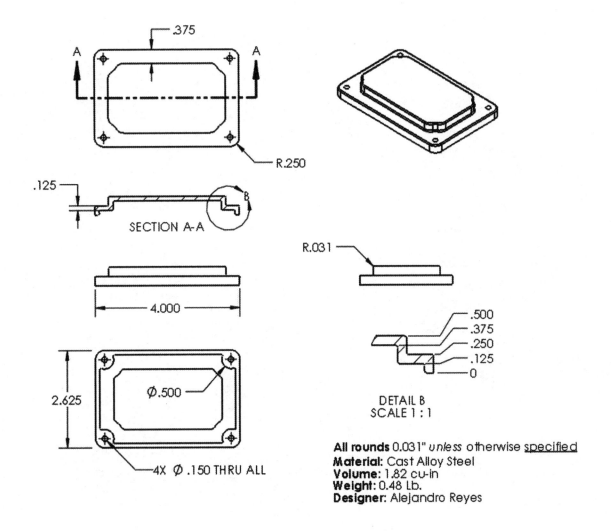

.375

A — A

R.250

.125

SECTION A-A

B

4.000

2.625

Ø.500

4X Ø .150 THRU ALL

R.031

.500
.375
.250
.125
0

DETAIL B
SCALE 1 : 1

All rounds 0.031" *unless* otherwise specified
Material: Cast Alloy Steel
Volume: 1.82 cu-in
Weight: 0.48 Lb.
Designer: Alejandro Reyes

Notes:

In this lesson we'll review previously covered material, add a section view using the "Slice section" option, we'll learn how to add Notes to our drawings, use ordinate dimensions, and how to add custom file properties to part and assembly files to be used in drawings and bills of materials. The drawing of the *'Top Cover'* will follow the next sequence.

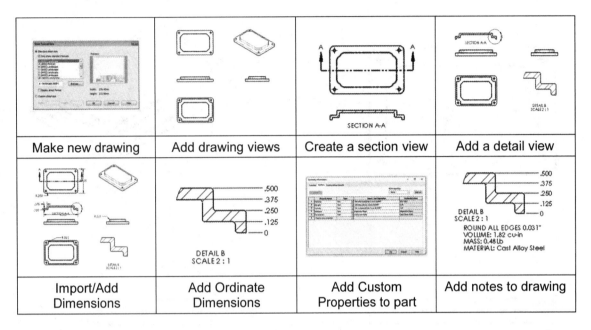

Make new drawing	Add drawing views	Create a section view	Add a detail view
Import/Add Dimensions	Add Ordinate Dimensions	Add Custom Properties to part	Add notes to drawing

19.1. - Open the *'Top Cover'* part and select the **"Make Drawing from Part/Assembly"** command as we've done before. Select the "A-Landscape" drawing size template without sheet format.

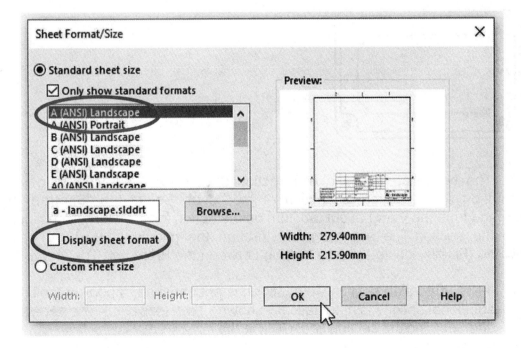

19.2. - The first step is to add the main views. Click-and-drag the Front view from the "**View Palette**" onto the sheet, and from it project Top, Bottom, Right and Isometric views. Change the display style to "Hidden Lines Removed" and "Tangent Edges Removed" for all views except the Isometric, which will have "Tangent Edges Visible."

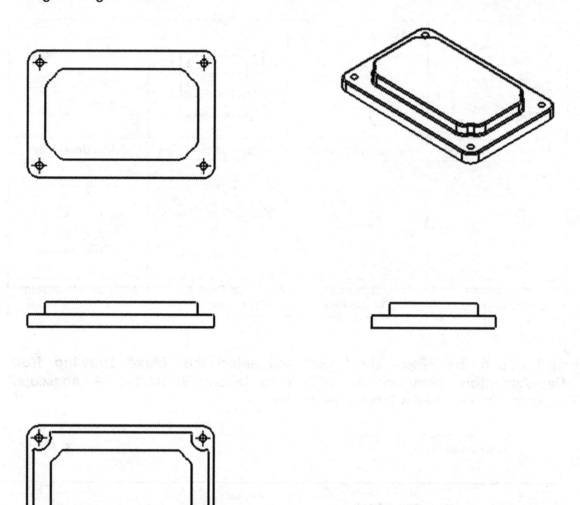

19.3. - The next step is to make a section through the Top view to get more information about the cross section of the cover. Select the "**Section View**" command from the View Layout tab, turn on the option for "Horizontal" section, and place the section line *approximately* through the middle of the Top view as indicated. Finally, locate the new section between the Front and Top views.

If the "Auto-start section view" option is not checked, a pop-up toolbar with additional options is automatically shown after locating the section line. Click in the green check mark to continue.

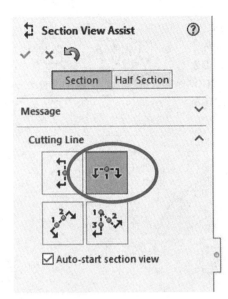
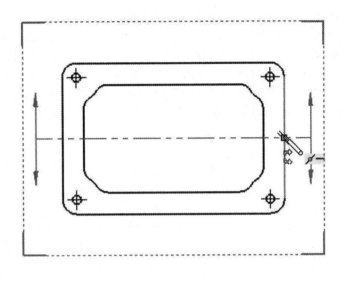

19.4. - After locating the section view turn ON the "**Slice Section**" option. By doing this the section view will only show the section's surface; using this option allows us to ignore the rest of the model behind the section line. To better illustrate this option, imagine we took a very thin slice of the model at the section line. Click OK when finished.

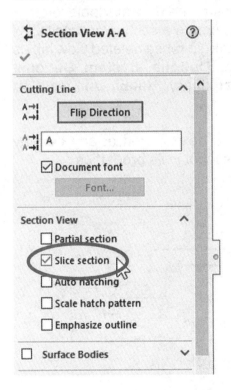
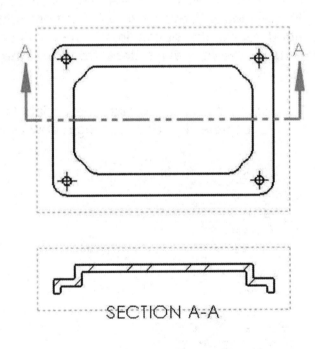

SECTION A-A

The direction of the section view can be reversed using the "Flip Direction" button or with a double click in the section line. If the section is reversed with the double click, the drawing needs to be rebuilt.

19.5. - To get a better view of the cross section, we'll need to make a detail of the right side of "Section A-A." Select the "**Detail View**" icon and draw a circle as shown; locate the detail view below the Right view and distribute the views evenly in the sheet. If center marks are automatically added, delete them for clarity.

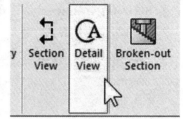

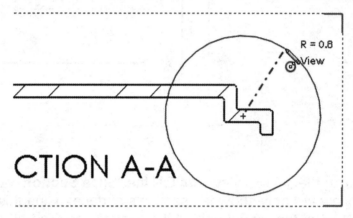

Every time a section or detail view is added, SOLIDWORKS increases the *new view label* value in the sheet's properties to the next available letter. If a view is deleted, say A-A, the next time we create a section or detail view it will be labeled B-B, even if view A-A is non-existent. To reuse deleted view labels we need to set the option in the menu "**Tools, Options, System Options, Drawings, Reuse view letters from deleted auxiliary, detail, and section views**."

Section, detail, or auxiliary view's labels can be changed at any time by selecting the view and entering a new view's label in its properties.

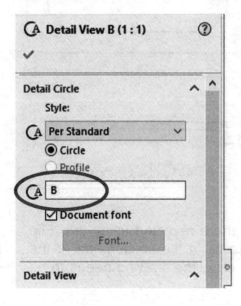

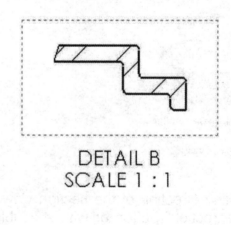

DETAIL B
SCALE 1 : 1

19.6. - Now that we have the needed views in the drawing, we need to import the model dimensions from the part. Go to the menu "**Insert, Model Items**" or the "**Model Items**" command in the Annotation tab of the CommandManager. Remember to select the options "Import items into all views" and "Entire Model," and include the "Hole Callout" dimensions. Make sure we are using English units with three decimal places.

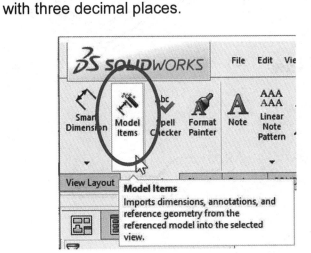

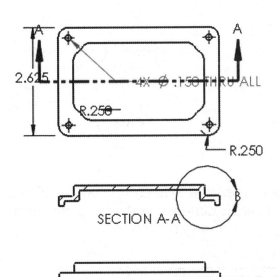

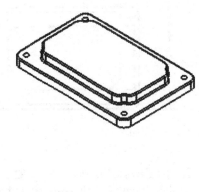

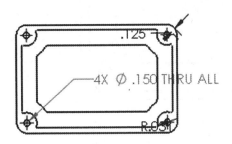

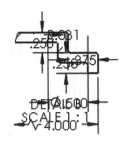

19.7. - Delete and arrange the dimensions as needed to match the next image.

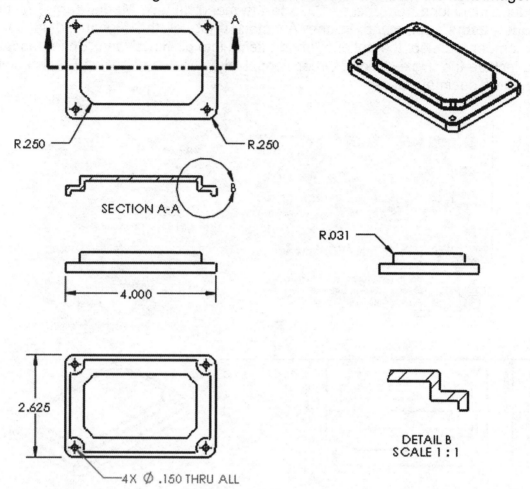

19.8. - In the bottom view, add a (reference) dimension to the round edge indicated using the "**Smart Dimension**" tool. When an arc is dimensioned a radial dimension is added.

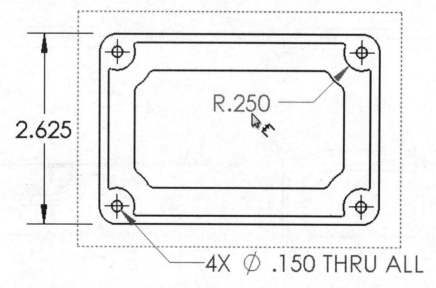

For our example we want to show a diameter dimension. To change the dimension, right-mouse-click on it, and from the pop-up toolbar select **"Display options, Display as Diameter**."

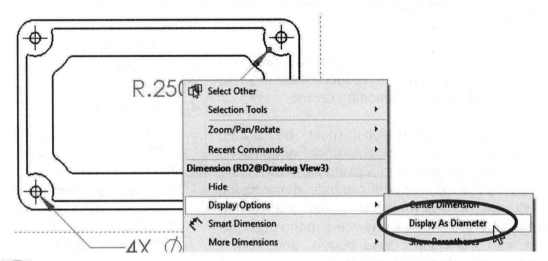

 To change the dimension to radial again repeat the previous step and select **"Display as Radius**."

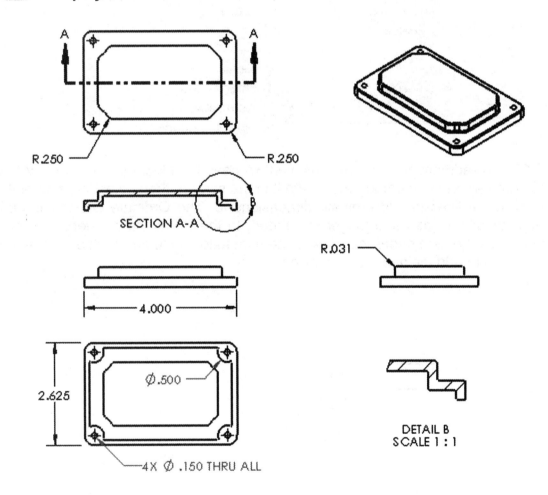

19.9. -"**Ordinate Dimensions**" are used to dimension multiple features from a common origin or datum, making drawings easier to read, they have a single leader line with the numerical value, and are automatically aligned to each other.

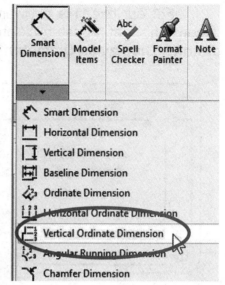

 Ordinate dimensions are favored and usually preferred to reduce detail drawing clutter and dimensioning errors.

Ordinate dimensions must be added manually, making them reference dimensions (unless they are added to a sketch and are being imported). To add vertical ordinate dimensions, select the drop-down-menu under the "Smart Dimension" tool, or click anywhere in the graphics area with the right mouse button, and from the pop-up menu select "**More Dimensions, Vertical Ordinate**."

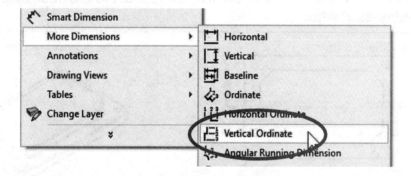

19.10. - To add ordinate dimensions, first we need to select the entity that will be the zero reference or datum and locate it in the screen, then select the rest of the edges or vertices to add ordinate dimensions to. The Ordinate Dimensions are automatically aligned and jogged if needed. In the Detail view, select the lower edge to be the zero reference, click to the right to locate it, and click on the rest of the edges to add the needed dimensions.

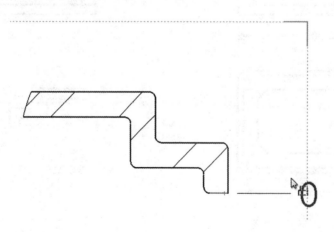

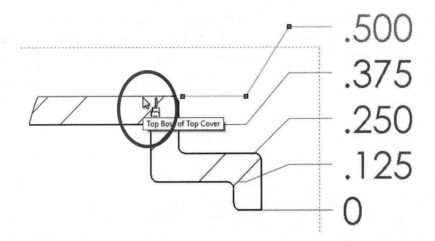

 If after adding ordinate dimensions a dimension is missed, we can right-mouse-click in any of the ordinate dimensions, select the option "**Add to Ordinate**" and click the missed entity to add the missing dimension.

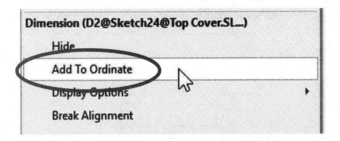

 By default, ordinate dimensions are automatically aligned to each other, but the alignment can be broken by selecting the "**Break Alignment**" command from the same right-mouse-click menu, and then moving the dimension.

19.11. – Repeat the previous step, but now use the "Horizontal Ordinate" dimension. Select the drop-down-menu under the "Smart Dimension" tool, or click anywhere in the graphics area with the right mouse button, and from the pop-up menu select "**More Dimensions, Horizontal Ordinate**."

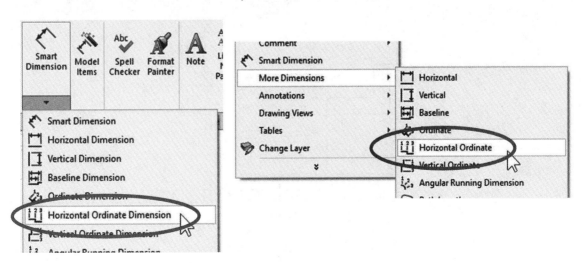

19.12. – Select the indicated edge to define the zero reference for the horizontal ordinate dimension, and then select the rest of the edges to dimension. After adding the dimensions, finish the command and move the Detail's information note for visibility.

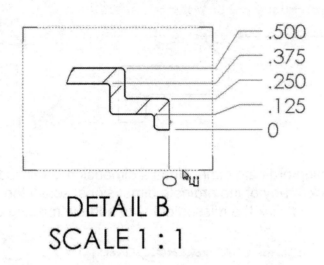

DETAIL B
SCALE 1 : 1

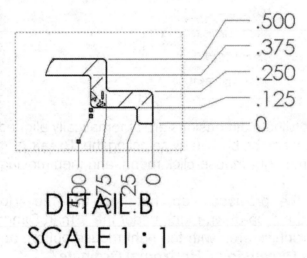

DETAIL B
SCALE 1 : 1

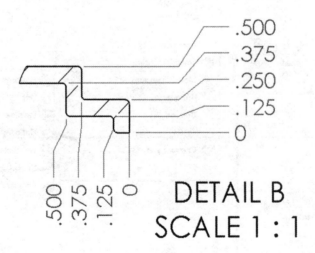

DETAIL B
SCALE 1 : 1

19.13. – After the dimensions have been added, we need to add notes to the drawing. Notes are used in drawings to communicate important information, such as the type of material, surface finish, dates, manufacturing information, company details, etc. To add notes to a drawing select the "**Note**" command from the Annotation tab or from the menu "**Insert, Annotations, Note**"...

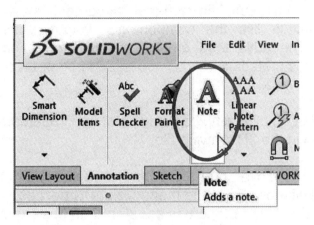

...and click below the detail view in the drawing to locate the note; the **Formatting toolbar** is automatically displayed next to the note (unless it's already visible). The text formatting toolbar works just like in other applications to change font, color, justification, and style.

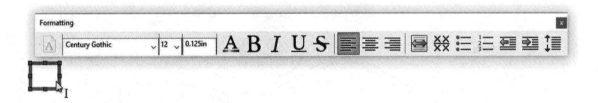

After the note is located in the drawing, we can type and format the text just as we would in a word processor. If multiple notes are needed in the drawing, click in the drawing to locate the next note. Feel free to experiment with different fonts and styles for this note. Click OK in the PropertyManager or press the "Esc" key when you are finished adding notes.

 To modify an existing note, double-click in it to activate the edit mode.

19.14. – Every Windows file can store additional file information known as metadata, which includes document or software specific information; for example, picture files include information such as picture height, width, color depth, camera information, etc. SOLIDWORKS uses "**Document Properties**" to store part, drawing or assembly specific information such as the designer's name, project, customer, part number, supplier, material, document version, etc. as well as parametric information like the component's weight, volume, surface area, dimensions, etc. The advantage of storing this information in the document properties is that this metadata can be accessed in other areas of the software to help us simplify our work.

To show this functionality we'll add Custom Properties to the 3D part file and then use them in the 2D drawing file.

Open the '*Top Cover*' part file (*Top Cover.sldprt*) or, if it's already open, make it the active window. From the menu "**File, Properties**" select the "Custom" tab; this is where document properties are stored. In this example we'll add three parametric properties linked to the part's Volume, Weight, and Material, and a user defined property with the designer's name.

 To open a part file directly from the drawing, click on the part in any drawing view and select the "**Open Part**" command from the pop-up menu.

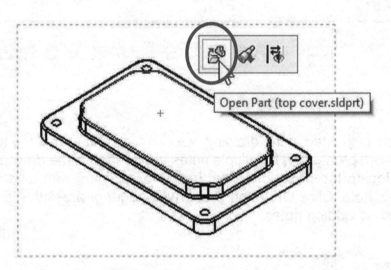

19.15. – In the "Property Name" column, click in the first cell and select "Material" from the drop down list; in the "Type" column "Text" is automatically selected; click in the "Value/Text Expression" cell, and select "Material" from the drop down list. SOLIDWORKS will automatically fill in the correct expression to make the property's value equal to "Cast Alloy Steel," which is the material we had previously assigned to this part.

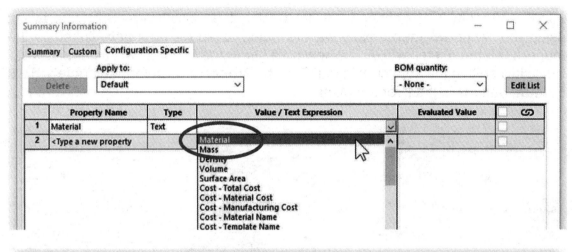

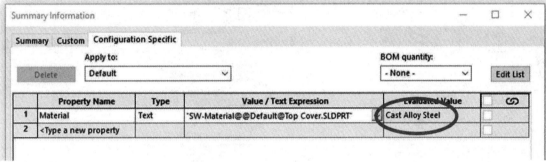

After adding a custom property, a new empty row appears. For the second custom property type the name "*Volume*" and select "Volume" from the value drop-down list; for the third property, select "Weight" for the property's name and "Mass" from the "Value" drop-down list. Finally, type "*MyName*" in the "Property Name" column and fill in your name for the value. If the "Evaluated Value" column is not updated, click OK to close the Custom Properties. After re-opening it all values are automatically calculated and filled in. When finished save the '*Top Cover*' part file.

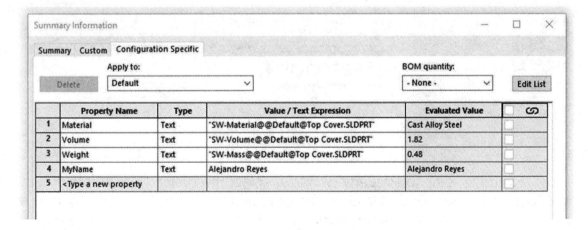

 Property names not listed can be typed in. If these values will be used for other documents, select the "Edit List" button to add them to the drop-down list of properties available for all components.

19.16. - After saving the changes, switch back to the '*Top Cover – Sheet1'* drawing using the "Window" menu. Now we are going to add a new note linked to the model's "Material" property. Add a new note below the first one and type "*Material:*" While still editing the note click in the "**Link to Property**" icon from the note's "Text Format" options in the PropertyManager.

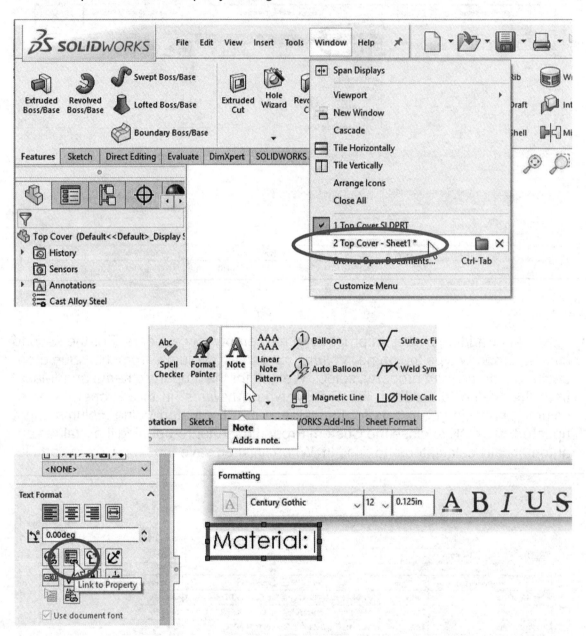

From the "Link to Property" window, select the option "Model found here" and "Drawing view specified in Sheet Properties" to use the Custom Properties of the model in the drawing sheet, then select "Material" from the "Property name:" drop down list. Notice the "Evaluated value:" field now shows "Cast Alloy Steel," which is the '*Top Cover*'s defined material. Click OK to add the Material's property value to the note and continue.

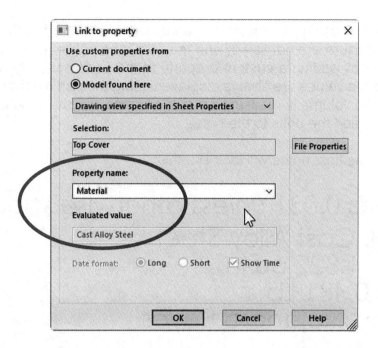

 Depending on how SOLIDWORKS is configured, you may see in the note "*$PRPSHEET: Material*"; this is the code used by SOLIDWORKS to read a part's properties. If you see this code while entering the note, it will be replaced by the component's property value after the note is finished.

Your note should now look like this:

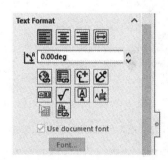 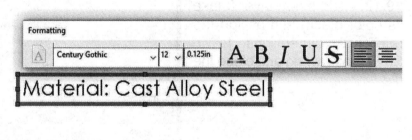

19.17. - The rest of the parametric notes can be added the same way. Press "Enter" to add a new row in the note; type "*Volume*:" and add a link to the "Volume" property's value; in the third row of the note type "*Weight*:" and add the link to the "Weight" property; finally, add a fourth row to the note, type "*Designer*:" and add a link to the "MyName" property.

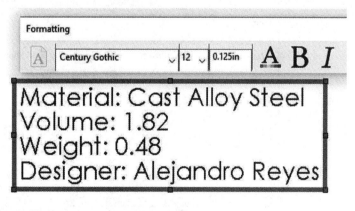

A drawback of using parametric notes that return a numeric value is that the units of measure are not listed, and they must be either added manually to the note or by adding a custom property to the part/assembly with the unit of measure, and the values are always displayed using the part's units of measure. In our example the volume is measured in cubic inches and the weight in pounds. We can manually add the units to the note.

After formatting, the final notes will look like this:

All rounds 0.031 *unless* otherwise <u>specified</u>
Material: Cast Alloy Steel
Volume: 1.822 cu-in
Weight: 0.481 Lb
Designer: Alejandro Reyes

When a model's (part or assembly) custom properties change, the drawing notes which use those properties will also be updated to reflect their new value. For example, if the 3D model changes in size, both its volume and weight will be different, and both the volume and mass custom property values will be updated accordingly in the detail drawing note.

19.18. - Add a new layer (make its color black) and assign the reference dimensions to it. Save the drawing and close the file.

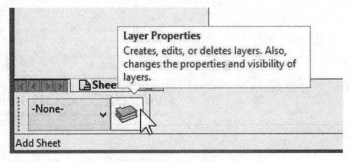

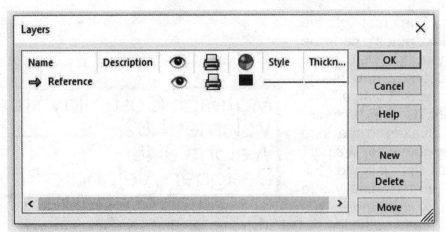

434

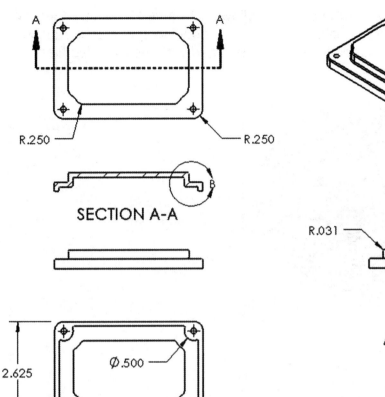

R.250

R.250

SECTION A-A

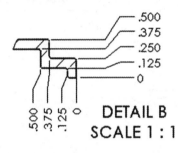

R.031

.500
.375
.250
.125
0

.500
.375
.125
0

DETAIL B
SCALE 1 : 1

2.625

Ø.500

4X Ø .150 THRU ALL

All rounds 0.031 *unless* otherwise specified
Material: Cast Alloy Steel
Volume: 1.822 cu-in
Weight: 0.481 Lb
Designer: Alejandro Reyes

435

Exercises: Make the detail drawing of the following Engine Project part that were done in the Part Modeling section to match the drawing previously supplied to make each part. High resolution images are included with the exercise files.

Cylinder Head

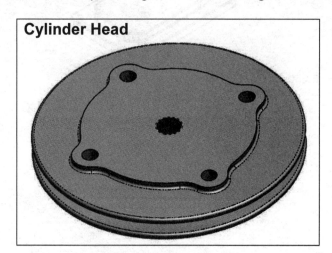

Drawing Four: The Offset Shaft

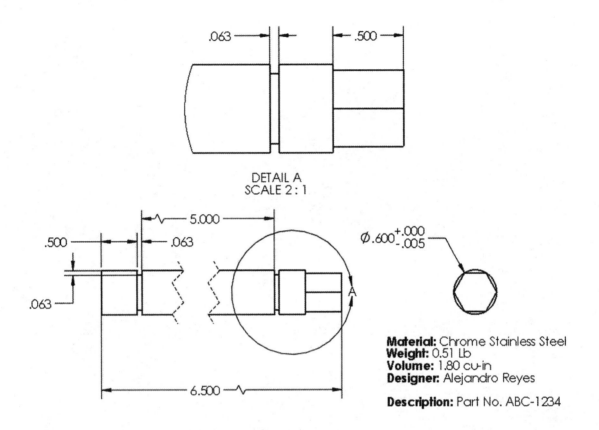

.063   .500

DETAIL A
SCALE 2 : 1

.500   .063   5.000

.063

6.500

$\varnothing.600^{+.000}_{-.005}$

Material: Chrome Stainless Steel
Weight: 0.51 Lb
Volume: 1.80 cu-in
Designer: Alejandro Reyes

Description: Part No. ABC-1234

Notes:

The *'Offset Shaft'* drawing, although a simple drawing, will help us reinforce previously covered commands including standard and detail views, importing dimensions, manipulating and modifying a dimension's appearance, notes and custom file properties, and we'll learn how to break a view with long objects to fit in the drawing sheet following the next sequence:

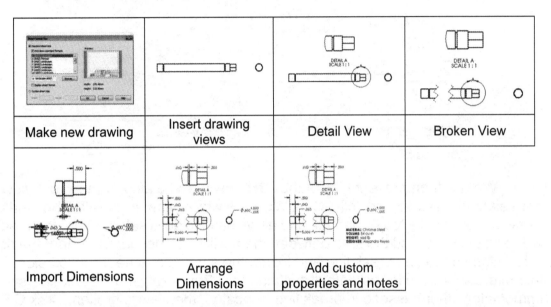

Make new drawing	Insert drawing views	Detail View	Broken View

Import Dimensions	Arrange Dimensions	Add custom properties and notes

20.1. - Since we have already done a few drawings, we'll ask you to make a new drawing using the "A-Landscape" sheet size and turn off the sheet format. In order to better fit the views in the sheet, right-mouse-click the sheet, select "Properties" and change the sheet's scale to 1:1. Add the Front view by dragging it from the View Palette and project the Right view. Finally, add a detail view of the right side of the shaft as shown, and if needed change the display mode to "Hidden Lines Removed." Set the drawing units to Inches with three decimal places.

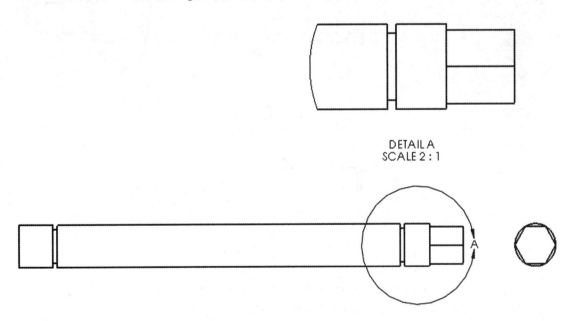

DETAIL A
SCALE 2 : 1

20.2. - For long slender elements like shafts, it's common practice to add a **Break** to shorten the view and save space in the drawing. To add a break in the Front view, select the "**Break View**" command from the View Layout tab in the CommandManager or the menu "**Insert, Drawing View, Break**."

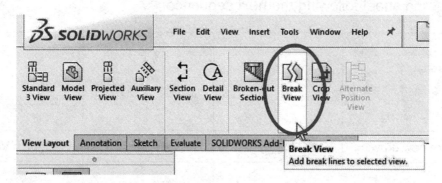

We are then presented with the "**Broken View**" dialog. If the Front view is not selected, click in it to tell SOLIDWORKS which view to break. The Vertical Break option is selected by default, and all we have to do is select where the breaks will be in the view, and the area between them will be removed. Click near the right side to locate the first break line and locate the second one near the left side. For this exercise we'll use the default settings for "Gap size" and "Break line style." Immediately after the second break line is located, the view is broken. Click OK to finish the command.

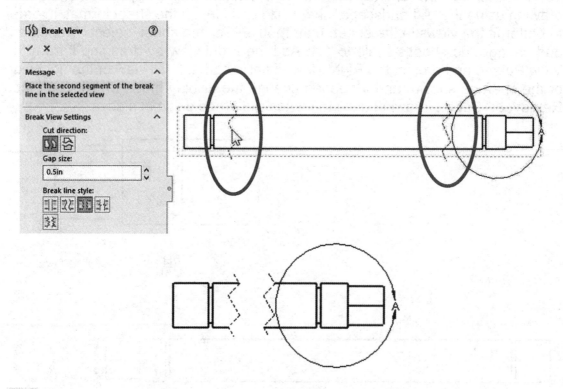

 To modify the broken view, click-and-drag the break lines. To delete the broken view (un-break) select one of the break lines and delete it.

20.3. -You probably know what we are going to do now... Yes, import the dimensions from the 3D model. Go to the menu "**Insert, Model Items**"; remember to select the "Import items into all views" and "Entire Model" options.

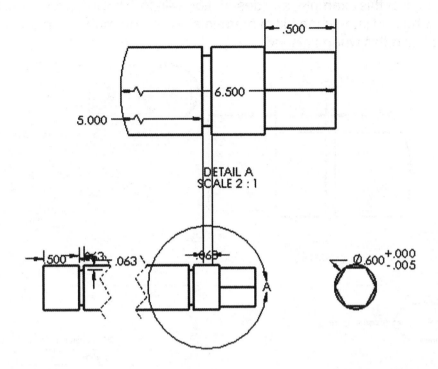

Arrange the imported dimensions as shown in the next image; remember to hold down the "Shift" key while moving dimensions from one view to another.

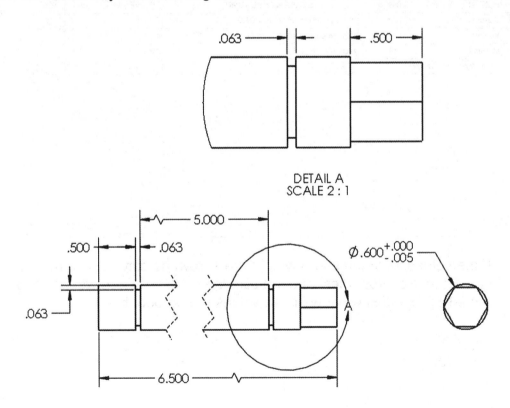

20.4. – Now we need to add custom properties to the *'Offset Shaft'*. Open the part and from the menu "**File, Properties**" add the same custom properties as we did in the *'Top Cover.'* Add a new property called "Description" and fill in a description for the shaft. For this example, any description will do. This information will be used in the drawing and later in the bill of materials. The "Summary Information" window should look like this when finished:

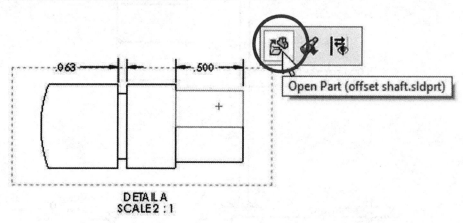

DETAIL A
SCALE 2 : 1

Open Part (offset shaft.sldprt)

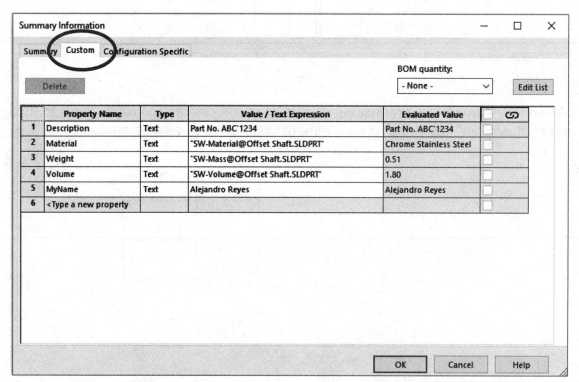

 If the values for volume and weight are shown as zeroes, you may have to close the file properties and rebuild the model. When you re-open the model's properties window, these values will be populated.

442

20.5. - Save the *'Offset Shaft'* part and go back to the drawing; add a note using the "**Note**" command from the Annotation tab, and type "*Material:*" "*Volume:*" "*Weight:*", "*Designer*" and "*Description*"; use the "Link to Property" command to add the corresponding property's value, and don't forget to add the units of measure.

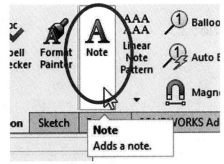

The finished note will be:

Material: Chrome Stainless Steel
Weight: 0.507 Lb.
Volume: 1.800 cu-in
Designer: Alejandro Reyes
Description: Part No. ABC-1234

20.6. - Save and close the drawing file.

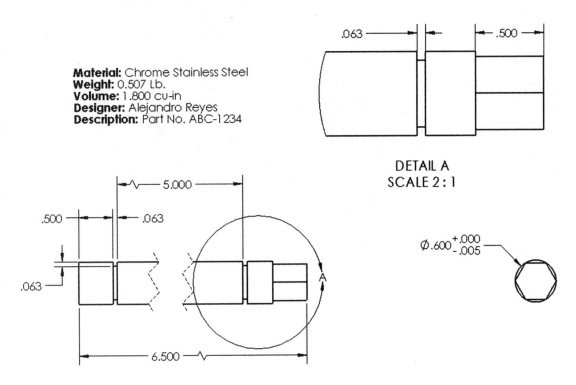

DETAIL A
SCALE 2 : 1

Notes:

Drawing Five: The Worm Gear

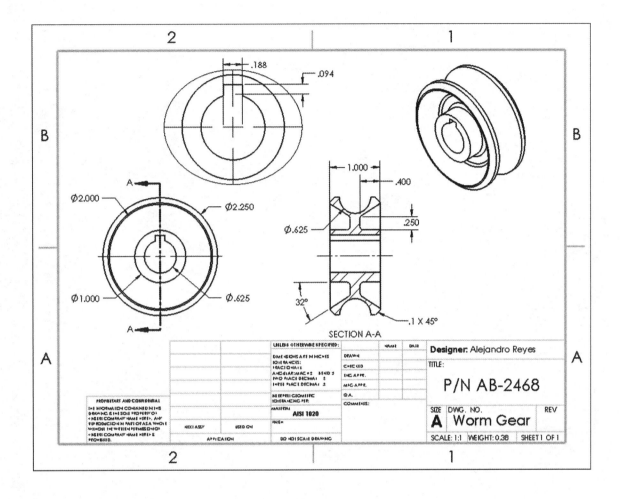

SECTION A-A

UNLESS OTHERWISE SPECIFIED:		NAME	DATE	**Designer:** Alejandro Reyes		
DIMENSIONS ARE IN INCHES TOLERANCES: FRACTIONAL± ANGULAR: MACH± BEND± TWO PLACE DECIMAL± THREE PLACE DECIMAL±	DRAWN			TITLE:		
	CHECKED					
	ENG APPR.					
	MFG APPR.			P/N AB-2468		
INTERPRET GEOMETRIC TOLERANCING PER:	Q.A.					
MATERIAL AISI 1020	COMMENTS:			SIZE **A**	DWG. NO. Worm Gear	REV
FINISH						
APPLICATION	DO NOT SCALE DRAWING			SCALE: 1:1	WEIGHT: 0.38	SHEET 1 OF 1

PROPRIETARY AND CONFIDENTIAL
THE INFORMATION CONTAINED IN THIS DRAWING IS THE SOLE PROPERTY OF <INSERT COMPANY NAME HERE>. ANY REPRODUCTION IN PART OR AS A WHOLE WITHOUT THE WRITTEN PERMISSION OF <INSERT COMPANY NAME HERE> IS PROHIBITED.

Notes:

In this lesson we'll review previously covered material and a couple of new commands like adding angular dimensions, changing a dimension's precision and adding chamfer dimensions as well as cropping a view and modifying the Sheet Format. The detail drawing of the *'Worm Gear'* will follow the next sequence.

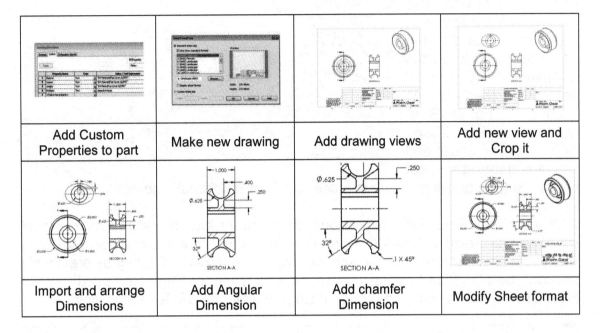

Add Custom Properties to part	Make new drawing	Add drawing views	Add new view and Crop it
Import and arrange Dimensions	Add Angular Dimension	Add chamfer Dimension	Modify Sheet format

21.1. - Just as we did in the previous drawing, we'll open the *'Worm Gear'* model and before we make the drawing, add the following custom properties. In the *'Worm Gear'* part select the menu "**File, Properties**" and complete it as shown in the next image. Save the part before continuing.

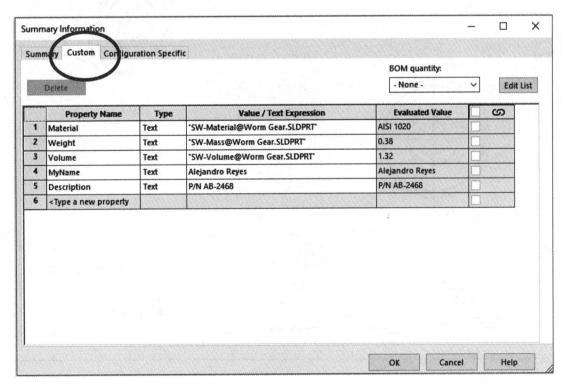

After adding the custom properties, select the "**Make Drawing from Part/Assembly**" icon. For the '*Worm Gear*' use the Drawing Template with an "A-Landscape" sheet size and leave the "Display Sheet Format" option checked in the "Sheet Format/Size" window. In this lesson we'll learn how to use and edit the drawing's sheet format. Set the drawing's units to inches and three decimal places.

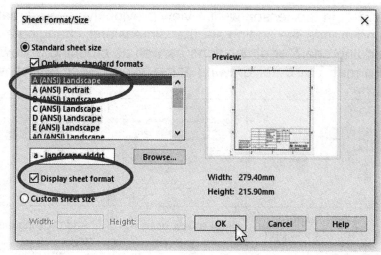

21.2. - Just as we've done in previous drawings, first we'll add the main model views to the drawing. Add the Front view from the "View Palette," an Isometric view, and the section view as shown. This is a review of material previously covered. Use "Hidden Line" display mode and "Tangent Edge Removed" for the Front and Section views and "Tangent Edge Visible" for the Isometric view.

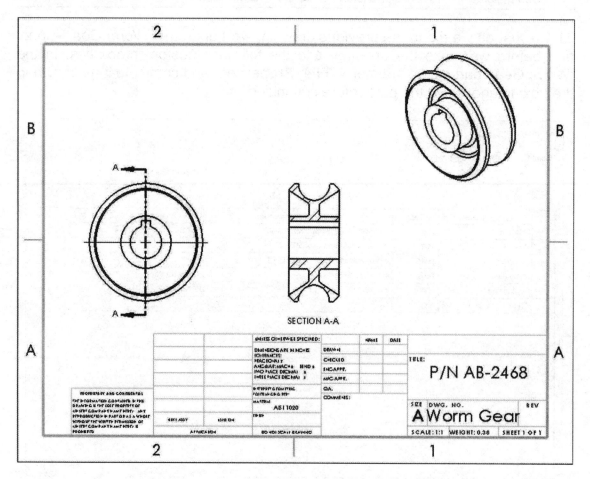

21.3. - To have a better view of the center of the part, we'll add a second Front view using the "**Model View**" command, and then we'll crop its center. The reason for not adding a Detail view is that adding another circle may make the Front view confusing.

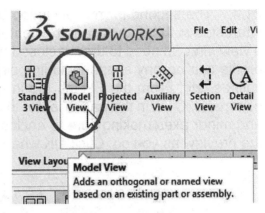

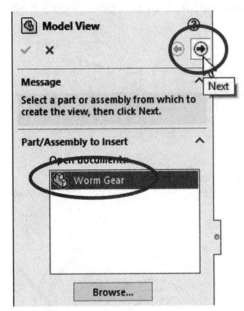

Use the "**Model View**" command from the View Layout tab, in the properties select the *'Worm Gear'* part from the list of currently open files (or Browse to select it) and click in the "Next" arrow at the top to go to the next step.

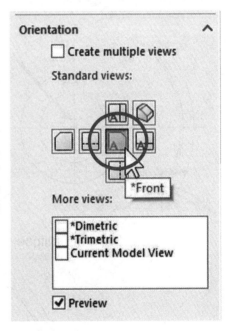

In the next screen select the Front view from the "Standard views," and leave the "Create multiple views" checkbox cleared, as we only need to add one view; to help us visualize the result turn on the "Preview" checkbox at the bottom.

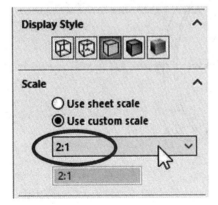

Before adding the view to the drawing, also change the view display style to "Hidden Lines Removed" and change the scale to 2:1 from the view's PropertyManager to make it two times larger.

Move the mouse pointer to the graphics area and locate the new Front view in the upper left corner. Click OK when done and change the new view's display to "Tangent Edges Removed."

21.4. - The second Front view is too big to fit in our page, and since we are only interested in the details near the center, we'll crop the view for a better fit. To crop a view, first we need to draw a closed profile using regular sketch tools. The closed profile can be any shape including circles, rectangles, polygons, ellipses, closed splines, etc. From the Sketch tab in the CommandManager select the **"Ellipse"** command. Click in the part's center to locate the ellipse, and then locate the major and minor axes making sure to enclose the center features of the part. You'll see the preview as you go. Click OK when finished.

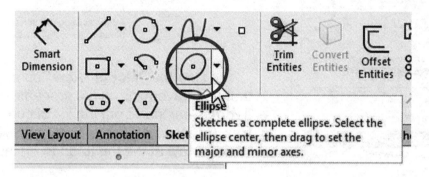

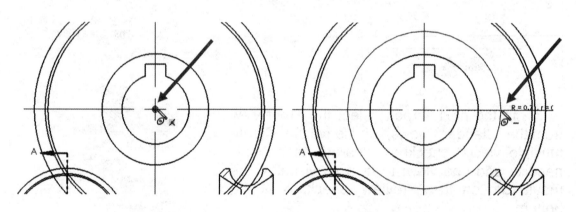

Start the ellipse   Locate major axis

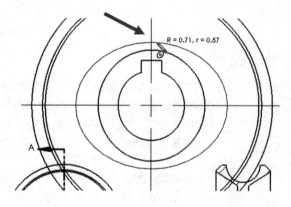

Locate minor axis to finish

21.5. – With the ellipse selected, click in the "**Crop View**" command from the View Layout tab. The view will be automatically cropped to the ellipse.

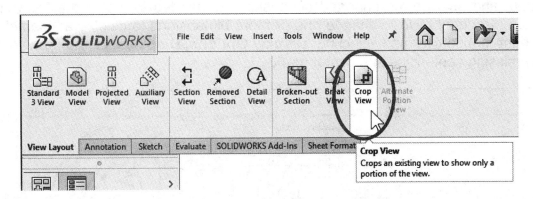

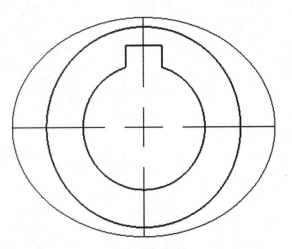

 To change or delete the crop from a view, select the cropped view in the FeatureManager (it will have a crop icon in it) or in the screen, click with the right mouse button and from the pop-up menu select "**Crop View, Edit Crop**" or "**Remove Crop**."

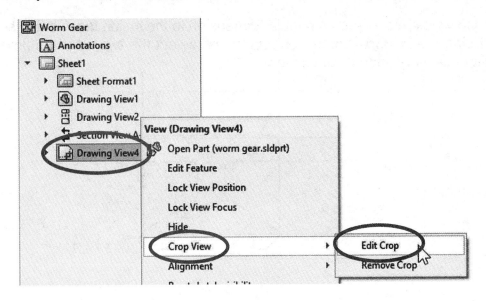

21.6. - After adding the main views, import the dimensions using "**Model Items**." Remember to select the options "Import items into all views" and "Entire Model." Delete and arrange the imported dimensions as needed to match the following image; you may have to change one or more dimensions to display as diameter by right-mouse-click and selecting "**Display Options, Display as Diameter**." Add any missing centerlines and center marks.

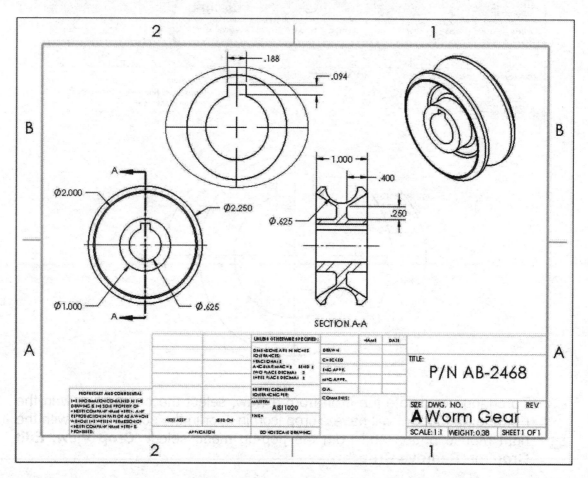

21.7. - Now we need to add an angular dimension to the inside faces in the section view. Using the "**Smart Dimension**" command select the two edges indicated to add the necessary angular dimension.

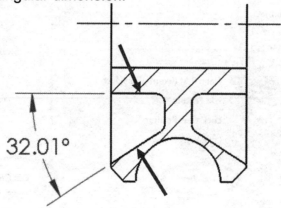

The dimension added uses the document's default precision of two decimal places, showing the dimension as 32.01°. In reality, a dimension like this should not be displayed using a precision of hundredths of a degree, since it is not needed or critical to the design, and more important is the fact that it would be extremely difficult to make; therefore, we need to change its precision.

To change the dimension's precision (Significant decimal numbers) select the dimension, move the mouse over the **"Dimension Palette"** and change the number of decimal places to "None" from the precision's pull-down menu. Click anywhere outside the Dimension Palette or hit the "Esc" key to finish.

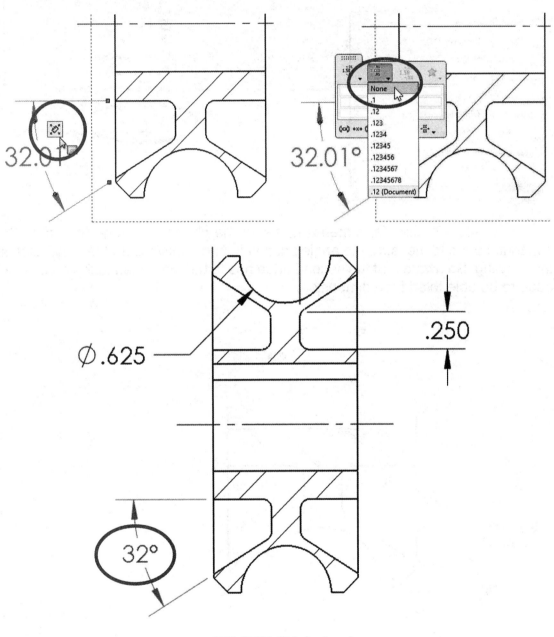

SECTION A-A

21.8. - The next thing we need to do is to dimension the chamfered edge. To add a "**Chamfer Dimension**" either click in the graphics area with the right mouse button, and from the pop-up menu select "**More Dimensions, Chamfer,**" or from the menu "**Tools, Dimensions, Chamfer**" or from the "**Smart Dimension**" drop-down icon.

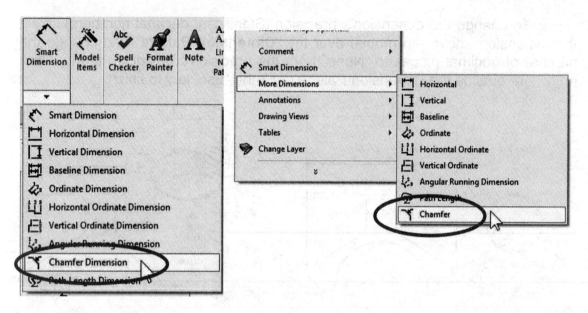

To add the chamfer dimension, select the chamfered edge first, then the horizontal edge to measure the angle against it, and finally locate the dimension in the drawing. Be aware that the second edge has to be connected to the chamfered edge to be able to add the dimension.

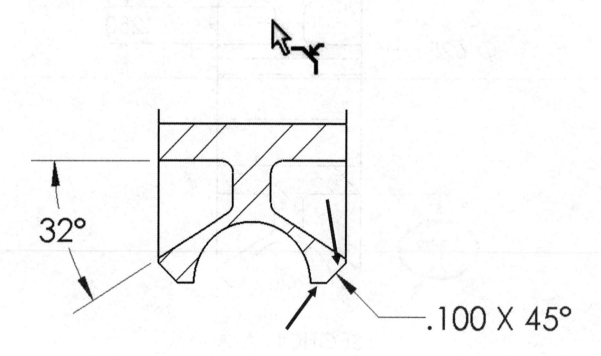

21.9. - The chamfer feature added to the model was intended to eliminate the sharp edges; therefore, it is a feature which doesn't require high precision, therefore we can change the dimension's precision. Select the chamfer dimension, and from the dimension's properties change the **"Tolerance/Precision"** to ".1" to change the dimension to show one decimal place. In the **"2ⁿᵈ Tolerance/Precision"** select "None" to remove the decimal places from the angle dimension.

 We have to make this change in the PropertyManager, because the **"Dimension Palette"** will change both values to the same precision and not individually.

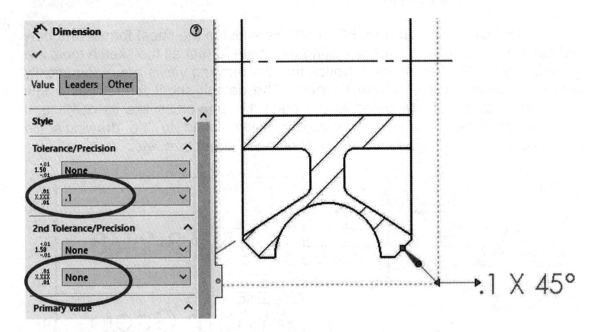

21.10. – When we started this drawing, we chose to include the title block in the drawing's template. The title block is the area of the drawing where important information is documented including company name, part number, material, fabrication and assembly notes, revisions, etc.

In SOLIDWORKS this is called the **"Sheet Format."** One characteristic of the sheet format is that it is "locked," and we cannot change it while we are working in the drawing views. Think of it as if our drawing views are in a transparent overlay on top of the Sheet Format. To change the title block, we need to activate it first. Right-mouse-click in the drawing area *or* in *"Sheet Format1"* in the Feature Manager, and from the pop-up menu select **"Edit Sheet Format."**

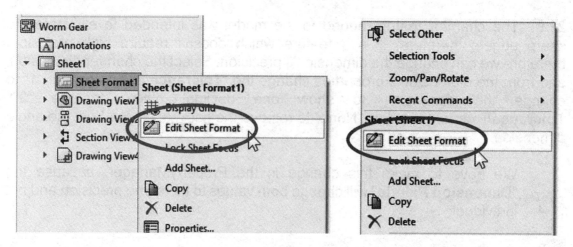

After selecting **"Edit Sheet Format"** we activate the sheet format and now we can make changes to it; while editing the sheet format all the sketch tools are available to modify it as needed. Notice that the drawing views are automatically hidden when editing the **"Sheet Format."** The default sheet format has a few parametric notes that are linked to the part's file properties like its name and material, and there are other parametric notes linked to the drawing's file properties, like the drawing's scale, sheet number, sheet size, etc.

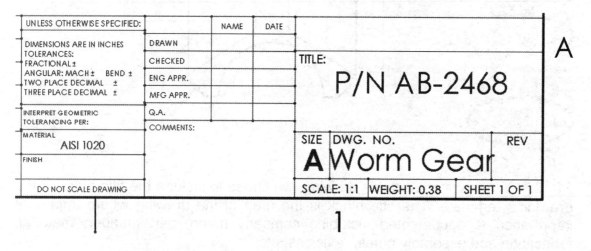

21.11. - Edit the notes by double-clicking them to change their font, size and style as needed, and add a new note linked to your name. Note that the "Material," "Weight," "Description" (*Title:*) and component name are automatically filled in from the 3D model's properties, and the scale and sheet number are filled in from the 2D drawing's properties (*this* document). These parametric notes are already included in the default sheet format. Format the notes to fit correctly in the spaces provided in the Sheet Format.

 Notes, like dimensions and other drawing elements, can also be assigned to a different layer to change their color.

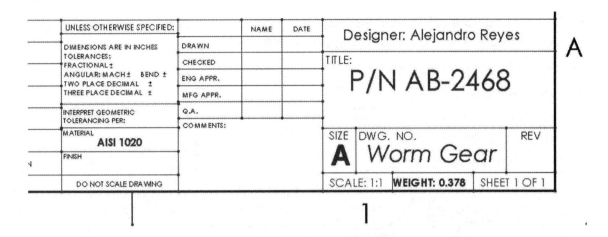

After we finish modifying the title block, exit the sheet format and return to editing the drawing sheet by selecting "**Edit Sheet**" from the right mouse button menu in the drawing area or clicking in the "**Sheet Format**" confirmation corner.

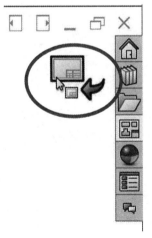

 If we modify an empty drawing's Sheet Format and save it as a template, the title block changes will be saved with the template and be available for new drawings based on this template. See the Appendix for more information on creating new templates and modifying existing ones.

21.12. - Change the reference dimensions (grey) to the "Format" layer to change their color to black. Save the drawing and close the file.

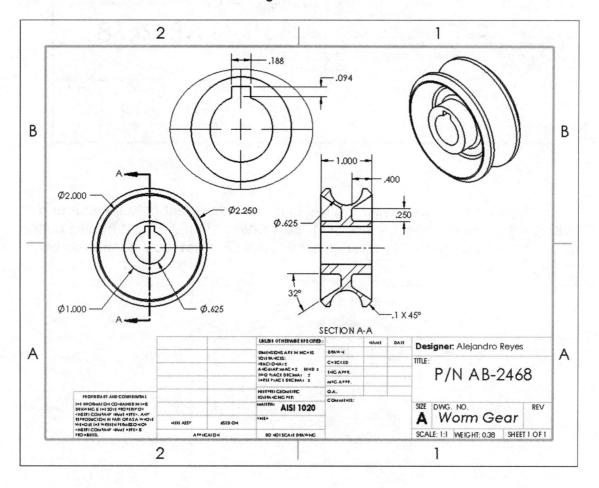

Exercises: Create a detail drawing of the following Engine Project parts made in the Part Modeling section and match the previously supplied drawings to make each part. High resolution images are included with the book's files.

Connecting Rod	Crankshaft
Con Rod Crankshaft Half Bushing	Bushing Top Con Rod
Oil Seal	Pin ConRod-Piston

Top Compression Ring	Middle Compression Ring
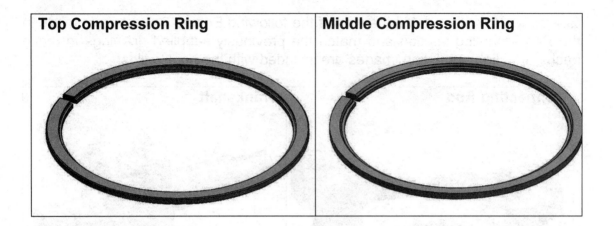	

TIP: Using the Section View command's default behavior makes a section the full width of the view; since we only need a section of one side of the ring to show its profile, after adding the Top view, draw a line using the sketch Line tool as indicated, and with the line selected, select the "Section View" command. By using a line that doesn't cut through the entire model the option "Partial Section" should be ON by default; additionally turn on the "Slice section" option to show only the section's surface.

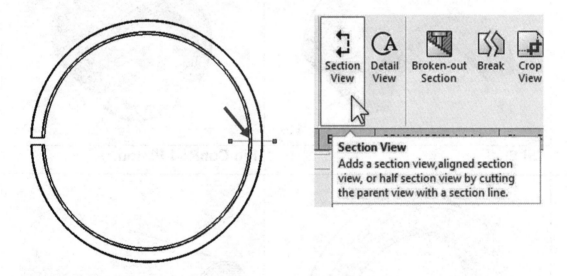

After making the section view it will be aligned to the Top view and have the same scale. To get a bigger view of the compression ring's profile, right-mouse-click in it and select "Alignment, Break Alignment" to *un-lock* the view's alignment and allow us to reposition it. Locate the section view centered above the Top view and change its scale to 8:1; continue adding dimensions and annotations as needed to complete the drawing.

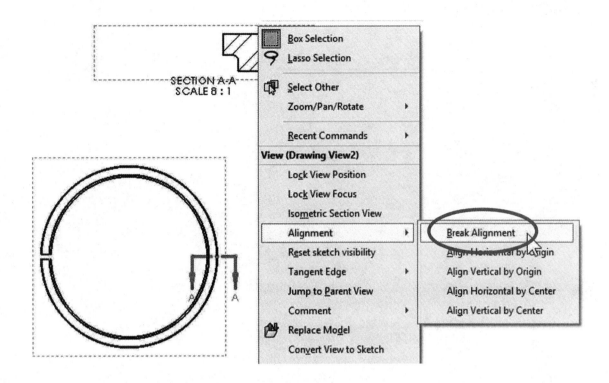

SECTION A-A
SCALE 8 : 1

▣	**B**ox Selection
ᓂ	**L**asso Selection
▦	**S**elect Other
	Zoom/Pan/Rotate ▸
	Recent Commands ▸
View (Drawing View2)	
	Lo**c**k View Position
	Lock View Focus
	Iso**m**etric Section View
	Alignment ▸
	Reset sketch visibility
	Tangent Edge ▸
	Jump to **P**arent View
	Comment ▸
	Replace Mo**d**el
	Con**v**ert View to Sketch

Break Alignment
Align Horizontal by **O**rigin
Align **V**ertical by Origin
Align Horizontal by Center
Align Vertical by Center

SECTION A-A
SCALE 8 : 1

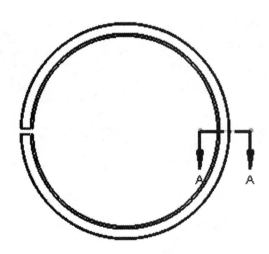

Notes:

Drawing Six: The Worm Gear Shaft

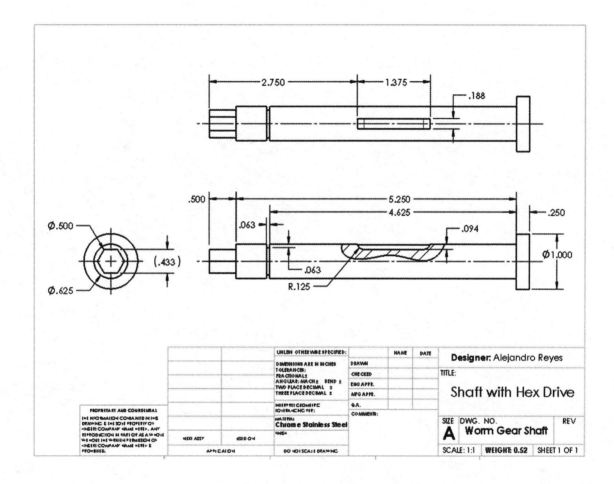

Notes:

Just as we did with the *'Offset Shaft'* drawing, we'll reinforce how to create new drawings, add views, import dimensions, move and format annotations, etc. We'll also learn how to use the Property Tab Builder to quickly and efficiently add custom properties, and a new type of view called "**Broken-Out Section**," used to look inside an area of the model without making a section view.

22.1. - The first thing we are going to do is to add several custom properties to the *'Worm Gear Shaft'* 3D model, including Material, Weight, Description ("*Shaft with hex drive*") and "MyName" as before.

A quick way to add custom properties to a model is by using the "Custom Properties Tab" located towards the bottom of the Task Pane. When used for the first time we need to create a new template with the properties we want to add to a model, drawing, or assembly file, but once we have made our custom templates, adding properties to a file is very quick.

Select the "Custom Properties Tab." To create a new template, press the "Create Now..." button to launch the "Property Tab Builder" program separately from SOLIDWORKS.

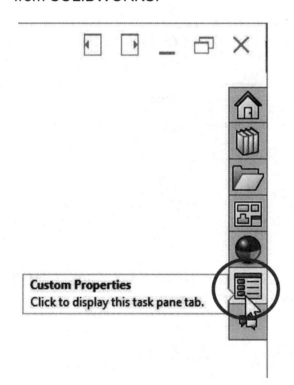

The Property Tab Builder is a stand-alone program to create forms containing the custom properties we want to add to our models using a drag-and-drop interface. In the "Control Attributes" section (Right side) we can define if this template is for a part, a drawing, an assembly, or a weldment. In this example we are going to make a Part template.

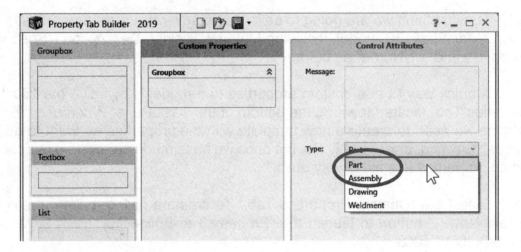

22.2.- The first thing we need to do is to select the "Groupbox" in the form and change its caption to "Exercise Properties" in the Control Attributes pane.

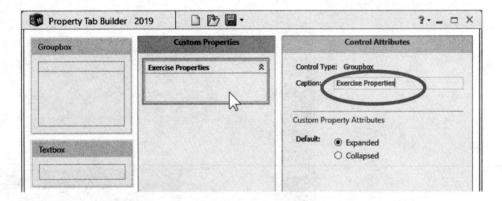

22.3.- The next step is to add a "Textbox" to the form by dragging it from the left side pane into the group box and then change its properties.

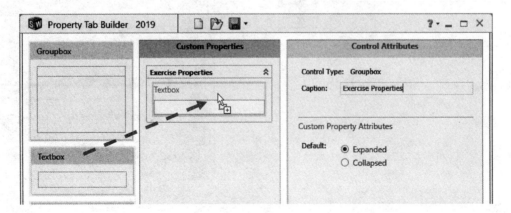

Type "Material" in the Caption to know what the property is. From the Name drop-down list select "Material"; this will be the name of the property. The Type of property is Text, and for the Value of the property select "[SW-Material]" from the drop-down list as we did when adding the property in a part file.

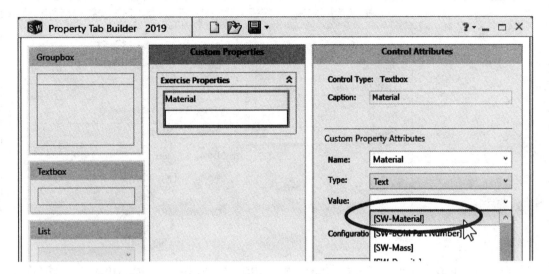

 One important detail to keep in mind is that Custom Properties can be configuration specific or model specific. At the bottom of the Control Attributes pane select the "Show on

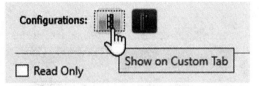

Custom Tab" option for all properties in this example, otherwise the properties will be added to the "Configuration Specific" tab and to use them later we would have to remember to call the configuration properties instead.

22.4.- Drag a second Textbox into the form; the other textbox will move to make room for it. Drop it right under the Material textbox at the bottom and fill in the control attributes.

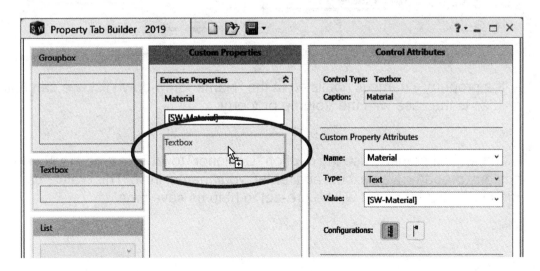

Type "Weight" in the Caption, select "Weight" from the Name drop-down list and "[SW-Mass]" from the Value drop-down list.

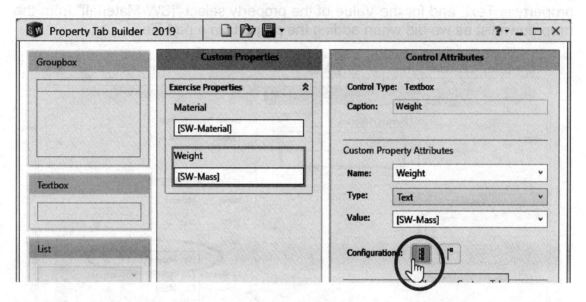

22.5. – Add three more text boxes for the "Volume," "Designer," ("*MyName*" property name) and "Description" as shown next.

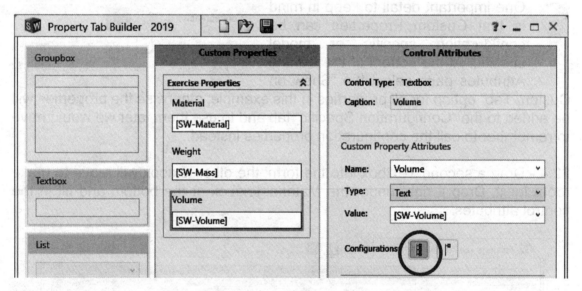

 If a property's name is not listed in the "Name" drop-down list, we can enter it, as is the case with the Volume property.

In the next property use the caption "Designer" to label it, type "MyName" in the custom property Name box and enter your name in it. This value can be changed later, but for now it will be pre-set to help us save time.

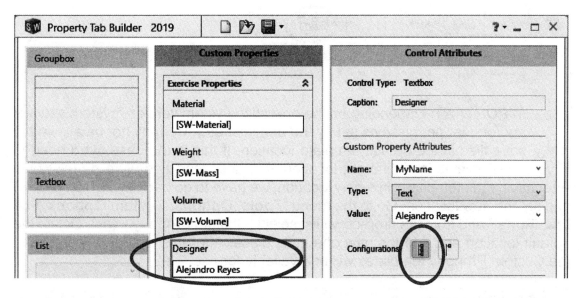

For the last property leave the Value field empty; this value will be entered when the custom properties are added to the part files later.

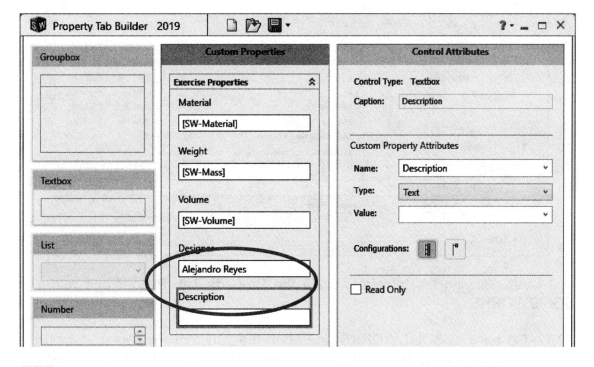

Other controls that can be added to a Property Tab template include drop-down lists (with typed items, linked to text files, excel spreadsheets, or databases), numbers using spin boxes, checkboxes and radio buttons for multiple selection items.

22.6. – After completing the template, we need to save it. Press the "**Save**" command and save it in the default custom property folder:

C:\Program Files\SOLIDWORKS Corp\SOLIDWORKS\lang\english

IMPORTANT: Depending on the operating system version, system settings and/or user permissions set by the administrator, we may not be allowed to save files in the default template location. If this is the case, we'll need to save our templates in a different location with user read and write permissions. After saving the templates in a new location, we have to go back to SOLIDWORKS and set the new location. Go to the menu "**Tools, Options, System Options, File Locations**," and from the drop-down list select "Custom Property Files." Delete the default location and add the new one in its place. After this change is completed the Custom Property templates will be available for use.

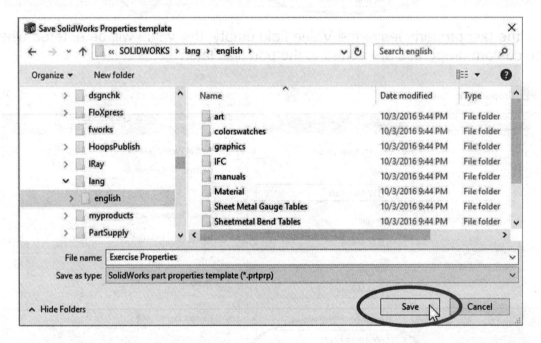

After saving the template, close the Property Tab Builder and return to SOLIDWORKS.

22.7. - Go back to SOLIDWORKS and open the "Custom Properties" tab. If the template we just created is not visible, click inside the tab and press "F5" to refresh and load it. The properties added to the template are ready for us to apply to the part. The "Material" property is already filled from the assigned part's material, "Weight" and "Volume" are automatically calculated and filled, "Designer" was preset with our name when we made the template (but can be changed if needed), and the only thing left to do is to type a description, which was intentionally left empty to enter a different value for each part. Type a description for the '*Worm Gear Shaft*'; even though the properties are already filled in the template, they have not yet been added to the 3D model. After pressing "Apply" the properties are added to the 3D model. Save the part to continue.

Pre-filled template   Completed template

22.8. – Now that we have learned how to create a custom properties template and added them to the '*Worm Gear Shaft*', make a new drawing using the "A-Landscape" drawing template including the sheet format, in inches and three decimal places.

Add the Front view from the View Palette, click above it to add a Top view, and to the left to add the Left view using "Hidden Lines Removed" display style for all 3 views. From the sheet's properties change the sheet scale to 1:1 and edit the Sheet Format to add the missing custom property notes and change their formatting to fit in the spaces provided.

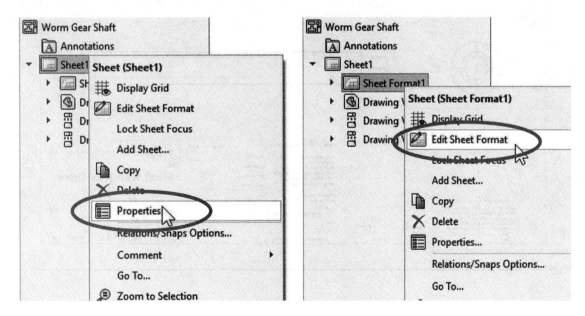

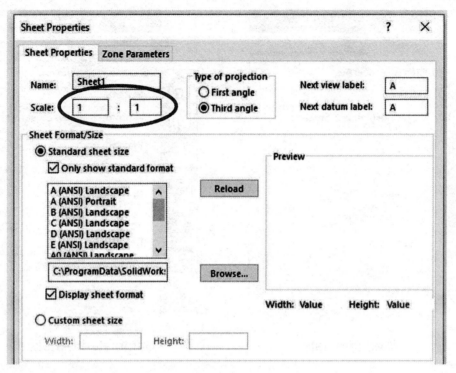

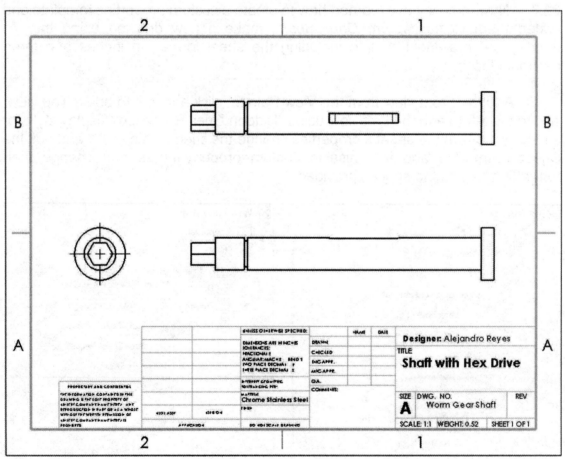

22.9. – Import the dimensions from the model…

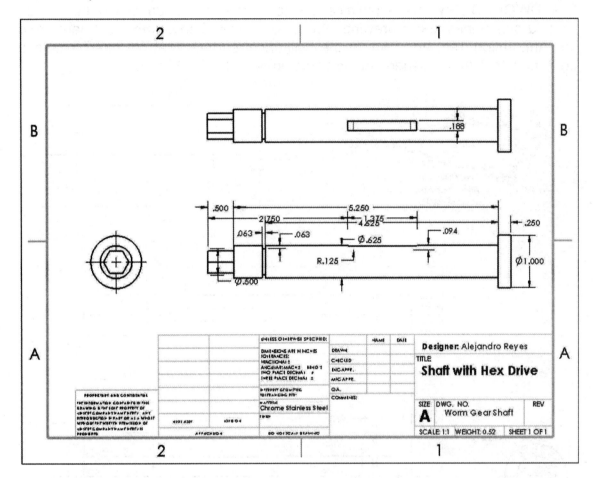

22.10. – And arrange them.

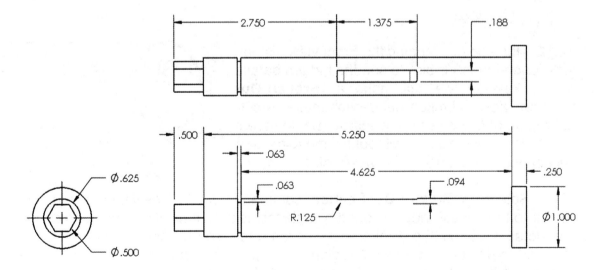

22.11. - One more thing we need to add to the drawing are the centerlines. SOLIDWORKS allows us to add a centerline to every cylindrical face of the model. To add centerlines to our drawing views, select the "**Centerline**" command from the Annotation Tab, and click in the cylindrical surface that we need to add a centerline to. When finished adding centerlines click OK to finish.

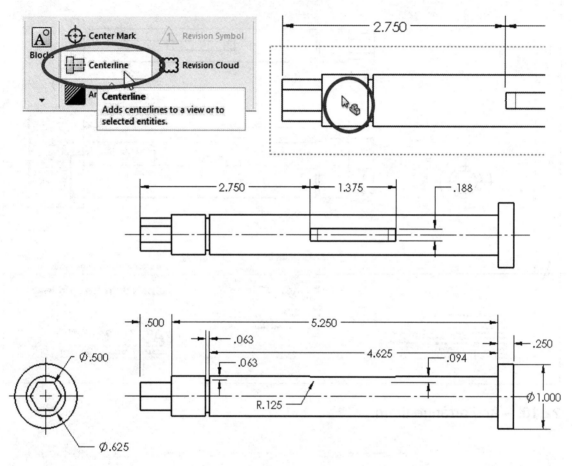

22.12. - Instead of changing the Front view display to "Hidden Lines Visible" to see the hidden details of the keyway, we will make a **Broken-Out Section**" view. A broken-out section makes a cut-out region of the view to a specific depth to reveal details otherwise hidden, without having to make a section view or changing the view's display style.

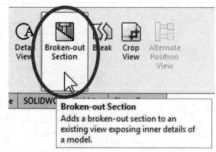

Select the "**Broken-Out Section**" command from the "View Layout" tab, and similar to the "Detail" view, where the circle tool is automatically selected to define the detail area, in the Broken-out-Section the "**Spline**" tool is automatically selected to define the break out region. The Spline is a smooth polynomial curve connected by multiple points. To draw a "**Spline**" add points approximately as shown around the keyway area in the Front view to surround it. Splines can be open curves, but in this case, we need a closed spline; to close the spline the last point has to be made coincident to the first one.

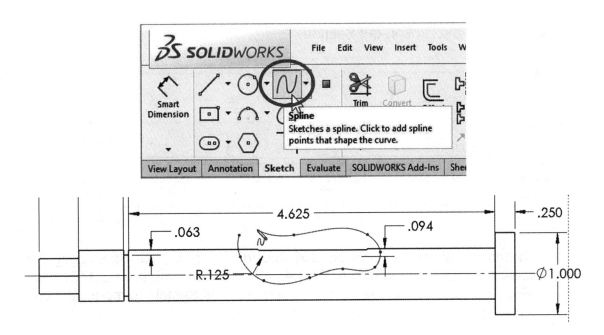

If we want to define the region for the Broken-Out Section using a different tool, such as lines, arcs, or ellipses, the only condition we have to meet is the profile has to be a closed profile and needs to be pre-selected before activating the "**Broken-Out Section**" command.

22.13. - When the last point of the **Spline** is added, we get the "**Broken-Out Section**" properties, where we are asked to enter a depth to make the cut or select an edge of the model to define the depth. Selecting the vertical edge indicated in the Top view to define the depth of the cut will cause the depth to be set at the edge's midpoint, making the cut go up to the center of the part. Activate the "Preview" option to see the resulting section.

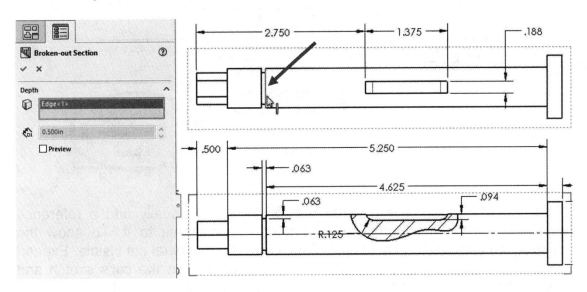

Click OK to finish the Broken-out Section.

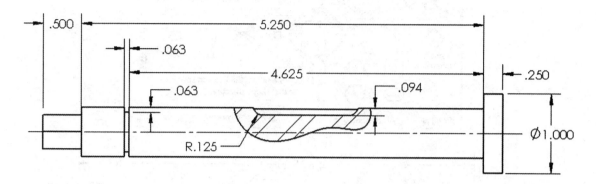

 To edit or delete the "**Broken-Out Section**," right-mouse-click <u>inside</u> the section's region and from the "**Broken-Out Section**" menu select "Delete," "Edit Definition" to change the cut's depth, or "Edit Sketch" to modify the cut-out area.

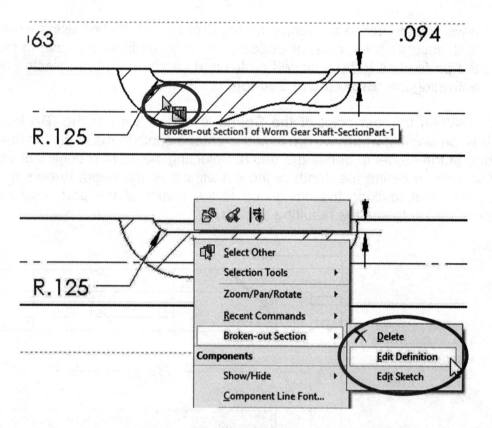

22.14. - In this drawing, it may be a good idea to manually add a reference dimension for the hexagonal cut and add a parenthesis to it. To show the construction circle we can make the sketch for the hexagonal cut visible. Expand the Left View in the FeatureManager, right-mouse-click in the cut's sketch and select "Show" from the pop-up menu. Save the drawing and close the file.

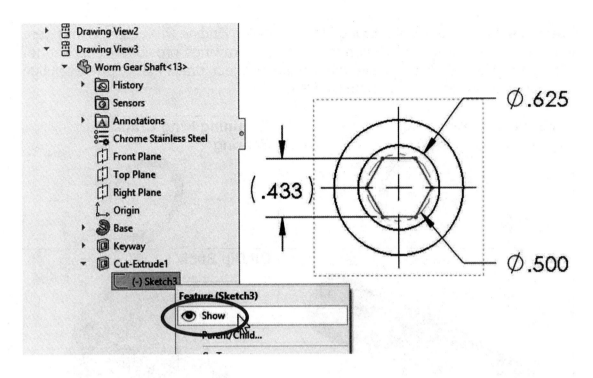

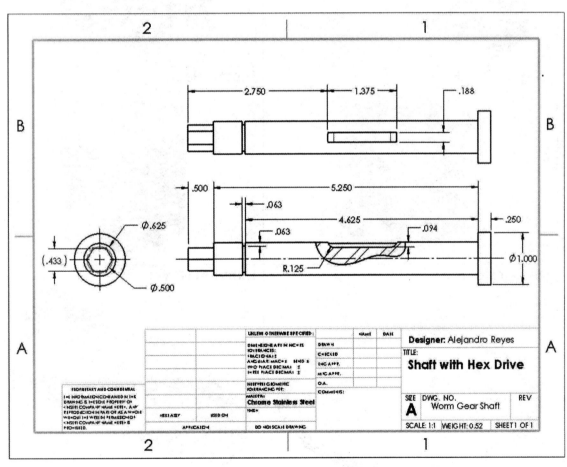

Exercises: Make the detail drawing of the following Engine Project parts that were done in the Part Modeling section to match the drawings previously supplied for each part. These will complete the Engine Project drawings. High resolution images are included with the exercise files.

Internal Retaining Ring	**Retaining Ring Crankshaft Bearing**
Muffler	**Oil Dip Stick**
Engine Block	**Piston Head**
Oil Control Ring	**Exhaust Cover**

Oil Pan

Crank Case Top

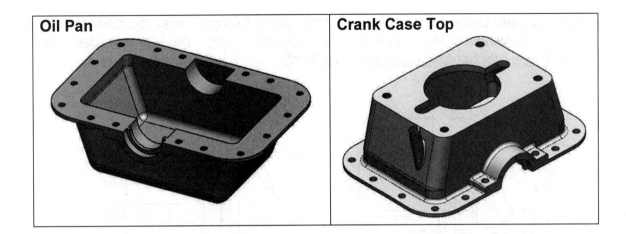

Extra Credit: Using the parts made for the gas grill project make the corresponding detail drawings. The finished components can be downloaded from http://www.mechanicad.com/download.html.

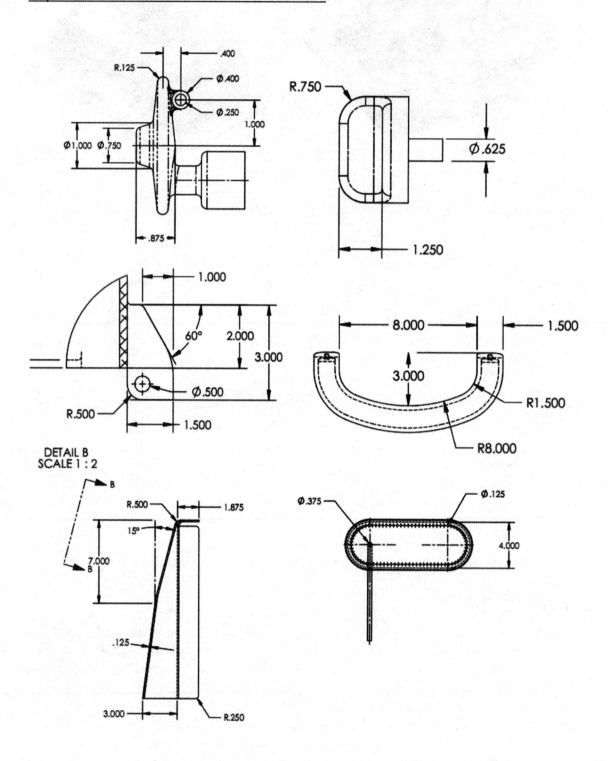

Chapter 5: Assembly Modeling

The next step in our design after making the parts and drawings is to get all the parts together into an assembly. The process of designing the parts first and then assembling them is known as "Bottom-Up Design." Think of it as buying a bicycle: when you open the box, you get all the pieces needed ready for assembly and some parts pre-assembled. A different approach known as "Top-Down Design" is where the parts are designed while working in the assembly; this is a very powerful tool that allows us to match parts to each other, changing a part if another component is modified. In this book we'll cover the Bottom-Up Design technique, since it is easier to understand and is also the basis for the advanced Top-Down Design, which is covered in the **Beginner's Guide to SOLIDWORKS Level II** book.

In general, it's a good idea to design the parts, make the assembly to make sure everything fits and works as expected (Form, Fit and Function), and *then* make the drawings of the parts and assemblies; this way the drawings are done at the end, when you are sure everything works correctly. In our case, we chose to make the drawings before the assembly to show how changes in the assembly are propagated to the part, and then to the drawing.

So far, we've been working on parts, single components that are the building blocks of an assembly. In an assembly we have multiple components, either parts or other assemblies (called sub-assemblies). The way we tell SOLIDWORKS how to relate components (parts and/or sub-assemblies) together is by using Mates (or relations) between them. Mates in the assembly are similar to the geometric relations in the sketch, but in the assembly, we use faces, planes, edges, axes, vertices and even sketch geometry from the component's features in order to mate them to each other.

To make an assembly we add components one at a time until we complete the design. In an assembly, every component has six degrees of freedom, meaning that they can move and rotate six different ways: three translations along the X, Y and Z axes, and three rotations about the X, Y, and Z axes. By mating components to each other we are essentially restricting how they move in relation to one another based on which degrees of freedom are constrained. This is the basis for assembly motion and simulation. Based on this, the first component we add to the assembly has all six degrees of freedom fixed by default. Therefore, it's a good idea to make sure the first component added to the assembly is one that will be a reference for the rest of the components. For example, if we make a bicycle assembly, the first component added to the assembly would be the frame. For the gear box we are designing, the first component added to the assembly will be the *'Housing',* since the rest of the components will be attached to it.

Notes:

The Speed Reducer Assembly

Notes:

In making the *'Speed Reducer'* assembly, we will learn about assembly tools and operations, including making new assemblies, adding components, adding Mating relations between them and later adding fasteners. We'll cover component design tables, interference detection, assembly exploded views, and how to change a part's dimensions while working in the assembly. The sequence to follow while making the *'Speed Reducer'* assembly is the following:

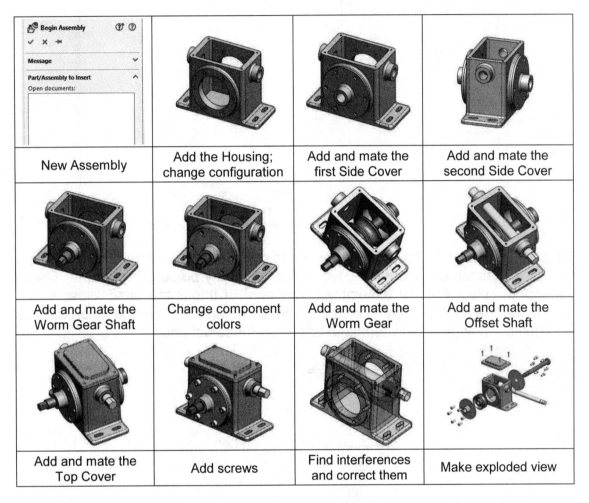

New Assembly	Add the Housing; change configuration	Add and mate the first Side Cover	Add and mate the second Side Cover
Add and mate the Worm Gear Shaft	Change component colors	Add and mate the Worm Gear	Add and mate the Offset Shaft
Add and mate the Top Cover	Add screws	Find interferences and correct them	Make exploded view

23.1. - One way to start a new assembly is by selecting the "**New**" document command, from the "**templates**" tab select the Assembly template, and click OK.

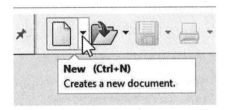

 Custom assembly templates are created just like part templates, the main difference is that we cannot assign a material to an assembly template because an assembly is a collection of individual parts where each part has a material assigned to it.

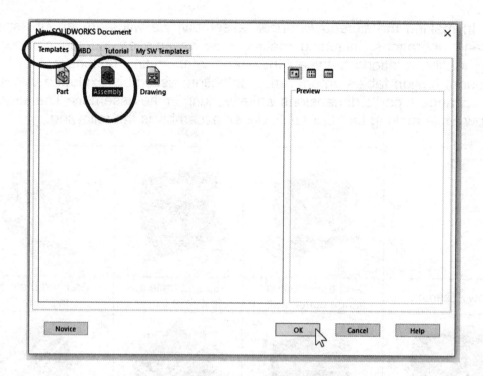

23.2. - If the option "Automatic Browse when creating new assembly" is enabled in the "Begin Assembly" command, we are immediately presented with the "Open" window to select the first assembly component. If this is the case, close it and cancel the "Begin Assembly" command. We are going to set our assembly options and make a new assembly template based on these options.

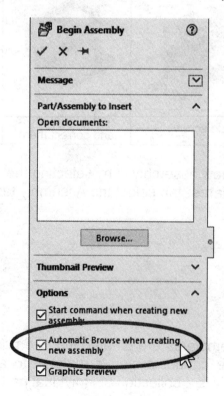

23.3. - Change the assembly's drafting standard to ANSI and its units to Inches with three decimal places.

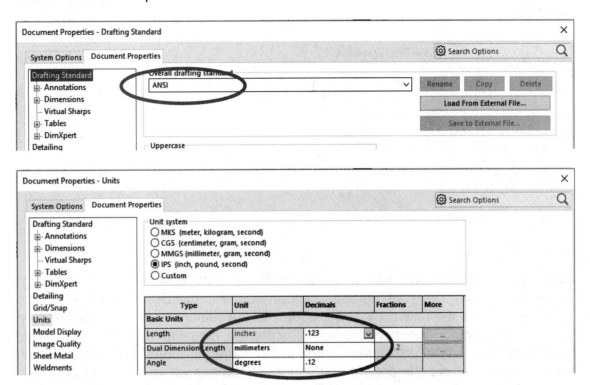

23.4. - From the menu "**File, Save As...**" select "Assembly Templates (*.asmdot)" from the "Save as type:" drop-down options and save this assembly as '*Assembly ANSI-Inch*' in the same folder where we previously saved our part templates and close it to continue.

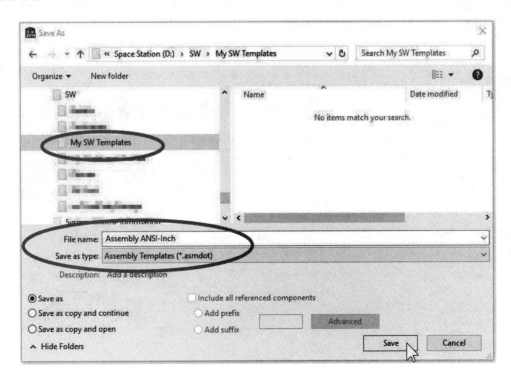

23.5. - After closing the new assembly template, select the "**New**" command, select the "My SW Templates" tab and using the newly created template make a new assembly.

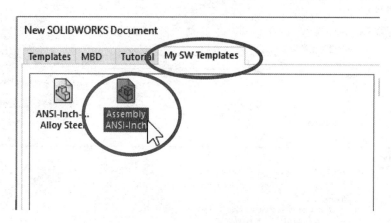

23.6. - The first thing we see when we make a new assembly is the "**Begin Assembly**" dialog to start adding components. As discussed previously, the *'Housing'* will be the first component added to our assembly. The "**Open**" dialog box is automatically launched; browse to locate the *'Housing'* and select it, and be sure to use the *"Machined"* configuration in the "Configurations" drop-down menu. If you cannot see the component that you want, make sure you are looking in the correct folder and have the "**Files of Type**" set to *"Part"* in the "Quick Filter." Click "**Open**" when done. If you have the "Graphics Preview" option box checked, you will see a preview of the component being added.

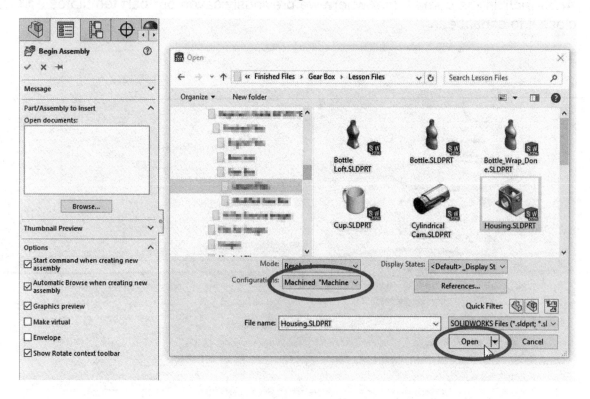

23.7. - After selecting the *'Housing'* we have to locate it in the assembly; when we move the mouse the part's preview follows it. What we want to do is to locate the *'Housing'* at the assembly's origin. If you cannot see the assembly's origin (which is hidden by default), turn it on by selecting the menu "**View, Hide/Show, Origins**" (this can be done while you are adding a component). The reason to locate the *'Housing'* at the assembly's origin is to have the *'Housing's* planes and origin aligned with the assembly's planes and origin. To add the *'Housing'* AND align it with the assembly's origin, click the OK button *or* move the mouse pointer to the assembly's origin and click on it. The cursor will have a double origin icon next to it.

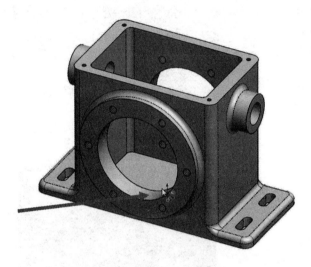

If the *'Housing'* is mistakenly added using the *"Forge"* configuration, do not worry, it can be changed. Select the "*Housing*" part in the FeatureManager, select the "*Machined*" configuration from the pop-up toolbar and accept the change with the green OK checkmark.

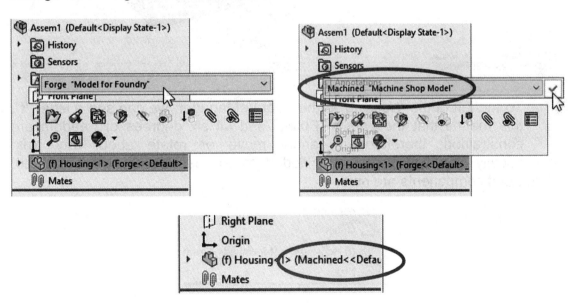

If the component added to the assembly has configurations, the selected configuration's name is shown next to the component's name in the FeatureManager in parentheses.

23.8. - Once in the Assembly environment, the toolbars are changed in the CommandManager; now we have an Assembly tab. Also notice at the bottom of the Assembly FeatureManager a special folder called "**Mates**." This is where the relations between components are stored.

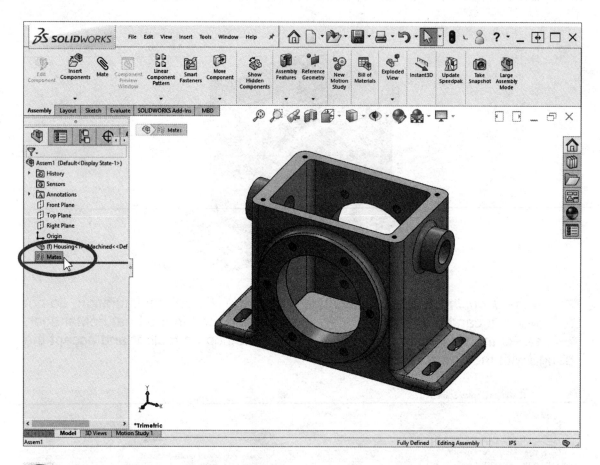

 To the left of the *'Housing'* in the FeatureManager, we can see a letter "**f**"; this means that the part is "**Fixed**" and all six degrees of freedom are constrained; therefore, it cannot move or rotate about any axis. Automatically, the first component added to an assembly is always fixed, subsequent components are not.

23.9. - To add the second component to our assembly, click the "**Insert Components**" command in the Assembly tab or the menu "**Insert, Component, Existing Part/Assembly**"; browse to the folder where the *'Side Cover'* part was saved, and open it. As with the '*Housing*', make sure the "Machined" configuration is selected before opening the '*Side Cover*'.

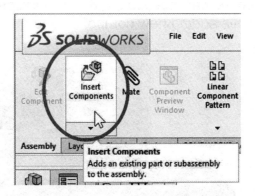

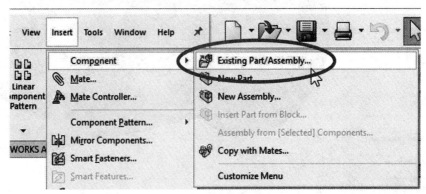

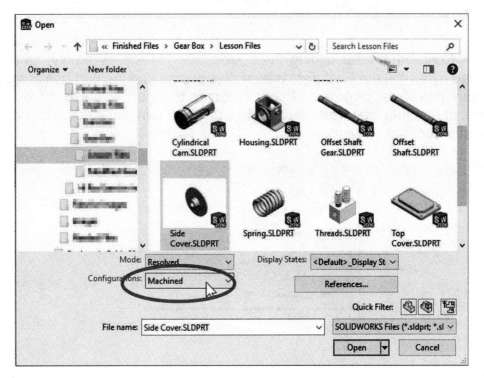

After opening the *'Side Cover'* we have a dynamic preview of the part in the assembly following the mouse cursor. Locate it next to the *'Housing'* as seen in the next image. Don't worry about the exact location; it will be located accurately in the next steps using mates.

 Keep in mind that we can manipulate the view (zoom, pan, rotate) just like in a part, even while adding new components to the assembly.

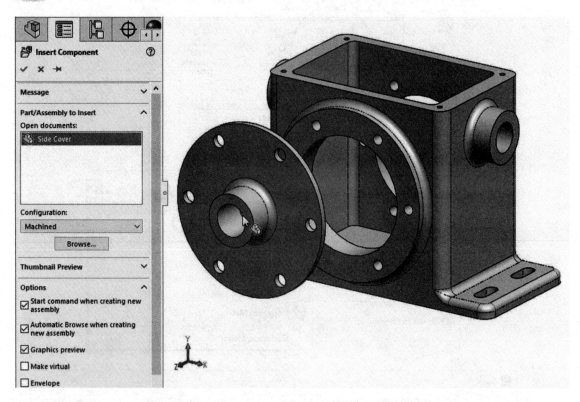

23.10. - Notice that after adding the *'Side Cover'* to the assembly, its name is preceded by a (-) in the FeatureManager; this means that the part has *at least* one unconstrained (free) degree of freedom. Since this part was just added, all six degrees of freedom are unconstrained (free) and the part can move in any direction and rotate about all three axes.

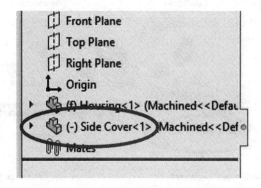

23.11. - Now that we have two components in the assembly, we are ready to add mates (Relations) between the *'Housing'* and the *'Side Cover'*. Mates allow us to

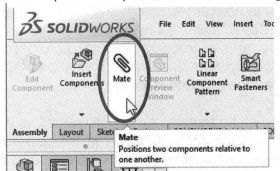

reference one component to another, locating and restricting the motion between them. As explained earlier, components can be mated using their faces, planes, edges, vertices, axes, and sketch geometry. Click the "**Mate**" icon in the Assembly tab or select the menu "**Insert, Mate**" to locate the *'Side Cover'* in reference to the *'Housing'*.

 When possible, select model faces as your first option for mates; most of the time faces are usually easier to visualize and select.

For the first mate, select the two <u>cylindrical faces</u> indicated in the next picture.

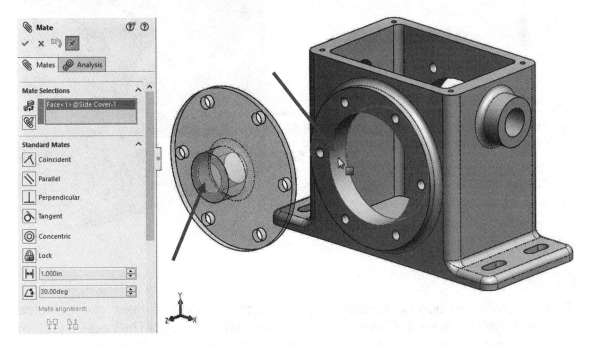

 After the first face is selected the component is automatically made transparent; this is an option that can be set at the bottom of the "Mate" command options.

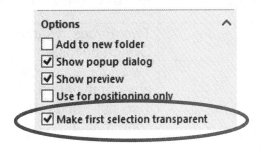

493

When the face of the second component is selected, SOLIDWORKS recognizes that both faces are cylindrical and automatically "snaps" them with a **Concentric Mate**. SOLIDWORKS defaults to concentric as it is the most logical option for this selection, and the *'Side Cover'* moves towards the *'Housing'* because it is the part with unconstrained degrees of freedom; in other words, it's the part that can move. Remember the *'Housing'* was fixed when it was inserted because it was the first component, and therefore, it cannot move.

The **Concentric Mate** is pre-selected and offered as an option; since this is what we intend to do, we have to accept it. Click the OK button in the pop-up toolbar or in the PropertyManager to add this mate between the two components.

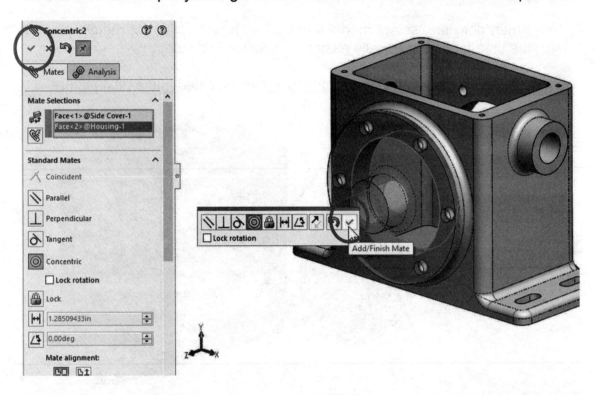

23.12. - Note the pop-up toolbar shows multiple options to mate the selected entities. (These options are listed in the PropertyManager, too.) The pop-up toolbar helps us to be more productive by minimizing mouse travel. The following table shows the standard (basic) mate options available:

Standard Mates	Entities that can be mated
Coincident	Two Faces, Planes, Edges, Vertices, Axes, Sketch points/endpoints, or any combination.
Parallel	Two flat Faces, Planes, linear Edges, Axes, or any combination.
Perpendicular	Two flat Faces, Planes, Edges, Axes, or any combination.

⟳ Tangent	Two cylindrical Faces; one flat Face and one cylindrical; a cylindrical Face and one linear Edge.
◉ Concentric ☐ Lock rotation	Two cylindrical Faces, two round Edges, two linear Edges or Axes, one cylindrical Face and one round Edge, one cylindrical Face and one linear Edge.
🔒 Lock	This option constrains all degrees of freedom of the component, locking it in place. Same effect as when the first component is added to the assembly.
↦ 1.000in ⊞ ☐ Flip dimension	Specify a distance between any two valid entities for Coincident or Parallel mates. "Flip dimension" reverses the direction to one side or the other. Using this mate on two flat Faces or Planes will also make them parallel.
⬔ 0.00deg ⊞ ☐ Flip dimension	Specify an angle between any two flat entities for Coincident, Parallel, or Concentric mates. "Flip dimension" will reverse the direction of the angular dimension.
Mate alignment: ⬚⬚ ⬚⬚ or ↗	Reverses the orientation of the components being mated. For two Faces, to look at each other or in the same direction.

After adding the first mate, the "**Mate**" command remains active allowing us to continue adding more mates to the assembly. It will remain active until we click **Cancel** or hit "Esc" on the keyboard. Notice that the mate just added is listed under the "**Mates**" box at the bottom of the PropertyManager. At this time, we'll add the rest of the mates to locate the 'Side Cover' and fully define it. Do not exit the "**Mate**" command yet.

23.13. - For the second mate, select the two cylindrical faces indicated, a bolt hole from the *'Side Cover'* and a screw hole from the *'Housing'*. SOLIDWORKS defaults again to a **Concentric** mate and rotates the *'Side Cover'* to align the holes; this mate will prevent the *'Side Cover'* from rotating. Remember we can use the **"Magnifying Glass"** (Shortcut "G") as we did to make selection of small faces easier. Click OK to add the mate. *IF* the *'Side Cover'* is *inside* the *'Housing'*, click-and-drag it with the left mouse button to move it out and add the remaining mates.

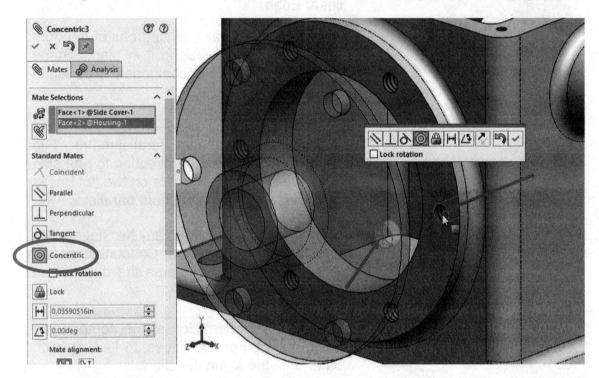

 The "Lock Rotation" option in the concentric mate prevents the mated parts from rotating. Checking this option in the first concentric mate would eliminate the need to add the second concentric mate.

In this case adding a second concentric mate works as expected because both holes are located <u>exactly</u> at the same distance from the center in both parts; in reality, if the holes were not exactly aligned, SOLIDWORKS would give us an error message letting us know that the mate cannot be added. In that case, it is better to align the two components using component Planes and/or Faces, with either a Parallel or Coincident mate, as will be shown later.

23.14. - The last mate will be a **Coincident mate** between the back face of the *'Side Cover'* and the front face of the *'Housing'* to make the faces touch; use the **"Select Other"** tool, or rotate the view to select the faces needed. (If the *'Side Cover'* is <u>inside</u> the *'Housing'*, click-and-drag to move it out.) After selecting the two flat faces, the cover will move until the selected faces touch; the **Coincident mate** option will be pre-selected as the concentric mate was preselected when selecting cylindrical faces. Click OK to add the last mate and close the **"Mate"** command by either clicking OK, Cancel or pressing the "Esc" key.

Right-mouse-click in the '*Side Cover*'   **Select back face of the '*Side Cover*'**

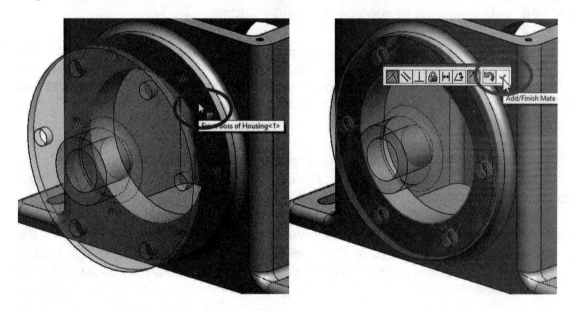

Select the front face of the '*Housing*'   **Click OK to add mate**

Using the "**Select Other**" tool allows us to select hidden faces without having to rotate the view. Hidden faces can be graphically selected in the screen, or from the list of faces located behind the face removed at the location of the right-mouse-click.

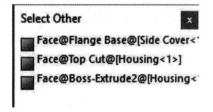

23.15. – Looking at the Feature-Manager we can see that the *'Side Cover'* is no longer preceded by a (-) sign; this means that all six degrees of freedom of the *'Side Cover'* have been constrained using mates; therefore, it cannot move or rotate about any axis anymore.

Notice that the "**Mates**" folder now includes the two Concentric mates and the Coincident mate just added; SOLIDWORKS adds the names of the mated components in each mate for reference. To get a better view of the component's names, we can change the width of the FeatureManager as needed by dragging its edge.

IMPORTANT: *If a (+) sign precedes a component's name in the FeatureManager, you probably also received an error message alerting you that the assembly had been **over defined**. If this is the case, it means you added conflicting mates that cannot be solved, or inadvertently selected the wrong faces or edges when adding mates.*

*The easiest way to correct this error is to either hit the **Undo** button or delete the last mate in the "Mates" folder; you will be able to identify the conflicting mate because it will have an error icon next to it and will be colored either red or yellow. If multiple mates have errors, start deleting the last mate at the bottom (the last mate added); chances are this is the cause of the problem. If you still have errors, keep deleting mates with errors from the bottom up until you clear all the errors. It's not a good idea to proceed with errors, as it will only get worse.*

23.16. – Now we are ready to add the second *'Side Cover'* to our assembly. Repeat the "**Insert Component**" command to add a second *'Side Cover'* and locate it on the other side of the *'Housing'* as shown. Remember to use the "Machined" configuration. Don't worry too much about the exact location; in the next step we'll move and rotate the part closer to its final location.

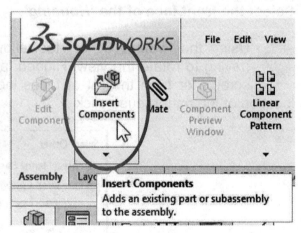

 A quick way to add a copy of a component is to hold down the "Ctrl" key and click-and-drag the part to be copied within the assembly's window.

An option to rotate the component and align it closer to its final orientation is by selecting the rotate buttons in the floating pop-up toolbar located near the bottom of the screen, where we can set the angle of rotation and the axis to rotate about. The part's preview will update dynamically before we click to locate the new component in the assembly.

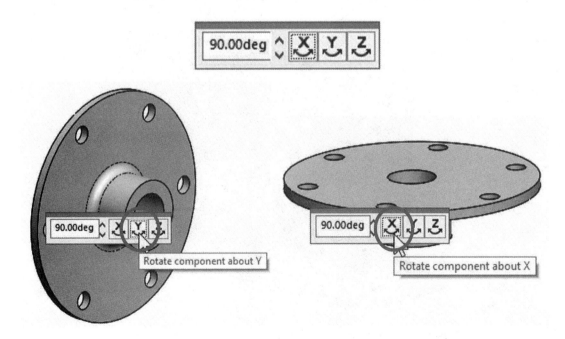

23.17. - When the second *'Side Cover'* is added, it has a (-) sign next to it in the FeatureManager. Remember, this means that it has <u>at least</u> one unconstrained Degree of Freedom (DOF). Since this part was just added to the assembly, all six DOF are unconstrained, and the component can be moved and rotated. Turn off the "Origins" if needed (menu "**View, Hide/Show, Origins**") and rotate the view just as we do with parts to look at the assembly from the back.

In order to "**Move**" a component within the assembly, click-and-drag it with the <u>left mouse button</u>. To "**Rotate**" it, click-and-drag it with the <u>right mouse button</u>. "Move" and "Rotate" the second *'Side Cover'* as needed to align it *approximately* as shown to add the necessary mates and locate it.

Move Component - Left Mouse Click-and-Drag

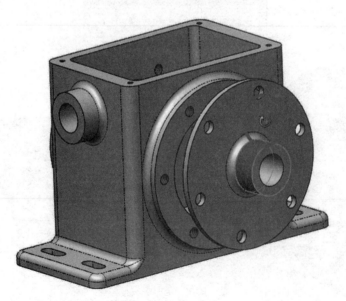

Rotate Component - Right Mouse Click-and-Drag

23.18. - Now we need to mate the second cover as we did the first *'Side Cover'*. Rotate the view (not the part) to get a better view of the faces to select. A different way to add mates is to select one of the faces to be mated, and from the pop-up toolbar select the "**Mate**" command; once the command is opened this face will be pre-selected and we only need to select the second face to add the next Concentric mate. Selecting the face of the *'Side Cover'* first makes the component transparent (the option is at the bottom of the "**Mate**" command), making it easier for us to make other selections through the transparent component.

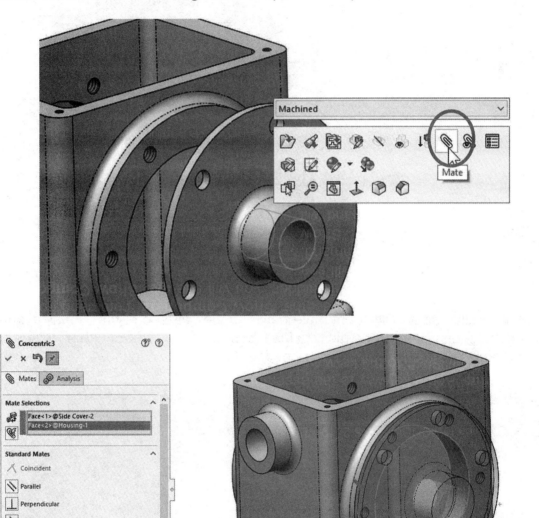

 Before adding the second concentric mate, left-mouse-click-and-drag the cover. Note it can move and rotate about its axis; under-constrained components are the basis for SOLIDWORKS to simulate motion.

23.19. - Now we need to add a new concentric mate to align the screw holes between the *'Side Cover'* and the *'Housing'*. Since the part is symmetrical, in this case it doesn't really matter which two holes are selected. If the cover had a feature that needed to be aligned, then orienting the part correctly would be important. Use the **"Magnifying Lens"** if needed to select small faces (the hole's edges can be used to add mates, too).

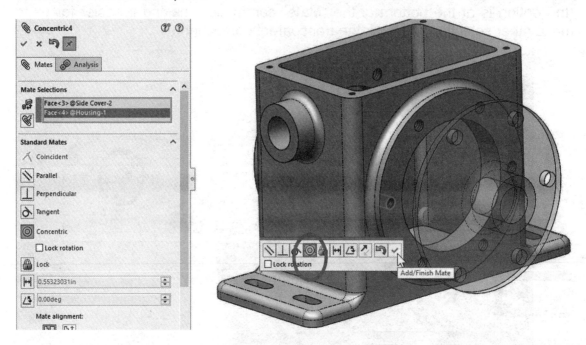

23.20. - Finally add a Coincident mate between the flat face of the *'Housing'* and the second *'Side Cover'* as we did with the first cover. Use **"Select Other"** to select the hidden face, or rotate the view as needed. Click OK to continue.

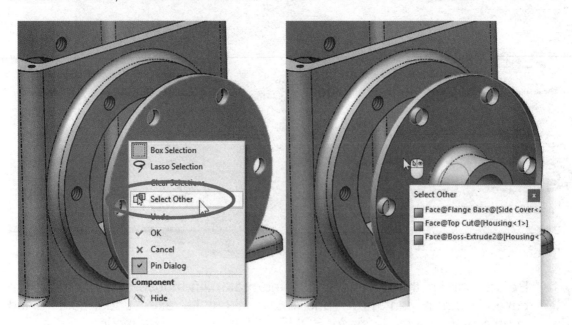

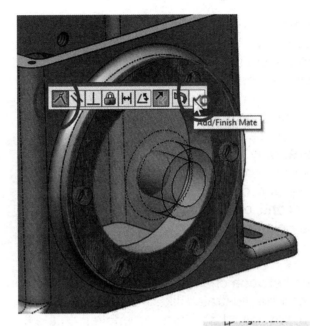

It is important to know that these mates could have been added in any order; we chose this order to make it easier for the reader to see the effect of each mate on the parts. Whichever order you select, you will end up with both *'Side Cover'* parts fully defined (no free DOF) and six mates in the "Mates" folder as listed next.

23.21. - Now that we have correctly mated both *'Side Covers'*, switch to an isometric view for clarity, add the *'Worm Gear Shaft'* using the "**Insert Components**" command as before, and locate it somewhere above the *'Housing'*.

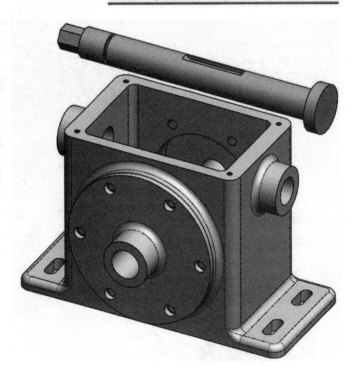

23.22. - With four different components in the assembly, it's a good idea to change their appearance to make it easier to identify them. To change a component's color, select the component in the FeatureManager or the graphics area, and from the pop-up toolbar select the "**Appearance**" icon. From the drop-down list, select the first option to change the part's color at the assembly level only. This means that the color will change only in the assembly and not in the part file. The second option changes the part's color at the part level.

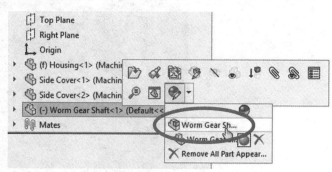

The difference between changing a part's appearance in the assembly or in the part file can be illustrated like this: We can paint the part *before* we assemble it, or we paint it *after* we assemble it. In the first case we assemble a painted part; in the second case the part is painted after it is assembled. In our example, we want to change the color only at the assembly level.

Now we are ready to change the part's color. Select the desired color from the color swatch selection box to change it and click OK to finish.

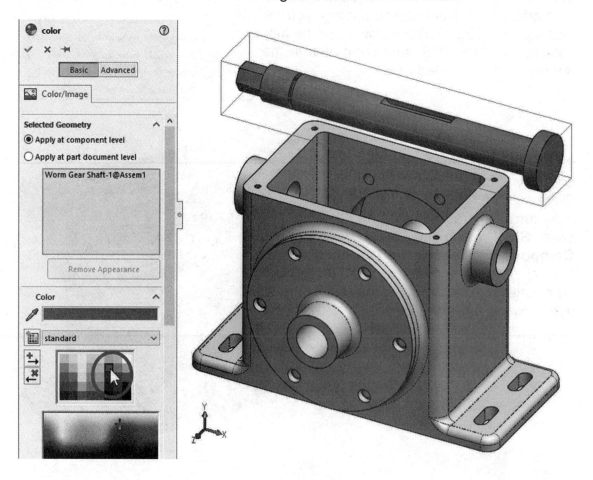

Using the same procedure, change the color of both covers to your liking before continuing to add mates. (It does look better in color!)

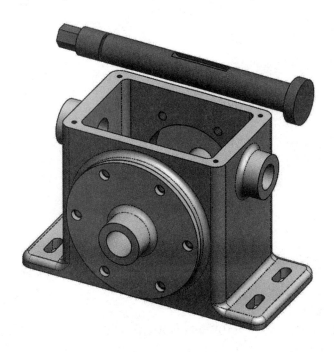

23.23. – For the next step, we want to locate the '*Worm Gear Shaft*' closer to its final location.

To move a part with more accuracy, we can use the "**Move with triad**" command. This is a very useful command to help us move components along a specific direction, on a plane, or to rotate about an axis.

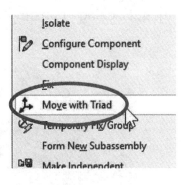

To move the '*Worm Gear Shaft*' using the triad, right-mouse-click on it and select "**Move with Triad**" from the pop-up menu; drag along the X, Y or Z axes, about the XY, XZ or YZ planes, or about the X, Y or Z ring's axis.

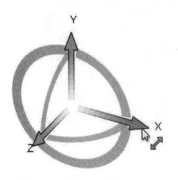
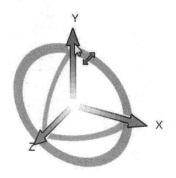
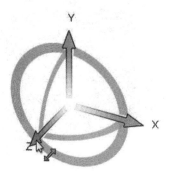

Move Along X axis   Move Along Y axis   Move along Z axis

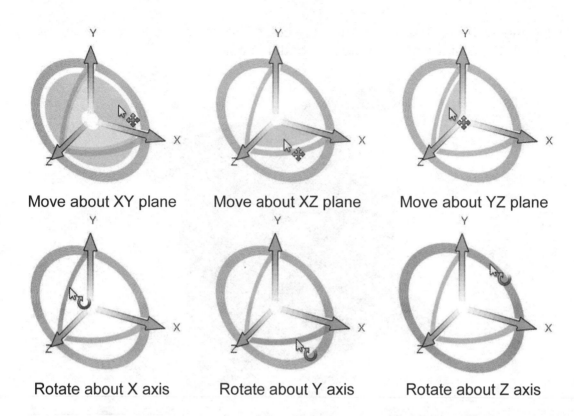

Move about XY plane   Move about XZ plane   Move about YZ plane

Rotate about X axis   Rotate about Y axis   Rotate about Z axis

Rotate the shaft 90° to the back about the Y axis; use the on-screen ruler as a guide. If the mouse pointer is near the ruler it will snap to the values. Feel free to explore the different options to manipulate models in the assembly using the triad. Clicking in an empty area of the screen removes the triad.

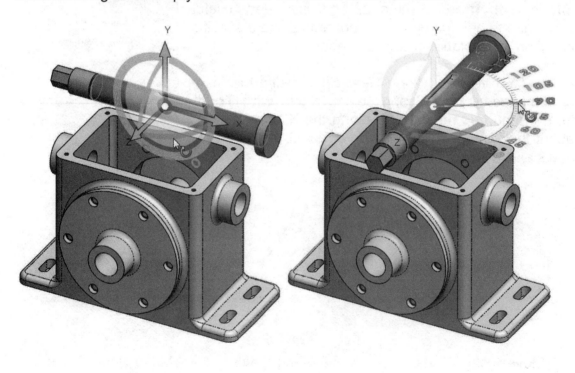

23.24. – Another way to add mates is by pre-selecting the faces to be mated and selecting the type of mate to add from the pop-up toolbar.

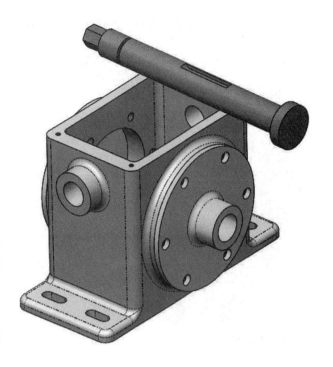

Only standard mates appropriate for the selected entities are available using this approach. Rotate the view to the back, and while holding the "Ctrl" key pre-select the two cylindrical faces to be mated; immediately after releasing the "Ctrl" key, the context toolbar will be displayed showing the options available.

Selecting the **Concentric** mate immediately mates the parts.

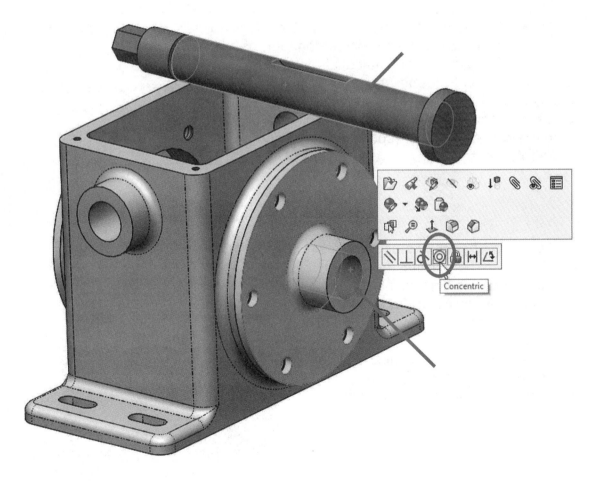

After adding the first concentric mate, repeat the same process but now pre-select the flat faces to add a Coincident mate. Use "**Select Other**" to select the hidden face if needed, or rotate the model.

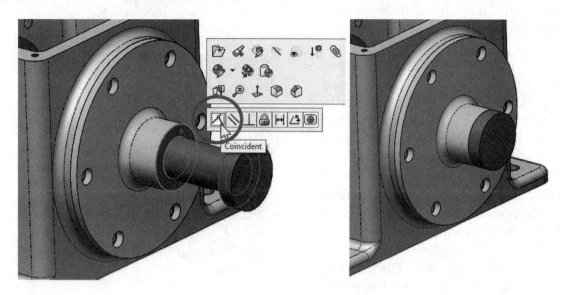

Notice that the *'Worm Gear Shaft'* is not fully defined yet; it still has a (-) sign before its name in the FeatureManager.

In this case, the only unconstrained or free DOF left is to rotate about its axis, and that is exactly what we want; the *'Worm Gear Shaft'* is supposed to rotate. If we click-and-drag it, we'll see it rotate. (Look at the keyway while rotating it.)

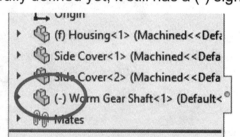

23.25. – In the next step add the *'Worm Gear'* part using the "**Insert Component**" command as before, and place it in the assembly above the *'Housing'* as shown.

In order to better differentiate between components in the assembly, change the part's color.

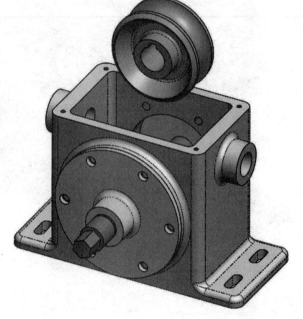

23.26. - To locate the *'Worm Gear'* in place use either of the two methods learned to add a concentric mate with the *'Worm Gear Shaft'*. Select the cylindrical inside face of the *'Worm Gear'* and the outside face of the *'Worm Gear Shaft'*. After adding the mate click-and-drag the *'Worm Gear'* to see how it moves along the shaft and rotates about it.

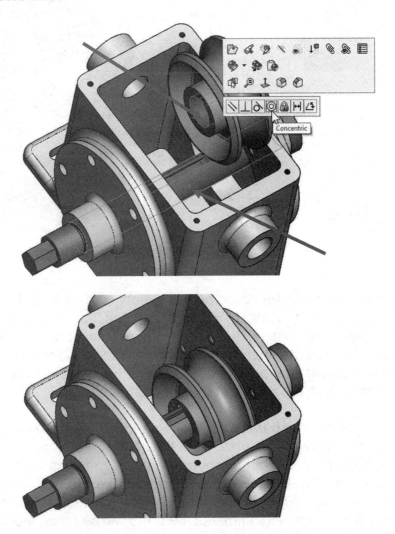

23.27. - Now we need to locate the *'Worm Gear'* in the middle of the *'Housing'*. To accomplish this, we could use the *"Front Plane"* of the *'Worm Gear'* and either the *"Front Plane"* of the *'Housing'* or the assembly, both of which are conveniently located in the center of the *'Housing'*; remember we made the *'Housing'* symmetrical about the origin, anticipating it could be helpful like in this case. ☺

In this step we can use the **Mate** command from the CommandManager or the context toolbar when we pre-select the planes. Add a **Coincident** mate using the *"Front Plane"* of the *'Housing'* (or the assembly) and the *"Front Plane"* of the *'Worm Gear'*. Select the **Mate** command from the CommandManager and expand the fly-out FeatureManager to make the selection. In the following image the *'Worm Gear'* has been moved slightly to the side for visibility purposes.

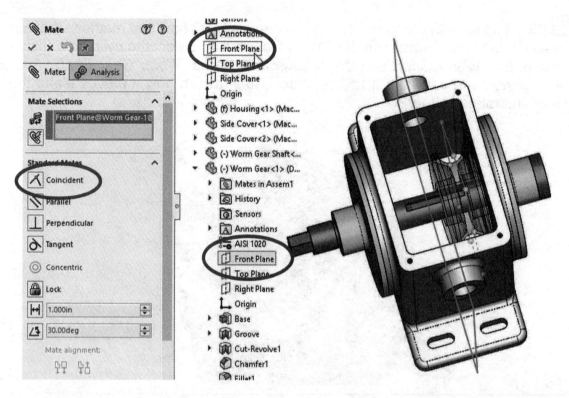

By adding this mate, the *'Worm Gear'* is now centered inside the *'Housing.'* Now we need to make the *'Worm Gear Shaft'* and the *'Worm Gear'* to rotate together as if they were assembled to each other. To accomplish this, we'll have to either add a **Coincident** mate using the corresponding planes from the parts, or a **Parallel** mate between the faces of the keyways. In this case we'll add a Parallel mate in this step and let the reader explore the other option.

23.28. – In order to add the parallel mate, we need to hide the *'Housing'* to have a better view of the keyway features. Exit the "**Mate**" command and select the *'Housing'* in the FeatureManager or anywhere in the graphics area; from the pop-up toolbar select the "**Hide/Show**" command to make the part invisible. Keep in mind that we are only hiding the part from view—it was not deleted from the assembly. When we finish adding mates the parts will be made visible again.

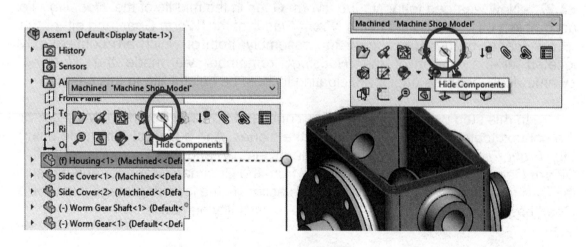

23.29. - As soon as the part is hidden it disappears from the screen and we can see inside without obstructions. The *'Housing'* icon in the FeatureManager changes to white, indicating the part is hidden.

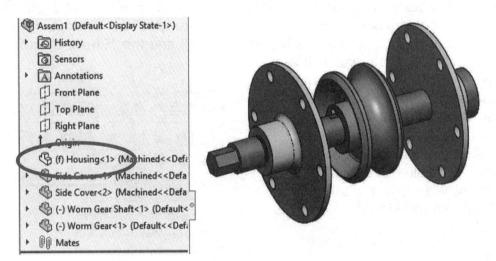

23.30. - Add a new "**Mate**" using the flat faces of the *'Worm Gear Shafts'* and *'Worm Gear's'* keyway. In this case, using a **Parallel mate** will help us absorb any small dimensional differences in case the mated faces are not exactly aligned, which would cause an error message. Making them parallel allows us to absorb those differences and still obtain the desired result. In general, a parallel mate is a more forgiving option and should be used in cases when other options could potentially cause a conflict. Click OK to add the parallel mate and continue.

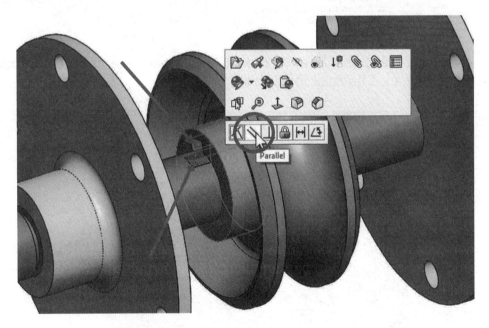

After adding the parallel mate, click-and-drag either the *'Worm Gear'* or the *'Worm Gear Shaft'*; see how both of them rotate at the same time, as if they had a keyway. When the *'Housing'* is hidden the rest of the components are still mated to them and behave accordingly.

Extra credit: Add a keyway to the assembly, delete the parallel mate and mate the *'Worm Gear Shaft'* to the keyway, and the keyway to the *'Worm Gear'*.

23.31. – After adding the mate, we need to make the *'Housing'* visible again. Select it in the FeatureManager and select the **"Show Component"** command from the pop-up toolbar. If a component is hidden, you will see the **"Show Component"** command; if it is visible, you will see the **"Hide Component"** command. Alternatively, selecting the **"Show Hidden Components"** command will hide all currently visible components, and all hidden components will become visible. Selecting a component will make it visible when the "Exit show-hidden" button is pressed.

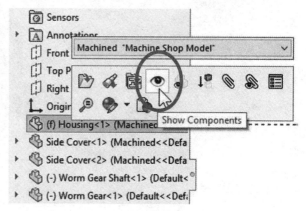

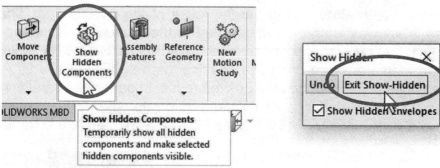

23.32. - Now add the *'Offset Shaft'* to the assembly using the **"Insert Component"** command and change its color for visibility.

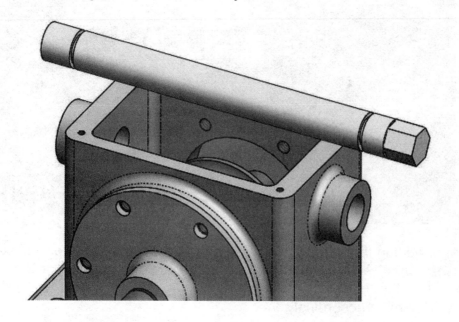

23.33. - Add mates by selecting one of the model faces to be mated, and from the pop-up toolbar, select the "**Mate**" icon. When the "**Mate**" command is displayed, the model face is pre-selected and we only need to select the other face to add a concentric mate, *or* pre-select both faces and select the "Concentric" mate from the pop-up menu. Do not use the "Lock Rotation" option; we need it to rotate.

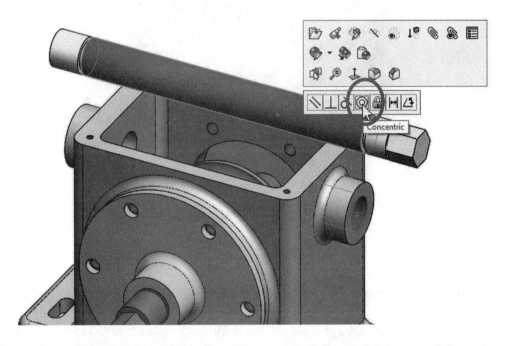

23.34. - Now we need to add a coincident mate to prevent the shaft from moving along its axis. We'll use the groove in the *'Shaft'* for this mate. In this case, we can use either the flat face or the edge of the groove; selecting the edge may be easier than selecting the face. Remember we can use the "**Magnifying Glass**" tool to zoom in and make our selections. Click OK to finish the mate.

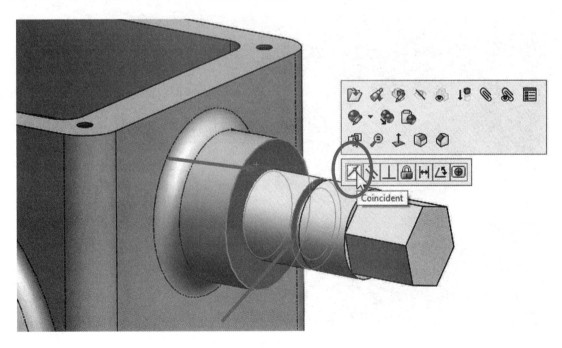

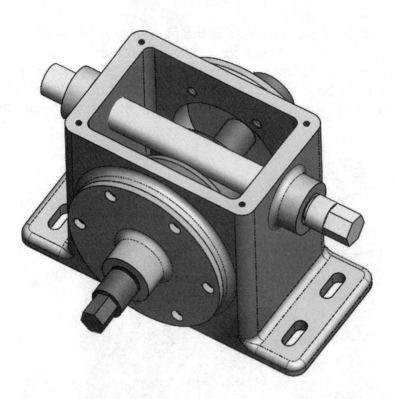

23.35. - The last component we're adding is the *'Top Cover'*. Add it to the assembly and change its color, too. The first concentric mate will align one of the holes of the cover to its corresponding hole in the *'Housing'*; press and hold the Ctrl key, select both indicated faces, and after releasing the Ctrl key select the concentric mate command from the pop-up menu. To add this mate, we can select cylindrical faces or circular edges. An interesting detail to know, by adding a concentric mate to a fully defined or fixed component, we are removing four DOF from the component: two translations and two rotations.

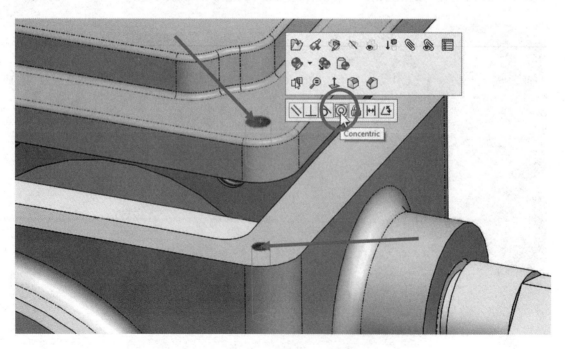

23.36. - Check the remaining two DOF by dragging the *'Top Cover'* with the left mouse button; it will rotate about the hole we mated and move up and down. By adding a coincident mate between the <u>bottom face</u> of the *'Top Cover'* and the <u>top face</u> of the *'Housing'* we'll remove one more DOF.

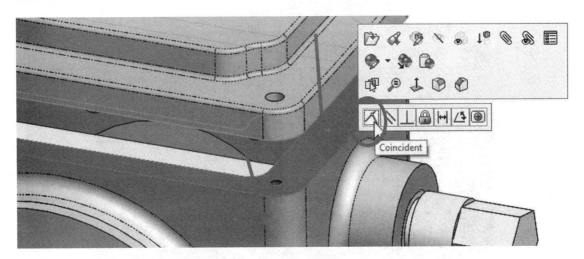

23.37. - If we click-and-drag the *'Top Cover'* it will turn; now we only have one DOF left. To finish constraining the *'Top Cover'* we'll add a **Parallel** mate between the *'Top Cover'* and the *'Housing'*. As we explained earlier, the reason for the Parallel mate is that sometimes components don't match exactly, and if we add a Coincident mate, we may be forcing a condition that cannot be met, over defining the assembly and getting an error message. The Parallel mate can be added between Faces, Planes, and/or Edges. In this particular case, we can use either a Coincident or Parallel mate since the parts were designed to match exactly. However, in real life they may not; that's why we chose a Parallel mate.

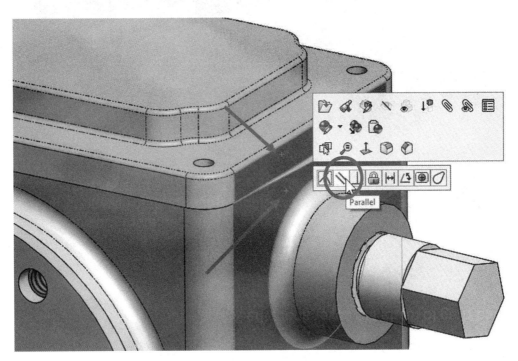

Your assembly should now look like this.

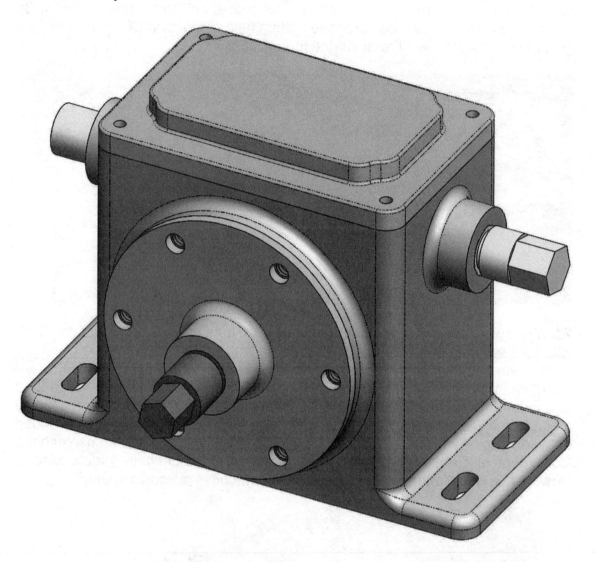

SmartMates

SmartMates are a quick and easy way to add basic mate types between components simply by dragging parts or assemblies onto each other using flat, cylindrical or conical faces, circular or linear edges, vertices, or temporary axes. To add SmartMates we need to hold down the "Alt" key while we drag a face, edge or vertex of the component to be mated *onto* the face, edge or vertex of the other component.

We'll re-create the entire assembly up to this point using the SmartMates approach, now that the concept of mates has been explained and we have a better understanding of the process. The next table shows the types of SmartMates available and their corresponding feedback icon.

 When SmartMates is enabled the icon will change to a clip attached to the face, edge, or vertex being dragged and the component being dragged will become transparent.

Entities to be mated/ Feedback icon	Resulting Mate
2 Flat Faces	**Coincident Faces**
2 Linear Edges	**Coincident Edges**

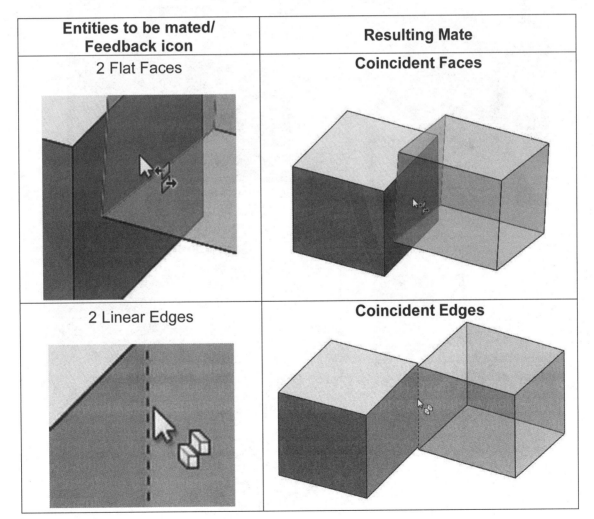

Entities to be mated/ Feedback icon	Resulting Mate
2 Cylindrical / Conical Faces	**Concentric**
2 Round Edges	**Peg-in-Hole**
	(Adds 1 Concentric mate between the cylindrical/conical faces and 1 Coincident mate between the flat faces)
2 vertices	**Coincident Vertices**

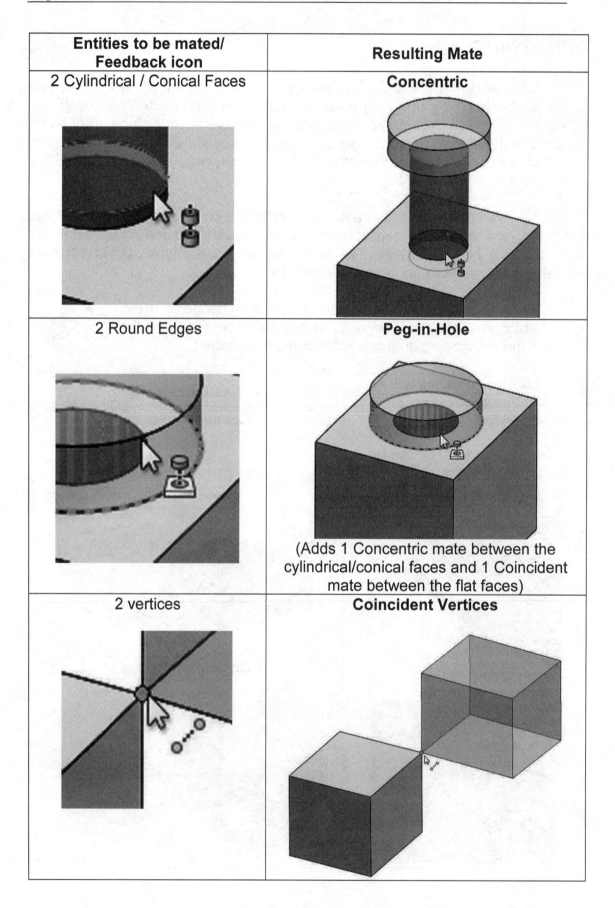

24.1. – Close the assembly without saving it; it will be recreated using an alternate process using the SmartMates functionality. Open the *'Housing'* part, and from the **"New"** document icon select **"Make Assembly from Part/Assembly"** or the menu **"File, Make Assembly from Part."** Make sure the "Machined" configuration is the active before making the assembly. Just as before, add the 'Housing' at the assembly's origin to align the part with the assembly's planes and origin. If the assembly's origin is hidden, turn it on using the menu **"View, Hide/Show, Origins"** and drop the part in the origin or click OK in the "Begin Assembly" command, this way the *'Housing'* will be automatically added at the assembly's origin.

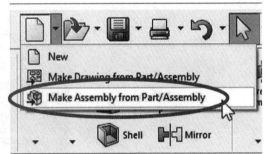

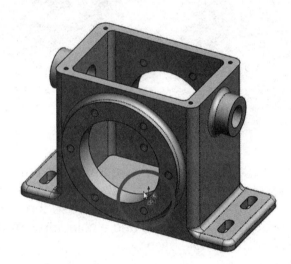

24.2. - Now, bring in the *'Side Cover'* with the **"Insert Component"** command and put it next to the *'Housing'* using the "Machined" configuration, just as we did before up to this point.

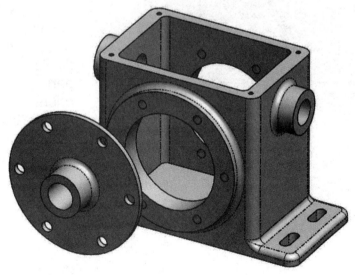

24.3. - Now we'll mate the parts using SmartMates. Press and hold the "**Alt**" key on the keyboard and, while holding it down, click-and-drag the *'Side Cover'* from the indicated edge. Notice that as soon as we start moving it, the "Mate" icon appears next to the mouse pointer, and the *'Side Cover'* becomes transparent.

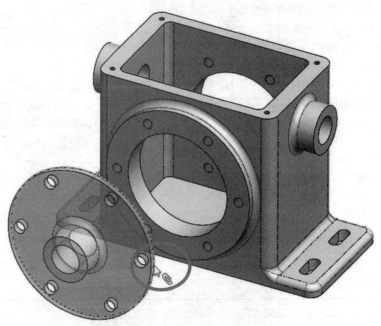

Keep dragging the *'Side Cover'* until we touch the flat face or the round edge in the front face of the *'Housing'*; at this time the *'Side Cover'* will 'snap' into place and will give us the "**Peg-in-Hole**" mate icon. Release the mouse button to finish. Using SmartMates we added a concentric and a coincident mate between the *'Side Cover'* and the *'Housing'* in a single step.

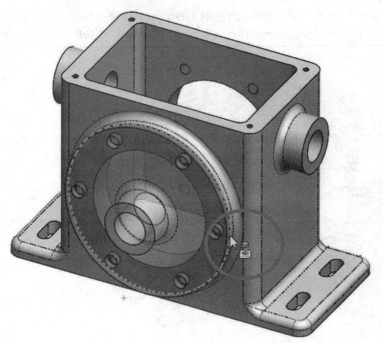

If the *'Side Cover'* is flipped pointing inside, release the "Alt" key while still holding the mouse button, and press the "Tab" key to flip its direction. When the part has the correct orientation release the mouse button to add the mates.

After expanding the "Mates" folder in the FeatureManager we can see a concentric mate and a coincident mate have been added. Selecting a mate shows the part's mated faces in the graphics area.

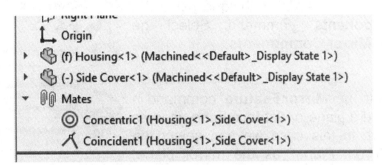

24.4. – To fully define the 'Side Cover' we'll add a concentric mate between two screw holes to prevent the '*Side Cover*' from rotating. Rotate the cover just enough to see both holes at the same time. Hold down the "**Alt**" key as before, then click-and-drag the face or edge of the hole in the '*Side Cover*' to the screw hole in the '*Housing*'. When the concentric mate icon is displayed, release the mouse button to add the mate. Now the '*Side Cover*' is fully defined and can no longer be moved.

24.5. – Now we need to add the second *'Side Cover'* to our assembly, but instead of manually adding it and then mate it using SmartMates, we are going to use a different technique using the **"Mirror Components"** command. Select the menu "**Insert, Mirror Components**."

Similar to the "**Mirror Feature**" command in a part, we need a plane or flat face to be used as a mirror plane. In this case we can select the assembly's "*Front Plane*" as the mirror plane, since it is located in the middle of the assembly, and in the "Components to Mirror" selection box, select the *'Side Cover'* either in the screen or the fly-out FeatureManager. After making your selections, click in the "Next" blue arrow to go to the next step in the "**Mirror Component**" command.

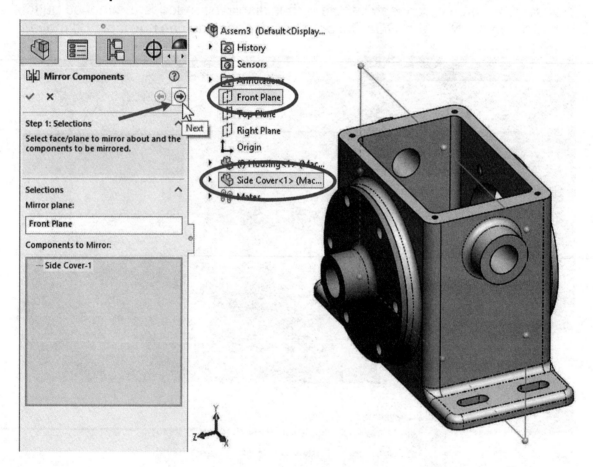

In the second step we get a preview of the mirrored part and we can optionally change its orientation and optionally create an "opposite hand" version of it. In this case, we don't need an opposite hand version, since the same part can be added on both sides. In case a mirrored component is not correctly oriented, select it and cycle through the different orientations until the desired orientation is shown using the "Reorient components" buttons at the bottom of the selection list. After finishing the command, a new instance of the *'Side Cover'* is added to the assembly and a *"Mirror Component"* feature is added to the FeatureManager.

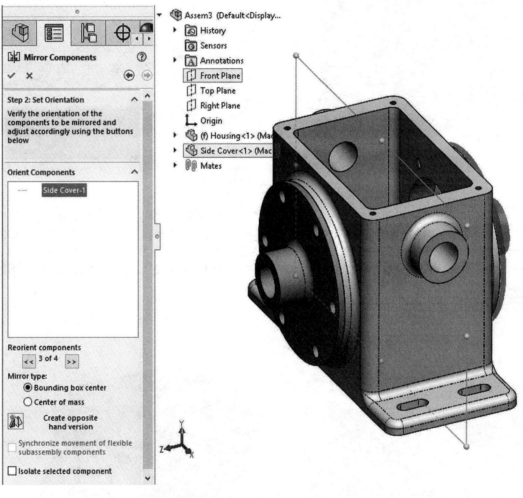

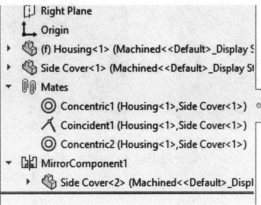

24.6. - Another way to add parts to an assembly is by dragging them directly from *Windows Explorer*. Open the folder where the assembly components are saved, and drag-and-drop the *'Worm Gear Shaft'* part directly into the assembly window.

24.7. - After adding the *'Worm Gear Shaft'* to the assembly, rotate it to a position that will allow us to see both of the edges to be mated; a limitation of using SmartMates is that we have to be able to see both of the entities to be mated. Hold down the "**Alt**" key, and click-and-drag the *'Worm Gear Shaft's'* edge to the *'Side Cover's'* hole to add a "Peg-in-Hole" mate.

 In this particular case, mating the outside edges of both the shaft and the cover would work exactly the same, since all outside edges are concentric to the inside edges.

24.8. - In this version of the assembly we'll use the '*Worm Gear Complete*' to learn a new type of mate later on. Another way to add components to an assembly is by arranging the part and assembly's windows side by side and dragging the part *into* the assembly.

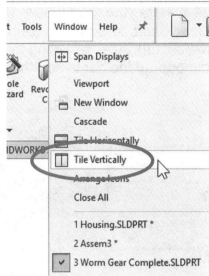

Open the '*Worm Gear Complete*' part, and using the menu "**Window, Tile Vertically**," tile the windows vertically. An advantage of using this approach is that if we click and drag a part's face into the assembly window the SmartMates functionality is automatically activated.

 We can drag parts into assemblies using faces, edges, or vertices, just as we would when using SmartMates. The only difference is that we don't have to press the "Alt" key while dragging from a part (or assembly) window into an assembly window.

If we click and drag the part from its name at the top of the FeatureManager, the SmartMates functionality is not activated and the part (or assembly) is added to the assembly but not mated.

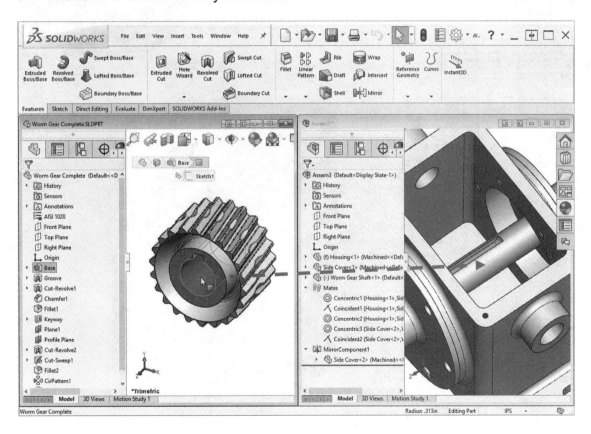

24.9. - Drag the *'Worm Gear Complete'* into the assembly's window over the *'Worm Gear Shaft'* to add a concentric mate using SmartMates (look for the concentric mate icon). After releasing the mouse Select the concentric mate and click OK in the pop-up toolbar to add the mate.

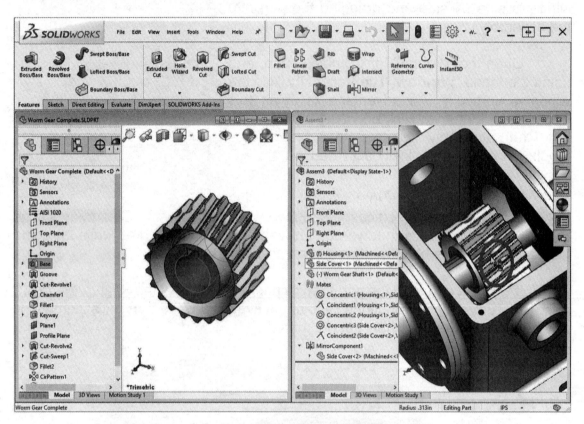

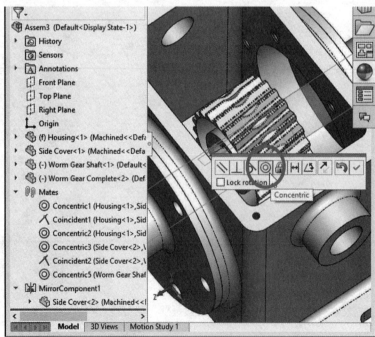

24.10. - The next step is to center the *'Worm Gear Complete'* inside the *'Housing'*. In the previous assembly we added a mate using the part's and assembly's planes. In this case we'll use a different mate type called "**Width**." The Width mate locates a component between two selections of another component.

Maximize the assembly window to use the entire screen, select the "**Mate**" command, expand the "**Advanced Mates**" section and select the "**Width**" mate. In the "Width selections:" select the two inside faces of the *'Housing'*; activate the "Tab selections:" box and select the two outside flat faces of the *'Worm Gear'* (the small round faces). The *'Worm Gear Complete'* will be automatically centered. Using the "Centered" constraint option the "**Width**" mate centers the "Tab" selections between the "Width" selections. Click OK to add the mate and finish.

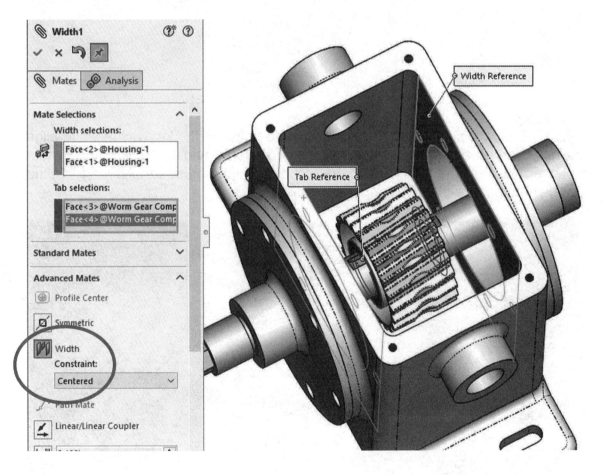

These are the "Constraint" options available in the Width mate:

a) *Centered* – The tab is centered between the width selections
b) *Free* – The tab can move freely between the width selections
c) *Dimension* – A side of the tab is located at a specific distance from either end of the width selections
d) *Percent* – The tab is located a percentage of the distance between the width selections

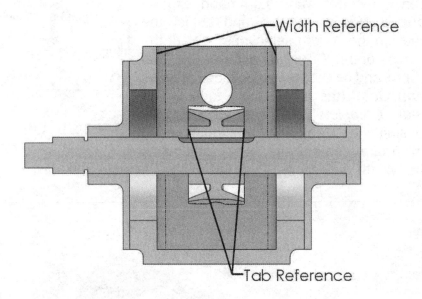

24.11. – Now we need to add the parallel mate between the keyway faces to align the shaft and the gear. Hide the '*Housing*' for clarity and add the parallel mate between the keyways using any of the techniques learned so far. After adding the parallel mate, show the '*Housing*' part again.

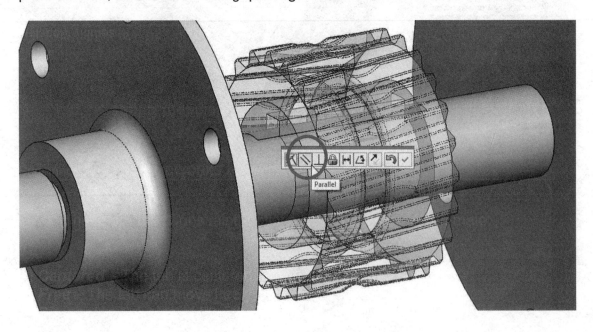

24.12. - Add the *'Offset Shaft Gear'* to the assembly using any method learned so far, and add a "Peg-in-Hole" mate using SmartMates using the indicated edge. **If the SmartMate is reversed (in the wrong direction)**, release the "Alt" key while still holding the left mouse button and press the "Tab" key to reverse the part's orientation. When the correct orientation is displayed release the mouse button.

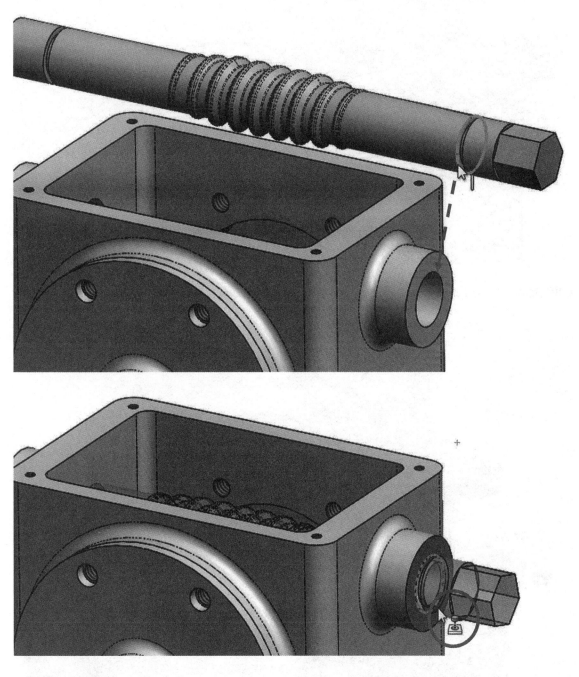

24.13. - Add the *'Top Cover'* to the assembly, and flip it upside down as shown using the "**Move with Triad**" command. Remember that for SmartMates to work we have to be able to see both of the entities to be mated, in this case two circular edges. Right-mouse-click in the cover and select "**Move with Triad.**"

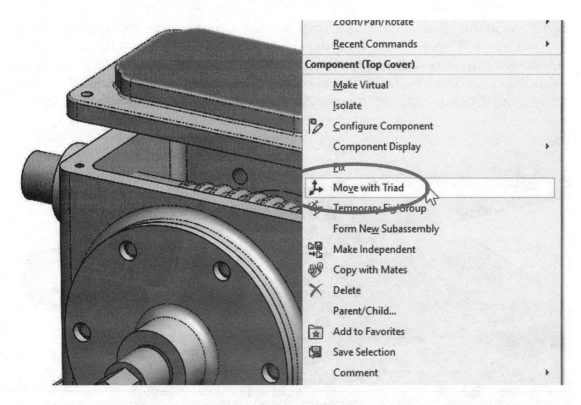

Click-and-drag to rotate the cover about the X axis enough to look at its underside.

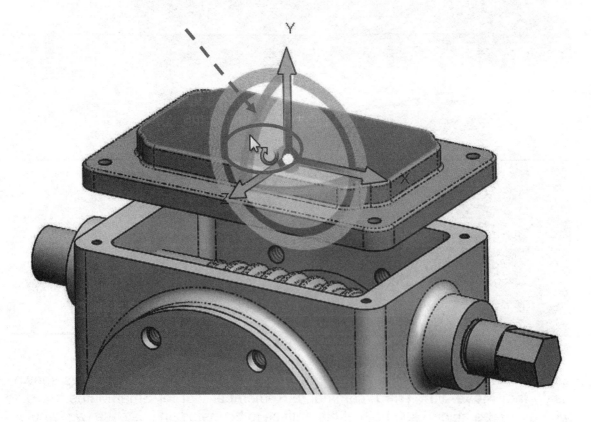

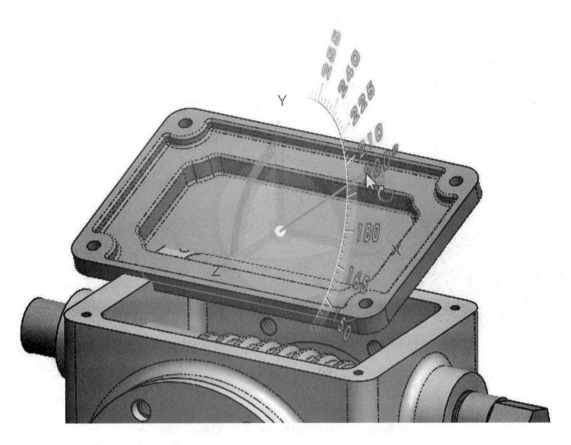

24.14. - Select the edge indicated, hold down the "**Alt**" key and drag it into the screw hole to add a **Peg-in-Hole** SmartMate.

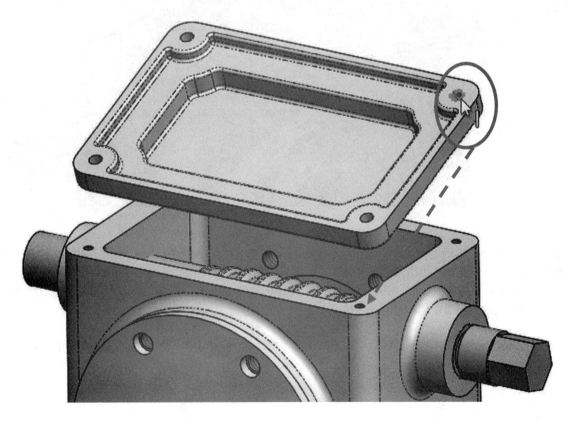

If the *'Top Cover'* is upside down in the preview, release the "**Alt**" key, and while still pressing the left mouse button press the "**Tab**" key once; notice how the preview changes by flipping the *'Top Cover'*. Pressing the "Tab" key again will flip the mate again. Make sure you have the correct orientation for the *'Top Cover'* before releasing the left mouse button to add the SmartMate.

One possible orientation; press the "**Tab**" key to flip.

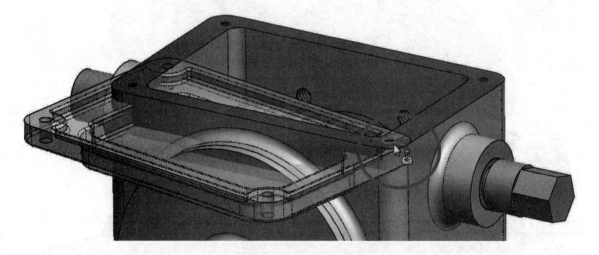

Orientation after pressing the "**Tab**" key

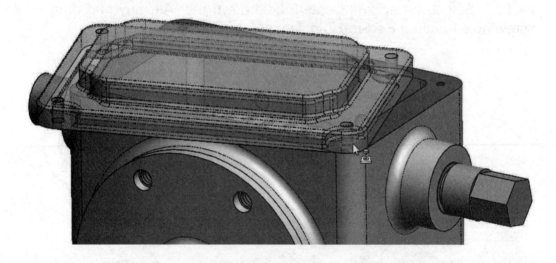

24.15. - Now add the final mate to the *'Top Cover'*. Hold down the "**Alt**" key and drag the indicated edge to the face (or edge) of the *'Housing'* to make them Coincident. Adding this mate will fully define the *'Top Cover'*. Change the color of the parts as before for easier visualization.

NOTE: The rest of the illustrations and exercises will be done using the full gear components added in this assembly.

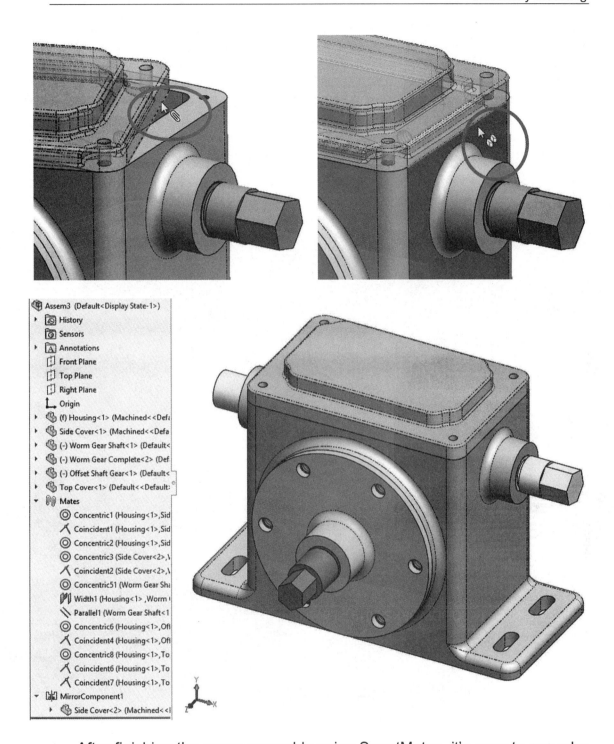

After finishing the same assembly using SmartMates, it's easy to see why using SmartMates is a good idea, since it speeds up the assembly process making it easier and faster.

Notes:

Fasteners

The commercial versions of SOLIDWORKS Professional and Premium, as well as the Educational Edition available to schools, include a hardware library called "SOLIDWORKS Toolbox" which includes nuts, bolts, screws, pins, washers, bearings, bushings, structural steel, gears, etc. in both metric and imperial unit standards.

25.1. - The "SOLIDWORKS Toolbox" is an add-in package that has to be loaded either using the menu "**Tools, Add-ins**" or the "SOLIDWORKS Add-Ins" tab in the CommandManager. Using the menu, we have to load both "SOLIDWORKS Toolbox Library" and "SOLIDWORKS Toolbox Utilities"; using the CommandManager we only have to activate the "SOLIDWORKS Toolbox" command.

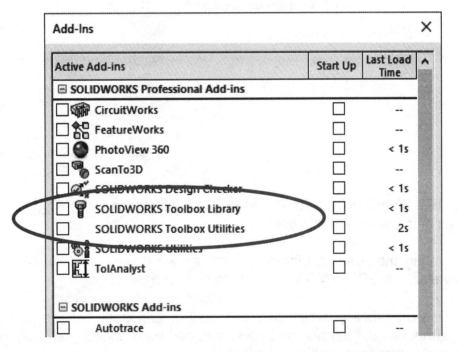

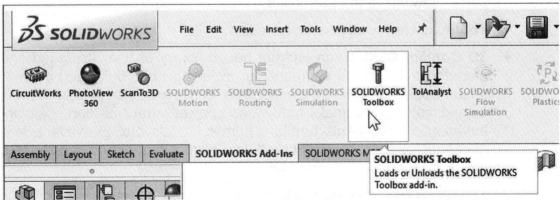

Using "SOLIDWORKS Toolbox" we can add hardware to our assemblies by dragging and dropping components, and SOLIDWORKS will automatically add the necessary mates, saving us time. To make it even more powerful and versatile, it can also be configured to add our own hardware.

25.2. - After activating the "SOLIDWORKS Toolbox" add-in, we can access from the **Design Library**. The Design Library is located in the Task Pane. If the Task Pane is not visible, go to the menu "**View, Toolbars, Task Pane**" to activate it.

 The "SOLIDWORKS Toolbox" add-in <u>was not included in the *Student Design Kit*</u> as of the writing of this book.

Activating the "Design Library" opens a fly-out pane that reveals all the libraries available in SOLIDWORKS. It contains four main areas:

- **Design Library**, which includes built-in and user defined libraries of annotations, features, and parts that can be dragged and dropped into parts, drawings, and assemblies.

- **Toolbox**, which was just described.

- **3D Content Central**, an internet based library of user uploaded and manufacturer certified components, including nuts, bolts and screws, pneumatics, mold and die components, conveyors, bearings, electronic components, industrial hardware, power transmission, piping, automation components, furniture, human models, etc., all available free of charge ready for drag and drop use. All that is needed to access it is an internet connection and log in.

- **SOLIDWORKS Content** allows the user to download weldments libraries, piping, blocks, structural members, etc.

The "Design Library" is a valuable resource for the designer, which can help us save time by not having to model components that are usually purchased, and in the case of the Supplier Certified library, components that are accurately modeled for use in our designs directly from the manufacturer.

25.3. - To add screws to our assembly, select the "Design Library" from the Task Pane and click in the (+) sign to the left of the "Toolbox" to expand it.

After expanding the "Toolbox" we can see the many options available. Depending on how "Toolbox" is configured, some standards may not be available. In this exercise we'll use "ANSI Inch, Bolts, and Screws" and select "Socket Head Screws."

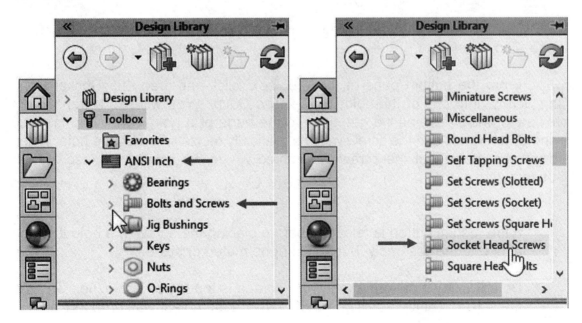

When we select the "Socket Head Screws" folder, we see in the lower half of the Design Library pane the available styles including Button Head, Socket

Head, Countersunk, and Shoulder Screws. For our assembly, we'll use "Socket Head Cap Screws (SHCS)."

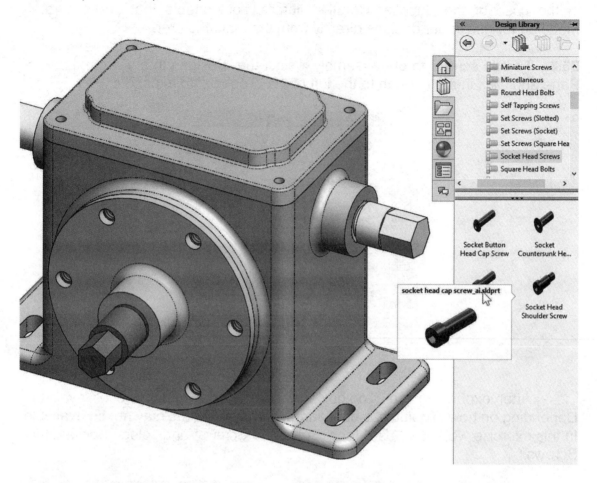

25.4. - The first screws we are going to add to the assembly are the #6-32 screws to the *'Top Cover'*.

From the bottom pane of the "Toolbox" click-and-drag the "Socket Head Cap Screw" into one of the holes in the *'Top Cover'*. You will see a transparent preview of the screw; as we get close to the edge of a hole it will snap in place using SmartMates and the screw will automatically resize to match the hole. When we get the preview of the screw assembled where we want it, release the left mouse button.

 If the mouse button is released before placing the screw in a hole it will be added to the assembly, but it will not be mated or sized.

 Do not worry if the preview in your screen is too big at first. When we drop the screw in place, it will be sized to match the hole it was dropped in.

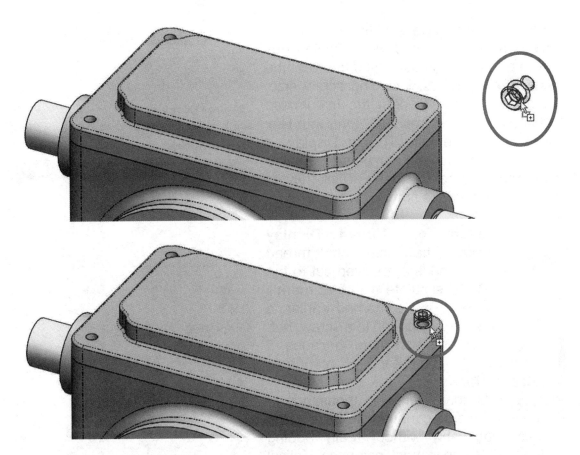

 Notice that the Design Library hides away as soon as we drag the screw into the assembly; if we want the Design Library to remain visible we need to press the "Auto Show" thumb tack in the upper right corner of the Task Pane.

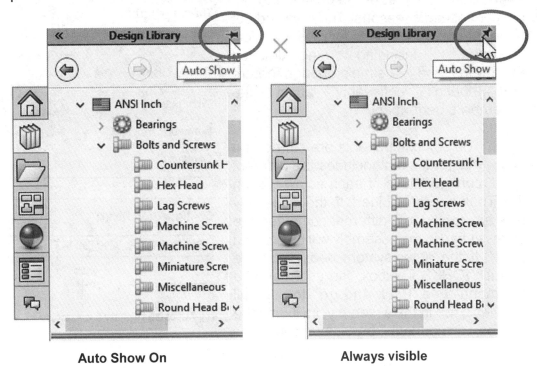

Auto Show On   **Always visible**

25.5. - As soon as we drop the screw in the hole, we are presented with a dialog box and a pop-up menu asking us to select the screw's parameters, including screw size, length, drive, thread length, etc. In our case we need a #6-32 screw, 0.5″ long with Hex Drive and Schematic Thread Display. If needed, a part number can be added at the time the fastener is made. Click OK to create and mate the first screw.

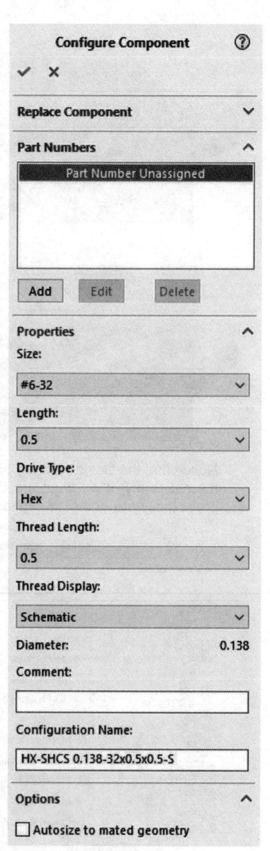

A word on Thread Display options: The "Schematic" thread selected adds a revolved cut to the screw to simulate a thread merely for visual effect. As we learned earlier, a helical thread can be added to a screw, but, in general, it's a bad idea because it consumes computer resources unnecessarily. For example, if an assembly has tens, hundreds or even thousands of fasteners it would certainly slow down the system noticeably without really adding value to a design in most instances. Helical threads are a resource intensive feature that is best left for times when the helical thread itself is a part of the design and not just for cosmetic reasons. The revolved cut gives a good appearance for most practical purposes, including non-detailed renderings, and is a simple enough feature that doesn't noticeably affect the assembly's performance.

Toolbox components are stored in a master file that includes the different configurations of each screw type in the Toolbox data folder. If the assembly files are copied to a different computer, the screws used in the assembly will be created there. If the other system does not have SOLIDWORKS Toolbox, the user can use the menu "**File, Pack and go**" to copy all the files used in the assembly, including the Toolbox components, parts, and drawings.

25.6. - After we click OK, the screw we specified is created with the selected parameters and mated in the hole where it was dropped. At this point we can add identical screws to the assembly if needed. In our example, we'll click in the other 3 holes of the *'Top Cover'* to add the rest of the screws. Notice the graphic preview of the screw when we move the mouse; as we get close to a hole the screw snaps to it adding a Peg-in-Hole SmartMate. After adding the four screws in the cover click "Cancel" to finish the command.

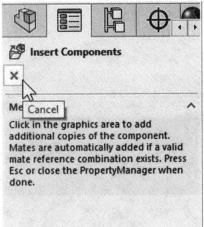

25.7. - Now we need to add *'¼-20 Socket Head Cap Screws'* to the *'Side Covers'*. Open the "Design Library" tab and, just as in the previous step, drag and drop the "Socket Head Cap Screw" in one hole of the *'Side Cover'*, but be careful to "drop" the screw in the correct location. If you look closely, there are two different edges where you can insert the screw: one is in the *'Side Cover'*, and the other is in the *'Housing'*. The screw will be mated to the hole you "drop" the screw in. Use the snap preview to help you identify the correct one.

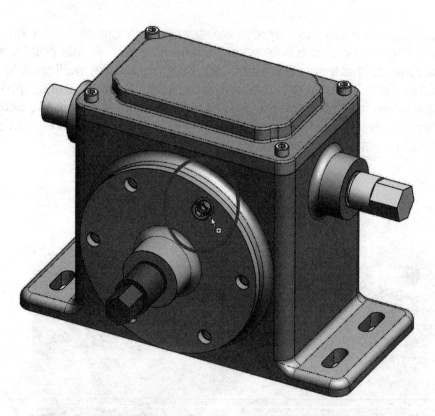

25.8. - After dropping the screw, the properties box is displayed. Select the ¼-20 x 0.5″ long screw with Hex Drive and Schematic thread display. Click OK to create the screw and mate it in place, and then close the "Configure Component" command; in this step we only need to add one screw.

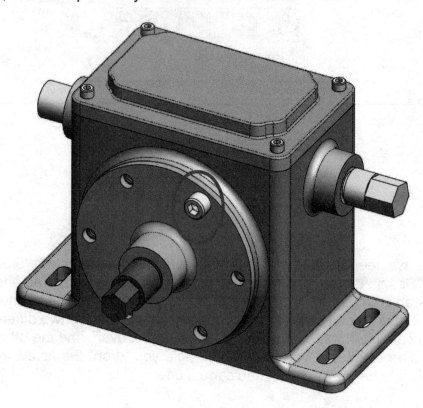

25.9. - Instead of manually adding the rest of the screws, we'll make a component pattern using the first screw. Just like a feature pattern in a part, we can add linear or circular **component patterns** in an assembly, but in this case we are going to use a different type of pattern that allows us to make a component pattern to match a part's *feature pattern*. Think of it this way: we make a pattern of bolts to match the pattern of holes in the *'Side Cove'*. This way, *IF* the pattern of holes changes (more or less holes), the pattern of screws in the assembly updates to match. In the Assembly tab, select "**Pattern Driven Component Pattern**" from the drop down menu in the "**Linear Component Pattern**" or the menu "**Insert, Component Pattern, Pattern Driven**."

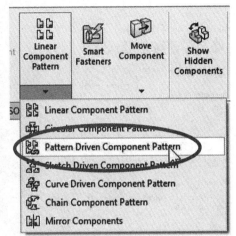

25.10. - In the "Components to Pattern" selection box select the ¼-20 screw previously added, and in the "Driving Feature" select any of the patterned holes in the *'Side Cover'* except the original hole. When selecting a hole's face, make sure the pop-up highlight shows the "*CircPattern1*" name; this way we'll know it's a patterned hole, otherwise we'll get a warning. After selecting the patterned hole's face we'll see the preview of the screws pattern; click OK to continue.

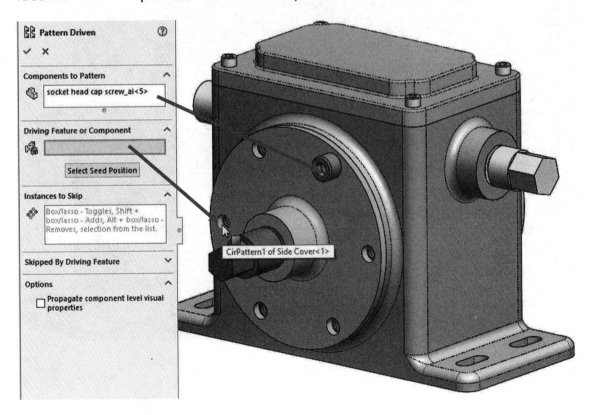

543

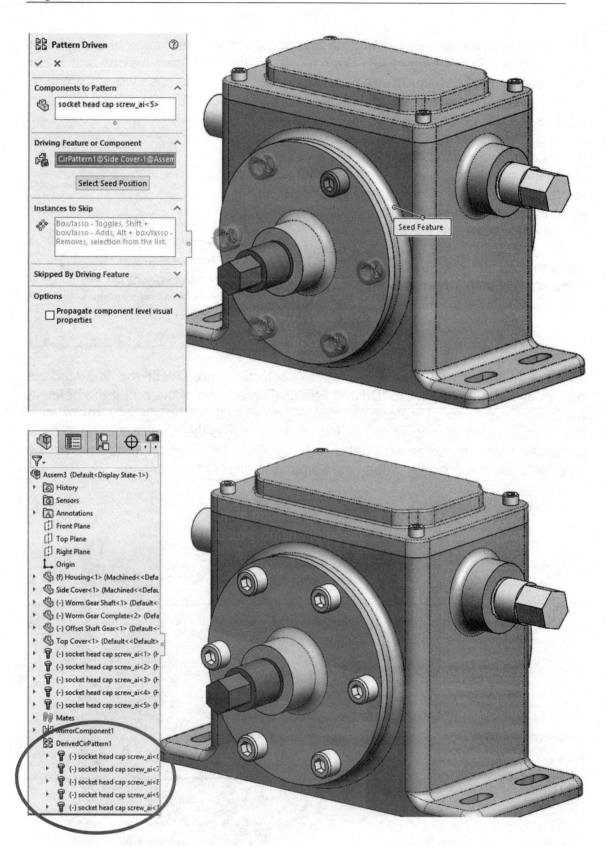

25.11. - Select the "**Mirror Components**" command from the menu "**Insert, Mirror Components**" or the drop-down menu in the "**Linear Component Pattern**" icon.

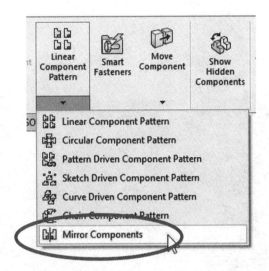

25.12. - Use the assembly's "*Front Plane*" as the "Mirror Plane" and in the "Components to Mirror:" selection box pick the ¼-20 screw and the "*DerivedCirPattern1*" feature (which will add all the screws in the pattern). Click in the "Next" arrow to continue to the next step and then OK to finish the mirror.

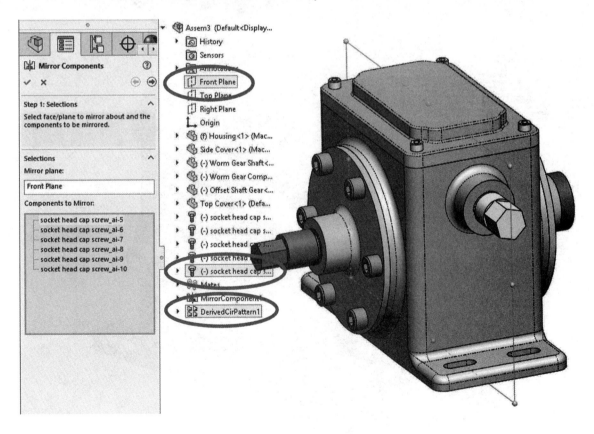

25.13. - Save the finished assembly as '*Gear Box Complete*'.

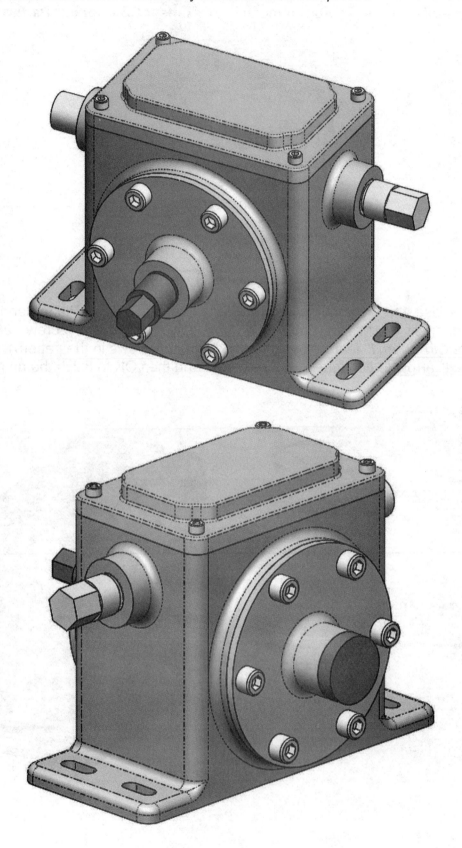

Configurations using Design Tables

Previously we covered how to make configurations of parts by manually suppressing features and changing dimensions. Configuring parts using this technique may be adequate when we have 2, maybe 3 configurations and a couple of configured features and/or dimensions, but it becomes very difficult to keep track of changes in the different configurations when we have 5, 10, 20, or more configurations with multiple features and dimensions.

In these cases, adding a Design Table is the best way to manage configurations. A Design Table is an Excel file embedded in (or linked to) a SOLIDWORKS part or assembly that controls dimension values, feature suppression states, configuration names, custom file properties, etc., allowing us to easily manage all the configurations in a single table.

 The use of Design Tables in SOLIDWORKS 2018 (Educational version 2018-2019) requires Microsoft Excel 2010, 2013 or 2016. No other spreadsheet software is supported.

26.1. - In this exercise we'll create a simplified version of a Socket Head Cap Screw, and then add multiple configurations using a Design Table. The first step is to create a model to configure using a template in inches. Make a new part and add the following sketch in the "*Front Plane.*" Dimensions are in inches. Remember to add the centerline that will allow us to add the doubled diameter dimensions.

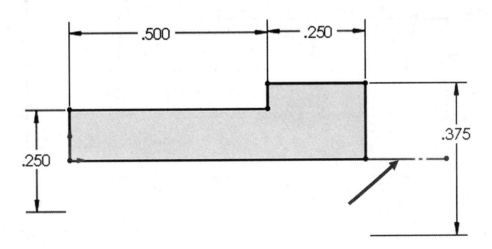

26.2. - SOLIDWORKS automatically assigns an internal name to all dimensions using the format *name@feature name*. We can temporarily see a dimension's name by resting the mouse pointer on top of it, or we can make them always visible (useful when making design tables) using the menu "**View, Hide/Show, Dimension Names**" or from the drop down View command icon. The dimension names in your case may be different from the book, based on the order they are added to the sketch.

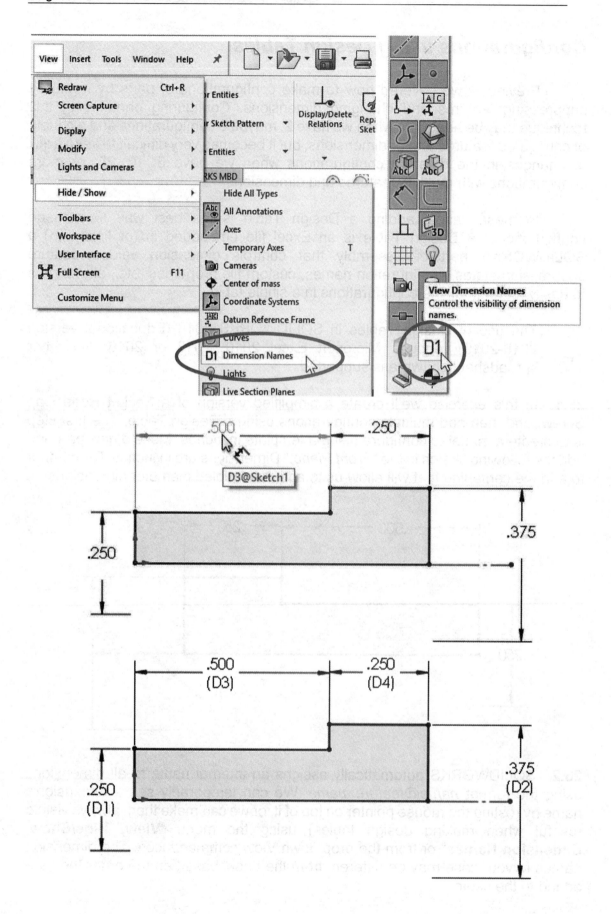

26.3. - When adding configurations with different dimension sizes, especially when we use a design table, it's a good idea to change the configured dimension's name and use one that allows us to easily identify them. To rename a dimension, select it in the screen and type a new name in its properties. When renaming dimensions, just add the name and don't worry about the *"@feature_name"* part; SOLIDWORKS adds this part automatically when we are finished. Turn on the Dimension Names and rename the sketch dimensions as shown.

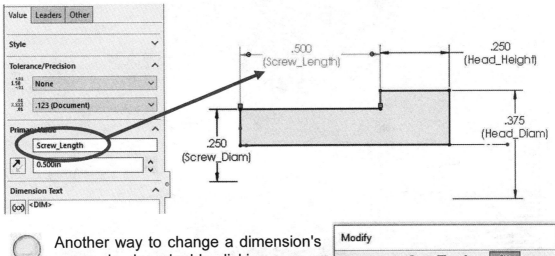

Another way to change a dimension's name is by double-clicking on a dimension, and then type a new name in the top field of the "Modify" dialog box.

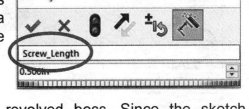

26.4. - After completing the sketch add a revolved boss. Since the sketch dimensions are automatically hidden after we make a feature, we need to make them visible while editing the part before creating the Design Table. Right-mouse-click in the "Annotations" folder in the FeatureManager, and activate the option "Show Feature Dimensions" from the pop-up menu.

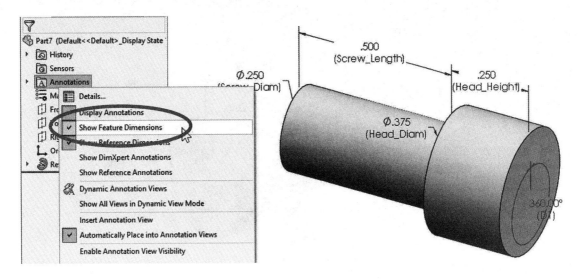

549

26.5. - Add a hexagonal cut in the head of the screw; make it 0.125″ deep.

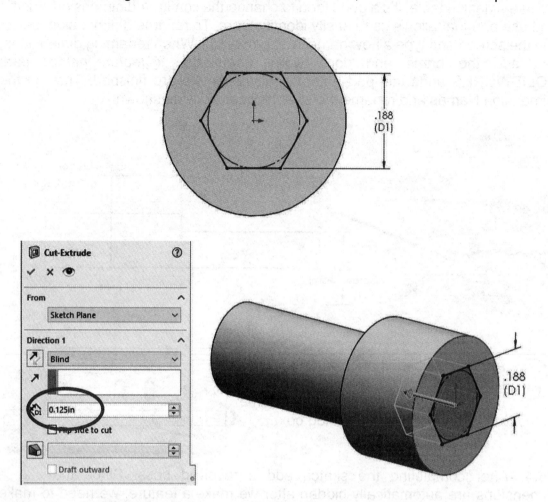

26.6. - After adding the extruded cut, change the cut dimension's names to "*Hex_Drive*" and "*Hex_Depth*." Note that dimensions added in the sketch are shown in black, and dimensions added by a feature are blue.

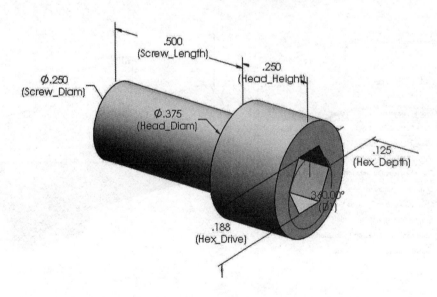

26.7. - Add a 0.015″ x 45 deg. chamfer to the head and tip of the screw as a finishing touch. Dimensions are hidden for better visibility.

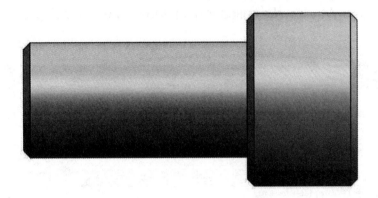

26.8. - When the part is finished, and renaming the dimensions, we are ready to add the design table. Go to the menu "**Insert, Tables, Design Table**."

Use the "Auto-create" option; it makes adding a design table easier. Leave the rest of the options to their default value.

When we click on OK, we are presented with a list of all the dimensions in the model; this is where we can select the dimensions that we want to add to the Design Table.

Model features and more dimensions can be added later to the table if needed. For this model, select the dimensions indicated. To make multiple selections, hold down the "Ctrl" key while selecting. This is the reason why renaming dimensions is handy; now we know what they are.

551

26.9. - After selecting the dimensions to configure, click OK. An Excel spreadsheet will be automatically embedded in the SOLIDWORKS part. Since we are using Excel embedded *inside* SOLIDWORKS, the menus and toolbars are changed to Excel according to Windows' application linking and embedding behavior.

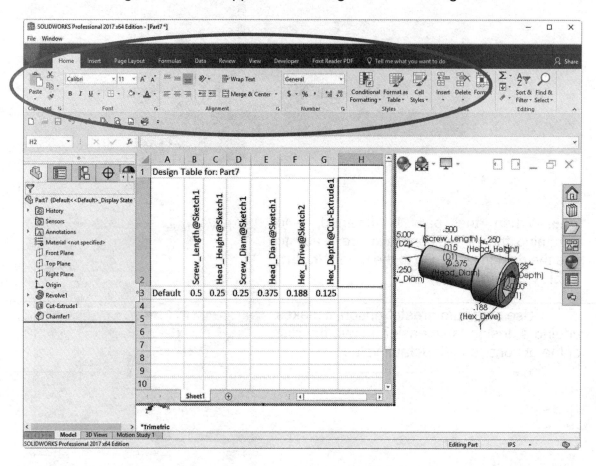

The normal behavior of embedded documents in Windows is to add a thin border around the embedded document. If needed, we can move the Excel file by dragging this border, or resize it from the corners. The borders are very small; be careful not to click outside the spreadsheet or you'll exit Excel and go back to SOLIDWORKS.

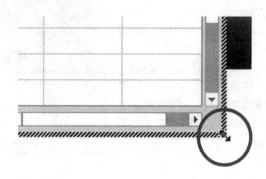

IMPORTANT: If you accidentally click outside the Excel spreadsheet, this is what will happen: you may be told that a configuration was created, or not, depending on whether the design table was changed or not. To go back to editing the design table, go to the ConfigurationManager, expand the "Tables" folder, make a right-mouse-click in the "Design Table," and select "Edit Table." This will get you back to editing the design table embedded in Excel.

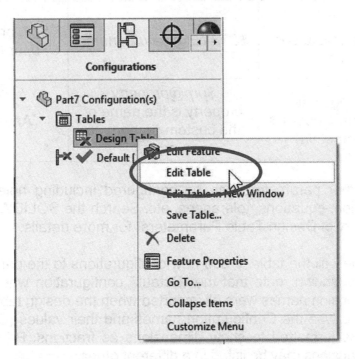

We can resize the table and columns if needed. As we always have at least one configuration, "Default" is listed in our Design Table with the corresponding values listed under each parameter. We can zoom in or out in the graphics area using the Shift + middle-mouse-button, rotate the model with the middle-mouse-button, or pan around with Ctrl + middle-mouse-button.

Making a left or right mouse click anywhere outside of the Excel window or the menus will exit Excel and send us back to SOLIDWORKS.

26.10. - The Design Table includes the configuration names, all the configured parameters, and their values for each configuration. The first row holds information about the part and is not used by the design table; we can type anything we want in this row, as the configuration's data starts in the second row.

Configurable parameters are listed in the second row starting at the second column, and configuration names are listed in the first column starting in the third row. Here is a list of some of the most commonly configured parameters in a part:

Parameter	Format in Design Table	Possible Values
Dimensions (To control a dimension's value)	*name@feature_name*	Any numerical value
Features (To control if a feature is suppressed or not)	*$STATE@feature_name*	*S, 1,* or *Suppressed* *U, 0,* or *Unsuppressed*
Custom properties (To add custom properties to a configuration)	*$prp@property* *property* is the name of the custom property to add	Any text string

 Many other parameters can be configured including hole wizard sizes, description, equations, tolerances, etc. Search the SOLIDWORKS help for "Summary of Design Table Parameters" for more details.

26.11. - Now we'll fill the table to add new configurations to the part. Edit the table and fill it out as shown; note that the "Default" configuration was renamed. The configured dimension names were all imported when the design table was created. We only need to type the Configuration names and their values (Cells A3 to G8). Column "F" was formatted to show dimensions as fractions. Be aware that the configured dimensions may be listed in a different order.

	A	B	C	D	E	F	G	H
1	Design Table for: Part1							
2		Screw_Length@Sketch1	Head_Height@Sketch1	Head_Diam@Sketch1	Screw_Diam@Sketch1	Hex_Drive@Sketch2	Hex_Depth@Cut-Extrude1	
3	6-32x0.5	0.5	0.138	0.226	0.138	7/64	0.064	
4	6-32x0.75	0.75	0.138	0.226	0.138	7/64	0.064	
5	10-32x0.5	0.5	0.19	0.312	0.19	5/32	0.09	
6	10-32x0.75	0.75	0.19	0.312	0.19	5/32	0.09	
7	0.25x0.5	0.5	0.25	0.375	0.25	3/16	0.12	
8	0.25x0.75	0.75	0.25	0.375	0.25	3/16	0.12	
9								
10								

Sheet1

After entering all the information in the Design Table, click anywhere inside the graphics area of SOLIDWORKS to exit the Design Table (and Excel). SOLIDWORKS will alert us of the newly created configurations. Click OK to acknowledge them and continue.

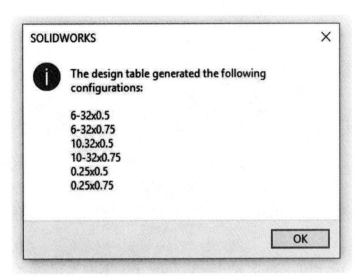

26.12. – Now the ConfigurationManager shows the configurations added in the Design Table. Note that configurations created with the Design Table have an Excel icon next to their name, and the dimensions configured in the table are displayed with a magenta color (default setting). Activate the different configurations to see the screw sizes created with each one.

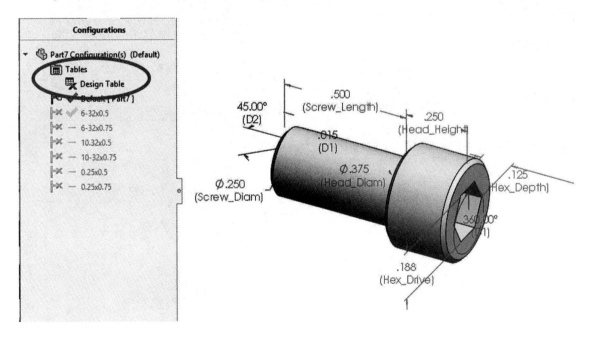

26.13. - The original *"Default"* configuration is not needed, so we need to delete it. Before deleting it, activate a different configuration first, since we cannot delete the currently active configuration. Right-mouse-click on *"Default"* and select "Delete."

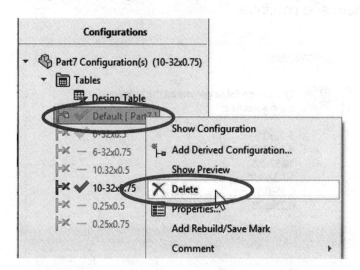

26.14. – Now we are going to learn how to include additional features or parameters in the Design Table. Right-mouse-click in "Design Table" and select "**Edit Table**." When asked to add Configurations or Parameters, click "Cancel" to continue, as we don't want to add any of the parameters shown to the Design Table.

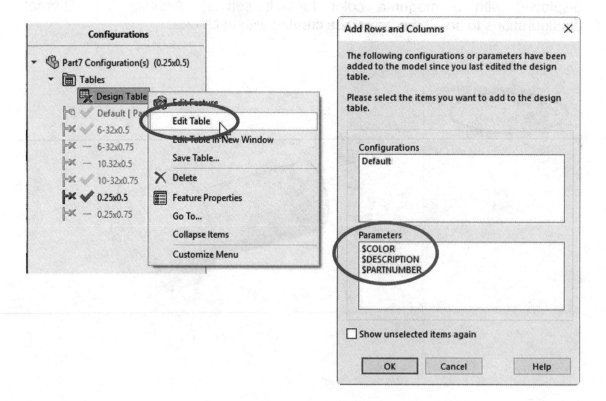

When the table is presented, the next available cell for adding parameters is pre-selected (in our case "H2"). To add the chamfer as a configurable parameter, we can type *$STATE@Chamfer1* directly in the cell, or select the FeatureManager tab to view the features and double-click in the "*Chamfer1*" feature.

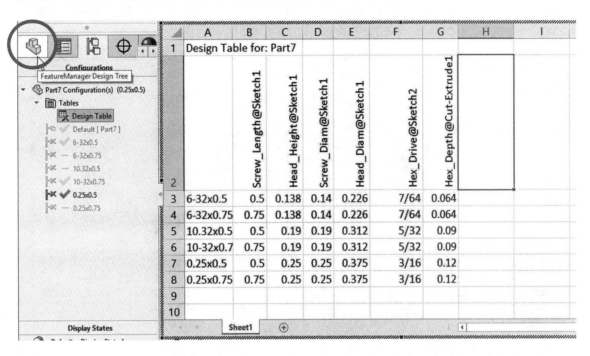

Double-click in the "*Chamfer1*" feature to add the correct nomenclature in the table and its current suppression state, in this case *Unsuppressed*. If the cell is left empty, the assumed value is *Unsuppressed*. Optionally we can type **S** for *Suppressed* or **U** for *Unsuppressed*.

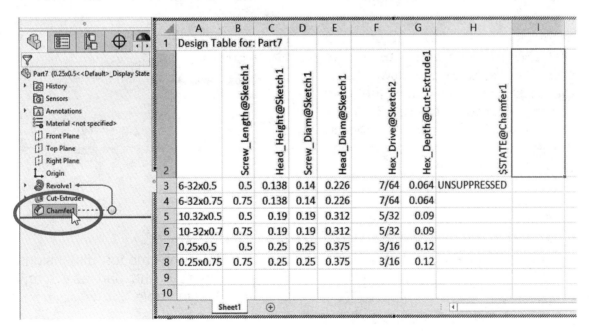

26.15. – Now we are going to modify the design table to add a new configuration. Copy the last configuration in the next row, and change its name by adding "*NoChamfer*" to the name (SOLIDWORKS does not allow duplicate names for configurations), and suppress the feature "*Chamfer1,*" just as an exercise. For short we used **U** and **S** for suppression states. Feel free to format the Design Table to your liking, as long as the layout isn't modified.

	A	B	C	D	E	F	G	H	I	J
1	Design Table for: Part7									
2		Screw_Length@Sketch1	Head_Height@Sketch1	Screw_Diam@Sketch1	Head_Diam@Sketch1	Hex_Drive@Sketch2	Hex_Depth@Cut-Extrude1	$STATE@Chamfer1		
3	6-32x0.5	0.5	0.138	0.14	0.226	7/64	0.064	U		
4	6-32x0.75	0.75	0.138	0.14	0.226	7/64	0.064	U		
5	10.32x0.5	0.5	0.19	0.19	0.312	5/32	0.09	U		
6	10-32x0.75	0.75	0.19	0.19	0.312	5/32	0.09	U		
7	0.25x0.5	0.5	0.25	0.25	0.375	3/16	0.12	U		
8	0.25x0.75	0.75	0.25	0.25	0.375	3/16	0.12	U		
9	0.25x0.75-NoChamfer	0.75	0.25	0.25	0.375	3/16	0.12	S		
10										

Sheet1

After finishing the changes in the Design Table, click in the graphics area to return to SOLIDWORKS as we did before; as soon as we leave Excel we get a message letting us know the new configuration was created. Click OK to acknowledge and continue. Switch through the different configurations to see the differences.

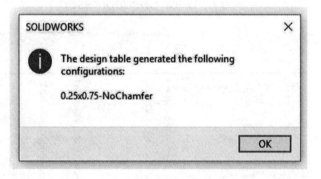

SOLIDWORKS ×

The design table generated the following configurations:

0.25x0.75-NoChamfer

OK

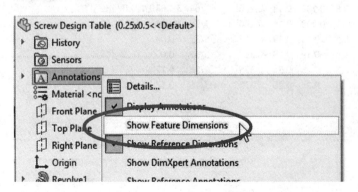

Hide the model dimensions and save the part as "*Screw Design Table.*"

26.16. – The finished configurations look like this:

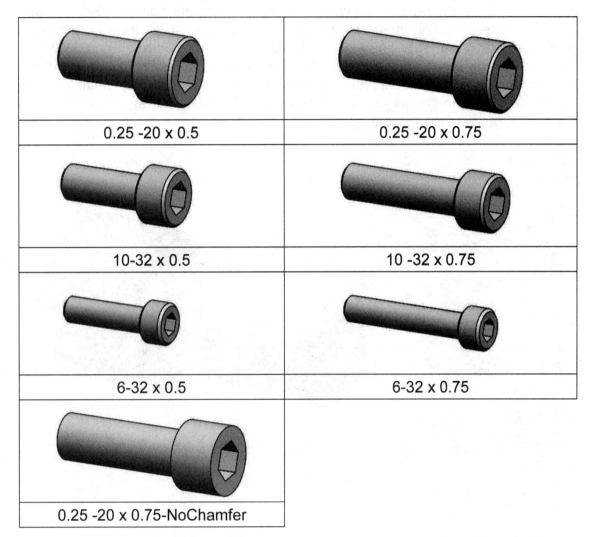

0.25 -20 x 0.5	0.25 -20 x 0.75
10-32 x 0.5	10 -32 x 0.75
6-32 x 0.5	6-32 x 0.75
0.25 -20 x 0.75-NoChamfer	

26.17. - For the screw we just made to be even more useful we must add it to our own components library, and give it the ability to auto-assemble, just like the SOLIDWORKS Toolbox components.

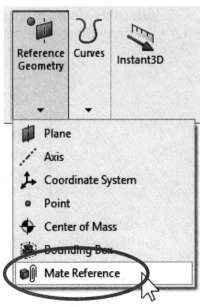

First, we need to add a new feature called **"Mate Reference"**; this feature tells the component what type of geometry to look for when we drag-and-drop it into an assembly, in this case, to auto-assemble using a "Peg-in-Hole" mate. Select the menu **"Insert, Reference Geometry, Mate Reference"** or from the drop-down menu in the "Reference Geometry" command.

In the "Primary Reference Entity" select the indicated edge. For the "Mate Reference Type" the only option we can use with a circular edge is "Default," and in "Mate Reference Alignment" leave the option to "Any"; this way we can flip the alignment using the "Tab" key if needed. Click OK to add the reference. With this mate reference, the screw will have the "Peg-in-Hole" behavior when it is drag-and-dropped from the "Design Library" into an assembly. If multiple mate references need to be added to a component, they can be renamed for easier reference, in this case, "Peg-In-Hole."

After the Mate Reference is added we can see it in the FeatureManager, where we can edit or delete it if needed. Save the part to continue.

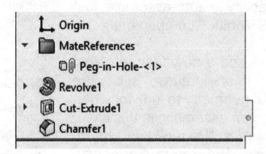

For simple mates we can use a single mate reference; for more complex mates we can use a second or even a third reference. Mate References can be defined using faces, edges, vertices, axes, planes or the origin. The following table shows the type of entity and mate options available. Using "Default" will try to add the default mate type for the entity in the assembly.

Entity		Mate Reference Types available
Cylindrical Face		• Default • Tangent • Concentric
Round Edge		• Default
Flat Face / Plane		• Default • Tangent • Coincident • Parallel
Axis		• Default • Concentric • Coincident • Parallel
Vertex/Origin		• Default • Coincident

26.18. – After adding the Mate Reference to the screw, the next step is to add it to the "Design Library." Open the "Design Library" tab, and press the "Keep Visible" push pin; maintaining the panel open is often helpful but not always required.

Expand the "Design Library" and select the "parts, hardware" folder. To add the screw to the library we can either:

- Select the "Add to Library" icon at the top of the Design Library

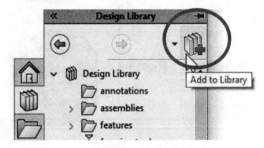

- Or drag-and-drop the screw from the top of the FeatureManager into the lower half pane or the folder where we want to save it. Be aware that for this method to work the Task Pane must remain visible.

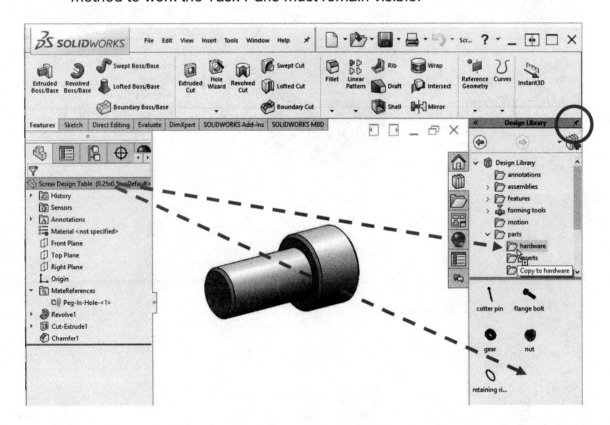

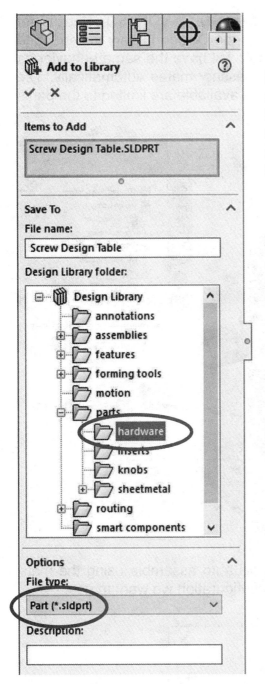

Both options will show the "**Add to Library**" command. The main difference between the two options is that using the drag-and-drop option pre-selects the part to add to the library, and using the button we have to select the part in the "Items to Add" selection box.

After selecting (or pre-selecting) the screw, select the folder where we want to store the library part and optionally add a description.

Leave the "File type" as Part. After pressing OK, the screw is added to the hardware folder and is available for use in assemblies.

Before using the screw from the library in an assembly, close the screw part; otherwise we'll be asked if we want to use the currently open file.

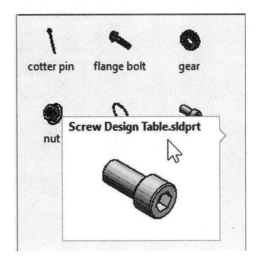

Alternate to SOLIDWORKS Toolbox: In case SOLIDWORKS Toolbox is not available, we can use the screw added to the Design Library instead, and dragging and dropping it into an assembly will have the same behavior as the screw from the Toolbox, adding all the necessary mates automatically. The major difference is that in this case the only sizes available are limited to the part's configurations.

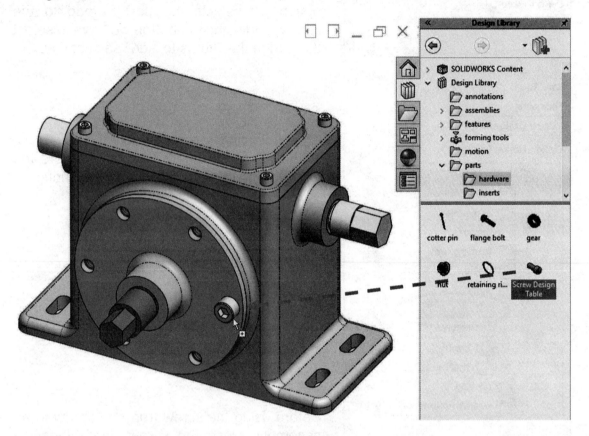

After we drop the screw in the hole it will auto assemble using the mate reference and then we'll be able to select the configuration we want to use.

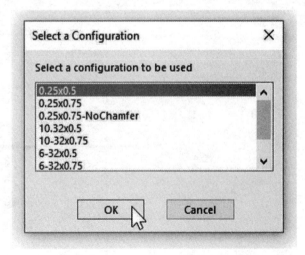

Interference Detection

27.1. - After the assembly is complete, one of the tools that can help us find out if we have problems in our assembly is the "**Interference Detection**" command. Go back to the '*Gear Box Complete*' assembly, and select the "**Interference Detection**" command from the "Evaluate" tab in the CommandManager or from the menu "**Tools, Evaluate, Interference Detection**."

The components we have made, once assembled, have a built-in interference by design, and its purpose is to help us learn how to use this tool. Select the "**Interference Detection**" command and click on "Calculate."

By default, the interference is calculated between all components in the assembly; however, in the "Selected Components" box we can limit the number of components used in the interference calculation, and/or optionally select which components to exclude from the calculation in the "Excluded Components" section, greatly improving performance, especially when working with larger assemblies.

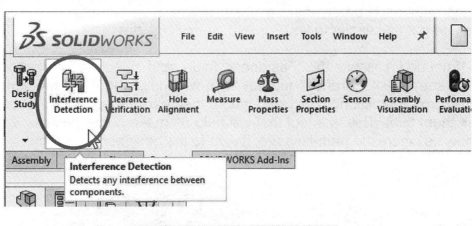

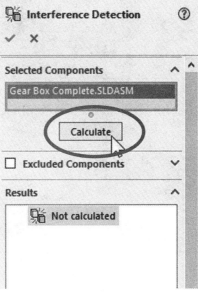

27.2. - After running the calculation, we see multiple interferences listed in the "Results" window. Most of these are fasteners because the holes they fit in are not threaded and the screw threads interfere with them. Since we are aware of those, we are going to turn on the "Create fasteners folder" option. When activated, fastener interferences will be grouped together, making it easier to identify problem areas that are not related to fasteners. The option "Make interfering parts transparent" is on by default. This option allows us to easily identify the interfering volumes.

NOTE: The fasteners folder option only works with Toolbox generated fasteners.

Another useful option is to hide non-interfering components; this way, the only components shown in the screen will be the ones interfering with each other, making the problem areas stand out and easier to find.

27.3. - If needed, collapse the "Fasteners" folder and select an interference from the list to see the interference; and now that we know where the problem is, we can take corrective action. In this step ignore the gear-on-gear interferences; we'll work on them later. This is a tool that will help us make better designs, but it will only show us where the problems are; it's up to the designer to fix them.

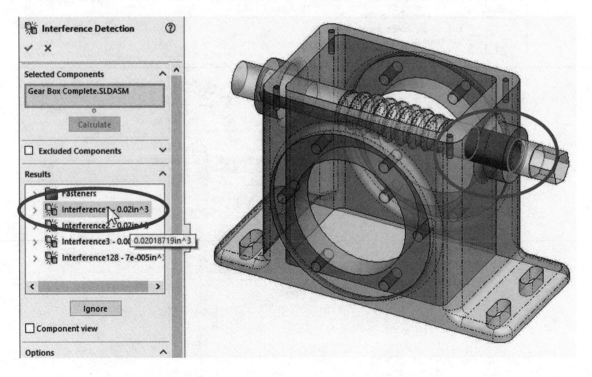

27.4. - To avoid checking for and displaying interferences we are not interested in, in this case gear-on-gear, select each of those interferences and click the "Ignore" button to hide them. In our example we have four instances; your case may be different depending on the position of each gear. Optionally we can rotate the gears to a position where they do not interfere with each other.

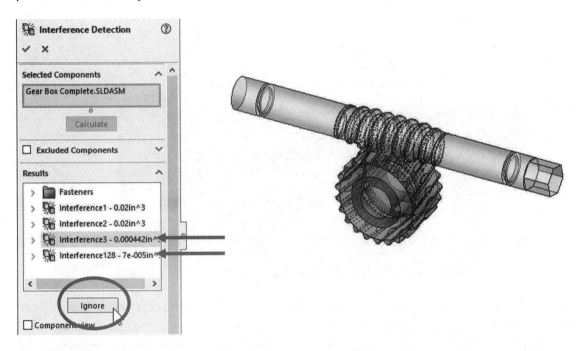

27.5. –To correct the interference in our design (since we know the hole in the '*Housing*' is smaller than the *'Offset Shaft's* diameter), we need to make the shaft's diameter smaller. To make this change, exit the "**Interference Detection**" command, and double-click in the shaft's cylindrical surface to reveal its dimensions.

Double-click the diameter dimension and change its value to 0.575″. Click the "Rebuild" icon as shown, and then OK to complete the "Modify" command. Rebuilding the model will tell SOLIDWORKS to update any models that were changed, like the *'Offset Shaft'* in this case.

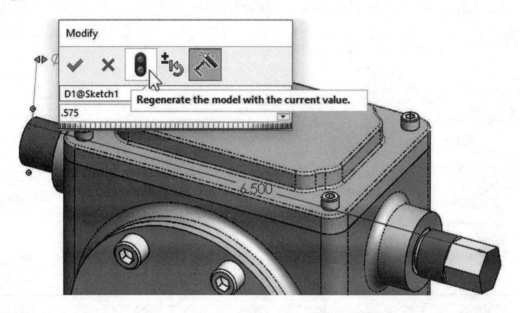

27.6. - After changing the shaft's diameter run the "**Interference Detection**" command again to confirm that we have resolved the interference between the shaft and the housing. The only interference displayed should be the fasteners. Be aware that we may see additional gear-on-gear interferences when the shaft's diameter changed; if this is the case, ignore them. Click OK to finish. In the next image non-interfering components are not hidden.

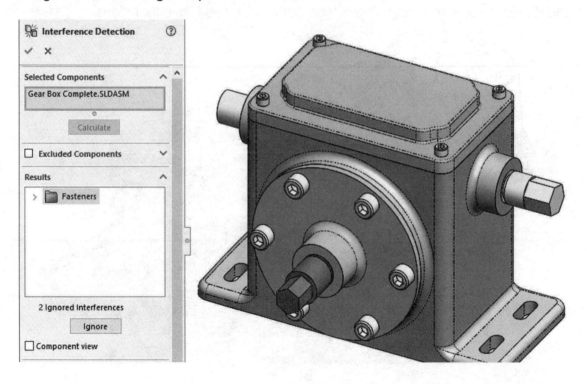

Assembly Configurations

28.1. - We have made two different versions of the *'Offset Shaft'*, one with a gear and one without it. Just as part configurations can show different but similar versions of a part, we can create assembly configurations to show different component configurations, parts, sizes, options, etc. While in the assembly, select the ConfigurationManager tab, add two new configurations and rename them "*With Gears*" and "*Simplified.*" After adding the new configurations rename the default configuration to "*No Gears.*"

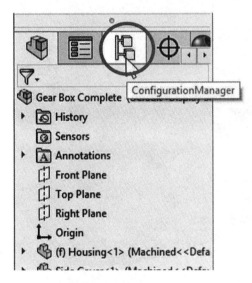

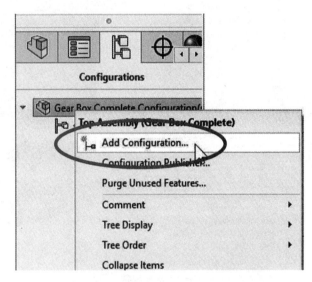

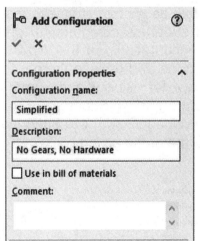

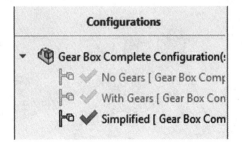

569

28.2. - Split the FeatureManager to show components and configurations, and activate the "*No Gears*" configuration.

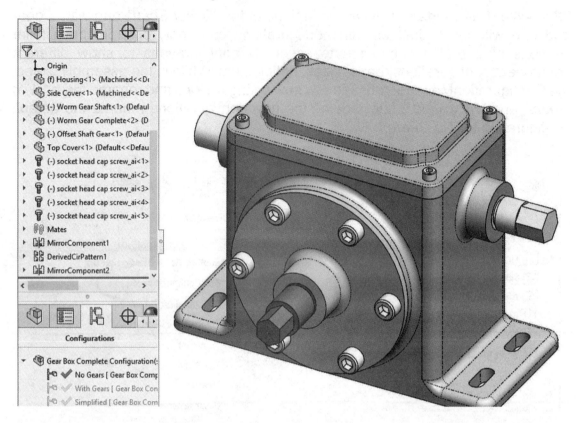

28.3. - Just as we can suppress features in a part's configuration, in an assembly we can suppress components (parts and sub-assemblies) and use a different part's configuration. For the "*No Gears*" configuration, we'll suppress the existing geared parts and replace them with the simplified version of each part. Hide the '*Top Cover*' for visibility, and suppress the '*Worm Gear Complete*' and '*Offset Shaft Complete*' parts. Suppressed components are shown in grey in the FeatureManager, and, like suppressed features in a part, they affect the weight and volume of the assembly.

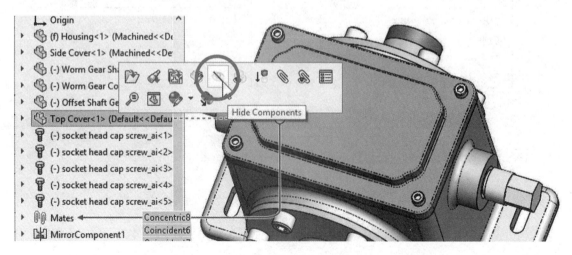

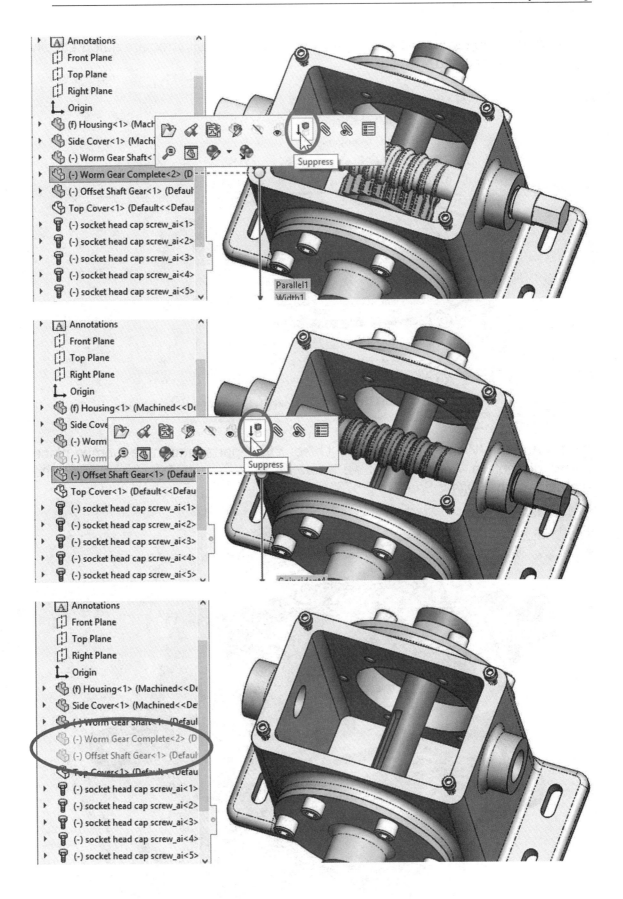

After suppressing a component in the assembly, its mates are automatically suppressed, too.

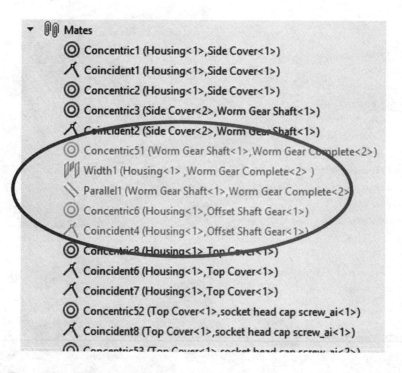

28.4. - Add the '*Worm Gear*' and the '*Offset Shaft*' parts, and mate them in place as we did before. And now that we know we have an interference, change the diameter of the '*Offset Shaft*' to 0.575".

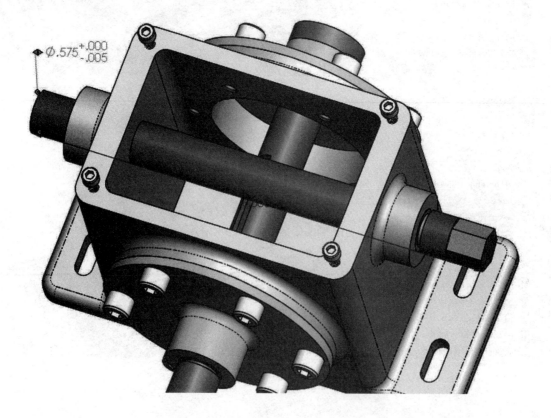

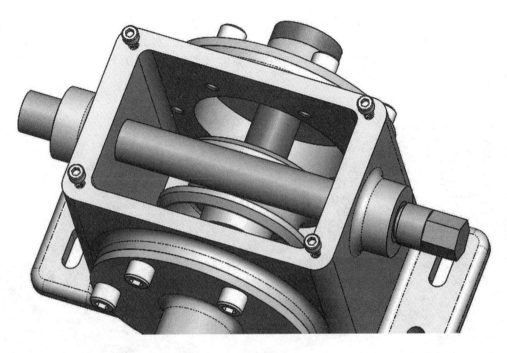

28.5. – After completing the "*No Gears*" configuration, switch to the "*With Gears*" assembly configuration. Notice the two versions of the offset shaft and worm gear overlapping each other. The reason is that newly added components to an assembly with multiple configurations are not automatically suppressed in the inactive configurations; therefore, we need to suppress the simplified '*Worm Gear*' and '*Offset Shaft*' that were just added in this configuration.

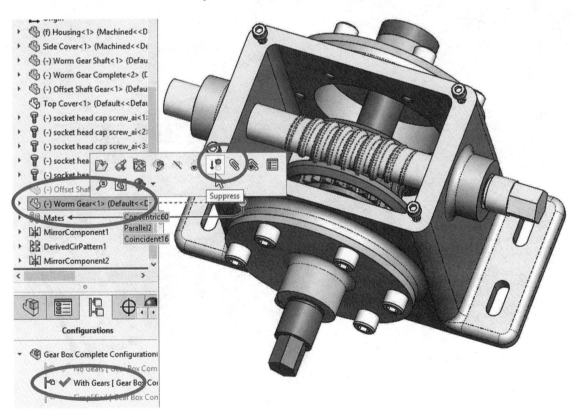

28.6. – Now switch to the "*Simplified*" configuration and suppress the geared components and all fasteners, making it similar to the "*No Gears*" configuration but without fasteners. Multiple components can be selected to suppress (or un-suppress) at the same time.

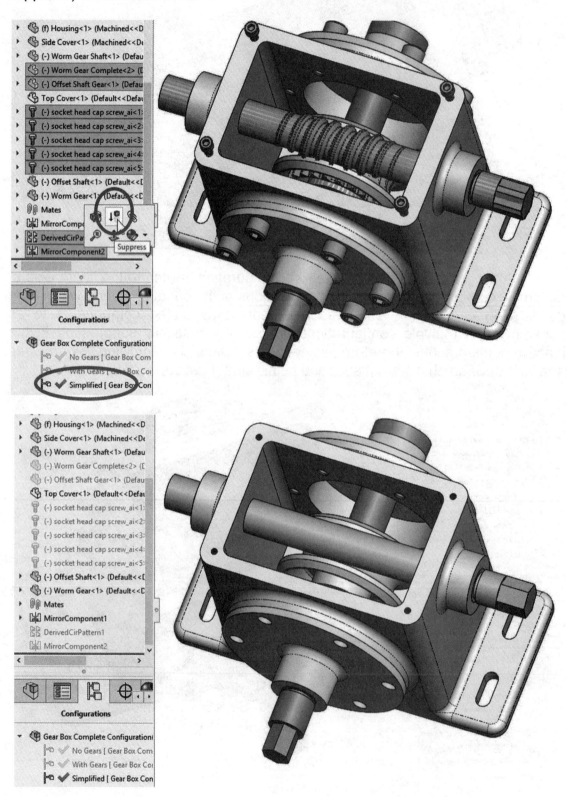

28.7. – When components are added to an assembly with existing configurations, by default, the mates added in the active configuration are suppressed in the other configurations. In our assembly, we added and mated the '*Worm Gear*' and '*Offset Shaft*' while the "*No Gears*" configuration was active; therefore, the mates added to locate these parts are suppressed in the '*With Gears*' and '*Simplified*' configurations.

Since the '*Worm Gear*' and '*Offset Shaft*' are used in the '*Simplified*' configuration, we need to Unsuppress their mates. At this point, we have added many mates to our assembly. To locate the mates we need to Unsuppress in the '*Simplified*' configuration, we have two options:

1. Select the '*Offset Shaft*' and the pop-up list will show the name of the mates used by this component; expand the mates folder and Unsuppress them.

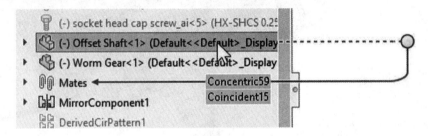

2. After selecting the '*Offset Shaft*' activate the PropertyManager tab at the top. The mates used by the selected component will be listed, and they can be Unsuppressed here. Unsuppress both mates to continue.

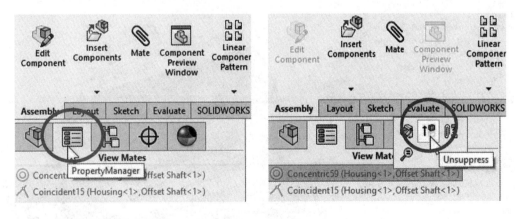

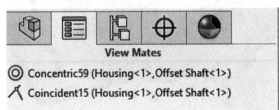

If these mates are not resolved (unsuppressed), the components will move out of place if dragged. Save the finished assembly with configurations.

After unsuppressing these mates, unsuppress the mates of '*Worm Gear*'.

The finished configurations for the Gear Box are as follows. The '*Top Cover*' has been hidden in all three configurations for visibility:

No Gears Configuration

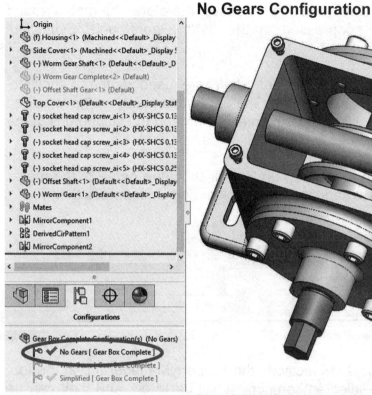

Simplified Configuration

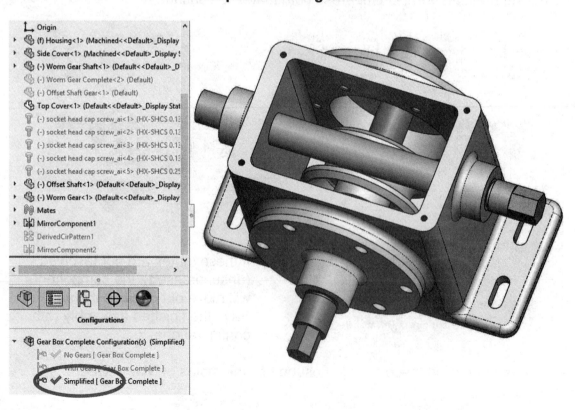

With Gears

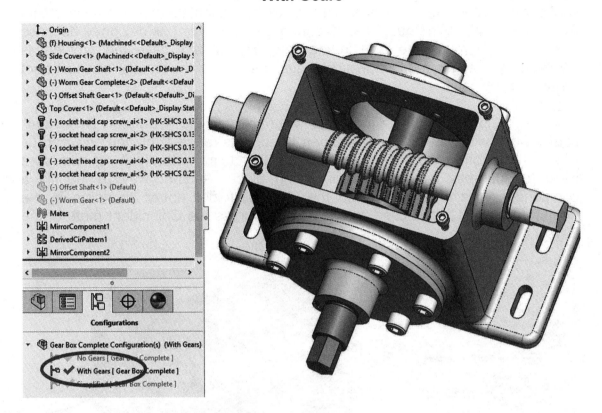

28.8. - When we change a part's dimension in the assembly, the change is propagated to the part and its drawing. After finishing the assembly and having changed the '*Offset Shaft*' diameter, open it's drawing to verify the drawing is updated. Save and close the drawing file. If asked to save modified files, click on "Save All" to save both the part and the drawing.

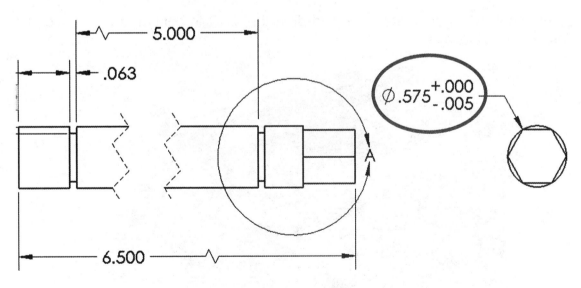

28.9. - After completing the assembly configurations activate the *"With Gears"* configuration. With the '*Top Cover*' hidden, we realize that there is another problem with our assembly: The geared shaft cannot be assembled as designed because it will not fit through the hole in the '*Housing*'. In this case the problem is intentionally exaggerated to make it obvious, but most of the time in real life it's not always this obvious.

To help us find these problems, we must use the **"Collision Detection"** command; it will allow us to move a part through its range of motion and alert us when it hits another component. To allow the shaft to move freely along its axis we need to suppress the mate holding it in place. Another way to identify a component's mates is to expand the **"Mates"** folder under its name in the FeatureManager. In this case, we need to suppress the Coincident mate of the '*Offset Shaft Gear*'.

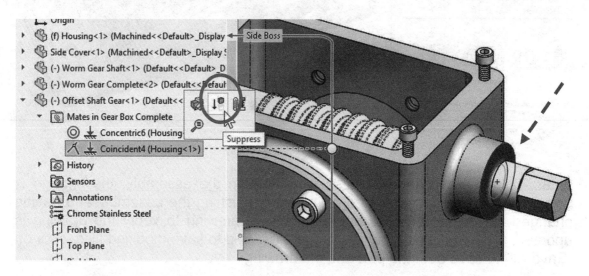

After suppressing the mate, click and drag the shaft to confirm it can move and rotate it about its axis.

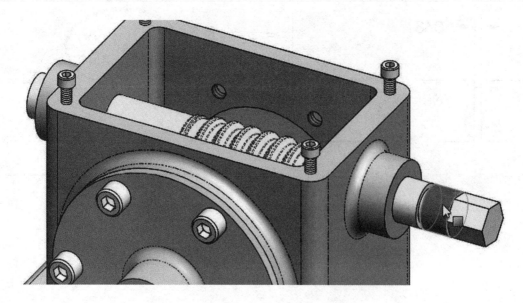

28.10. – When we work with assemblies, often we need to simplify the view by hiding the components not directly related to the task at hand; in this case we are interested in the interferences between the '*Housing*' and the '*Offset Shaft Gear*'. A quick way to isolate the components is to hold down the "Ctrl" key, pre-select the shaft and the housing, and make a right-mouse-click to select the **"Isolate"** command. Every component not pre-selected will be temporarily hidden; a toolbar with an **"Exit Isolate"** button will become visible. Pressing the "**Exit Isolate**" button will show the hidden components and return to the previous view state.

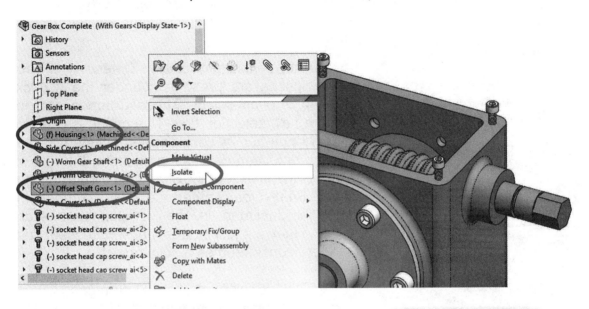

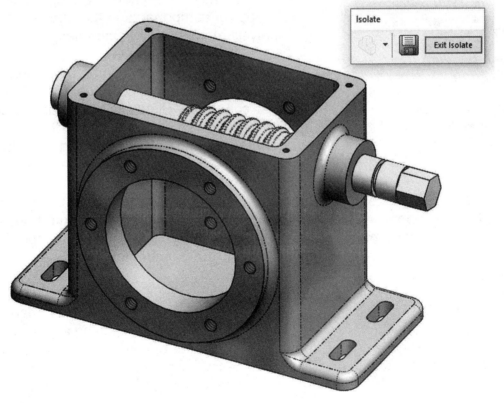

28.11. – Now we are going to run the collision detection. From the Assembly tab select the "**Move Component**" command. This command is essentially the same as dragging a component in the graphics area; the main difference is with this command we have more options, including "Collision Detection" and "Physical Dynamics."

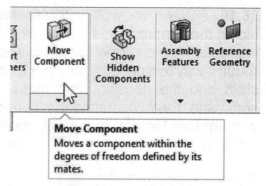

Move Component
Moves a component within the degrees of freedom defined by its mates.

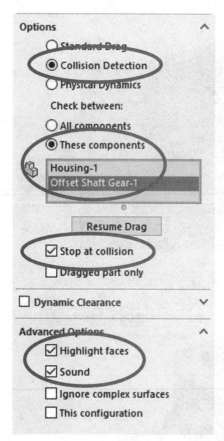

After selecting "Collision Detection" a new set of options are presented; under the "Check between:" option select "These Components" to reveal a new selection box. This option is used to limit the number of components included in the collision detection calculation, ignoring the rest of the parts in the assembly. Select the *'Housing'* and *'Offset Shaft Gear'*, otherwise every component would be included in the collision calculation because they are just hidden, not suppressed, and therefore considered as a solid object.

The collision detection command is a resource intensive command; therefore, limiting the number of components in the analysis makes the analysis run faster and smoother.

After selecting the components press the "Resume Drag" button to continue. Make sure the "Stop at collision," "Highlight faces" and "Sound" options are also selected; this way the parts will stop when they hit each other, and we'll hear a sound and see the colliding faces.

Using the "All components" option can be used if the assembly only has a few components and we have a fast computer. The maximum number of components depends on geometry complexity, mates, computer speed, etc. and therefore it's more convenient to limit the number of components in the analysis.

If we start moving the parts with the "Collision Detection" command and there is an interference between any two components being analyzed, we'll get a warning letting us know about it and the option "Stop on collision" will be disabled, but the sound and highlight faces will continue to work. This is another good reason to limit the number of components included in the analysis.

28.12. - Click and slowly drag the shaft, notice how the movement is slower than usual, and, depending on the PC's hardware, may be sluggish. When the helical thread hits the housing's wall, it will highlight the colliding faces, alert us with a sound, and stop per the options selected.

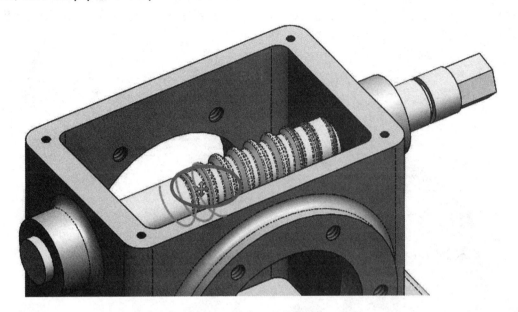

28.13. - Collision and interference detection tools can help us identify and fix design errors early in the process before manufacturing a product. After identifying potential problems click "Exit Isolate" to return to the previous view state. Close the "**Move Component**" command, and Unsuppress the coincident mate to return the shaft to its original location.

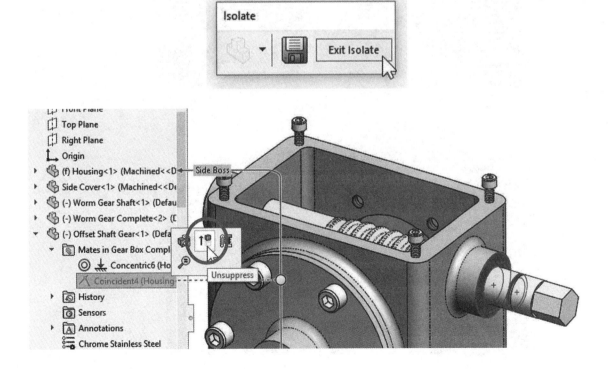

28.14. - Now we will modify the *'Housing'* to allow the shaft to be assembled. We need to make the housing's hole for the shaft larger and add bushings. Open the *'Housing',* double click the "*Side Boss*" to show its dimensions and change the diameter from 1.000″ to 1.250″. Then show the shaft's hole diameter and change it from 0.575″ to 0.775″. Rebuild the model and return to the assembly to continue.

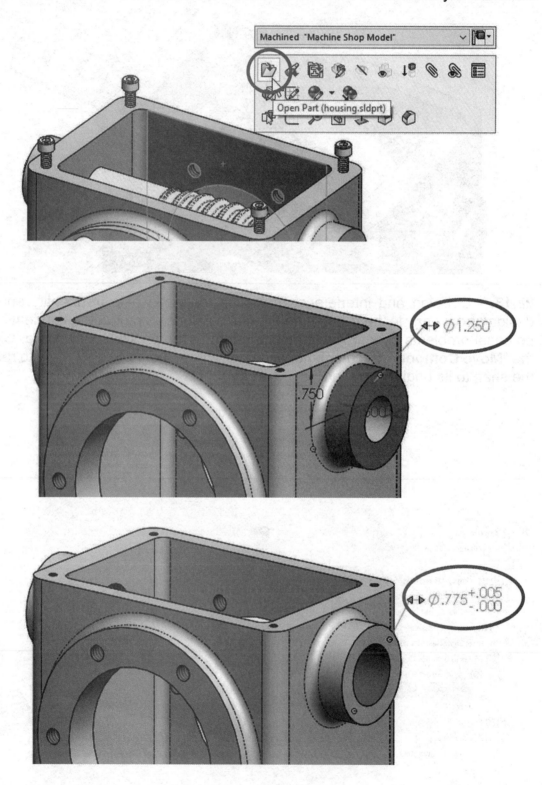

28.15. – Make a new part for the bushing. Add this sketch in the Right plane and extrude it 0.875″; remember to add the tolerances to the diameter dimensions as indicated. Change the material to "Brass" and save it as *'Bushing'*.

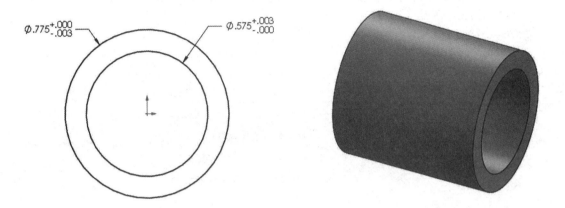

28.16. – Return to the assembly, and add two bushings and mate them at each side of the *'Offset Shaft Gear'*. We can see the assembled bushings after making the *'Housing'* transparent.

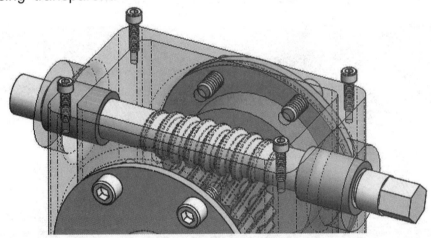

To make a component transparent, select it in the FeatureManager or the graphics screen and select "**Change Transparency**" from the pop-up menu. To make the component opaque again repeat the process and turn transparency Off.

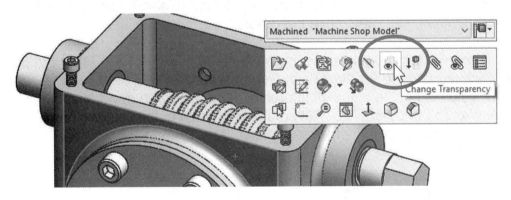

28.17. – As a final step to finish the assembly we'll add a new type of mate called "Gear." The Gear mate allows us to define a rotational ratio between two cylindrical components, even if the two components don't have any type of contact or alignment. Adding this mate will allow us to simulate a gear driving another gear.

While still in the "With Gear" configuration, select the *'Worm Gear Shaft', 'Worm Gear Complete'* and *'Offset Shaft Gear'* from the Feature-Manager, right-mouse-click and select "Isolate" to show these components only.

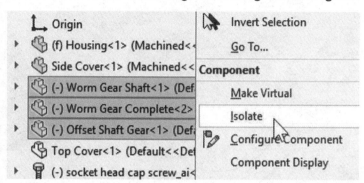

28.18. - Before adding the Gear mate, move the components to a non-interfering position, as seen in the next image. Rotate the view to see the clearance between the gears and move them to approximately this position. While this step is not required for the gear mate to work, it makes the simulation look better.

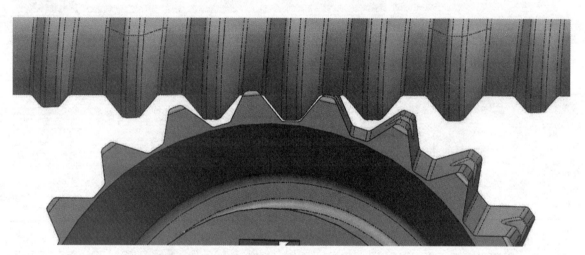

28.19. - Select the **"Mate"** command, expand the "Mechanical Mates" group near the bottom, and select the "Gear Mate." For this type of mate we can select a cylindrical face, a round edge or an axis of the *'Offset Shaft Gear'* and another from either the *'Worm Gear Complete'* or the *'Worm Gear Shaft'*; since both are connected, moving one will move the other.

After making the selections, in the teeth/diameter of the *'Offset Shaft Gear'* enter a value of 1, and for the *'Worm Gear'* or *'Worm Gear Shaft'* enter a value of 22. This value is not the actual diameter, but the gear ratio at which one will rotate with respect to the other. The "Reverse" option inverts the rotation of the gears if needed. Click OK to add the mate and finish the command.

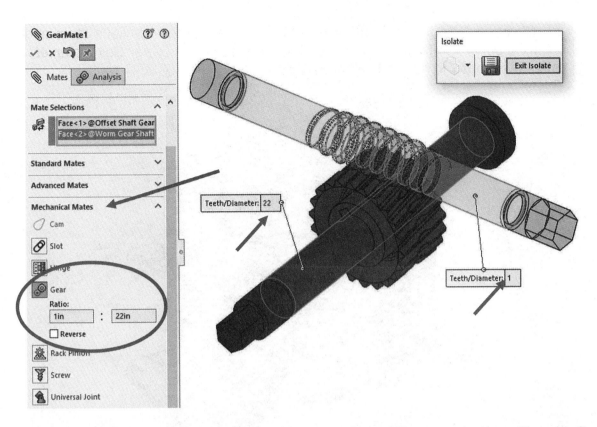

28.20. - To test the gear mate click and drag with the left mouse button either shaft to see the effect. Dragging the '*Worm Gear Complete*' will cause the '*Offset Shaft Gear*' to turn fast and vice versa.

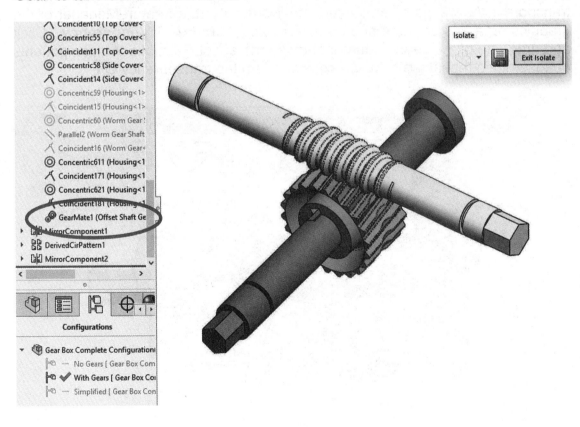

Click the "Exit Isolate" button to return the view to the previous state and show the 'Top Cover' to finish the assembly. Your assembly 'Gear Box Complete' is finished and should now look like this.

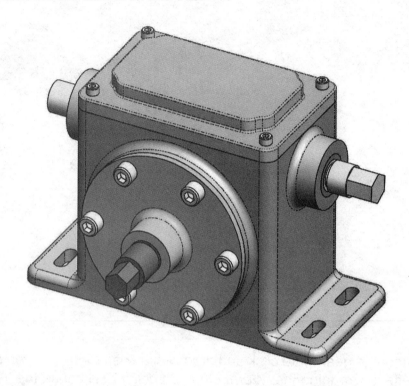

28.21. – Instead of hiding a component to view inside or behind it, we can make it transparent. As we did before, select the 'Housing' in the FeatureManager or the graphics area and from the pop-up menu select "**Change Transparency**." Click and drag one of the gears to turn it and see the effect. To make the components opaque again repeat the same process and turn the transparency Off.

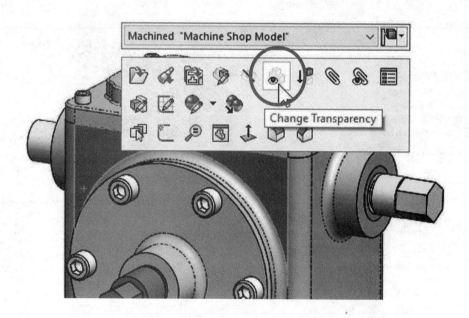

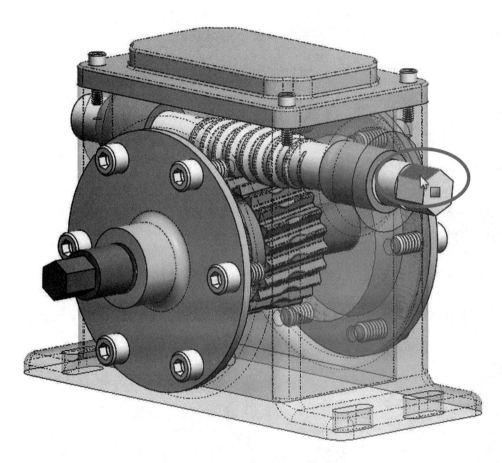

28.22. - An easy way to manipulate component's display mode in an assembly is by using the "Display Pane." The "Display Pane" can be accessed at the top of the FeatureManager. If the arrow is not visible, make the FeatureManager wider to reveal it. It will be located at the top-right. Click to expand and collapse as needed.

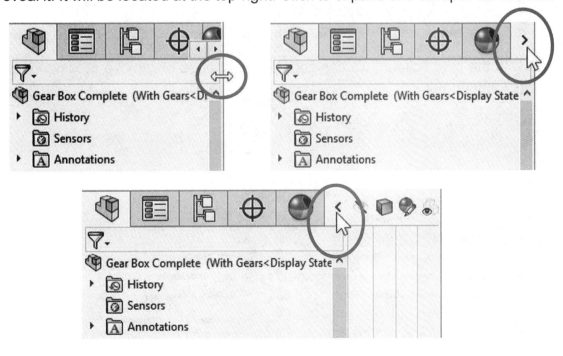

In the "Display Pane" (Default shortcut "F8") we can:

- Click in the first column to Hide/Show a component.

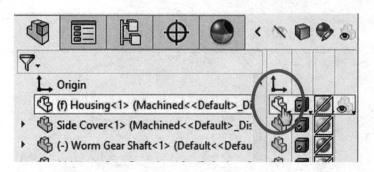

- In the second column to change a component's display style.

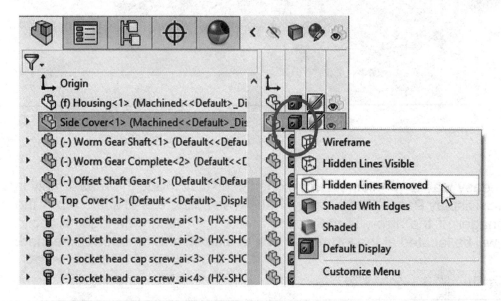

- In the third column, we can copy/paste, change or remove a component's appearance.

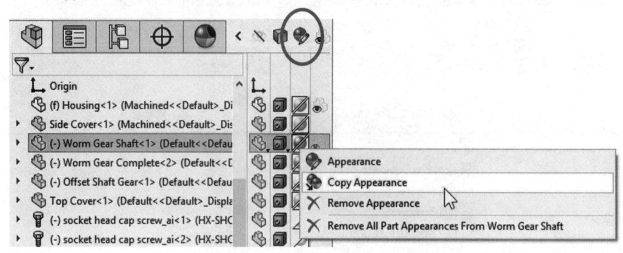

- In the fourth column to make the component transparent/opaque.

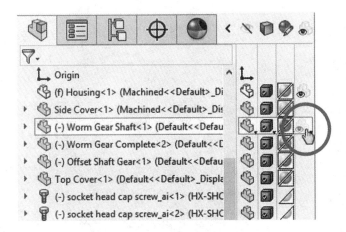

Feel free to explore the different display options for the assembly; often we need to change a component's display to be more efficient.

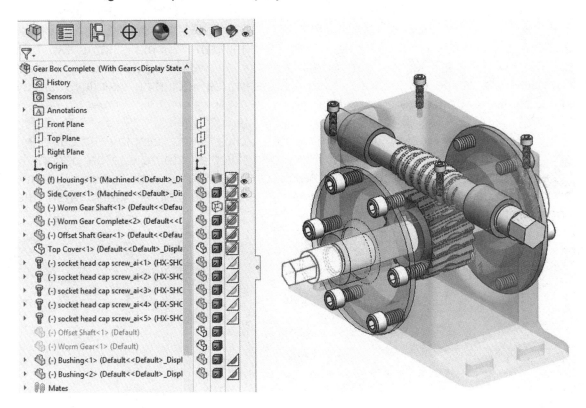

When selecting faces in a transparent component, *IF* there is an opaque component behind the selection, the opaque component is selected. If there are no opaque components, the top-most transparent face is selected. To select the transparent face if an opaque component is behind it, we need to hold down the "Shift" key while selecting. This is the default setting. If the option "**Tools, Options, System Options, Display/Selection, Allow selection through transparency**" is turned off, the top-most face will always be selected. Notice the outline of the face to be selected in each case.

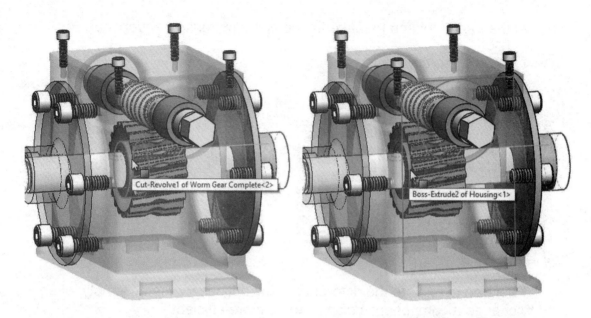

Default selection behavior   Selection holding "Shift" key

28.23. - At the bottom of the ConfigurationManager is the "Display States" tab, and just like configurations, we can add as many as we need to show different combinations of the component's visual appearance, including visibility, display style, color and transparency settings.

To add new display states, right-mouse-click in the "Display States" pane and select the "Add Display State" command. Display states can be optionally linked to configurations by activating the checkbox at the bottom; this way we only have one display state per configuration and it is shown only when the configuration is changed.

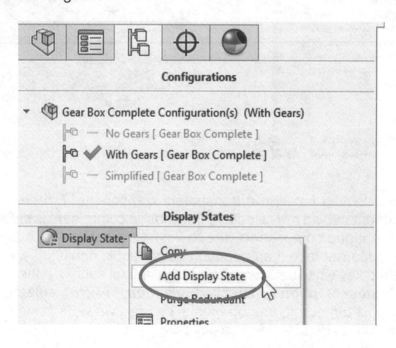

After adding the new display state, rename it "*All Visible*" and change all the components back to visible, no transparency, and default display style. To activate a different display state, we need to double click in it.

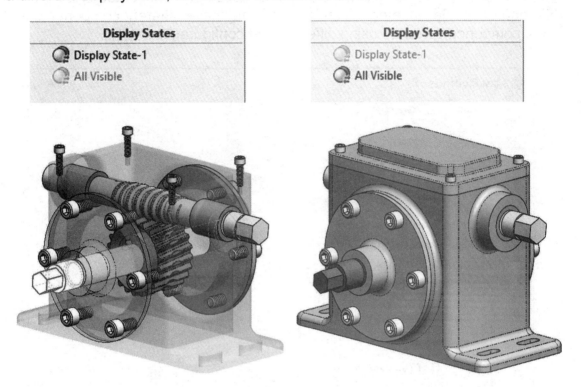

28.24. - An assembly's weight and mass properties can be calculated the same way as in a part using the "**Mass Properties**" command. The difference is that in the assembly, the weight will be the combined weight of the individual components based on the materials they are made of. This is why it's a good idea to always assign a material to each component.

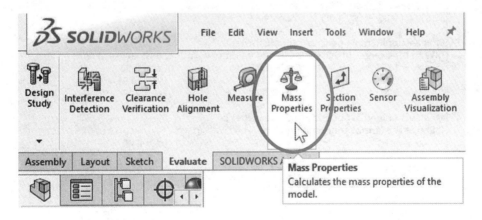

Optionally, mass properties (weight, center of mass, and moments of inertia) can be overridden if needed by selecting the "**Override Mass Properties…**" button. Keep in mind that each assembly configuration may give us different results because each configuration may have different components and some components may be using different part configurations.

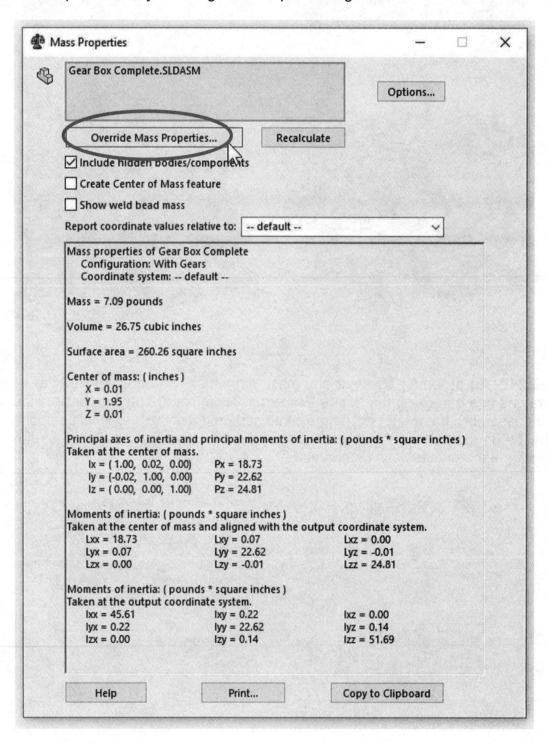

Exploded View

29.1. - After completing the assembly, our next step is to make an exploded view. Exploded views are used for documentation, assembly instructions, training, sales presentations, etc. Activate the *"With Gears"* configuration, and select the **"Exploded View"** command from the Assembly tab in the CommandManager or from the menu **"Insert, Exploded View."**

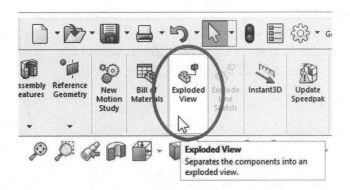

In the Explode PropertyManager we can see the "Settings" selection box is active and ready for us to select the component(s) that will be exploded.

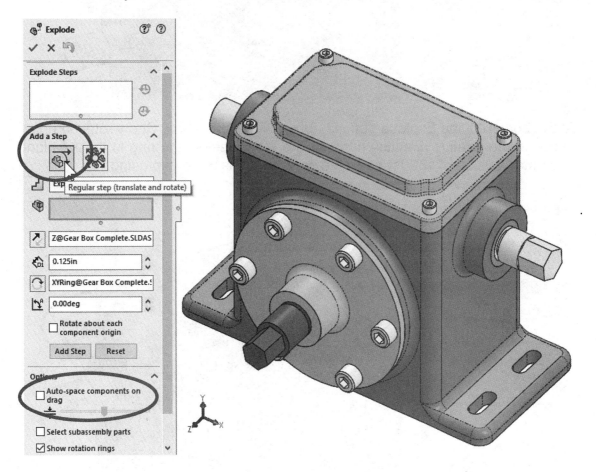

29.2. - There are two ways to add explode steps: the first option is to select a component and drag the tip of a direction axis to the desired explode distance using the on-screen ruler as a guide, and the second option is to manually define the specific explosion parameters for direction and rotation of a component. The main difference is we can only explode *and* rotate a component at the same time using the second option.

The first step will be to individually explode and rotate the four screws in the *'Top Cover'*; therefore, we'll use the second option. Select the first screw in the screen; the yellow triad will be displayed showing the directions available for explosion. Activate the "Explode Direction" field, select the "Y" direction axis in the screen and enter a value of 0.5″ for the explode distance.

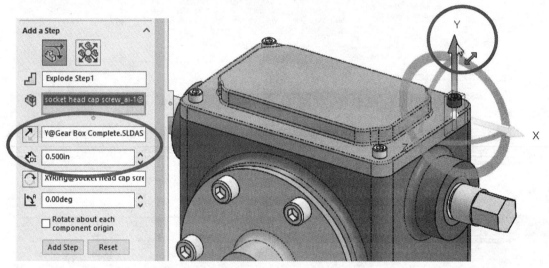

At the bottom of the Explode options, turn ON the checkbox "Show rotation rings" to reveal the rotation direction rings for the screw, select the ring in the XZ plane and enter a value of 1800° to turn the screw 5 times while exploding.

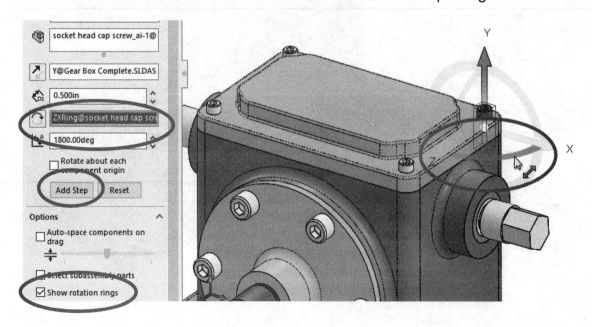

After all the explode step information is entered, press the "Add Step" button to add the exploded view step.

 If needed, select the "Explode Step1" in the "Explode Steps" box to adjust the step's settings and press "Done" to continue.

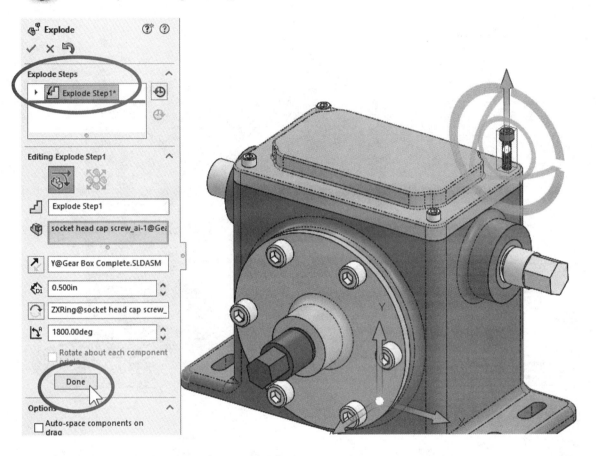

Now we are ready to start the second step.

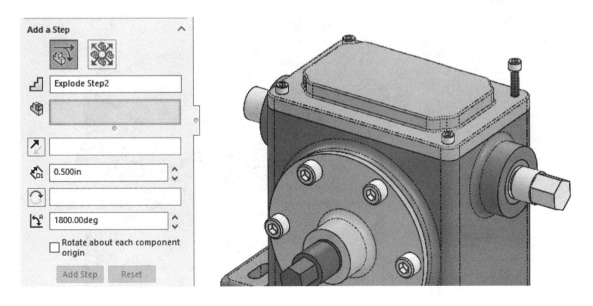

29.3. - Now we need to repeat the same process for the remaining three screws on the top. Select the second screw, select the "Y" direction arrow and the XZ plane of rotation and press "Add Step;" the distance and rotation values remain the same.

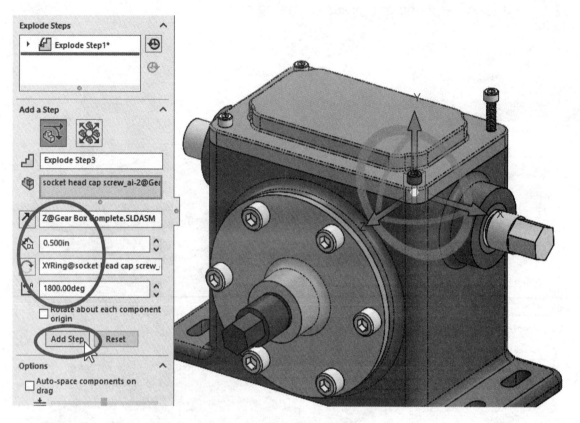

29.4. - Repeat for the other two screws.

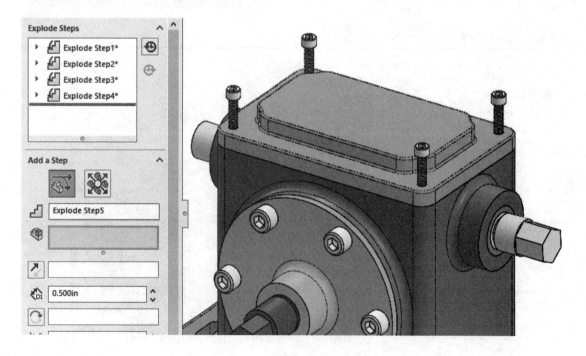

29.5. - Now we need to explode the four screws upwards at the same time. In this case select the four screws and click-and-drag the yellow "Y" distance direction arrow approximately 3 inches up. When we explode the components this way, the step is completed automatically and we are ready for the next step.

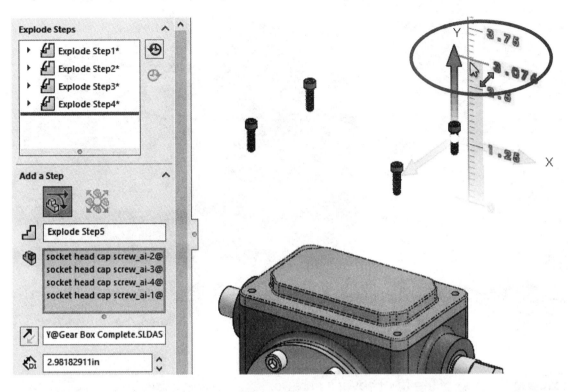

29.6. - Now select the '*Top Cover*' and drag the "Y" direction arrow up approximately halfway between the '*Housing*' and the screws.

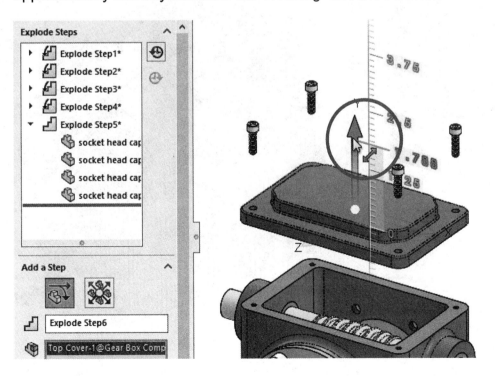

29.7. - Next are the front cover's screws. Just like the top screws, explode and rotate them individually 0.5″ in the "Z" direction and rotate 5 turns (1800°) about the XY plane.

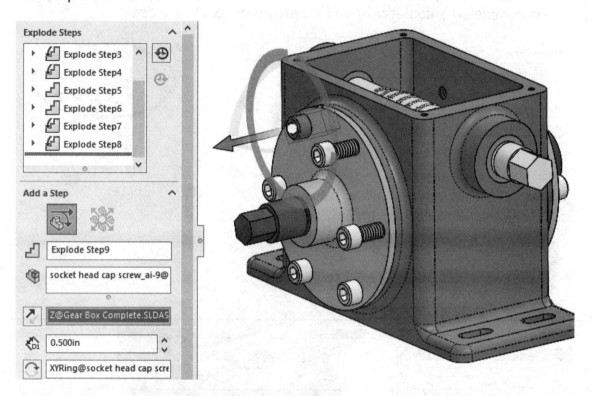

Explode the rest of the front cover's screws.

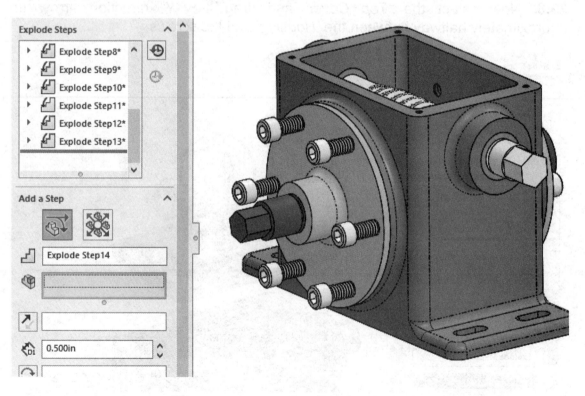

29.8. - Select all the front cover's screws, and click and drag the "Z" direction arrow approximately 6 inches.

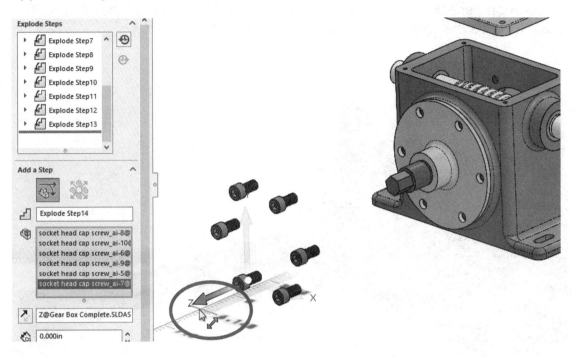

29.9. - Select the 'Side Cover' and drag it in the "Z" direction.

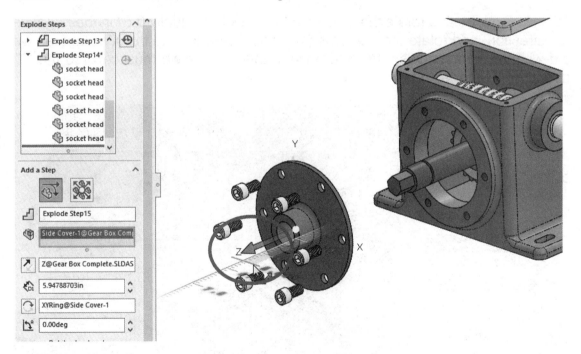

 In case we need to edit an exploded step, select it in the "Explode Steps" list and drag the arrow's tip, or double-click in the explode step and enter a new distance value to change the exploded distance and/or rotation.

29.10. - The rest of the explode steps are done the same way. Select the component(s) and either drag the tip of the desired direction arrow, or pick a direction and specify the distance and/or rotation. Now explode the *'Worm Gear Shaft'* towards the back approximately 8" dragging the "Z" arrow.

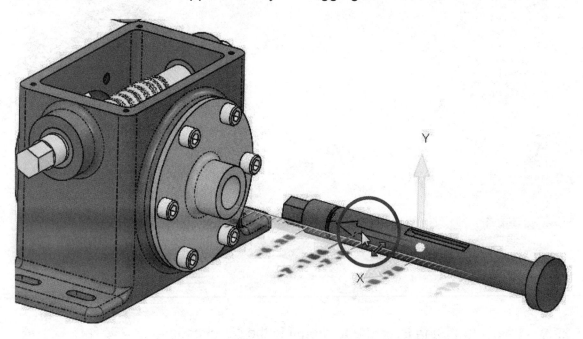

29.11. - Explode and rotate the screws in the back individually 0.5″ in the negative "Z" direction, and rotate them 5 turns (1800 degrees) around the XY plane as we did with the front screws. The difference in this case is we must reverse the "Z" direction or enter a negative value.

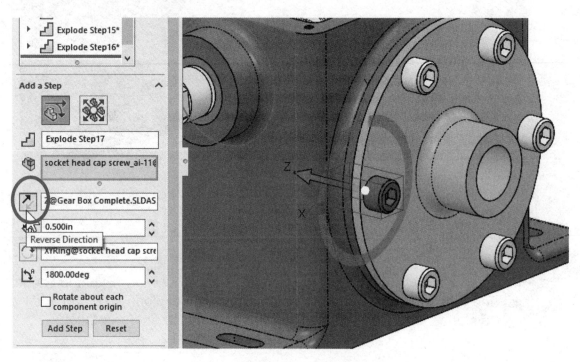

29.12. - Explode the rest of the screws in the rear cover.

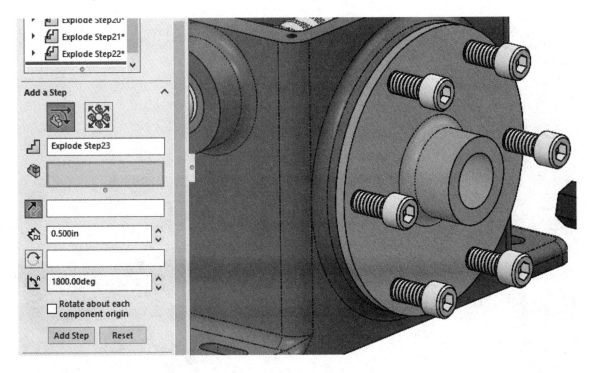

29.13. – Now select all six rear screws and explode them back about 7 inches.

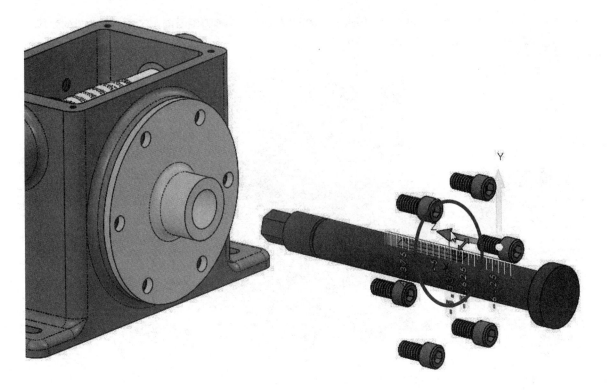

29.14. – Add a new step to explode the rear *'Side Cover'*.

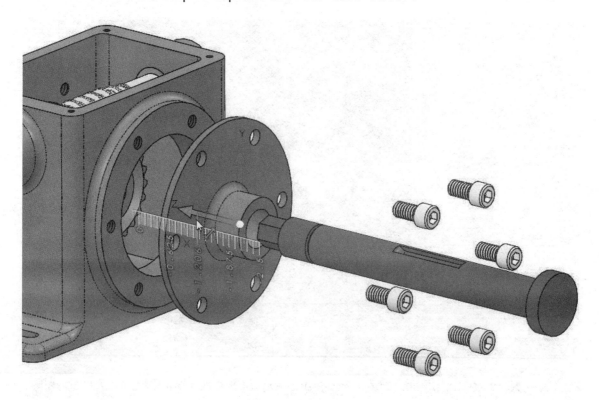

29.15. – Explode the *'Worm Gear Complete'* towards the front between the *'Housing'* and the front *'Side Cover'*.

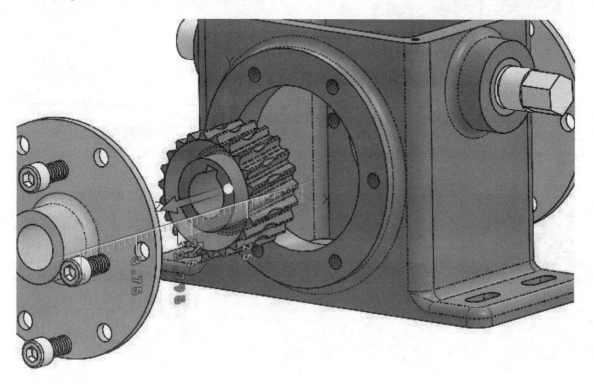

29.16. – Add additional steps to explode the bushings and the '*Offset Shaft Gear*'. Explode the right side bushing out of the 'Housing' first, then to the back. When making exploded views, try to group components that will be exploded the same distance and direction at the same time to reduce the number of steps needed. After adding the last step click OK to finish the **Exploded View** command.

The finished exploded view should look approximately like this:

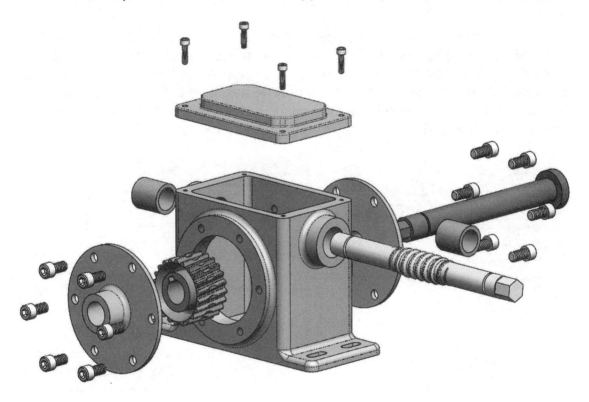

 Multiple exploded views can be added per assembly configuration.

29.17. – After adding an exploded view to our assembly, we can show it in the exploded state or collapsed as needed. To collapse an exploded assembly, click with the right mouse button at the top of the FeatureManager in the Assembly name and select "**Collapse**" from the pop-up menu, or in the ConfigurationManager right-mouse-click in "*Exploded View1*" directly under the configuration with the exploded view, and select collapse. Another way is to double click "*Exploded View1*" to expand or collapse the assembly.

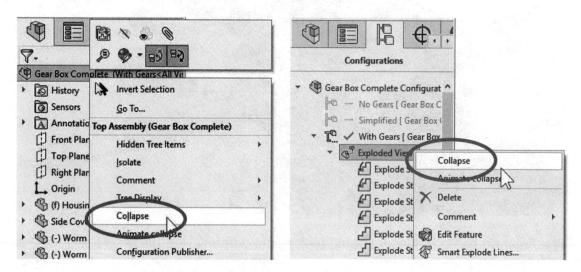

29.18. - To explode the assembly again, make a right-mouse-click at the top of the FeatureManager and select "**Explode**," or in the ConfigurationManager double click "*Exploded View1*" or right-mouse-click in the exploded view and select "Explode." Be aware that this option will only be available if the assembly has been exploded. If we need to edit the exploded view steps, select the "**Exploded View**" command or select "**Edit Feature**" from the pop-up menu.

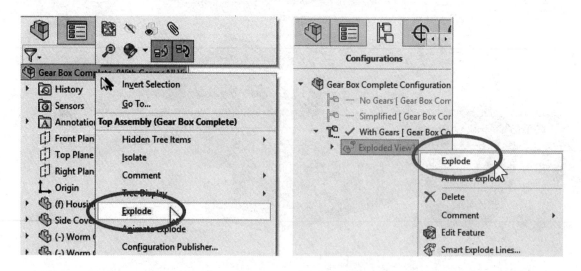

 If a configuration has two or more exploded views, exploding the assembly from the FeatureManager will use the last exploded view used.

29.19. - From the same pop-up menu we can select "**Animate collapse**" if the assembly is exploded, or "**Animate explode**" if the assembly is collapsed. This brings up the Animation Controller to animate the explosion and optionally save a video with the exploded view animation.

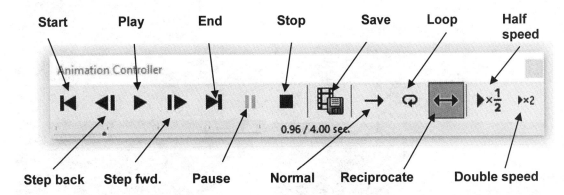

29.20. - After completing the exploded view, we can add sketch lines to show the path the components will follow for assembly. To add them, select the "**Explode Line Sketch**" command from the "Exploded View" drop-down menu.

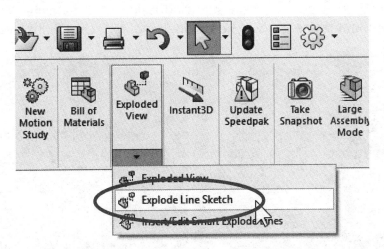

If not already active, the Explode Sketch toolbar is automatically displayed showing two commands: "**Route Line**" and "**Jog Line**." The route line is used to add lines connecting faces, edges, vertices, axes, etc. of one component to another in the exploded view, and the Jog Line will add a jog to an existing explode sketch line. The Explode Lines we are adding are a 3D Sketch connecting the selected entities. For the first line, select the round face at the front of the '*Housing*', then the cylindrical face inside the '*Worm Gear Complete*' and finally a cylindrical face of the '*Side Cover*'.

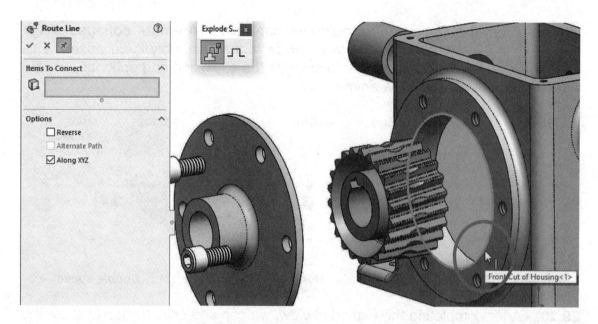

If the route line is pointing in the wrong direction, click in the direction arrow head or the "Reverse" checkbox in the options to reverse it. The option "Along XYZ" is active by default, and helps us create the lines along the assembly axes, otherwise it would be made using a 2D sketch located at an angle.

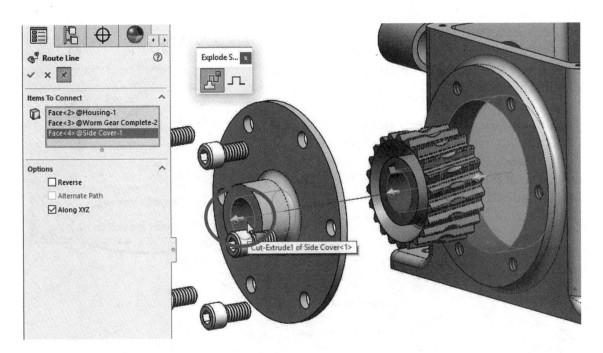

29.21. - When the selections for this line are completed, click OK to add it and start a new one. Notice the line is drawn using a broken line style and is automatically hidden when passing behind components.

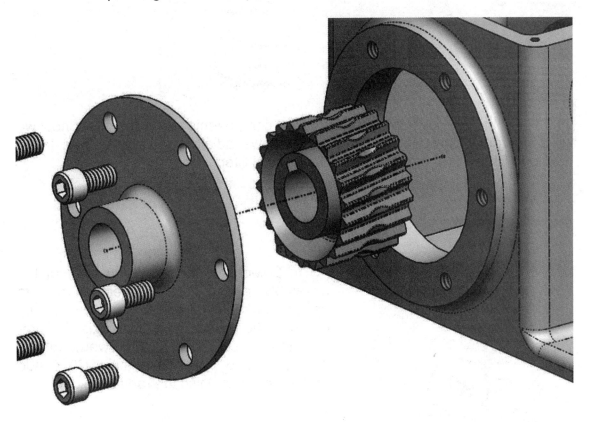

29.22. - Before adding the next explode line, we'll add another step to the exploded view. Exit the "**Explode Line Sketch**" command to exit the 3D Sketch.

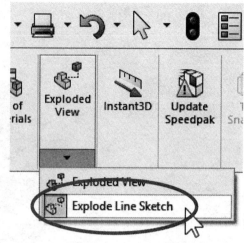

To modify the exploded view, expand the configuration in the ConfigurationManager, right-mouse-click in "*Exploded View1*" and select "Edit Feature." Notice the "*3DExplode1*" feature added to the exploded view; this is the 3D sketch with the exploded lines.

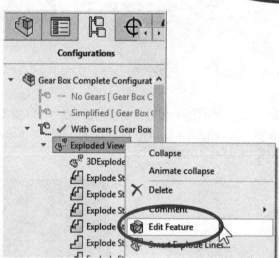

29.23. - In the previous steps we exploded components along the axes; now we will add a radial explode step to move the screws radially away from an axis. In the "Explode Step Type" group activate the "Radial Step" option.

Select the six screws in front of the assembly, and optionally select an axis to define the radial direction. In our case, by default, the assembly's "Z" axis is pre-selected, which is exactly what we need. Set the distance to 1" and 0° rotation. Click "Add Step" and OK to finish the radial explode step.

608

If we need to explode components radially about a different direction, activate the "Explode Direction" box, and select the new axis, a cylindrical face or a linear edge.

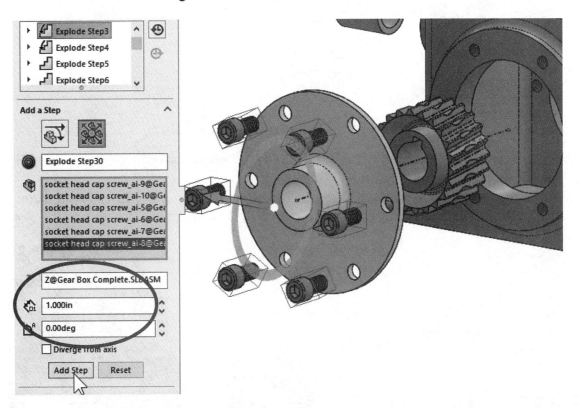

When finished, exit the "**Explode**" command to continue.

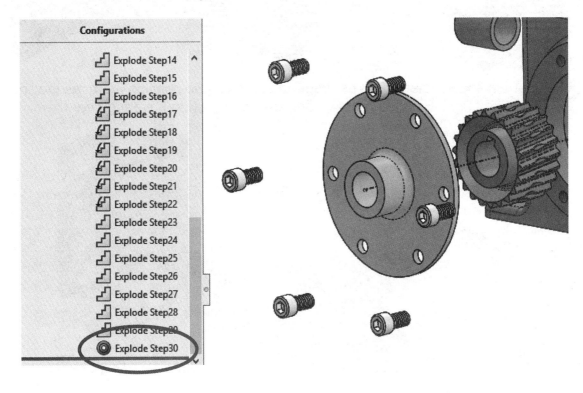

29.24. - Activate the "Explode Line Sketch" command again, and turn off the "Along XYZ" option.

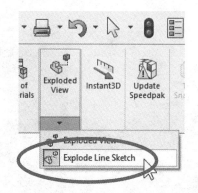 

Select the indicated screw hole, side cover hole and screw. By having the "Along XYZ" option off, the line is created on a rotated 2D sketch. Click OK to add the line.

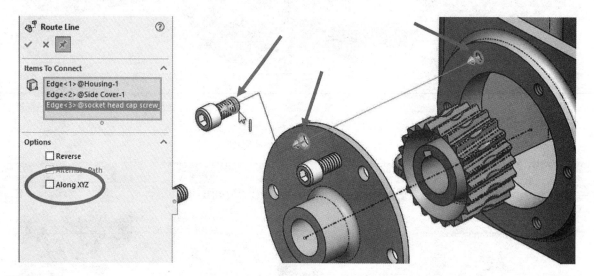

 Certain line segments can be dragged to resize and position them, as the line indicated next. Editable lines will show a handle when hovering over them.

29.25. - After finishing this line, the "Along XYZ" checkbox is automatically turned On again. To see the effect of this option, now add a line to the screw to the right of it, selecting the holes and screw in the same order. Feel free to drag the line segments to your liking.

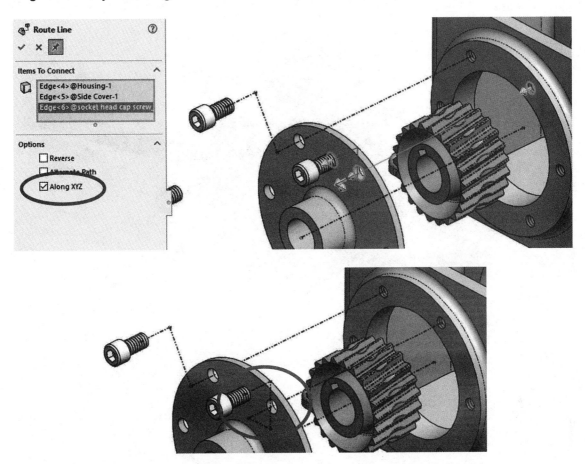

In case an exploded sketch line needs to be deleted, exit the **"Route Line"** command, and while the **"Explode Sketch line"** is active, select the lines to be deleted, and delete them as we would in a sketch. If the **"Explode Sketch Line"** command is active, we are essentially editing a 3D sketch.

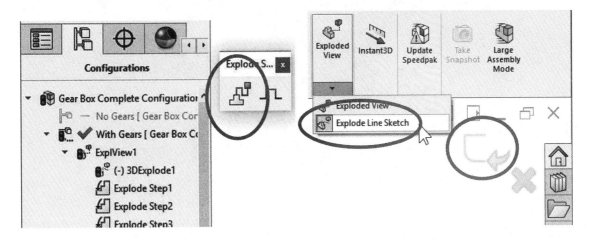

29.26. - Add a few more explode lines to illustrate the assembly order and exit the "**Explode Line Sketch**" command.

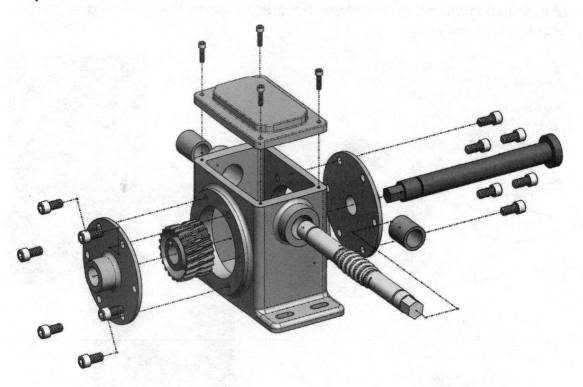

Engine Project: Assemble the engine using the parts built in the Part modeling exercises, and make an exploded view. First make the following sub-assemblies as indicated (shown in exploded view) before adding them to the final assembly.

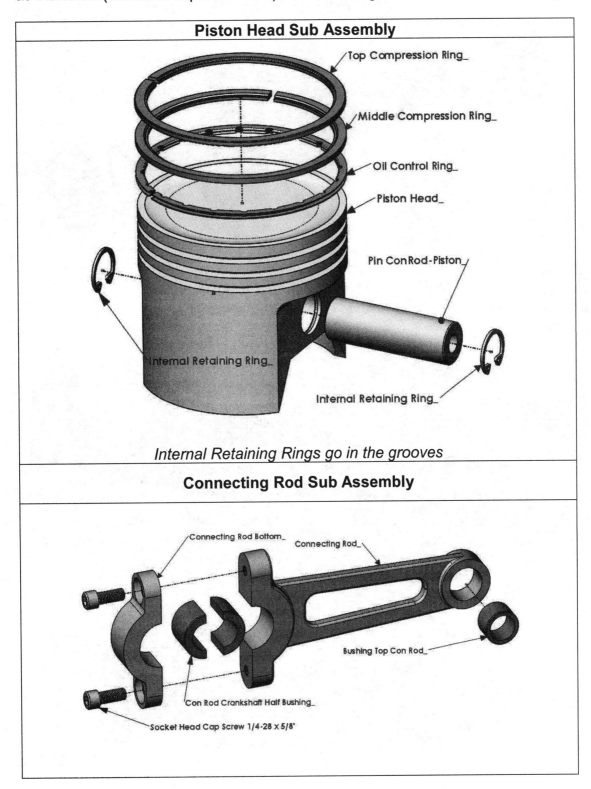

Piston Head Sub Assembly

Top Compression Ring_

Middle Compression Ring_

Oil Control Ring_

Piston Head_

Pin Con Rod - Piston_

Internal Retaining Ring_

Internal Retaining Ring_

Internal Retaining Rings go in the grooves

Connecting Rod Sub Assembly

Connecting Rod Bottom_

Connecting Rod_

Bushing Top Con Rod_

Con Rod Crankshaft Half Bushing_

Socket Head Cap Screw 1/4-28 x 5/8"

Crankshaft Sub Assembly

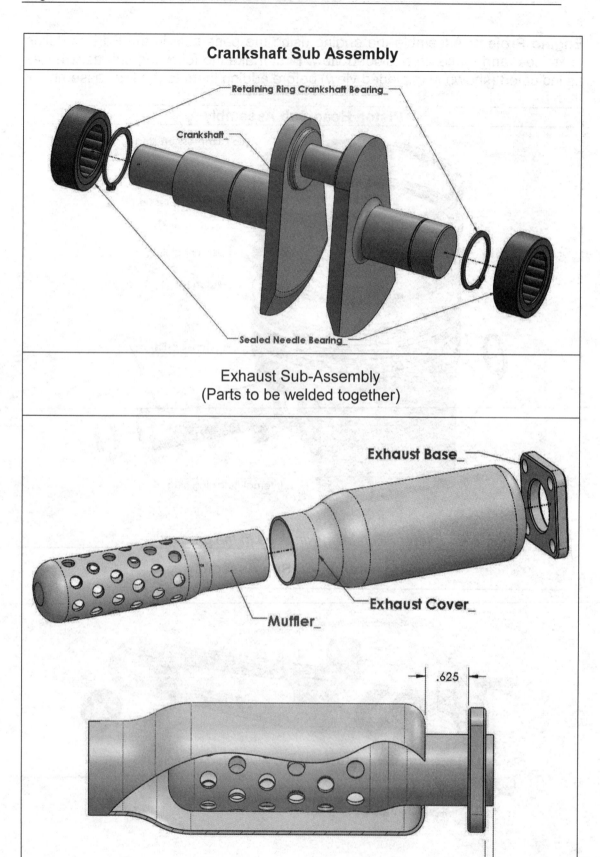

Retaining Ring Crankshaft Bearing_

Crankshaft_

Sealed Needle Bearing_

Exhaust Sub-Assembly
(Parts to be welded together)

Exhaust Base_

Exhaust Cover_

Muffler_

.625

.115

Sequence of components to assemble the engine.

Oil Pan	Crankshaft Sub Assembly	Shaft Seal Gasket	Oil Seal
Oil Pan Gasket	Connecting Rod Sub Assembly	Crank Case Top	Cylinder Gasket
Piston Head Sub Assembly (only one Concentric mate to not over define it)	Engine Block	Make the Piston Head **concentric to** the Engine Block	Head Gasket
Cylinder Head	Shaft Seal Cover	Exhaust	Intake
Oil Dip Stick	Socket Head Cap Screws: ¼-20 x ½", ¼-20 x 5/8", 6-32 x ½"		

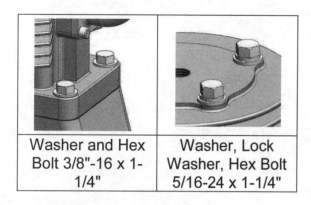

Washer and Hex Bolt 3/8"-16 x 1-1/4"	Washer, Lock Washer, Hex Bolt 5/16-24 x 1-1/4"

As an optional finishing touch before making the exploded view, add a spark plug. You can download one from the 3D Content Central website (free), but you will have to sign up to access it. The link to the spark plug we used is:

http://www.3dcontentcentral.com/secure/download-model.aspx?catalogid=171&id=86052

After completing the engine assembly, make a section view, and move the crankshaft. If the assembly was made correctly the crankshaft will rotate and the piston will move up and down as it should.

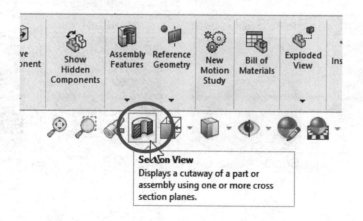

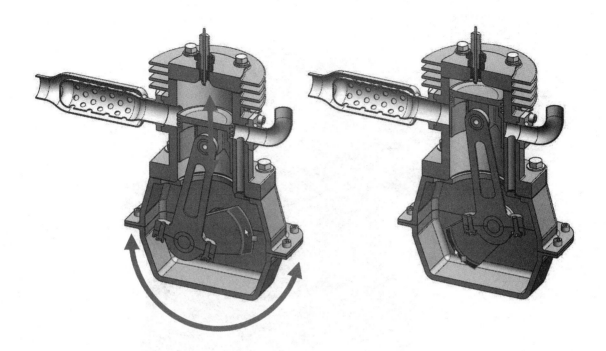

After adding all the fasteners, the FeatureManager will be *very* long. To make it easier to navigate it, we can group multiple components together using folders. Select all the fasteners in the FeatureManager, right-mouse-click on them, select "**Add to New Folder**" and rename it "*Fasteners*." This way we can hide, suppress, delete, or isolate all of them at the same time. After adding a folder, we can drag-and-drop components to and from it to add or remove items from the folder.

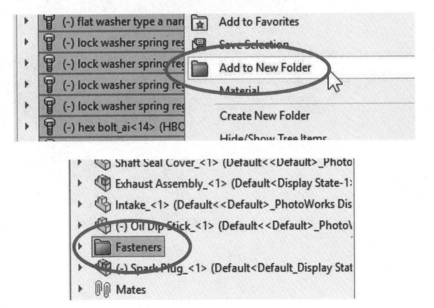

Image created using RealView graphics

For the engine's exploded view we can use the option "Select sub-assembly's parts" to explode sub-assembly components individually. If this option is OFF the entire sub-assembly will be selected and moved as a single part.

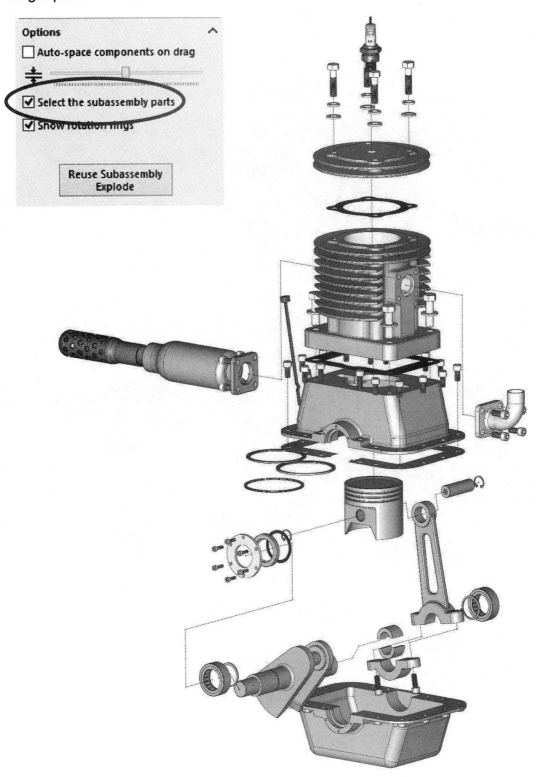

Extra Credit: Make the assembly of the *Gas Grill* using the parts created in the part modeling extra credit project, and then add an exploded view for documentation purposes. The grill's cover must be able to open and close. Close the cover using collision detection.

Chapter 6: Assembly and Design Table Drawings

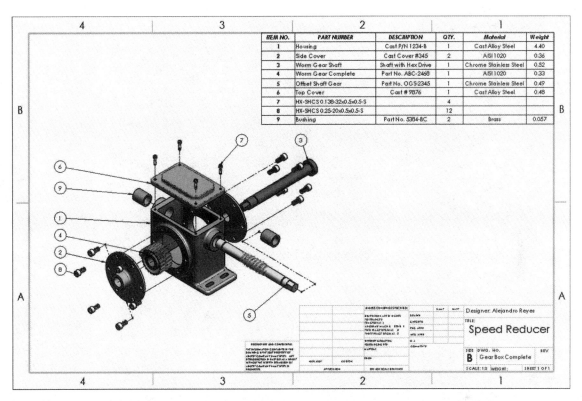

ITEM NO.	PART NUMBER	DESCRIPTION	QTY.	Material	Weight
1	Housing	Cast P/N 1234-B	1	Cast Alloy Steel	4.40
2	Side Cover	Cast Cover #345	2	AISI 1020	0.36
3	Worm Gear Shaft	Shaft with Hex Drive	1	Chrome Stainless Steel	0.52
4	Worm Gear Complete	Part No. ABC-2468	1	AISI 1020	0.33
5	Offset Shaft Gear	Part No. OGS-2345	1	Chrome Stainless Steel	0.49
6	Top Cover	Cast # 9B76	1	Cast Alloy Steel	0.48
7	HX-SHCS 0.138-32x0.5x0.5-S		4		
8	HX-SHCS 0.25-20x0.5x0.5-S		12		
9	Bushing	Part No. S384-BC	2	Brass	0.057

Designer: Alejandro Reyes
TITLE: Speed Reducer
SIZE B DWG. NO. Gear Box Complete REV
SCALE: 1:2 WEIGHT: SHEET 1 OF 1

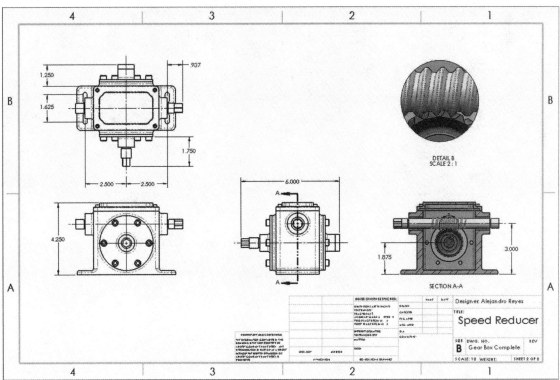

DETAIL B
SCALE 2:1

SECTION A-A

Designer: Alejandro Reyes
TITLE: Speed Reducer
SIZE B DWG. NO. Gear Box Complete REV
SCALE: 1:2 WEIGHT: SHEET 2 OF 2

621

Notes:

Assembly Drawing

30.1. - After finishing the assembly, we are going to make an assembly drawing. In this case, the drawing will include an exploded view, Bill of Materials (BOM), and identification balloons, main views with overall dimensions, section and detail views. To make the drawing, open the assembly '*Gear Box Complete*' and activate the *"With Gears"* configuration; it doesn't matter if it's exploded or collapsed. Select the "**Make drawing from part/assembly**" icon as we did before, make a drawing using the "B-Landscape" sheet size and include the sheet format ("Display sheet format" checked).

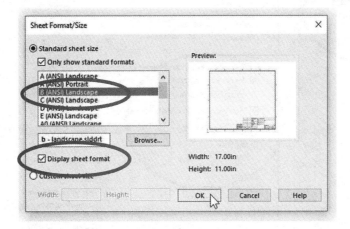

From the "View Palette," drag the "Isometric Exploded" view to the sheet. Change the Display Style to "Shaded with Edges." If the sheet's scale is not 1:2 change it in the sheet properties.

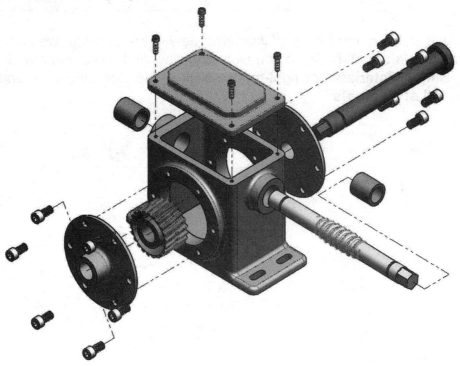

623

When an assembly has an Exploded View, we can change the drawing view to show it either Exploded or Collapsed afterwards. In the drawing view properties we can change to a different configuration, to show it exploded or not, its display state, and, if the configuration has multiple exploded views, a different exploded view. If needed, right-mouse-click in the view, select "Properties" and turn on/off the option "**Show in Exploded State**."

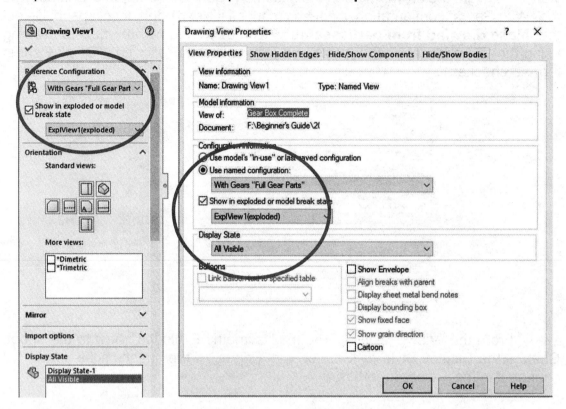

30.2. - Once we have the exploded isometric view in the drawing, we need to add a **Bill of Materials** (BOM). Select the isometric view and select the menu "**Insert, Tables, Bill of Materials**," or right-mouse-click in the isometric view and select "**Tables, Bill of Materials**."

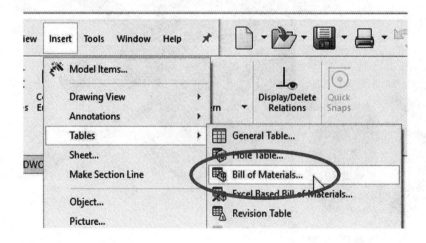

In the Bill of Materials PropertyManager use the following options:

Table Position: "Attach to anchor point" unchecked. This way we can locate the BOM table anywhere in the drawing, otherwise the table will be attached to the BOM anchor point defined in the drawing template.

BOM Type: "Parts-only" to list all the parts in the assembly, regardless of sub-assemblies.

Configurations: "With Gears," the assembly configuration we are making a drawing of.

Part Configuration Grouping: "Display configurations of the same part as separate items"; by using this option, the screws, which are different configurations of the same part, will be displayed as different items in the BOM.

Keep Missing Item: OFF. This option will include removed/replaced items from the assembly in the BOM.

Item Numbers: Start at: 1, Increment: 1.

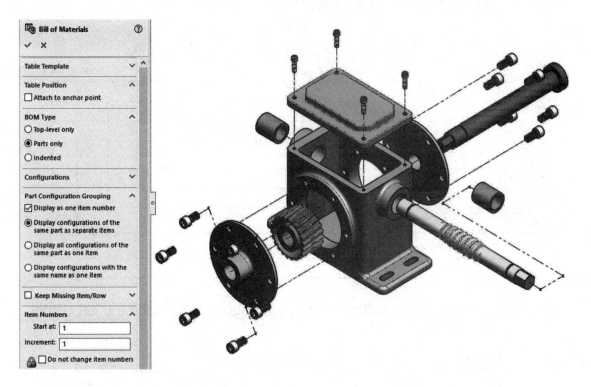

30.3. - After clicking OK, move the mouse and locate the BOM in the top right corner of the drawing. While locating the table, notice the table will snap to the corner. The column's width and row's height can be adjusted by dragging the table lines just as in MS Excel.

Since we added the "Description" custom property to the *'Worm Gear Shaft'* it is automatically imported into the Bill of Materials. We added a "Description" property to the *'Worm Gear'* and *'Offset Shaft'*, but it was to the simplified version of the parts, not the full gear parts added to this assembly.

ITEM NO.	PART NUMBER	DESCRIPTION	QTY.
1	Housing		1
2	Side Cover		2
3	Worm Gear Shaft	Shaft with Hex Drive	1
4	Worm Gear Complete		1
5	Offset Shaft Gear		1
6	Top Cover		1
7	HX-SHCS 0.138-32x0.5x0.5-S		4
8	HX-SHCS 0.25-20x0.5x0.5-S		12
9	Bushing		2

Resize column width to fit the PART NUMBER

	A	B	C	D
1	ITEM NO.	PART NUMBER	DESCRIPTION	QTY.
2	1	Housing		1
3	2	Side Cover		2
4	3	Worm Gear Shaft	Shaft with Hex Drive	1
5	4	Worm Gear Complete		1
6	5	Offset Shaft Gear		1
7	6	Top Cover		1
8	7	HX-SHCS 0.138-32x0.5x0.5-S		4
9	8	HX-SHCS 0.25-20x0.5x0.5-S		12
10	9	Bushing		2

…and re-position the Bill of Materials table by dragging it from the top left corner.

	A	B	C	D
1	ITEM NO.	PART NUMBER	DESCRIPTION	QTY.
2	1	Housing		1
3	2	Side Cover		2
4	3	Worm Gear Shaft	Shaft with Hex Drive	1
5	4	Worm Gear Complete		1
6	5	Offset Shaft Gear		1
7	6	Top Cover		1
8	7	HX-SHCS 0.138-32x0.5x0.5-S		4
9	8	HX-SHCS 0.25-20x0.5x0.5-S		12
10	9	Bushing		2

30.4. - It is possible to customize the Bill of Materials by adding more columns with information imported from the components' custom properties.

To add columns to the table, right-mouse-click in the "QTY" cell in the table (or any other column), and from the pop-up menu select "**Insert, Column Right**" (or Left…) to locate the new column. For the "Column type" value select "CUSTOM PROPERTY," and from the "Property name" drop down list, select "**Material**."

 For components with a custom property assigned, its corresponding value will be automatically filled in the Bill of Materials table.

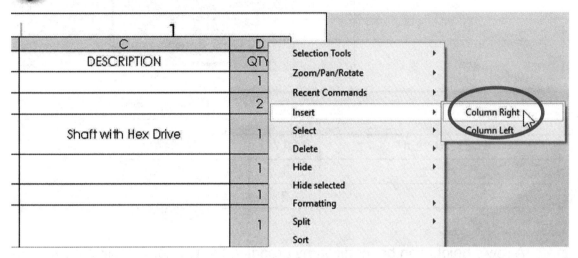

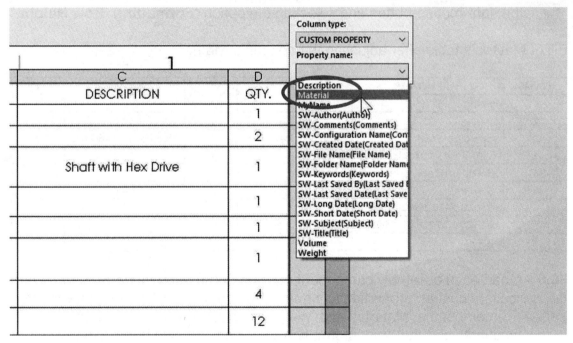

Repeat the process to add another column to the right with the custom property "**Weight**"; when finished adjust the column's width and relocate the table in the drawing.

Column type:

CUSTOM PROPERTY

Property name:

Description
Material
MyName
SW-Author(Author)
SW-Comments(Comments)
SW-Configuration Name(Con
SW-Created Date(Created Dat
SW-File Name(File Name)
SW-Folder Name(Folder Name
SW-Keywords(Keywords)
SW-Last Saved By(Last Saved
SW-Last Saved Date(Last Save
SW-Long Date(Long Date)
SW-Short Date(Short Date)
SW-Subject(Subject)
SW-Title(Title)
Volume
Weight

	C	D	E	
	DESCRIPTION	QTY.	Material	
		1		
		2		
	Shaft with Hex Drive	1	Chrome Stainless Steel	
		1		
		1		
		1	Cast Alloy	

	C	D		E	F
	DESCRIPTION	QTY.		Material	Weight
		1			
		2			

 A row's height can be modified by dragging the row, or selecting a row with the right-mouse-button and selecting the option "**Formatting, Row Height**."

Bill of Materials table after adjusting the column's width.

ITEM NO.	PART NUMBER	DESCRIPTION	QTY.	Material	Weight
1	Housing		1		
2	Side Cover		2		
3	Worm Gear Shaft	Shaft with Hex Drive	1	Chrome Stainless Steel	0.52
4	Worm Gear Complete		1		
5	Offset Shaft Gear		1		
6	Top Cover		1	Cast Alloy Steel	0.48
7	HX-SHCS 0.138-32x0.5x0.5-S		4		
8	HX-SHCS 0.25-20x0.5x0.5-S		12		
9	Bushing		2		

30.5. - Open each assembly component missing information in the BOM, and add the necessary custom properties using the "Custom Properties" template created earlier. Remember the "Material" and "Weight" properties will be automatically filled in the template, and the only information we need to enter is the description of each part. Remember to press "Apply" after adding the properties to a part and save it. After the missing custom properties have been added to each part, return to the assembly drawing. If needed, rebuild the drawing to update the table.

To open a part from within the drawing, select the part in the drawing and click in the "**Open Part (…)**" icon from the pop-up menu, or right-mouse-click the component's name in the Bill of Materials and select "Open."

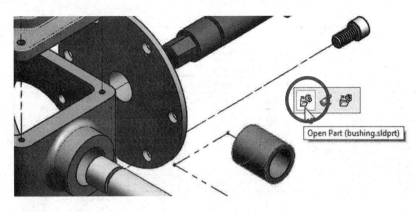

ITEM NO.	PART NUMBER	DESCRIPTION	QTY.	Material
1	Housing		1	
2	Side Cover		2	
3	Worm Gear Shaft	Shaft with Hex Drive	1	Chrome Stainless Ste
4	Worm Gear Complete		1	
5	Offset Shaft Gear		1	
6	Top Cover		1	Cast Alloy Steel
7	HX-SHCS 0.138-32x			
8	HX-SHCS 0.25-20x0			
9	Bushing			

Selection Tools
Zoom/Pan/Rotate
Recent Commands
Open top cover.sldprt
Insert
Select

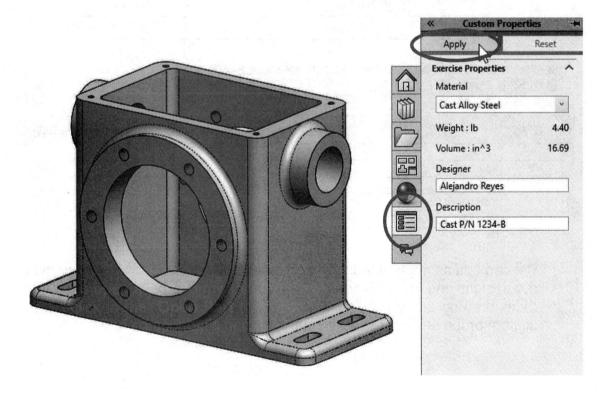

30.6. - When finished, your table should look like this. Feel free to add your own "Description" to each component.

ITEM NO.	PART NUMBER	DESCRIPTION	QTY.	Material	Weight
1	Housing	Cast P/N 1234-B	1	Cast Alloy Steel	4.40
2	Side Cover	Cast Cover #345	2	AISI 1020	0.36
3	Worm Gear Shaft	Shaft with Hex Drive	1	Chrome Stainless Steel	0.52
4	Worm Gear Complete	Part No. ABC-2468	1	AISI 1020	0.33
5	Offset Shaft Gear	Part No. OGS-2345	1	Chrome Stainless Steel	0.49
6	Top Cover	Cast # 9876	1	Cast Alloy Steel	0.48
7	HX-SHCS 0.138-32x0.5x0.5-S		4		
8	HX-SHCS 0.25-20x0.5x0.5-S		12		
9	Bushing	Part No. 5384-BC	2	Brass	0.057

Cells, rows, columns or the entire table can be formatted using the pop-up formatting toolbar, just like MS Excel. By default, the entire table uses the drawing document's font. To modify the first row's font, select the row's header, (or column, cell or table) and turn Off the "Use Document Font" option to display the formatting toolbar. Modify the table's headers to your liking.

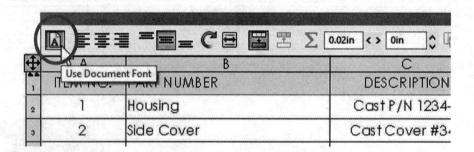

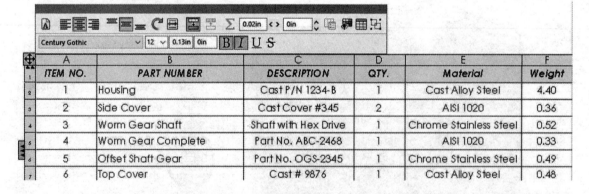

We can add as many custom properties as needed to accurately document our designs and, as we learned, these properties can be used in the part's detail drawing and the Bill of Materials. Assembly files can also be given custom properties just like individual components.

The final Bill of Materials table should look something like this:

ITEM NO.	PART NUMBER	DESCRIPTION	QTY.	Material	Weight
1	Housing	Cast P/N 1234-B	1	Cast Alloy Steel	4.40
2	Side Cover	Cast Cover #345	2	AISI 1020	0.36
3	Worm Gear Shaft	Shaft with Hex Drive	1	Chrome Stainless Steel	0.52
4	Worm Gear Complete	Part No. ABC-2468	1	AISI 1020	0.33
5	Offset Shaft Gear	Part No. OGS-2345	1	Chrome Stainless Steel	0.49
6	Top Cover	Cast # 9876	1	Cast Alloy Steel	0.48
7	HX-SHCS 0.138-32x0.5x0.5-S		4		
8	HX-SHCS 0.25-20x0.5x0.5-S		12		
9	Bushing	Part No. 5384-BC	2	Brass	0.057

30.7. - In the next step we need to visually identify the assembly components. The Auto Balloon command automatically adds an identification balloon to each component of the assembly. To Start, either make a right-mouse-click in the assembly drawing view and select "**Annotations, Auto Balloon**" from the pop-up menu, or select the assembly view and go to the menu "**Insert, Annotations, Auto Balloon**."

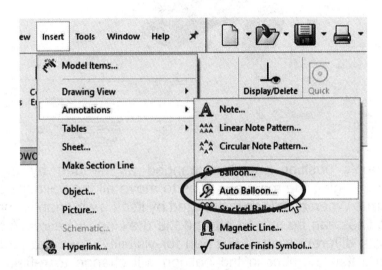

When the Auto Balloon's PropertyManager is displayed, select the option "Follow Assembly Order" to make the balloons and BOM to follow the component order in the assembly's FeatureManager, in "Balloon Layout" use the "Square" pattern to locate the balloons around the assembly view, turn on the "Ignore multiple instances" button to avoid adding balloons to duplicate components, check the "Insert magnetic lines" to align the balloons using an invisible line that annotations are attached to when we move them close to it (helpful to align annotations).

In the "Balloon Settings" section select the "Circular" option to define the balloon's style, in the balloon's size use the "2 Characters" option, and in the "Balloon text:" option select "Item Number" to identify the components using the BOM item numbers.

30.8. - If we click-and-drag any of the balloons before we finish the "**Auto Balloon**" command, all the balloons will move in or out at the same time. This way we can locate them closer to the view. When satisfied with the general look of the balloons, click OK to add them to the drawing view. If needed, move the assembly view to make space for the identification balloons.

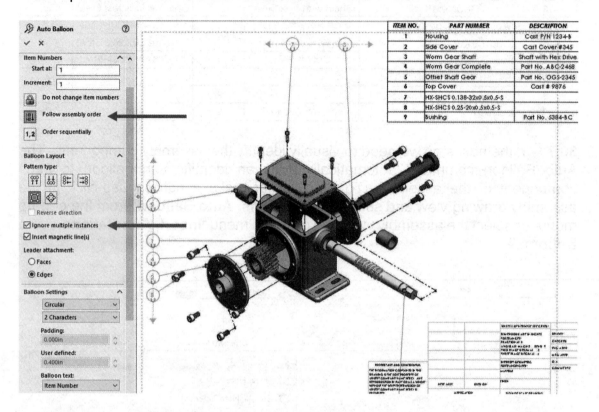

ITEM NO.	PART NUMBER	DESCRIPTION
1	Housing	Cast P/N 1234-b
2	Side Cover	Cast Cover #345
3	Worm Gear Shaft	Shaft with Hex Drive
4	Worm Gear Complete	Part No. ABC-2468
5	Offset Shaft Gear	Part No. OGS-2345
6	Top Cover	Cast # 9876
7	HX-SHCS 0.138-32x0.5x0.5-S	
8	HX-SHCS 0.25-20x0.5x0.5-S	
9	Bushing	Part No. 5384-BC

30.9. - A balloon's position can be changed as needed by dragging them individually or dragging the magnetic lines to move all the balloons attached to it at the same time. When a balloon is dragged by itself it will snap to magnetic lines, and magnetic lines can be moved around the drawing. A balloon's arrow tip can be dragged to a different area of the part for visibility; if we drag it to a different component, the item number in the balloon will change to reflect the part the balloon is attached to. Magnetic lines will be visible after selecting a balloon. Arrange the balloons as needed to improve readability.

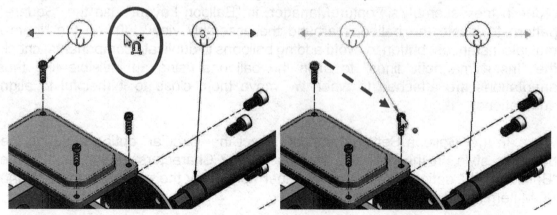

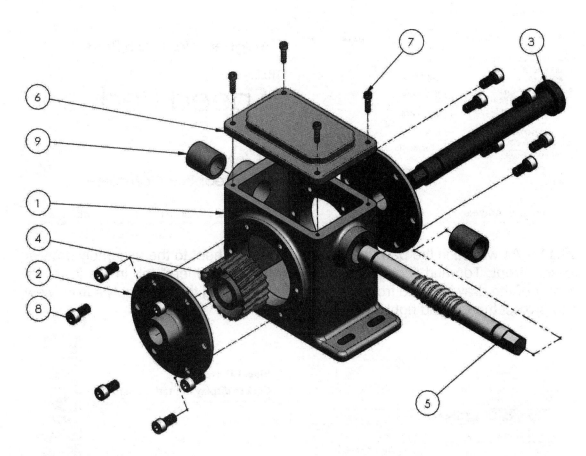

30.10. – Edit the Sheet Format to make corrections and add the missing notes in the title block.

ITEM NO.	PART NUMBER	DESCRIPTION	QTY.	Material	Weight
1	Housing	Cast P/N 1234-B	1	Cast Alloy Steel	4.40
2	Side Cover	Cast Cover #345	2	AISI 1020	0.36
3	Worm Gear Shaft	Shaft with Hex Drive	1	Chrome Stainless Steel	0.52
4	Worm Gear Complete	Part No. ABC-2468	1	AISI 1020	0.33
5	Offset Shaft Gear	Part No. OGS-2345	1	Chrome Stainless Steel	0.49
6	Top Cover	Cast # 9876	1	Cast Alloy Steel	0.48
7	HX-SHCS 0.138-32x0.5x0.5-S		4		
8	HX-SHCS 0.25-20x0.5x0.5-S		12		
9	Bushing	Part No. 5364-8C	2	Brass	0.057

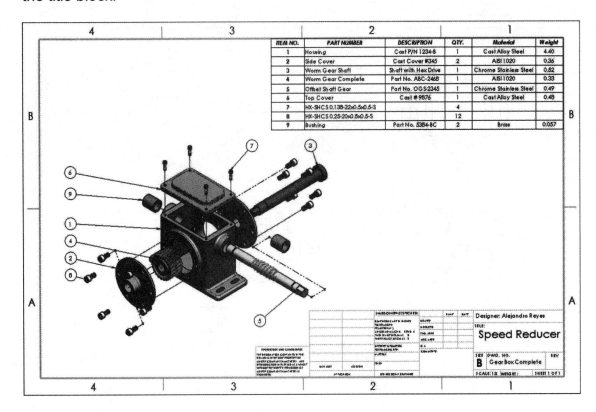

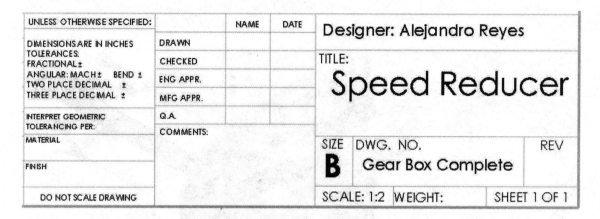

UNLESS OTHERWISE SPECIFIED:		NAME	DATE	Designer: Alejandro Reyes		
DIMENSIONS ARE IN INCHES	DRAWN					
TOLERANCES: FRACTIONAL±	CHECKED			TITLE:		
ANGULAR: MACH± BEND ± TWO PLACE DECIMAL ± THREE PLACE DECIMAL ±	ENG APPR.			Speed Reducer		
	MFG APPR.					
INTERPRET GEOMETRIC TOLERANCING PER:	Q.A.					
MATERIAL	COMMENTS:					
				SIZE **B**	DWG. NO. Gear Box Complete	REV
FINISH						
DO NOT SCALE DRAWING				SCALE: 1:2	WEIGHT:	SHEET 1 OF 1

30.11. - As we did in the part drawings, add a new sheet to the assembly drawing to add Front, Top and Right views. Click in the "Add Sheet" tab in the lower left corner of the drawing, and from the "View Palette" drag the assembly's front view, and project the top and right views from it.

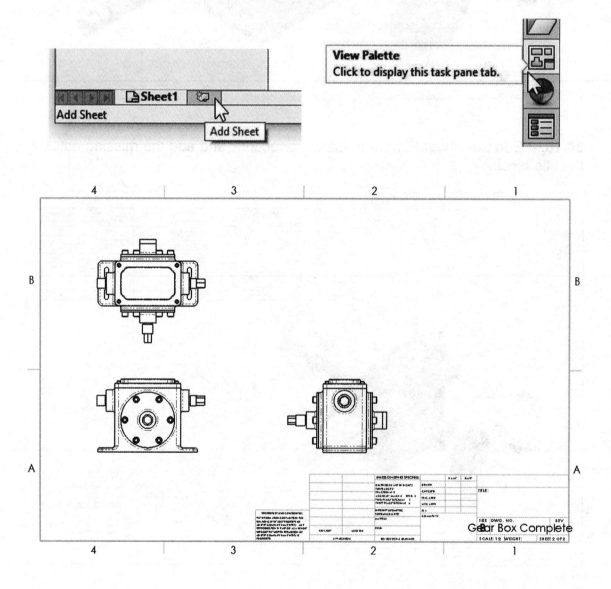

30.12. - Adding a section view to an assembly is just like in a part; the exception is that in the assembly we have the option to exclude selected components from the section. Select the "**Section View**" command and add a vertical section through the middle of the right view. Immediately after locating the section line the "Section Scope" dialog box asks us to select the components that will not be cut by the section view.

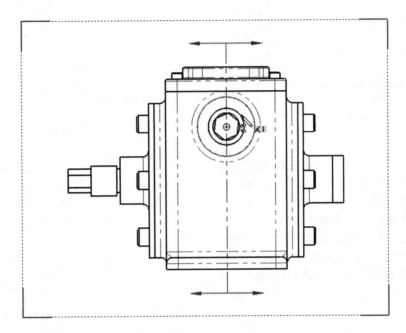

When asked to select the components excluded from the section view, select the 'Offset Shaft Gear' in the right view, and click OK to finish the selection, and locate the section view. If a section line cuts fasteners, we can turn ON the "Exclude fasteners" option to automatically exclude all Toolbox components. Locate the section view to the right side. Notice the 'Offset Shaft Gear' is not cut.

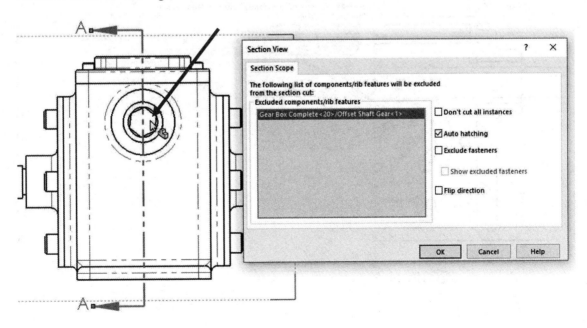

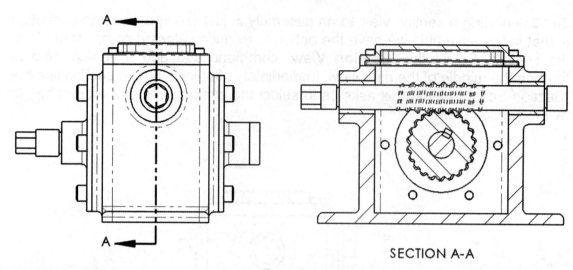

SECTION A-A

30.13. - To exclude additional components from the section view, right-mouse-click in the section view, select "Properties" from the pop-up menu, go to the "Section Scope" tab and select the '*Worm Gear Complete*'. Click OK to continue and change the section view style to "Shaded with edges."

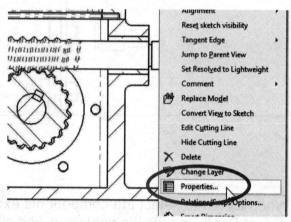

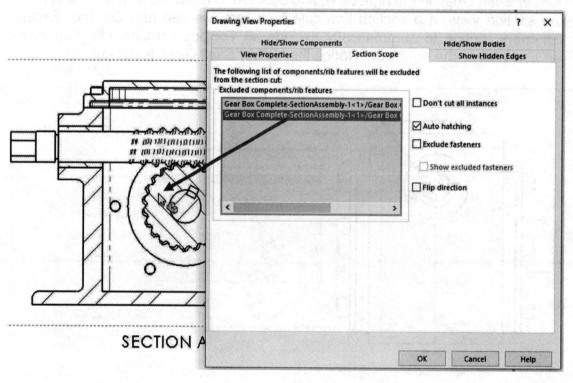

SECTION A

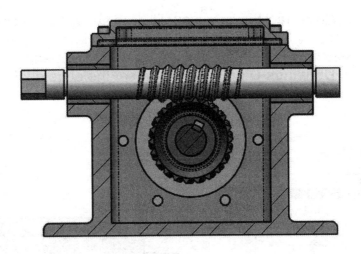

SECTION A-A

If the section view (or any view in a drawing) is not accurately displayed, for example, circles are shown as polygons, we can increase the image quality by going to the menu "**Tools, Options, Document Properties, Image Quality**" and increase the image quality. Be aware that using a high quality image may impact your PC's performance.

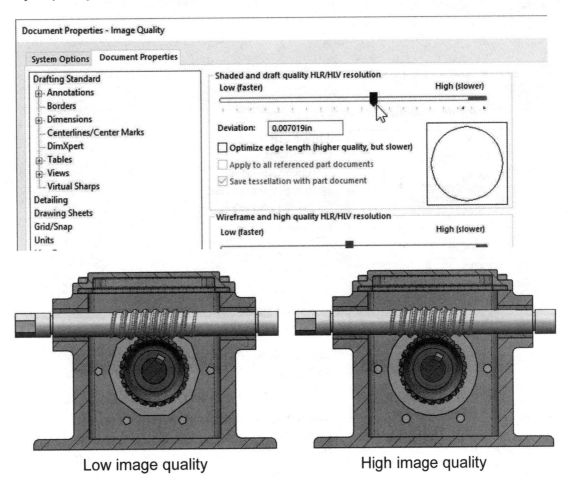

Low image quality   High image quality

Another way to correct the image quality is to select the "High quality" option in the "Display Style" section of the view's properties in the PropertyManager. This option can be pre-set in the menu "**Tools, Options, System Options, Drawings-Display Style**" to set the edge quality for wireframe or shaded edge views. The default setting for shaded with edges views is draft quality.

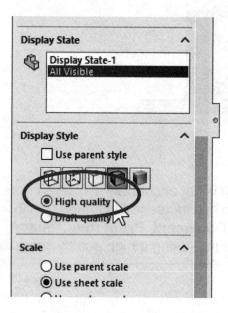

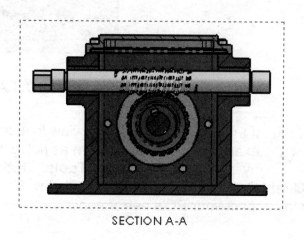

SECTION A-A

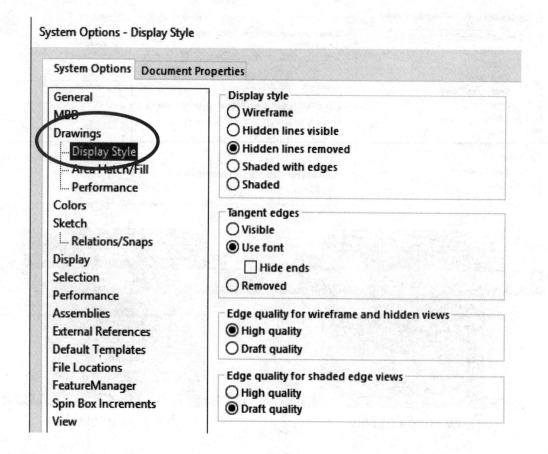

30.14. - Add a new detail view of the region where both gears mesh and change the detail's scale to 2:1. Holding down the "Ctrl" key while drawing the detail circle (or any sketch entity) will temporarily disable automatic sketch relations.

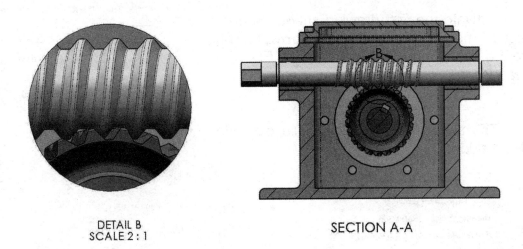

DETAIL B
SCALE 2 : 1

SECTION A-A

30.15. – After completing the drawing views, since the assembly does not have any dimensions, we cannot import any annotations from the assembly to the drawing. In this case, we have to manually add the overall assembly dimensions to the drawing using the "**Smart Dimension**" command. These will be reference dimensions, and will be displayed in grey. To show them in black we need to move them to the "FORMAT" layer, or select the "FORMAT" layer first, *and then* add the reference dimensions. Add the necessary centerlines and centermarks where needed.

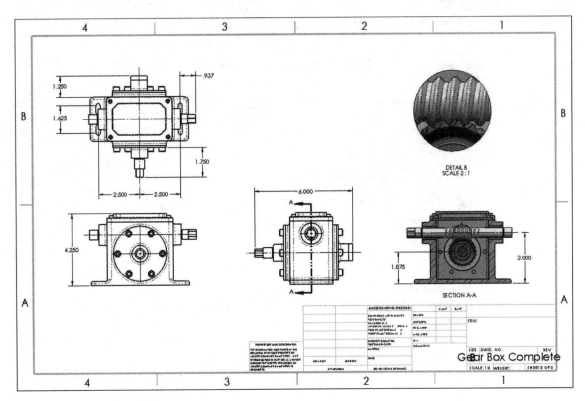

30.16. - We can modify the sheet format in the second sheet to match the first sheet, but instead we'll save the sheet format of the first page and apply it to the new sheet.

Activate the first drawing sheet, "Sheet1," select the menu "**File, Save Sheet Format…**" and save the sheet format using the name '*Exercises b – landscape*'.

 The sheet format uses the file extension *.slddrt

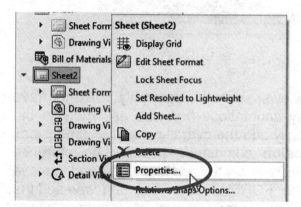

After saving the sheet format, go back to "Sheet2." Right-mouse-click either in the drawing sheet (not a view), *or* the FeatureManager in "*Sheet Format2,*" and select "Properties." Browse to locate the new sheet format and click "Apply Changes" to load it.

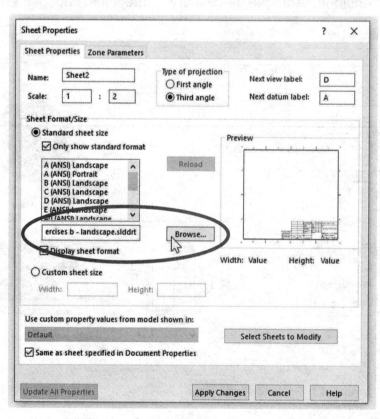

After loading the shaved sheet format, every note, modification, and custom property added in the saved sheet format will be added to the new drawing sheet. Save the drawing and close the file.

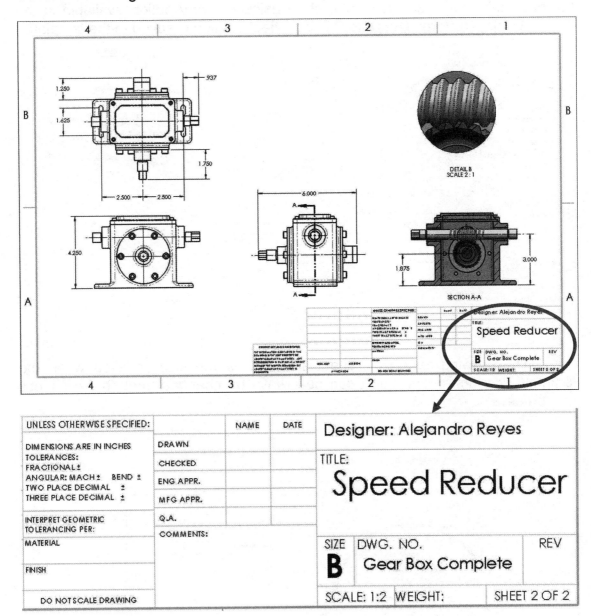

To create a drawing template, start a new drawing without adding any drawing views. The sheet format can be modified, change units of measure, drafting standard, add and/or modify layers, line styles and thickness, define the drawing's image quality, scale, etc. Basically, every option available in the menu "**Tools, Options, Document Properties**" section is saved to the template.

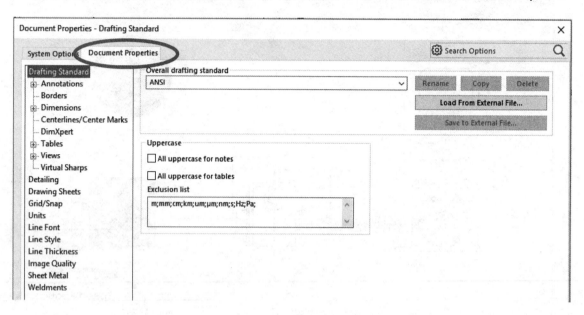

The drawing template's sheet size will be the size selected when the template was first made, but it can be always be changed through the sheet's properties.

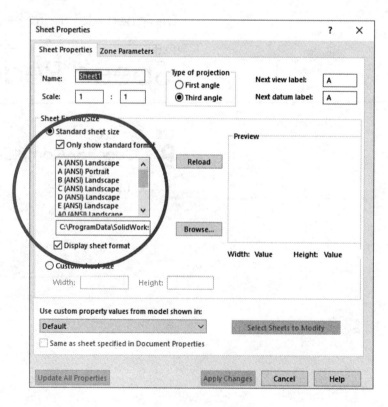

By default, if more sheets are added to a drawing using this template, the sheet format will not be copied into the new sheet, only the first sheet will have the sheet format saved with the template. To have a sheet format loaded with a new sheet, go to the menu "**Tools, Options, Document Properties, Drawing Sheets**", activate the "Use different sheet format" option and browse to the desired sheet format's location; this way when we add a new sheet using this template a sheet format will be loaded.

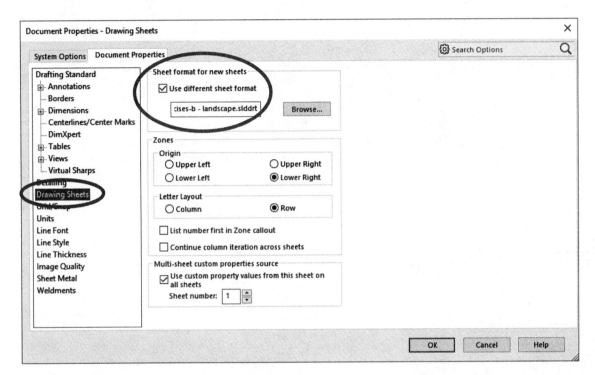

Make as many drawing templates as needed using different combinations of units, drafting standards, sheet formats, etc.

Notes:

Design Table Drawing

31.1. - A drawing of a part with a design table is essentially the same as any other drawing, with the only difference that we can add the Design Table to the drawing.

Make a new drawing of the *'Screw Design Table'* part using the A-Landscape template. Add Front, Right and Isometric views, import and arrange dimensions, and modify the sheet format as shown. Make the Isometric view's scale 2:1. Use the *"0.25x0.5"* configuration.

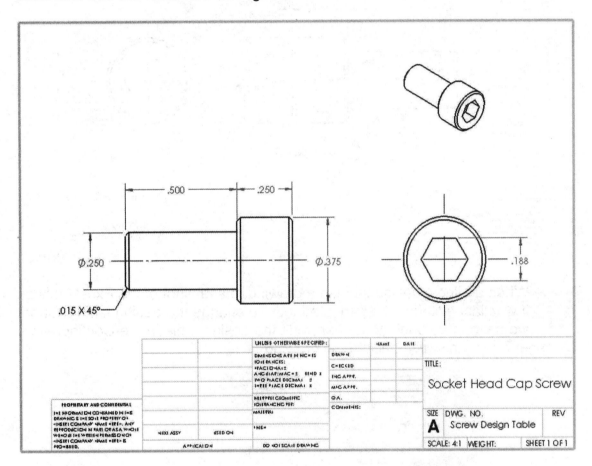

31.2. - Right-mouse-click in a part's view and select "**Tables, Design Table**" from the pop-up menu to import the part's design table; it will be immediately added to the drawing.

Once the Design Table is added to the drawing we can optionally turn on dimension names to help identify the configured dimensions using the menu "**View, Hide/Show, Dimension Names.**"

645

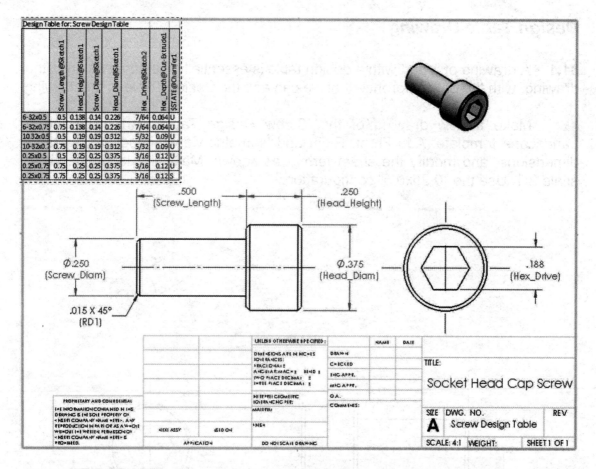

Design Table for: Screw Design Table

	Screw_Length@Sketch1	Head_Height@Sketch1	Screw_Diam@Sketch1	Head_Diam@Sketch1	Hex_Drive@Sketch2	Hex_Depth@Cut-Extrude1	$STATE@Chamfer1
6-32x0.5	0.5	0.138	0.14	0.226	7/64	0.064	U
6-32x0.75	0.75	0.138	0.14	0.226	7/64	0.064	U
10-32x0.5	0.5	0.19	0.19	0.312	5/32	0.09	U
10-32x0.7	0.75	0.19	0.19	0.312	5/32	0.09	U
0.25x0.5	0.5	0.25	0.25	0.375	3/16	0.12	U
0.25x0.75	0.75	0.25	0.25	0.375	3/16	0.12	U
0.25x0.75	0.75	0.25	0.25	0.375	3/16	0.12	S

When adding a design table to a drawing, a snapshot of the Excel table as it was last edited in the part is added. To change the design table's format we have to go back to the part, edit the design table in the Configuration-Manager and then switch back to the drawing.

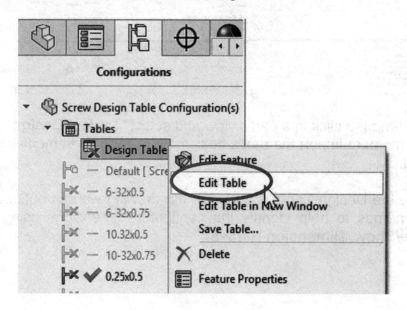

646

31.3. – The "Hex_Drive" dimension is shown in the design table using a fractional dimension, and the imported dimension in the drawing is in decimal format. To show the dimension as fractional, select the dimension and from the dimension's properties select the "Other" tab, activate the "Override Units" checkbox and select "Fractions." Set the denominator to "16" and activate the option "Round to nearest fraction," this way the dimension will be rounded to the nearest 1/16th of an inch.

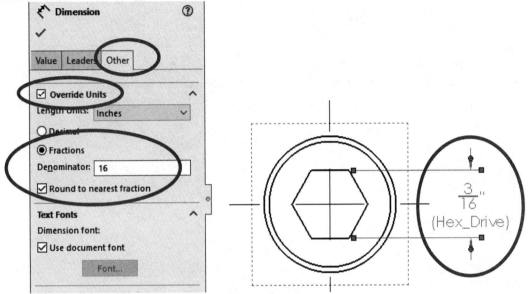

Save and close the drawing.

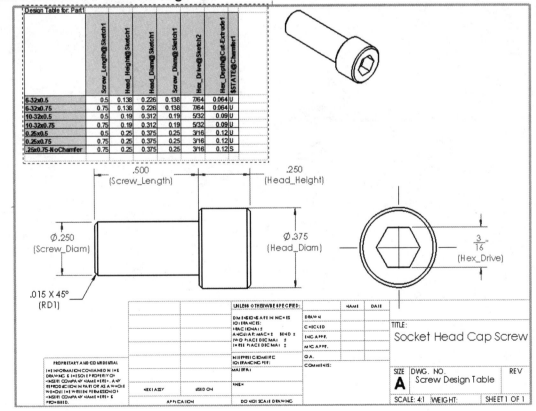

Notes:

Engine Project: Make an assembly drawing of each sub-assembly and one of the exploded Engine Assembly. Use the following images as a guide.

For the "Connecting Rod Sub Assembly" use the "A-Portrait" template, and for the "Engine Assembly" drawing use the "C-Landscape" template to better accommodate the large assembly. Feel free to format the BOM to your liking.

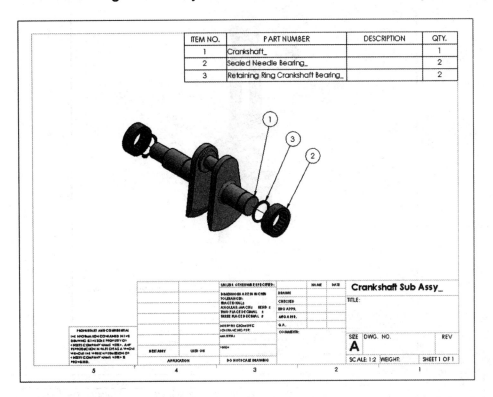

ITEM NO.	PART NUMBER	DESCRIPTION	QTY.
1	Crankshaft_		1
2	Sealed Needle Bearing_		2
3	Retaining Ring Crankshaft Bearing_		2

For the Exhaust assembly add a Broken-Out section to show the interior.

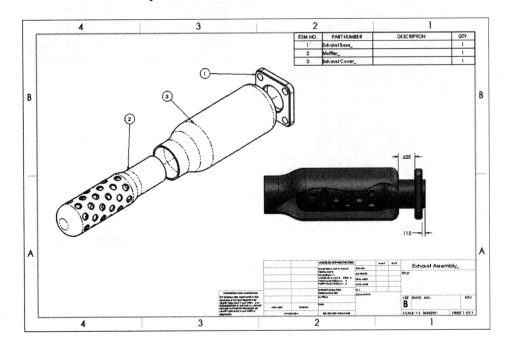

ITEM NO.	PART NUMBER	DESCRIPTION	QTY.
1	Exhaust Base_		1
2	Muffler_		1
3	Exhaust Cover_		1

ITEM NO.	PART NUMBER	DESCRIPTION	QTY.
1	Connecting Rod_		1
2	Connecting Rod Bottom_		1
3	Con Rod Crankshaft Half Bushing_		2
4	Bushing Top Con Rod_		1
5	HX-SHCS 0.25-28x0.625x0.625-S		2

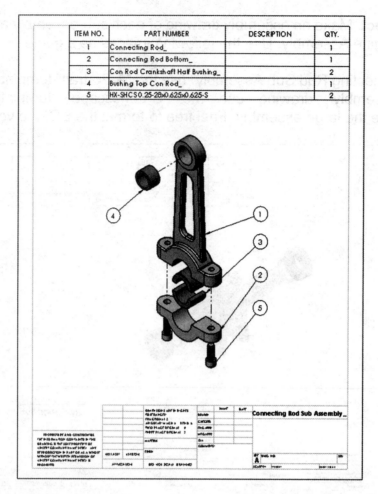

Connecting Rod Sub Assembly_

ITEM NO.	PART NUMBER	DESCRIPTION	QTY.
1	Piston Head_		1
2	Pin ConRod-Piston_		1
3	Internal Retaining Ring_		2
4	Top Compression Ring_		1
5	Middle Compression Ring_		1
6	Oil Control Ring_		1

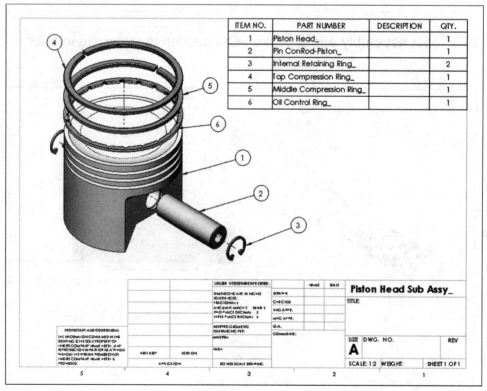

Piston Head Sub Assy_

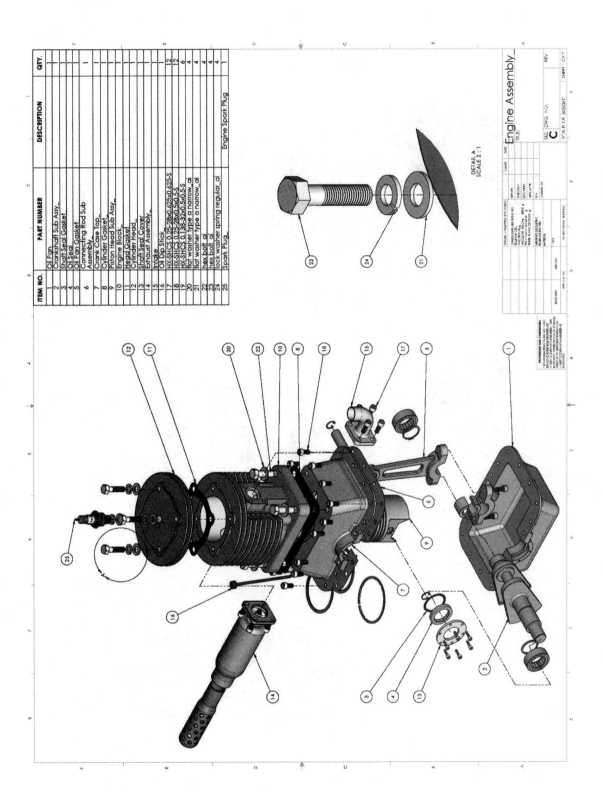

ITEM NO.	PART NUMBER	DESCRIPTION	QTY.
1	Oil Pan		1
2	Crankshaft Sub Assy.		1
3	Shaft Seal Gasket		1
4	Oil Seal		1
5	Oil Pan Gasket		1
6	Connecting Rod Sub Assembly		1
7	Crank Case Top		1
8	Cylinder Gasket		1
9	Piston Head Sub Assy.		1
10	Engine Block		1
11	Head Gasket		1
12	Cylinder Head		1
13	Shaft Seal Cover		1
14	Exhaust Assembly		1
15	Intake		1
16	Oil Dip Stick		1
17	HX SHCS 0.25-28x0.625x0.625-S		12
18	HX SHCS 0.25-28x0.5x0.5-S		12
19	HX SHCS 0.138-32x0.5x0.5-S		6
20	flat washer type a narrow_ai		4
21	flat washer type a narrow_ai		4
22	hex bolt ai		4
23	hex bolt ai		4
24	lock washer spring regular_ai		4
25	Spark Plug	Engine Spark Plug	1

DETAIL A
SCALE 2 : 1

Engine Assembly

TITLE:

SIZE C   DWG. NO.

SCALE 1:9   WEIGHT: SHEET 1 OF 2

651

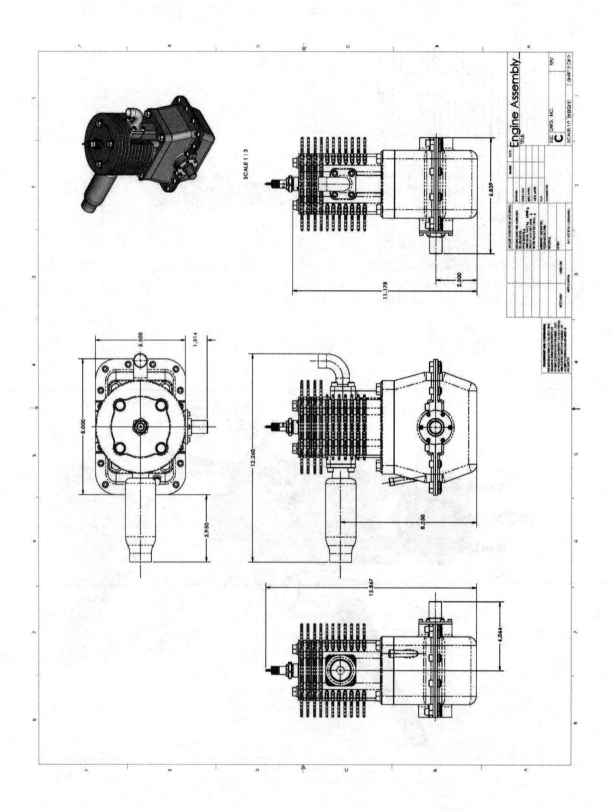

Chapter 7: Animation and Rendering

SOLIDWORKS Professional and Educational editions include PhotoView 360, an integrated software used to generate photo realistic rendered images of parts and assemblies. Photo realistic images generated before a product or design is finished can be very useful; for example, we can show a potential customer how the product would look, prepare advanced marketing campaigns ahead of manufacturing, promotional videos, etc.

The resulting images can be of such a high quality and realism that they can (and often are) confused with pictures taken in a professional photo studio. PhotoView 360 allows us to select component materials, colors, backgrounds, light sources, shadow settings, reflections, and many advanced lighting settings. To master the rendering process and obtain high quality images takes time and patience, since we need to experiment with many settings, render a test image, adjust, and repeat.

To obtain high quality, photo realistic rendering results, the user will usually have to make multiple iterations, changing and fine-tuning different settings in each pass, especially lighting. In this lesson, we'll cover the basic and most commonly used settings to help us understand their effect and obtain acceptable rendering results.

The animation is used to show assemblies in motion, to help us see and understand how they work in real life, animate exploded views and collapse them, for example, to explain how to assemble or take a product apart. Also, not only can we animate an assembly, we can change a component's visual properties including transparency, appearance, display settings, hide and show them, etc. to better illustrate their operation, inner workings or highlight features.

After an animation is completed it can be saved to a file as an *.AVI video or a series of pictures to be used in third party video editing software. At the time of saving the video we can choose to create the video using the SOLIDWORKS screen or a photo realistic animation using PhotoView 360. In the second case, the result will take considerably longer to generate, as each frame of the animation will be photo realistically rendered, and depending on the rendering settings, the length of the animation, and the size of the image, this can take a long time to produce.

To practice, we'll produce renderings and animations of the parts and assemblies of the gear box and the engine we have completed so far.

653

Notes:

PhotoView 360

32.1. - To practice rendering open the '*Gear Box_PV*' assembly from the exercise files or use the gear box assembly we have made so far.

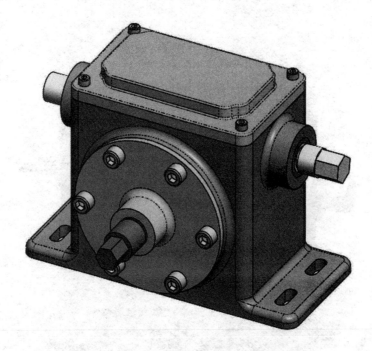

32.2. - To start rendering, we need to load PhotoView 360 in SOLIDWORKS. Go to the menu "**Tools, Add-ins…**" and select "PhotoView 360," or in the Command Manager, select the SOLIDWORKS Add-Ins tab and activate "PhotoView 360." We can also turn on the other add-ins included in SOLIDWORKS, including "SOLIDWORKS Toolbox."

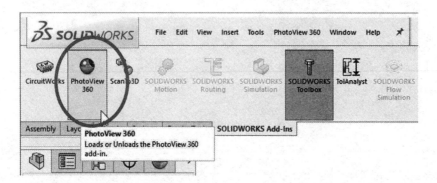

After loading PhotoView 360 a new "PhotoView 360" menu and a "Render Tools" toolbar is added to SOLIDWORKS.

32.3. - The first thing we need to do to prepare for a render is to activate the "DisplayManager" tab; this is the area where we can add, edit, and delete a component's appearance, decals, light sources, cameras, and scenes. Make the FeatureManager wider if needed or press the default shortcut F8. After the tab is activated we can see the different sections for:

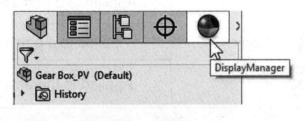

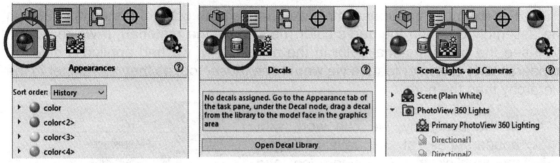

Appearances **Decals** **Scenes, Lights, Cameras**

After selecting "View Appearances" we can see all the appearances used in the assembly (or part) listed in chronological order (History), Alphabetical order or by Hierarchy. History lists the appearances in the order they were added, and Hierarchy lists the appearances at the level at which they are added.

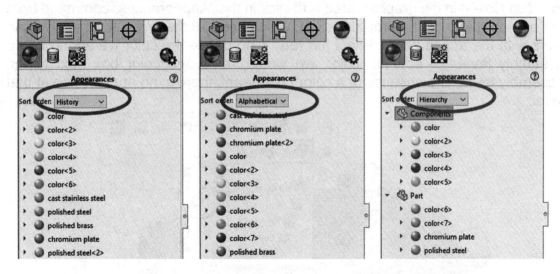

Appearances will be shown depending on the level at which they are added, and higher hierarchy appearances will override lower level appearances. This is the hierarchy for appearance display:

	Component	Applied to a part or sub-assembly in the assembly.
	Face	Applied to a part's face.
	Feature	Applied to a part's feature.
	Body	Applied to a part's body.
	Part	Applied to the entire part.

For example, in our assembly we changed the color of the components in the assembly covering the part's color (defined by the material). If we remove it we'll see the part's material color in the assembly; if we had applied a different appearance to a face or feature we would see those, too, as they are higher in the hierarchy than the part.

In the assembly, we can work with appearances at all levels. Right-mouse-click in an appearance and select the option to Add, Edit, Copy, Paste or Remove an appearance to a Component, Face, Feature, Body or Part.

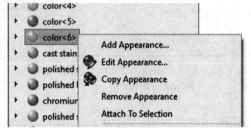

32.4. – Appearances can be added, removed or edited in the DisplayManager or the graphics area. In this step, we'll remove the color from the '*Top Cover*'. Select the '*Top Cover*' in the graphics area and click in the "Appearances" command from the pop-up menu. In the drop-down list, we can see all the levels at which we can work with the appearance. Select the red "X" to delete the color we added at the assembly level. To edit the color, we would select the color box next to the assembly level. The absence of a color box means there is no appearance at that level.

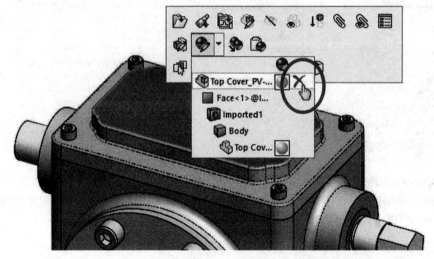

After removing the appearance, the '*Top Cover*' now looks like this, showing the material's appearance added at the part level.

32.5. - For the next step we are going to add a decal to our assembly. Decals are used to show logos, labels, markings, etc. in a component. Open the '*Top Cover*' part (if using the supplied exercise files for the PhotoView 360 exercise, you may be asked "*Do you want to proceed with feature recognition?*" If this is the case, select "No" to continue). After opening the part, select the DisplayManager tab and go to the "View Decals" section.

Any decals applied to a component will be listed in this section. Click in "Open Decal Library" to show the library decals available in the "Appearances, Scenes, and Decals" section of the task pane.

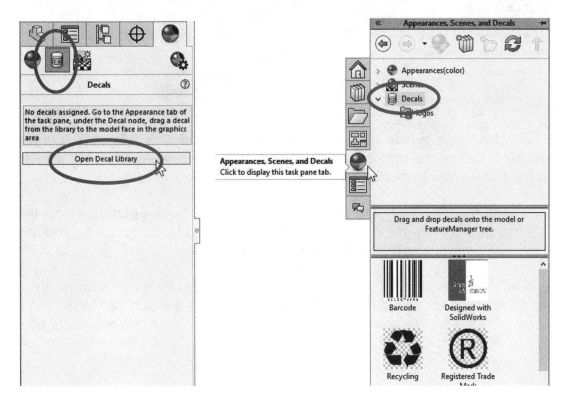

To create a new decal right-mouse-click in the DisplayManager pane, select "Add Decal" and follow the instructions to define the custom decal's parameters.

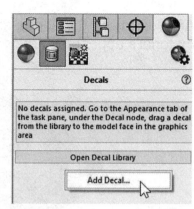

32.6. - From the "Decals" library in the Task Pane, scroll down to the "Recycling" decal and drag it onto the top face of the cover. As soon as we drop it, the decal's properties are displayed. In this page, we can also select a different image if needed. Notice the recycling logo is rotated in this view.

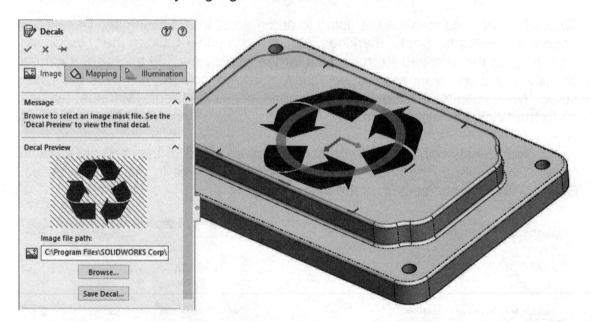

32.7. - Just as an exercise we are going to rotate the logo in this step. It is not necessary because the part is symmetrical and can be rotated, but we'll do it to show the reader how to modify the decal.

Select the "Mapping" tab in the properties. Here we can change how the label is mapped in the surface (flat, cylindrical, spherical), its location, size, rotation and optionally mirror it vertically and/or horizontally.

Set the label's size to 1" and mirror it both vertically and horizontally, or rotate it 180 degrees. Leave the location in the center of the part (X=0, Y=0). We can also drag the decal's corners to resize it, the ring to rotate and anywhere else to move it. Click OK to add the decal and save the '*Top Cover*'. Close the part file and return to the assembly.

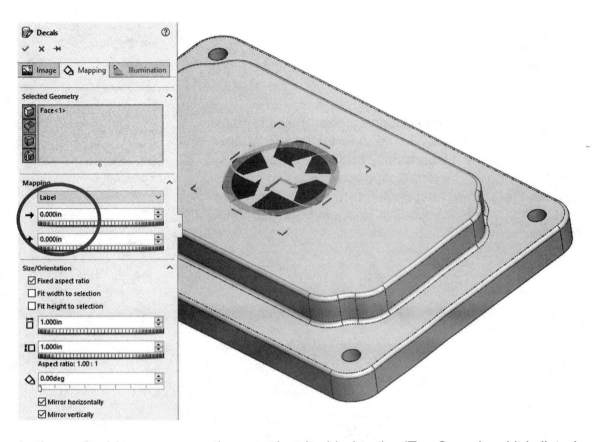

In the assembly we can see the new decal added to the *'Top Cover'* and it is listed in the "Decals" section in the assembly.

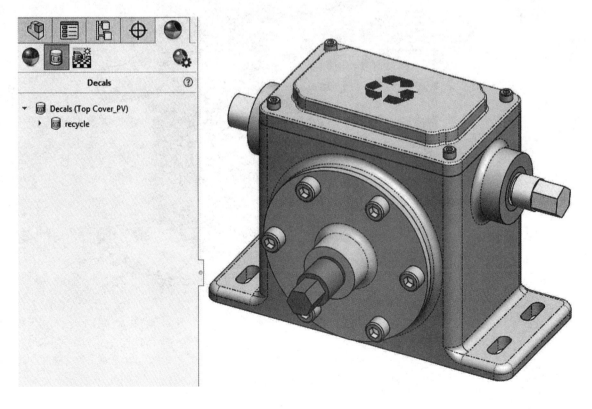

32.8. – Our next step will be to define the scene and light settings for the assembly before rendering the final scene. Select the "Scene, Lights, and Cameras" section.

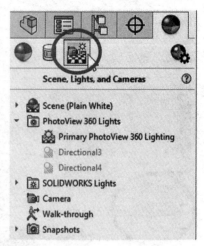

If the hardware supports it, RealView graphics can be activated in the View toolbar. RealView will show a high-quality real time render using the component's materials, appearances, and scenes added. RealView is not required to render an image but helps when composing images before a final render.

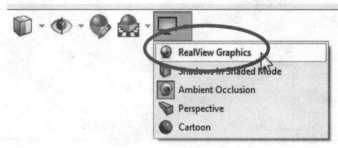

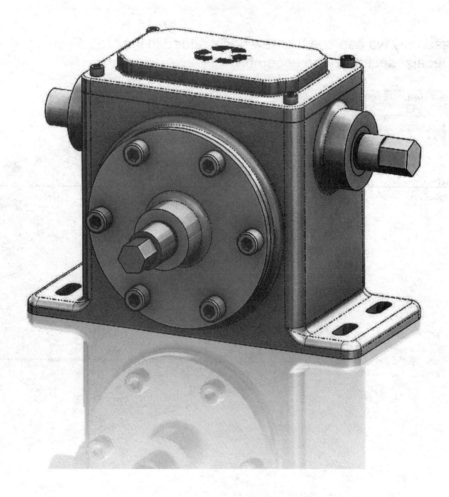

32.9. – The SOLIDWORKS built-in scenes have predefined light settings, backgrounds, and floor textures. To apply a scene to our assembly (or a part), open the "Appearances, Decals and Scenes" tab in the Task pane, scroll down to "Scenes, Studio Scenes" and select the "Reflective Floor Checkered" scene. To add it to our assembly we can:

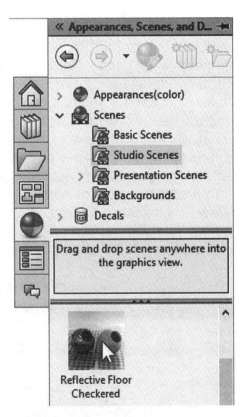

- Double click in it
- Right-mouse-click and select "Add Scene to part," or
- Drag it to the graphics area.

After adding a scene, we can see the new scene parameters in the Display Pane. To fine tune the scene, right-mouse-click in Scene, Floor, Background, and Environment sections to modify to our liking. The images with floor reflections were made with RealView active.

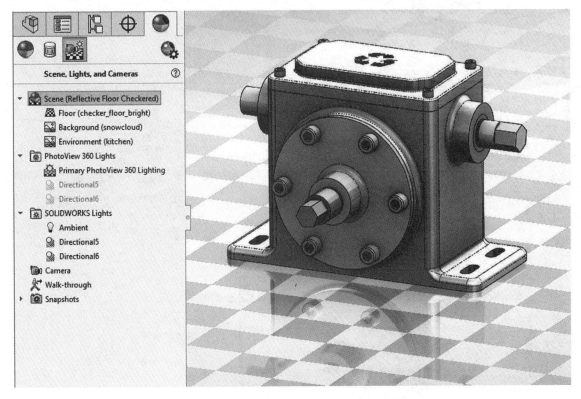

32.10. - Pre-defined scenes include light settings to match the scene. Turn on the "**Preview Window**" command in the Render Tools tab in the CommandManager. A preview window will appear to help us adjust our settings before rendering the final image. Be aware that computer resources will be used while the preview image is being dynamically rendered, and overall performance will be impacted.

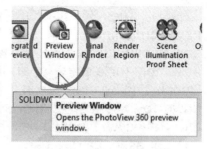

After turning on the preview, we may be asked if we wish to turn on a camera, turn on perspective view or continue. For our example, select "**Turn on Perspective View**." By selecting this option, objects in the scene will look more realistic; all lines going "*into the screen*" will change to converge into a vanishing point in the distance, much like objects are perceived by the eye.

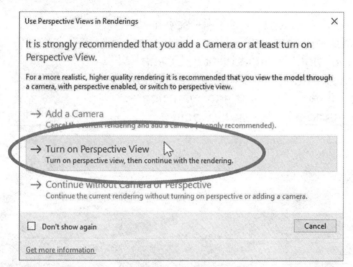

 The Perspective View can be turned on at any time in the "**View Settings**" drop-down command or the menu "**View, Display, Perspective**."

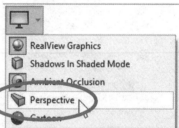

In the preview window, we can see the light is (in this case) too bright. To change the lighting, expand the "PhotoView 360 Lights" section in the DisplayManager. Right-mouse-click "Primary PhotoView 360 Lighting" and select "Edit Primary PhotoView 360 Lighting."

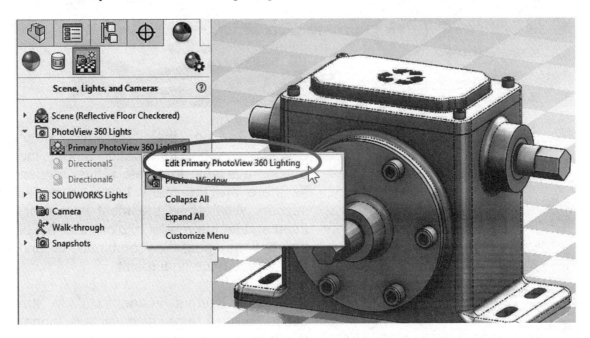

This is one of the settings we need to tweak to obtain better results. Change the illumination parameters until the preview's image improves; click OK to finish.

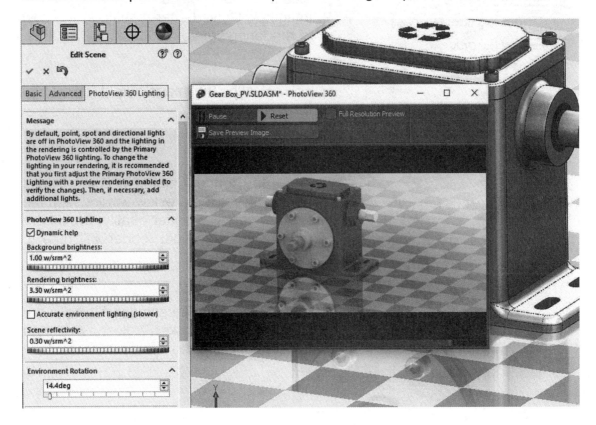

32.11. – After the illumination settings are complete, select the "**PhotoView 360 Options**" to define the final render options.

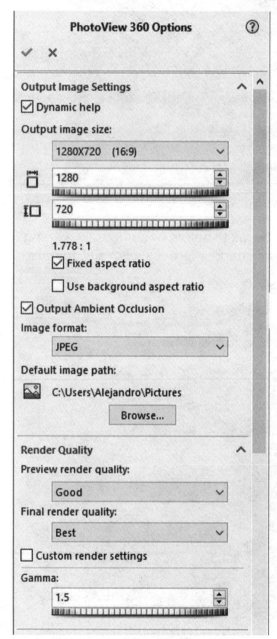

Turn on the "Dynamic Help" checkbox to get information about the options available to create a final render, as there are too many to cover in this section. We'll show the most commonly used and let the reader explore the rest of the options, possible combinations and their effect in the final result.

Set "Output image size" to the desired size. A larger size will take longer to render but will provide a bigger image to resize and print. We'll use 1280x720 in our example for speed purposes; larger images will give better results, at the cost of rendering time.

Define the image output format and the path to save it.

In "Render Quality" we can change the quality for the preview window and the final render. The better quality, the longer the preview and final render will take to complete.

The gamma settings will make the scene lighter or darker. Adjust as needed to obtain the desired results.

Generally, when creating a final rendered image, start with a small image size, turn off the advanced lighting effects which add rendering time (Bloom, contour/cartoon rendering, Direct caustics and Output ambient occlusion) and lower the final render quality to evaluate the resulting image, change positions, lights, colors, environment, etc. before committing to a time-consuming high-quality final render.

The PhotoView 360 rendering engine takes full advantage of multi-core processors; therefore, the faster the processor speed and the more processor cores available, rendered images will complete faster.

32.12. – To evaluate multiple scene illumination options simultaneously we can use the "**Scene Illumination Proof Sheet**" command to render multiple small images with various lighting settings, making it easier to get the best result.

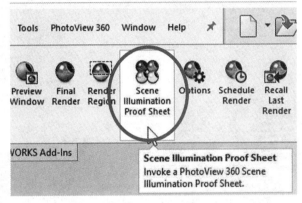

Selecting an image will position it in the center, and the rest of the images will be updated to show darker and lighter lighting options.

Moving the increment slider to "Fine" will make small lighting adjustments, and "Coarse" will make larger adjustments. When the desired image is selected, click OK to apply those settings to our scene.

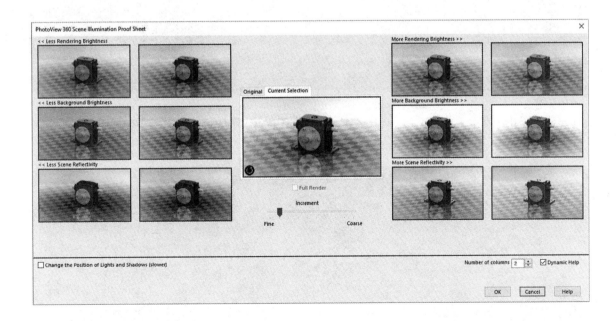

32.13. - Once the render options are set, select the "**Final Render**" command from the Render Tools tab.

A new window with the final render progress will appear. Each processor's core will process a small area at a time; in this case the processor has eight cores, reducing the time needed to render an image.

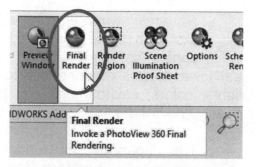

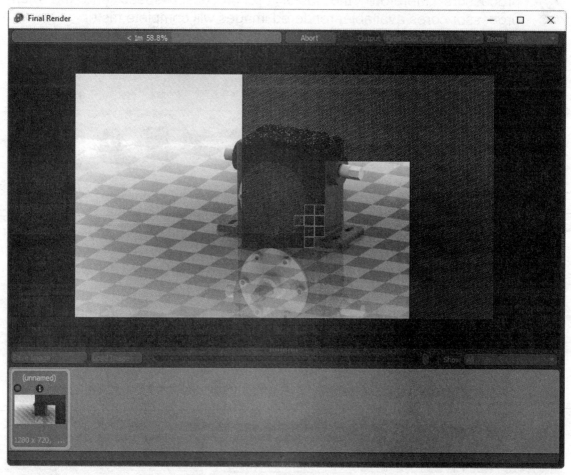

32.14. - After the render is finished, select the thumbnail at the bottom and click "Save Image," set the location and save it.

To add realism to the final renderings, remove the colors assigned to the rest of the components and activate the "Perspective" view in the View toolbar.

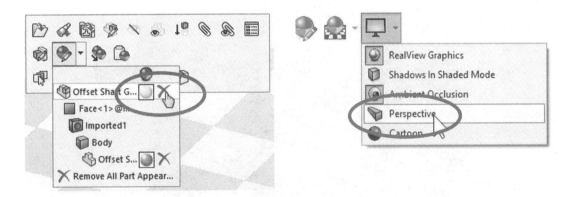

Feel free to explore different options, lighting, and environments.

Engine Project: Make a render of the engine using the knowledge acquired so far.

The previous image was made using a "**Cut Extrude**" assembly feature. This is a cut at the assembly level that is not reflected at the part level; think of it as cutting the parts *after* they are assembled together. Add a sketch in the assembly selecting a face just as we do in a part.

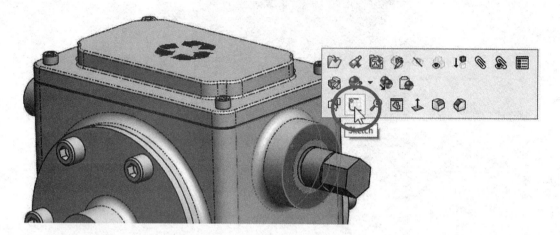

After starting a sketch in the assembly, we get a warning letting us know that the sketch is being created inside the assembly and not in a part.

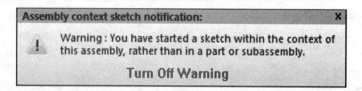

Draw a rectangle (or any other profile) to make the cut. Make sure it covers the areas of the assembly you want to cut.

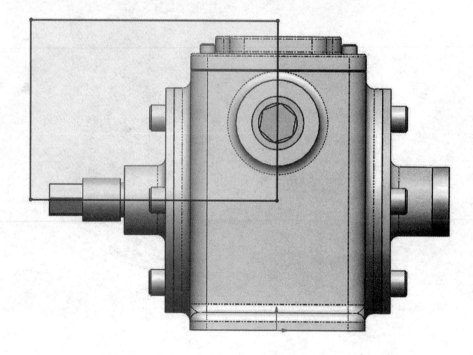

Select "**Extruded Cut**" from the drop-down menu under "**Assembly Features**" in the Assembly tab of the CommandManager.

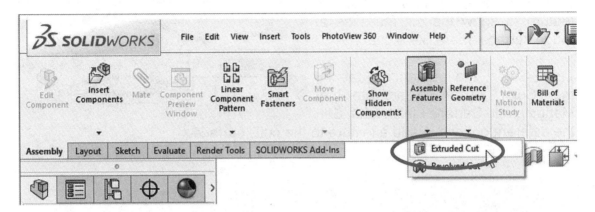

Make the cut using the "Through All" option (only a suggestion). Under the "Feature Scope" uncheck the "Auto-select" option and select the components that will be included in the cut feature; this is the same behavior as the assembly drawing's section view. If a selected component is completely enclosed by the cut (like some of the screws), it is deleted from view. Click OK to finish.

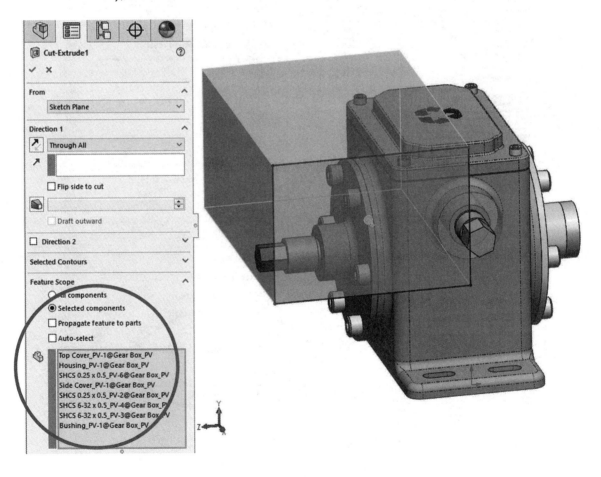

As a finishing touch before rendering, we can add a different texture to the assembly's cut-extrude feature to show the cut faces as a rough surface, more like a cut.

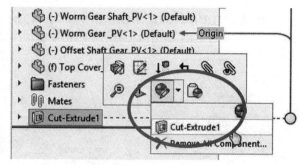

Select the "Advanced" tab in the "Color" option to activate the Mapping and Surface Finish tabs. Set the different options to add a texture to the cut's surface.

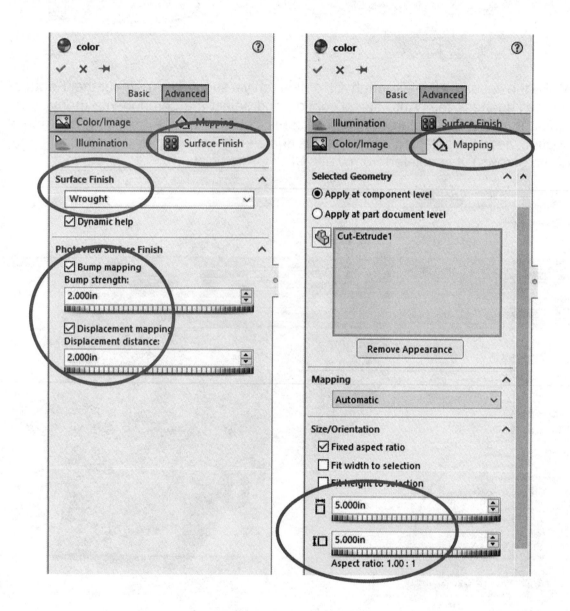

After mastering the basics, explore different options, textures, lighting effects, and the multiple settings available to create photo realistic images of your projects and designs.

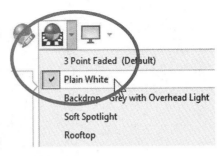

The following image was created using the "Plain White" scene from the "Apply Scene" in the View toolbar.

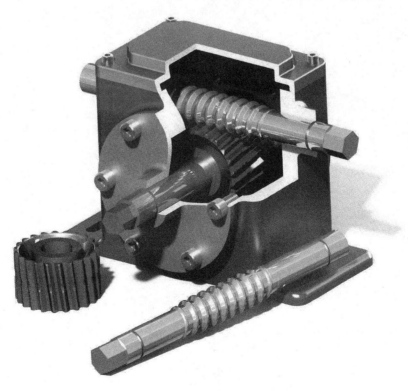

Notes:

Animation

Notes:

33.1. – With the built-in tools in SOLIDWORKS we can create an animation of a part or an assembly; in the latter case, the animation can show the assembly in motion, how to assemble the components (collapse), how to disassemble them (explode), etc. using a simple interface.

To learn how to use the animation tools open the '*Gear Box Animation*' assembly from the exercise files, or use the assembly made in the previous lessons. To start an animation, select the "Motion Study 1" tab at the bottom left corner of the assembly window to access the MotionManager.

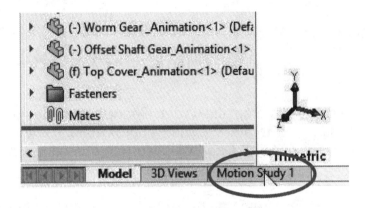

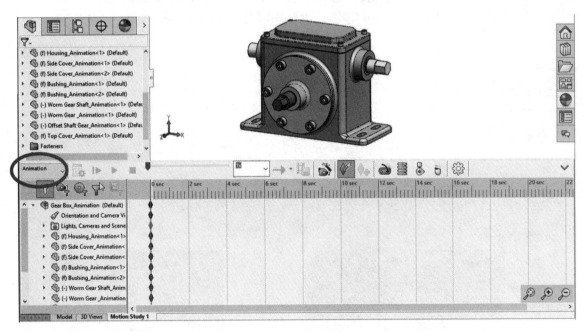

SOLIDWORKS has two different types of motion studies. The first one is "Animation," where we move components manually or with a motor to a new position using a timeline. The second type is "Basic Motion"; the difference is that a basic motion study considers physics, mass, gravity, etc. and the components are driven by motors, gravity, contact between them, and/or springs. In this lesson, we'll only cover Animation.

33.2. – For our first animation we'll manually move components to a new position and record it in the timeline. First, we need to move the time bar or click in the timeline at the four seconds mark to define the length of the first animation step.

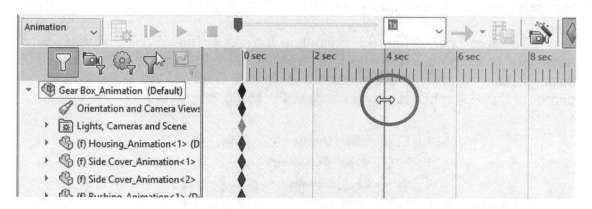

33.3. – Now move (click-and-drag) the '*Worm Gear Shaft*' to the position it will have at the four seconds mark; rotate the shaft approximately ¼ to ½ a turn in the screen. Notice the '*Offset Shaft Gear*' will also rotate because of the gear mate added previously.

 By default, the "**Autokey**" command is ON. For this animation step turn it OFF to learn the manual option first and the automatic option later.

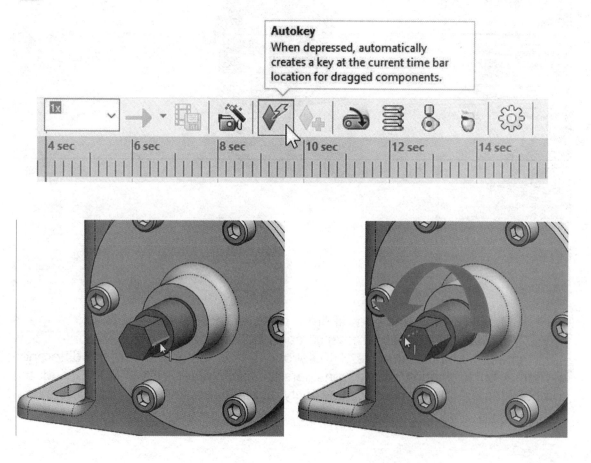

After rotating the '*Worm Gear Shaft*', click the "**Add/Update Key**" command at the top of the MotionManager, or right-mouse-click in the time bar at the '*Worm Gear Shaft's* level and select "**Place Key**." This key will mark the position of the component at four seconds in the animation.

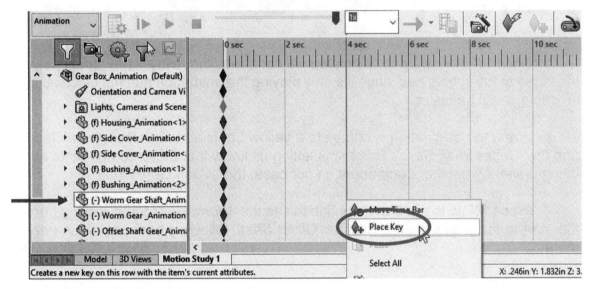

33.4. - After locating the first key, the animation's length is four seconds, and a green bar is added between 0 seconds and 4 seconds in the '*Worm Gear Shaft*'.

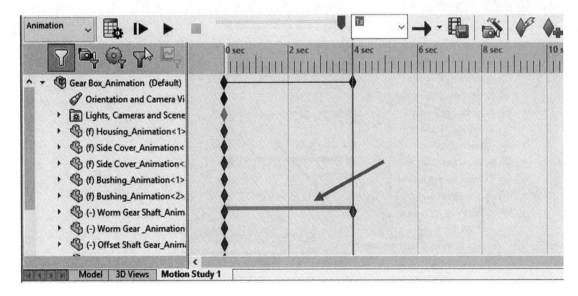

33.5. - Once the end of the first animation step is defined, SOLIDWORKS needs to calculate the motion of the components.

Click either in the "**Calculate**" or "**Play**" command; while the animation is calculated for the first time, the playback may be slow, depending on the number of moving components and/or visual effects being animated.

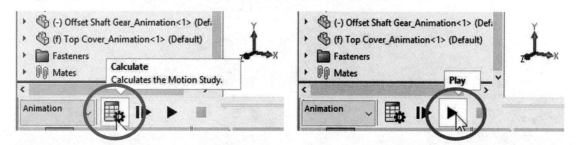

Once the animation is calculated we can press the "**Play**" command to see the video at full speed, and since it's only playing the animation and not calculating anything, it will be faster.

When the calculation is complete a yellow bar is added to the '*Worm Gear*' and the '*Offset Shaft Gear*'. This bar is letting us know that these components are being driven by another component, in our case, the '*Worm Gear Shaft*'.

Press "**Play**" to run the animation to see the '*Worm Gear Shaft*' turning from the start to the final position, and the '*Offset Shaft*' following because of the gear mate.

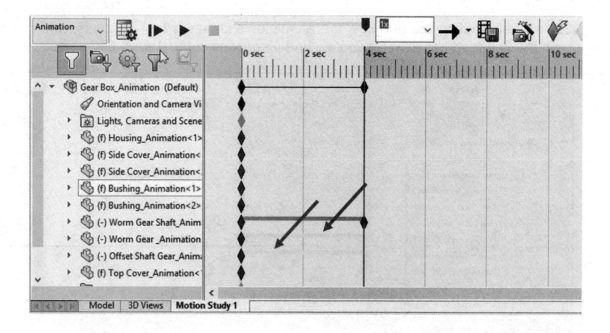

33.6. - During an animation we can also change a component's appearance. In this step, we'll make the '*Housing*' transparent. Move the time bar to the 2 second position.

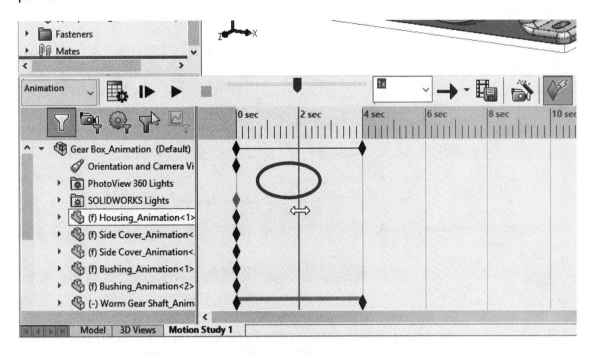

Right-mouse-click in the '*Housing*' and select "Change Transparency." A magenta line will be added between 0 and 2 seconds in the timeline.

 You may need to expand the pop-up menu at the bottom to reveal the "Change Transparency" command.

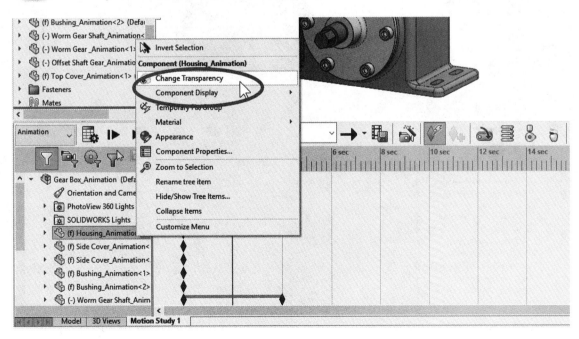

Expand the '*Housing*' in the animation manager to see the different aspects of the component that can be animated. The magenta line was added in the Appearance row; this is letting us know the component has a visual change during the specified timeframe.

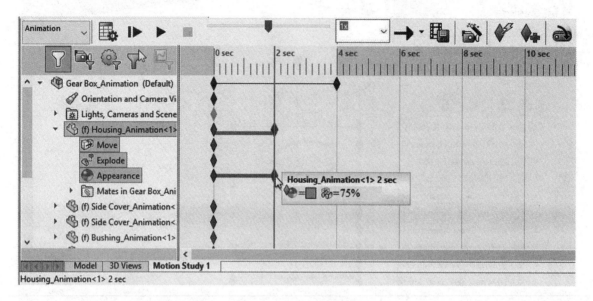

33.7. - After pressing play the '*Housing*' starts to fade at 0 seconds and becomes transparent at 2 seconds, letting us see the gears moving inside.

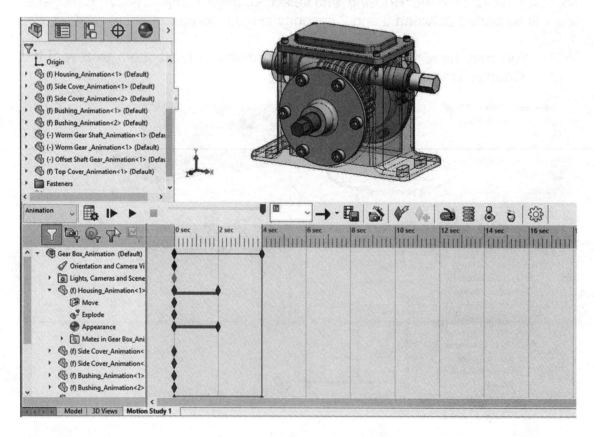

33.8. - For the next step we'll hide the '*Top Cover'.* Move the timeline to 4 seconds, click in it in the FeatureManager and select "**Hide**." A new animation step is added, also in magenta because it's an appearance change, going from 0 to 4 seconds.

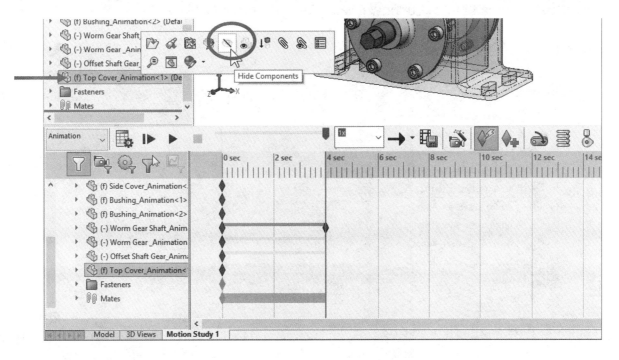

Pressing "Play" will show the animation, but now the '*Top Cover'* will be completely hidden at 4 seconds.

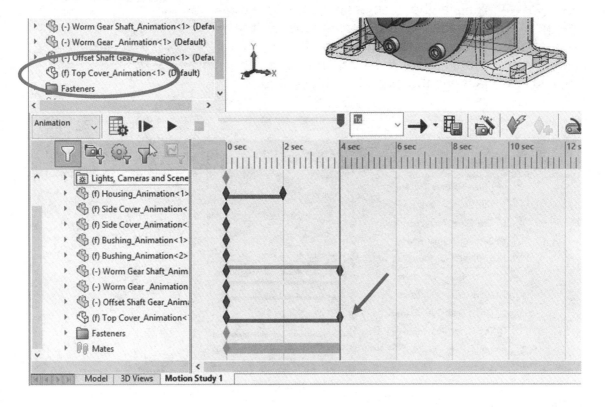

33.9. – We can also modify the markers in the timeline to change the animation. Just as an example, click in the '*Worm Gear Shaft*' motion marker at 4 seconds and drag it to the 6 second mark. By making this change the shaft will rotate the same distance as before, but now it will take 6 seconds instead of 4.

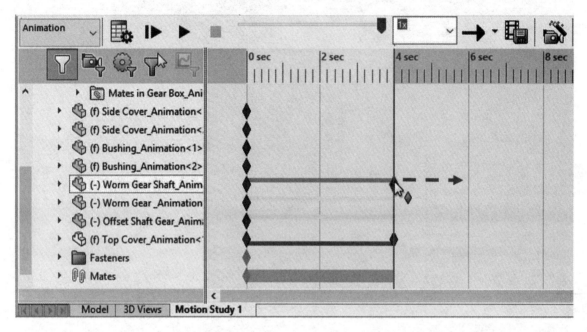

The yellow background at the top of the timeline is now hashed to let us know that we need to re-calculate the animation because changes were made to the movement of components. Click the "**Calculate**" button to recalculate the animation; the background is solid yellow again and the animation is updated.

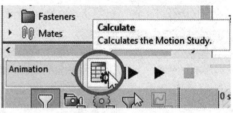

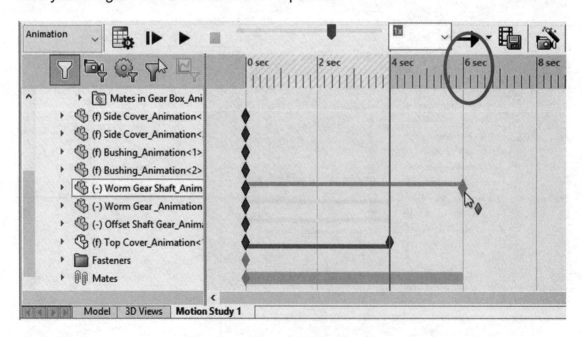

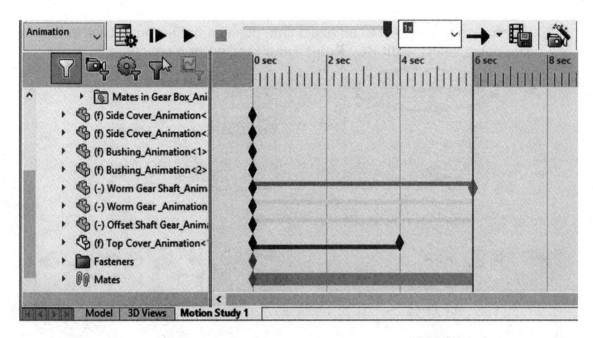

33.10. – In the next step we'll change the time to hide the '*Top Cover*'. Expand the '*Top Cover*' in the animation manager to see its animation keys. Click and drag the starting key to the 3 second marker in the timeline.

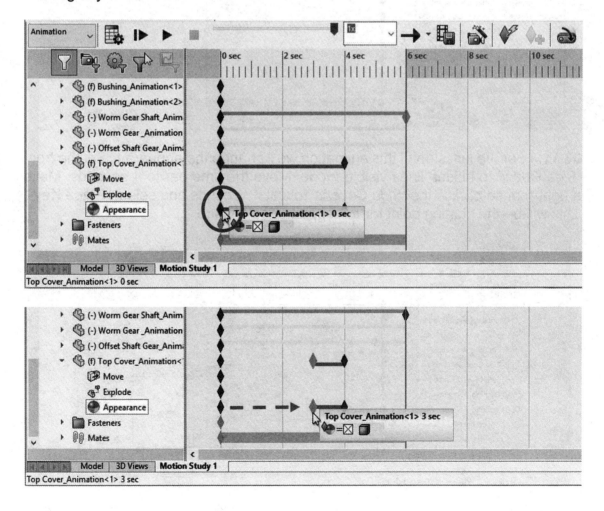

687

To change a key point to an exact time position right-mouse-click in the key point, select "**Edit Key Point Time**" and enter a new time.

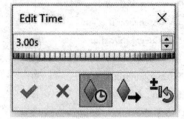

Our animation manager now looks like:

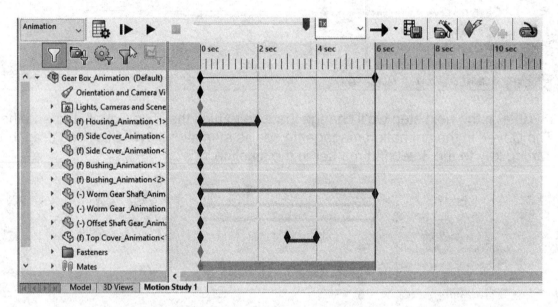

33.11. - For the last step of this animation we'll change the appearance of the front '*Side Cover*' to hidden lines visible mode. Move the time bar to 4 seconds. Make a right-mouse-click in the '*Side Cover's*' row at 4 seconds and select "**Place Key**." This will be the starting point for the appearance change.

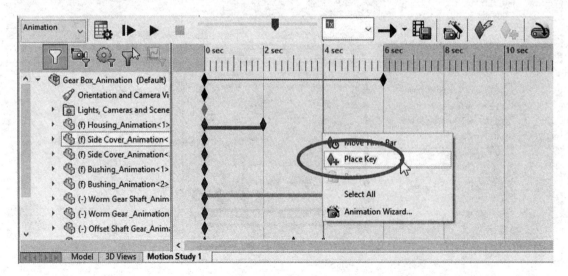

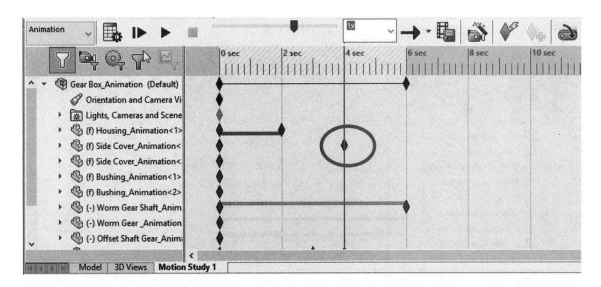

Move the time bar to 6 seconds, right-mouse-click in the 'Side Cover' and select "**Component Display, Hidden Lines Visible.**" A new key is added, and a magenta line shows that the component's display will change between these time keys.

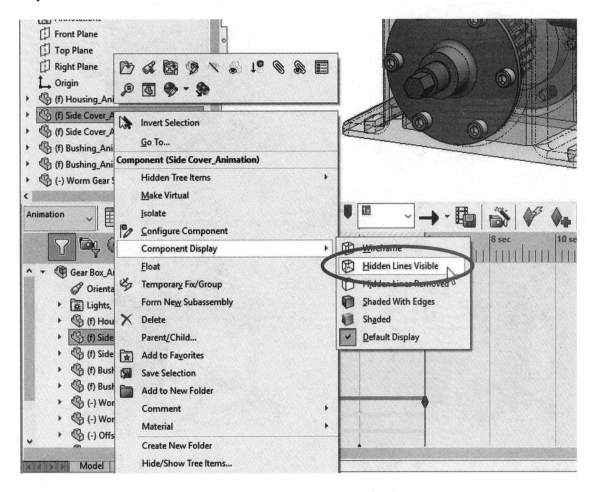

If needed, recalculate to finish the animation.

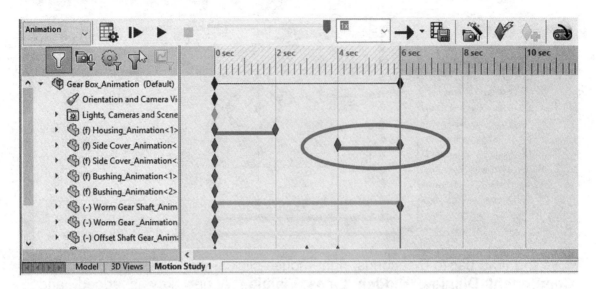

33.12. - Play the animation to see the result.

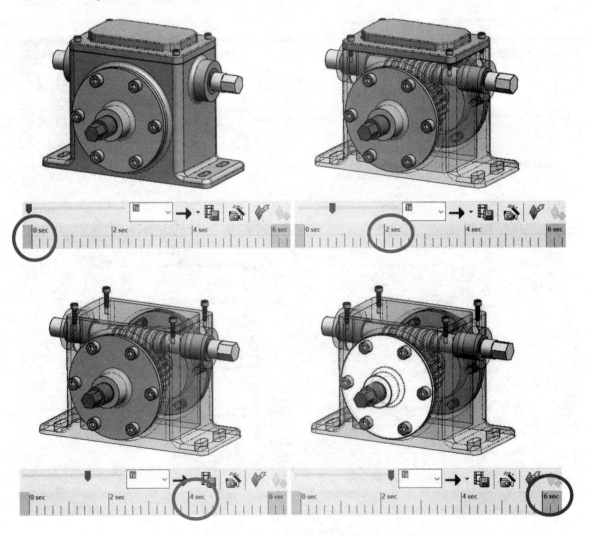

33.13. - The view orientation is automatically locked when we start a new motion study, and that's why it always goes back to the same orientation when the animation is played, even if we rotate the view. To play the animation using a different view orientation, right-mouse-click in "*Orientation and Camera Views*" and select "**Disable Playback of View Keys**." After disabling it we can play the animation using any view orientation/zoom we wish.

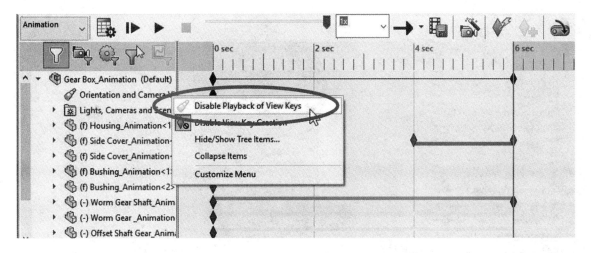

33.14. - To add view orientation changes to the animation, we must meet the following three conditions:

- "**AutoKey**" must be ON.
- "**Disable Playback of View Keys**" must be OFF.
- "**Disable View Key Creation**" must be OFF.

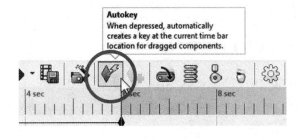

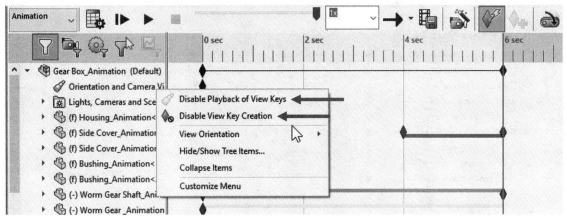

The "**AutoKey**" command automatically adds key points in the animation at the location of the time bar when we move a component or change the view orientation if the previous conditions are met.

Move the time bar to 0 seconds and set the view orientation to the starting view for the animation.

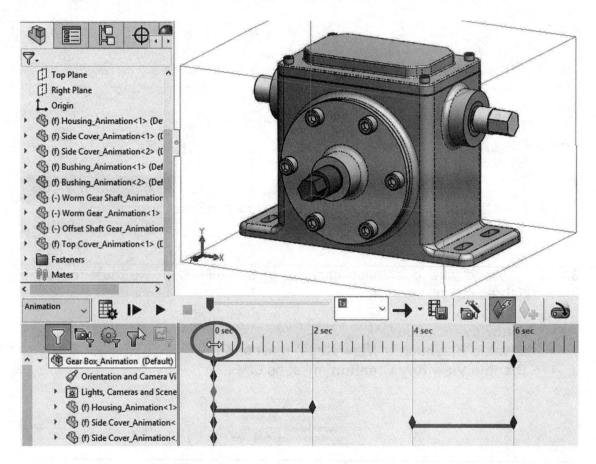

33.15. - Move the time bar to 2 seconds and add a key in the "*Orientation and Camera Views*" row.

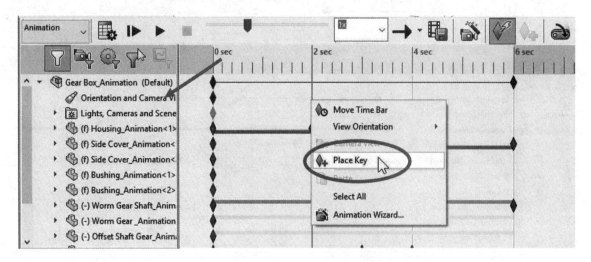

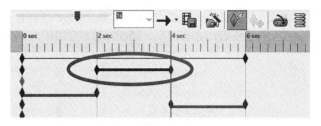

Move the time bar to 4 seconds and change the view's orientation. The new animation step is added to the view orientation automatically between the 2 seconds key and the 4 seconds key and will remain at that orientation.

If the "**AutoKey**" command is used for component movement, it's a good idea to turn the "**Disable View Key Creation**" option ON to prevent accidentally adding view orientation keys to the animation. Add view orientation steps later to have more control and get the desired results. After finishing the view orientation animations, turn the "**Disable View Key Creation**" option back ON. The Final animation sequence is this. Feel free to explore animating more components to see their effect. We'll see explode and collapse animation in the next step.

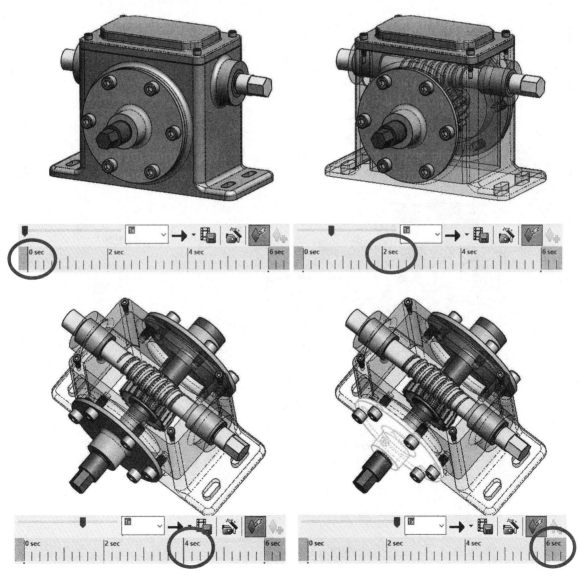

33.16. - To create a new animation right-mouse-click in the "*Motion Study 1*" tab and select "**Create New Motion study**" or the menu "**Insert, New Motion Study**."

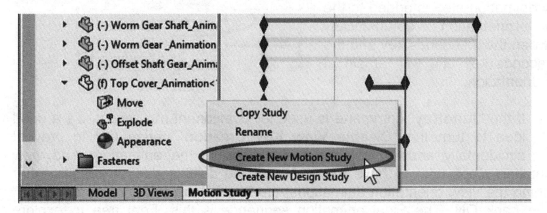

In the new motion study click the "**Animation Wizard**" command, select the "Explode" option and click "Next" to continue.

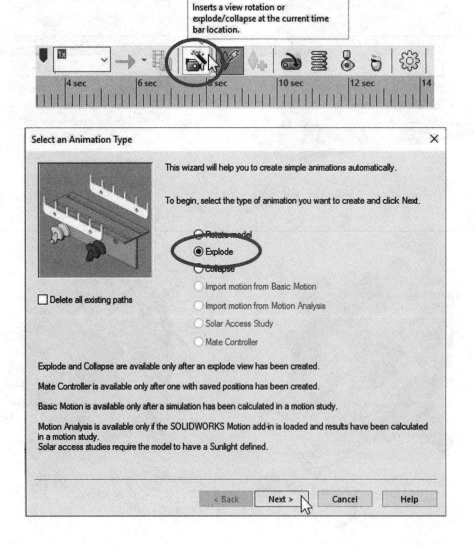

When we animated the exploded view previously, the animation's length was fixed at 8 seconds, causing the exploded view to run very fast. Using the animation wizard, we can make the explosion any length we want. In the following step set the animation's duration to 30 seconds and leave the start time at 0 seconds. Click "Finish" to complete the animation.

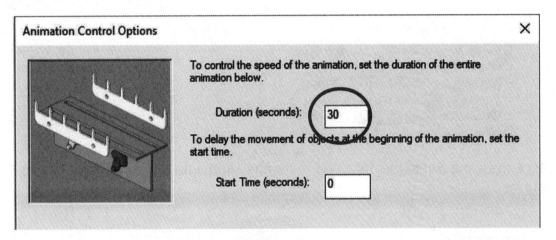

All the explode steps are automatically added to the timeline. Press "Play" to see the result.

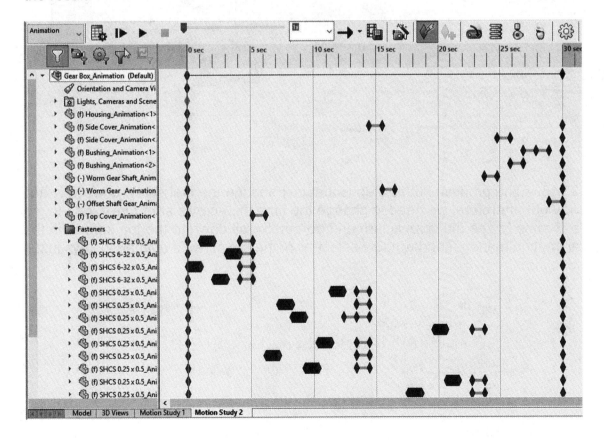

33.17. – After the exploded view steps are added to the animation, turn OFF the option "**Disable View Key Creation**" and add orientation view animations to zoom it to the screws when they are exploding out.

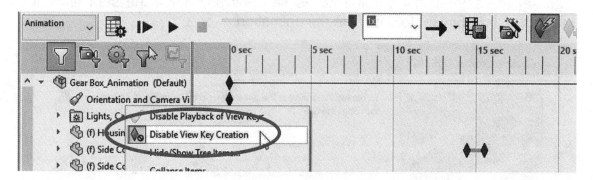

Move the Time Bar to the 1 second mark and zoom into the '*Top Cover*' to see the screws.

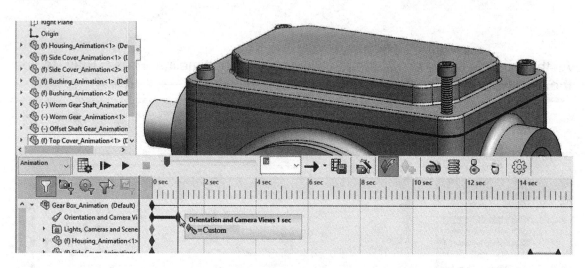

33.18. - The problem in this step is that in 1 second the first screw is already half way out; therefore, we need to change the time the screws start to explode to get a chance to see the screws turn and go out. Scroll down to the top screws in the animation manager and window-select all of the top screw's animation keys in the timeline.

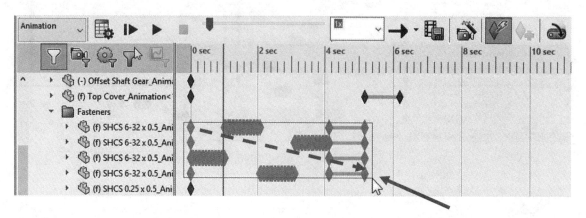

After selecting all the keys right-mouse-click in any one of them and select "**Edit Key Point Time**." By pre-selecting all the keys they will be shifted in the time line simultaneously.

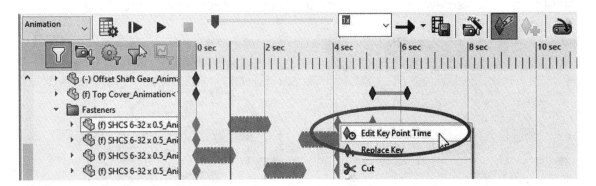

 Individual key points can be dragged in the timeline to make quick time adjustments.

Enter a value of +1 second to move the screws' keys to start moving 1 second later. After re-calculating the animation, the viewpoint will zoom to the '*Top Cover*' and then the screws will start turning and going out.

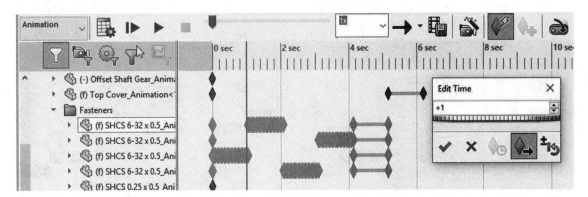

 To change the speed of the animation's playback, select a multiplier from the "**Playback Speed**" drop-down box. Be aware that changing the playback speed will also change the speed when saving the video.

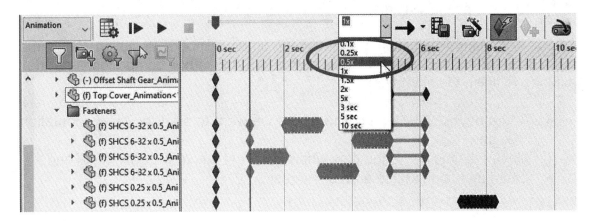

33.19. – The screws start moving one after another. To make the animation more "fluid," we need to overlap the screw's motion instead of starting to move one after another. Select the second screw's keys and shift its start time -0.5 seconds before, the third screw -1 second, and the fourth screw -2 seconds.

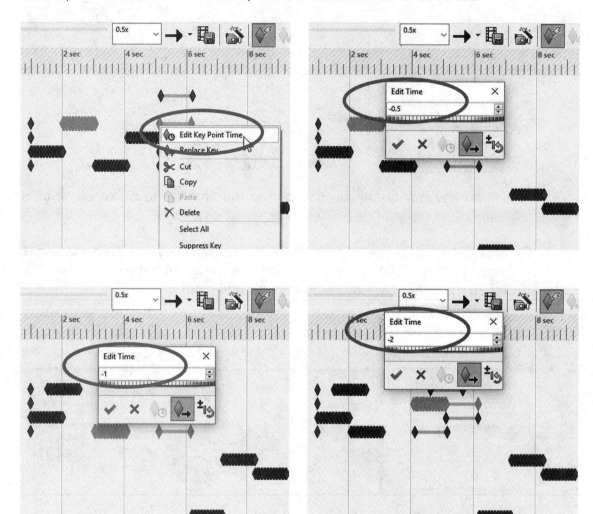

 Multiple animation keys can be selected and moved at the same time to a different animation time step.

33.20. – Continue adjusting the timing of other components using the same approach. Add similar keys to view and zoom into other components as they explode to finish the animation.

- Add a new key in the timeline where we want to start the view orientation change.
- Move the time bar to the time where we want the view transition to end.
- Zoom and pan into the area of interest.

After setting the view orientation, the new animation step is automatically added between the keys.

Notice that the next timeline does not have a gap between some view orientation steps. This is done by adding a new key, moving the time bar to the new location, changing the view orientation and repeating the process a second time. The final effect is that the view keeps moving from one orientation to the next without stopping.

 Remember to turn "**Disable View Key Creation**" ON when done manipulating views and avoid accidentally changing the view orientation and zoom settings of the animation.

33.21. - As a final step, we are going to add a motor to make the gears move for a few seconds. Move the timeline to the beginning, and from the Animation toolbar, select the "**Motor**" command. We can add linear or rotary motors to any component which can move along its free degrees of freedom.

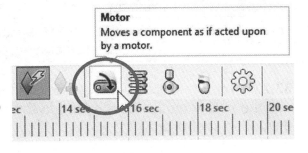

For this example, we'll add a rotary motor to the '*Worm Gear Shaft*', which will also move the '*Worm Gear Complete*' and the '*Offset Shaft Gear*' thanks to the gear mate. In the "Motor Type" properties, select the "Rotary Motor" option. For the "Component/Direction" select the outside face of the shaft. A preview will show the direction of rotation. If needed (or wanted), we can reverse the direction. In the "Motion" section select the "Distance" option and make the displacement 1800 degrees (5 full turns), starting at 7 seconds, with a duration of 8 seconds (finishing at 14 seconds). Click OK to add the motor.

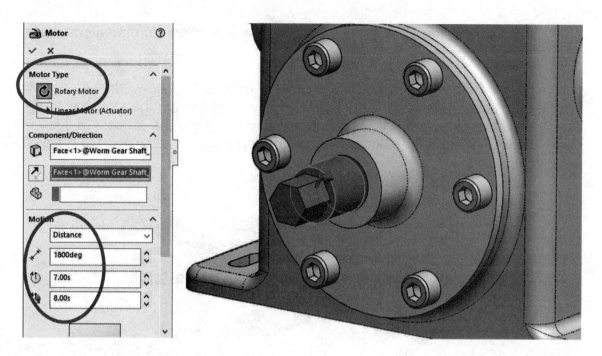

Other options to define a motor include:

Constant Speed	Components will move at a constant speed for the length of the animation
Distance	Specify a distance within a given time frame
Oscillating	Oscillating motion defining the angular displacement, frequency and phase shift
Segments	Define multiple motion segments using functions for displacement, velocity, and/or acceleration
Data Points	Define motion segments by specifying displacement, velocity and/or acceleration data points
Expression	Specify a mathematical function to describe displacement, velocity, acceleration, etc.

33.22. - After re-calculating the animation to account for the added motor, we can see the yellow motion bars added to the components moved by the motor in the animation controller.

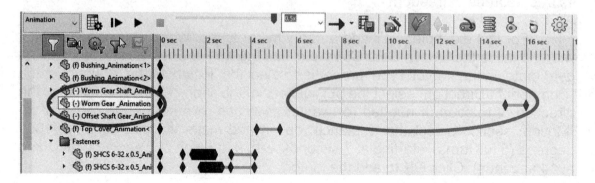

"Disable Playback of View Keys" must be turned OFF before saving the video, otherwise the animation will not show view orientation changes.

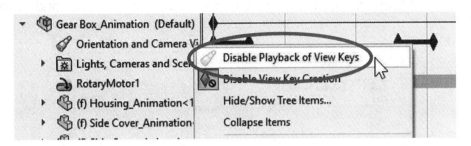

33.23. - The last step is to save the animation to a file. Select the "**Save Animation**" command.

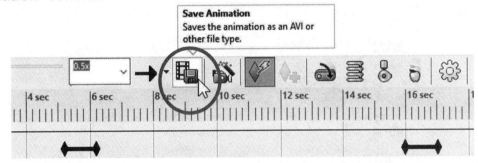

In the "Save As" dialog, browse to select the location to save the video. In the "Save as type:" we can select to save as different video formats, or series of images, also in different formats that can be imported into video editing software. Select the "MP4 Video file" format and set the speed to 30 frames per second for smooth video playback.

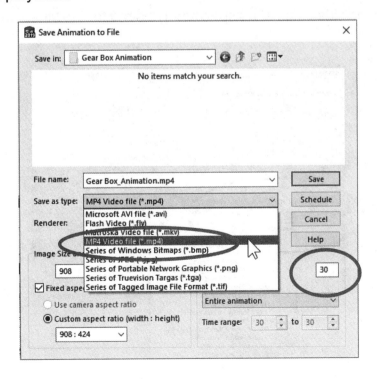

In the "Renderer" selection box the default selection is "SOLIDWORKS screen." In this case the animation is made using screen shots of the SOLIDWORKS graphics area. Select the image size, aspect ratio, and frames per second.

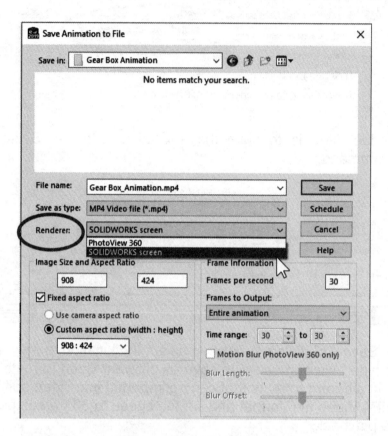

Finally, select the video size and aspect ratio desired for the video and click "Save" to continue.

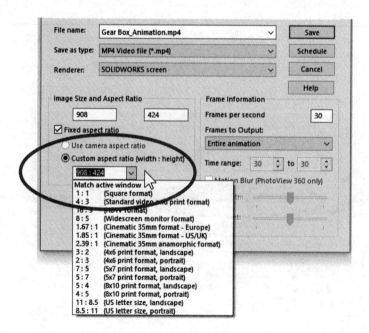

The window size used for the video includes the FeatureManager's area but with the FeatureManager hidden. Press "**F9**" or click to hide the FeatureManager to get a better idea of the area included in the video.

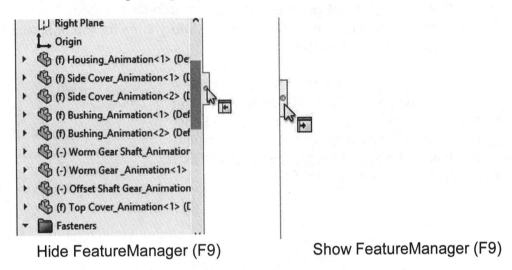

Hide FeatureManager (F9)   Show FeatureManager (F9)

If PhotoView 360 is available and loaded, the animation can be rendered using PhotoView to obtain a photo realistic video. Be aware that selecting PhotoView will render each frame of the animation.

For example, if the animation is 20 seconds long, and we make an animation with 15 frames per second (standard video is 30 frames per second), PhotoView will render 300 frames. If the average frame takes 2 minutes, depending on the hardware used, PhotoView settings, image size, etc., saving this video with *half* the desired frame rate will take approximately 10 hours to complete.

Considering the excessive amount of time required to make a photo-realistic video, we want to be sure the result is acceptable, if possible, in the first attempt. To minimize the risk of having to wait a very long time only to find out the resulting animation is not what we expected, here are a few suggestions when saving a video using the PhotoView renderer:

- Make sure the animation runs smoothly, including component movement, component's display, and view orientations.
- In PhotoView adjust all the necessary settings to obtain the desired image quality in the final render and write them down.
- Save a video using the SOLIDWORKS screen renderer to make sure the image size is correct, and the animation is moving correctly.
- Go back to PhotoView and lower the quality settings to get a fast render for a test.
- Save a new animation using the PhotoView renderer (using the low-quality settings), a small (proportional) image size, and low frame rate. This way we'll be able to quickly produce a video to make sure we are getting the desired effect and lighting conditions.

- Adjust whatever needs to be adjusted as needed until the result is acceptable using the lower quality settings.
- When the animation is moving correctly using these settings, set the PhotoView settings to the optimal parameters recorded before.
- Set the image size and frame rate to the desired values for the final render.
- Wait for the render to complete.

At the bottom of the PhotoView options there is a section for "Network Rendering." What this option does is to use multiple clients using a shared network directory to split rendering jobs to finish faster. Multiple SOLIDWORKS licenses are required for this option.

Exercises:

Generate an animation of the engine assembly using the knowledge acquired in this lesson. This is only a suggestion; it's time to get creative. ☺

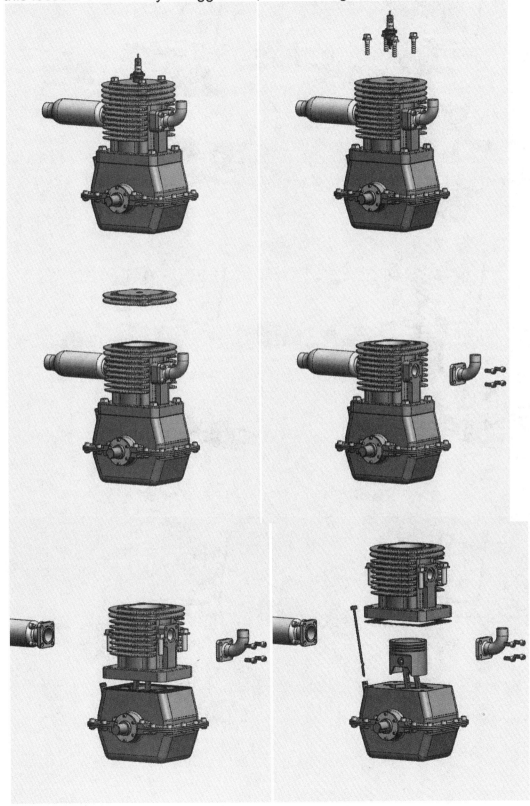

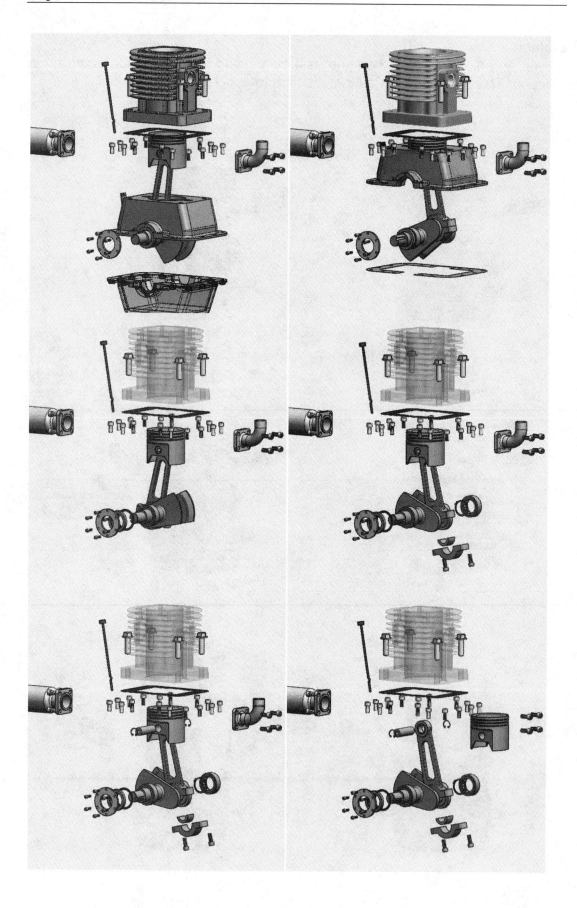

Chapter 8: Analysis: SimulationXpress

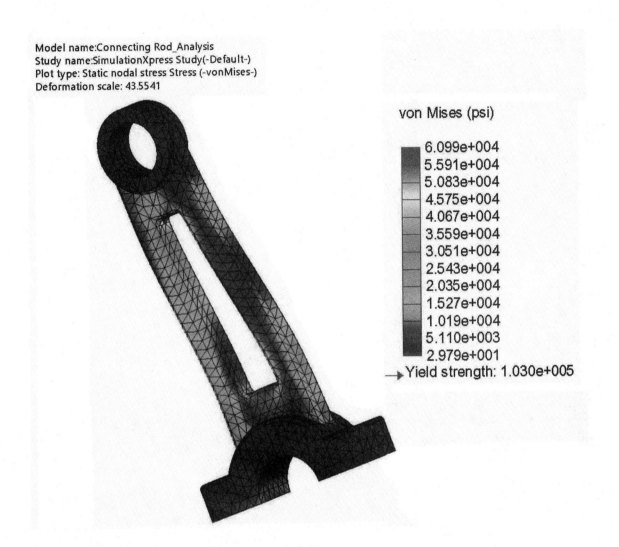

Model name:Connecting Rod_Analysis
Study name:SimulationXpress Study(-Default-)
Plot type: Static nodal stress Stress (-vonMises-)
Deformation scale: 43.5541

von Mises (psi)

6.099e+004
5.591e+004
5.083e+004
4.575e+004
4.067e+004
3.559e+004
3.051e+004
2.543e+004
2.035e+004
1.527e+004
1.019e+004
5.110e+003
2.979e+001

→ Yield strength: 1.030e+005

Notes:

Background: Why Analysis?

Simulation is a very important tool in engineering. Computers and software have come a long way since the early analysis tools first became available, and modern tools have made simulation a lot easier to use, faster, more accurate and more accessible than ever, enabling designers and engineers to check their design, make sure it's safe, understand how it will deform, if it will fail and under what circumstances, or how it will perform in any given environment (temperature, pressure, vibrations, etc.).

The biggest advantage, by far, when analyzing a design is that we will have a safe design, and at the same time, save money by making decisions as to what materials to use, component sizes, features and even appearance early in the design process, when it is still all in "*paper*" (maybe a more appropriate term now would be '*in Bytes*' ☺) and cheaper to modify. As the product development process advances beyond the design stage, making design changes to a product is increasingly more expensive as we approach the manufacturing stage.

Picture it this way: Imagine we design the newest must-have gadget and it looks really nice. We make molds, tooling, order parts, and set up an assembly line. Soon our gadget is in the stores, and a month later we start receiving customer complaints: when a button is pressed hard, the battery cover falls off. Then, *after* we know we have a problem, we run an analysis and find out that *we should have* had a thicker *this or that*, with a *widget* in between the *thingamajig* and the *thingamabob*, and those changes would solve the problem. Now, all we need to do is to change the design, the tooling, the assembly line, marketing, and above all, convince every customer that it is fixed. At this point, our customers' perception of our company and credibility are destroyed. This is usually the most expensive part. It's easy to see why making analysis of our products early in the design process will help us design better products and save money and resources.

With that said, it must also be noted that just because we made a simulation does not guarantee that our designs will be safe or successful, as there are many factors involved. Reasons for product failure include using the product beyond its designed capacity, abuse, material imperfections, fabrication processes and things and circumstances we could have never thought of. This is where the designer needs to consider every possible scenario, and of course, simulate it as realistically as possible with the correct analysis tools.

SOLIDWORKS includes a _basic_ structural analysis package called "**SimulationXpress**." As its name implies, it has limited functionality that allows only certain scenarios to be analyzed. To understand what these limitations are, first we need to learn a little about how analysis works. A general overview of the inner workings of analysis is as follows:

Analysis, also known as "Finite Element Analysis," is a mathematical method where a component's geometry is divided into hundreds or thousands of small pieces (*elements*), where they are all connected to one or more neighboring elements by the vertices (called *nodes* in analysis). The elements have simple geometry that can be easily analyzed using basic stress analysis formulas. Elements are usually tetrahedral or hexahedral for most solid models. There are other types of elements including shell and linear elements for different types of geometry like thin walled 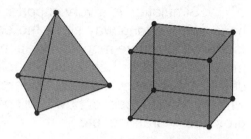 or long slender components. SimulationXpress only uses tetrahedral elements.

To make a simulation, we need to know the material our component is made of (material's physical properties), how it's supported (Restraints), and the forces acting on it (Loads). The next step is to break down the model into small elements; this process is called meshing. The resulting group of interconnected elements is called the **mesh**; in SimulationXpress the only parameter we can change, if we choose to, is to define the average element size, and SimulationXpress automatically breaks down the geometry into elements to create the mesh.

The next step is to define how the component is supported; in other words, which faces, edges or vertices are restrained. Each node has six degrees of freedom, meaning that each node can move along X, Y and Z and rotate about the X, Y, and Z axes. Restraints limit the degrees of freedom (DOF) of the nodes in the face or edge that they are applied to. SimulationXpress limits the user to restraining (Fix) all DOF in faces. In other words, they are completely immovable.

This approach is enough for simple analyses, but to obtain a more accurate and reliable solution, a more realistic simulation must be used that allows selective displacement and rotation of element nodes.

After we know how the model is supported, we need to define the Loads (forces) that act upon it. A limitation of SimulationXpress is that it only allows Forces and Pressures to be applied to model faces. We can define the Loads to be normal to the face they are applied to or perpendicular to a reference plane.

Another limitation of SimulationXpress is it can only make **Linear Static Stress Analysis**. Stress is defined as the internal resistance of a body when it is deformed and is measured in units of Force divided by Area. For example, if we load a bar with a 1 in² cross section area with a 1 Lb. load, we'll have a stress of 1 Lb/in². The stress depends on the forces and the geometry regardless of the material used. The material properties will make a difference as the model will deform more or less.

The "Static" part means the model is immovable; in other words, the restraints will not let the model move when the loads are applied; loads and restraints are in equilibrium, otherwise the analysis would be "Dynamic."

The "Linear" part implies the deformation of the model is proportional to the forces applied; twice the force, twice the deformation. These deformations are generally small compared to the overall size of the part and occur in the **Elastic** region of the "Stress-Strain" curve of the material used. If we remove the force, the model returns to its original shape like a spring.

In general terms, the **Yield Stress** is the point where the stress-strain curve is no longer linear. If a material is stressed beyond the Yield Stress it will be permanently deformed. In this case, we will have **Plastic Deformation**. Thinking of the spring, if we pull the spring too far, it will be permanently deformed, meaning it had plastic deformation. **If the material's yield stress is exceeded with the applied loads, the SimulationXpress analysis results will be invalid.** If this is the case, we will need to change our design, its geometry, the loads applied, or material used to have stress results below the material's yield stress.

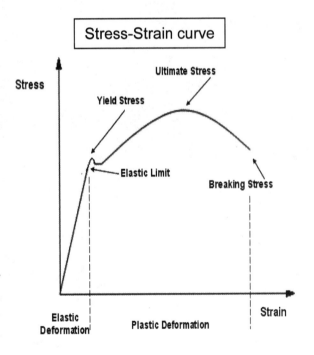

711

Notes:

Engine's Connecting Rod Analysis

34.1. - To show the SimulationXpress functionality, open the '*Connecting Rod_Analysis*' part from the exercise files included. This is a slightly different version of the part made for the engine project.

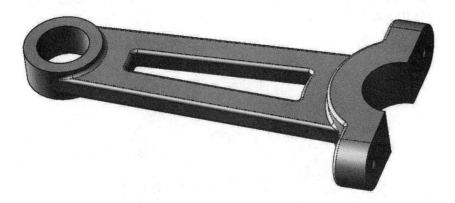

For our analysis, we are going to pretend the engine's crankshaft is blocked during normal operation, resulting in a high stress in the connecting rod. We'll assume it happens at the point where the crankshaft is horizontal and the connecting rod is at the maximum angle with the vertical, causing the maximum torque as shown in the next image.

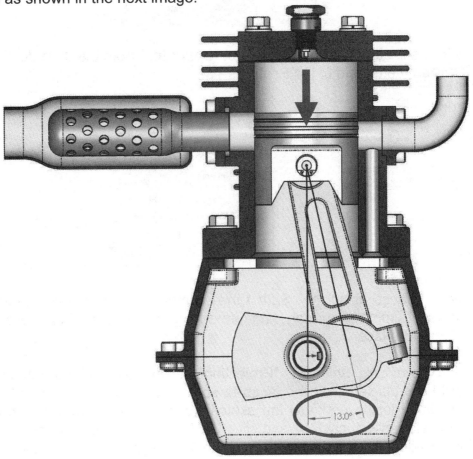

713

34.2. - To properly simulate the downward force, we must divide (split) the cylindrical face of the connecting rod in two, where the bottom half will carry the load. To split the face, we will use the "**Split Line**" command. The first thing we need to do is to add a sketch in one side of the model. Using the "**Midpoint Line**" command draw a line starting in the center, up to the inside edge at an angle. Add a vertical construction line and dimension the angle 103°. The force will be applied perpendicular to this line.

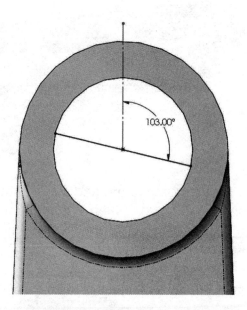

34.3. - While still editing the sketch, select "**Split Line**" from the drop-down menu in the "**Curves**" command of the Features Tab.

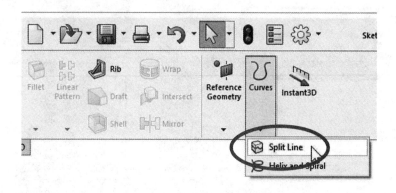

 Dividing faces using the "**Split Line**" command is a common practice in analysis to more accurately simulate a load in a small part of a larger area in a model's face.

In "Type of Split" select the "Projection" option; this means that the current sketch will be projected over the faces to split. In the "Selections" box "Current Sketch" is listed letting us know the sketch we are working on is pre-selected. Select the inside cylindrical face and click OK to split it. Notice the face now has two edges dividing the face in two. The force will be applied to the bottom half.

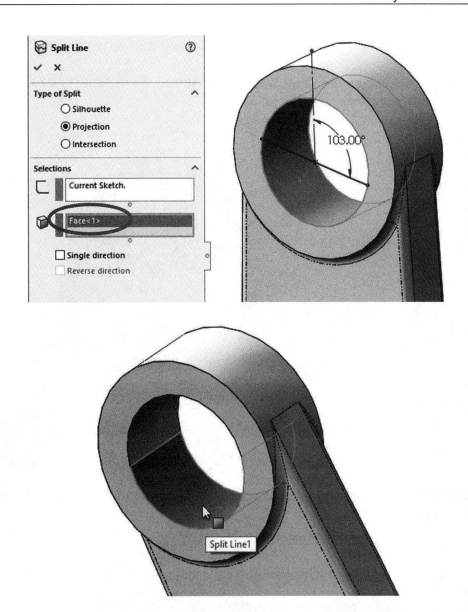

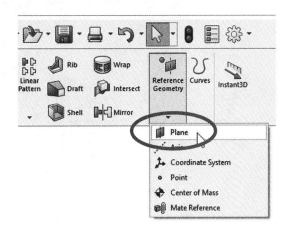

34.4. - Now we need to create an auxiliary plane that will be used to define the direction of the force. In the Features tab, select the "**Plane**" command from the "**Reference Geometry**" drop-down menu.

To define this plane, we have two options. Use either one to continue.

a) Select the edge of one split line and a vertex of the other split line (Edge and vertex definition)

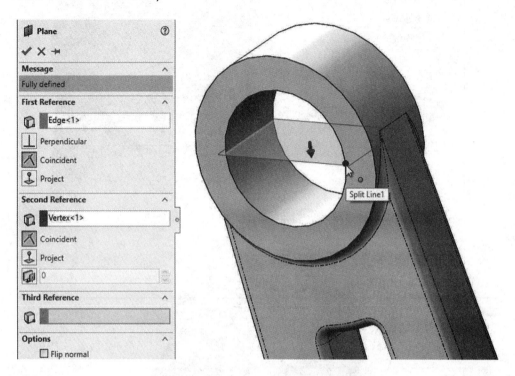

b) Select the two vertices of one split line's edge, and one from the second split line's edge (3 vertices definition)

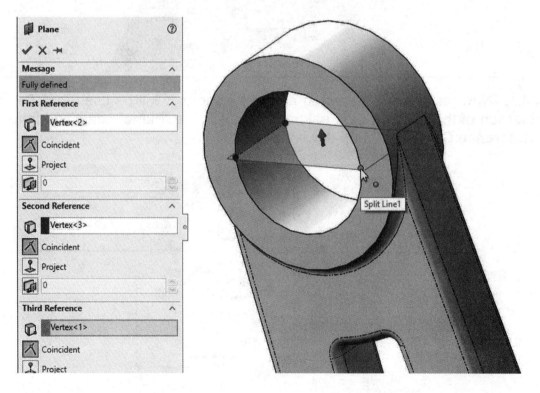

34.5. - Now our model is prepared for analysis; activate the SimulationXpress wizard using the menu "**Tools, Xpress Products, SimulationXpress**" or in the Evaluate tab of the CommandManager.

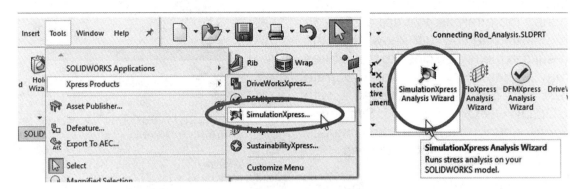

The SimulationXpress wizard is integrated in the Task Pane and is automatically displayed. Select "Options" to set units to English, turn ON the checkbox "Show annotation for maximum and minimum in the result plots," and leave the results location to the default 'temp' folder. Click OK to accept the options and then "Next" to continue.

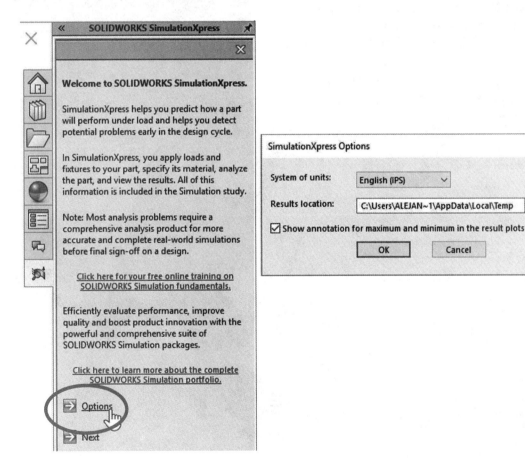

34.6. - The first step will be to define the "Fixtures"; this means the faces that will be constrained (all six degrees of freedom restrained) and will not move during the analysis. We are also given a warning letting us know that the results in the region close to the fixtures may be unrealistic, and more accurate results can be achieved with *SOLIDWORKS Simulation Professional*, where we can simulate more realistic restraint conditions.

Select "Add a Fixture" to define how the part will be supported. In the Fixture selection box select the two faces at the bottom of the connecting rod and click OK to finish. In this example these two faces will be assumed to be immovable. Remember that the analysis results may not be accurate near the fixed faces but will give us a good approximation in the middle region of the connecting rod, which is what we are interested in analyzing.

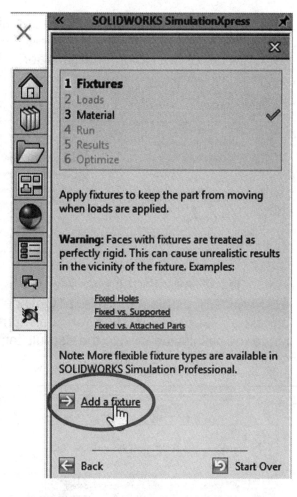

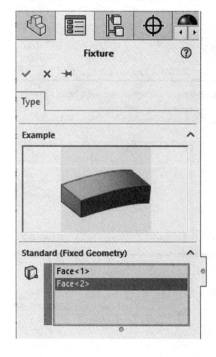

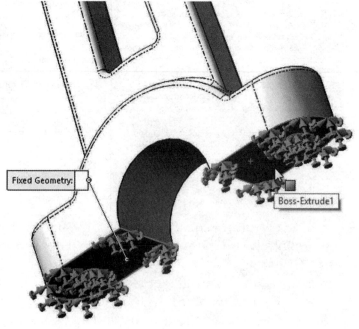

If more Fixtures were required, select "Add a Fixture" as needed until all the necessary restraints are in place. In our case, these two faces are the only ones to be fixed. The SimulationXpress Study now shows the "*Fixed-1*" condition and "Fixtures" shows a green checkmark letting us know this step's minimum requirement has been met. Click Next to continue.

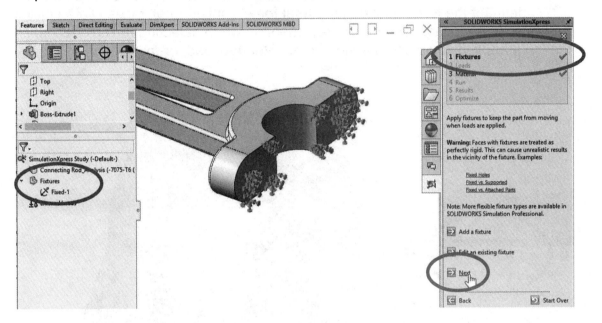

34.7. - In the next step of the analysis we need to define the force(s) and/or pressure(s) acting in our model. The load we are going to apply will be a force of 2,500 lbf in the lower half of the previously split face, perpendicular to the auxiliary plane we made.

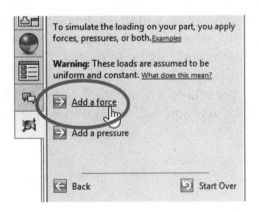

Click "**Add a Force**" and select the bottom half of the split face to apply the force uniformly in the entire surface. The preview shows the force vectors perpendicular to the surface in a radial fashion, but since we need it to be perpendicular to the auxiliary plane, use the "Selected direction" option and select the plane previously made. Note that now the direction vectors are perpendicular to "*Plane1*," as explained at the beginning of the exercise. Make the units "English" and enter a value of 2,500 lbf. If needed, click "Reverse direction" to make the force to point down. Click OK to add the force and continue.

719

If we don't give the force a direction, it will be normal to the selected face, making the force radial to the cylindrical surface.

 In our example, we are adding the force to one face only, and the "Per Item" option is the only option available. If two or more faces are selected, we can define if the force added is the total force applied to all faces (Total) or if it is applied to each face (Per Item). Pay attention when selecting multiple faces, as the force applied could be a fraction of the intended load, or multiple times more.

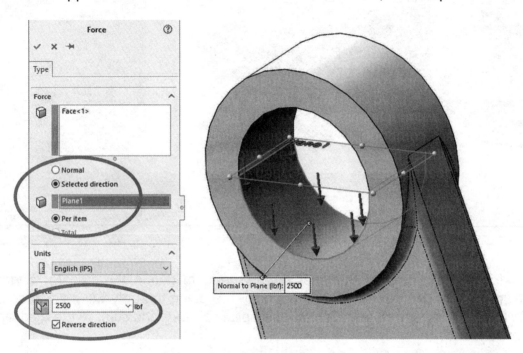

34.8. - The "*Force-1*" condition is added to the SimulationXpress study where "Loads" shows a green checkmark letting us know this step's minimum requirement (1 force or pressure) has been met.

If more forces or pressures are needed, they can be added using "Add a Force" or "Add a Pressure" as required, or we can edit an existing load. Click Next to continue.

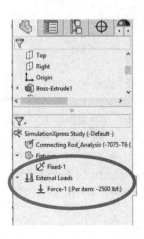

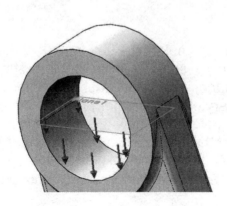

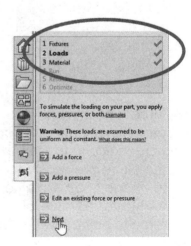

34.9. - The next step is to define the material that will be used for the part. This is important because as we explained earlier, for the analysis we need to know the physical properties of the material.

Since a material had been previously assigned to the part before starting the analysis (7075-T6 aluminum alloy), this step is already marked as complete and we don't need to do anything else. We can see the green checkbox next to "Material" in the SimulationXpress wizard as well as the Young's Modulus and Yield Strength values for reference.

If a material has not been assigned before starting the analysis, we can select "Change Material" and pick one from the materials list. Click "Next" to continue.

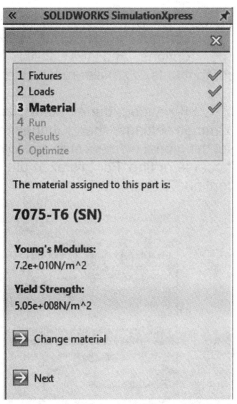

34.10. - Now that we have given Simulation-Xpress the minimum information required to make an analysis (Fixtures, Forces and Material), we are ready to make the mesh (break the model into smaller 'finite' elements) and run our simulation. At this point we have an option to run the analysis with the default settings for the mesh or change them.

In general, using a smaller mesh size generates more elements that lead to a more accurate solution; however, increasing the number of elements also increases the time and computing resources needed to run the simulation. SimulationXpress allows us to change the mesh density or the general size of the elements.

In this example reducing the element size from the default size (increasing the mesh density) will not significantly impact the time to solve the problem or its accuracy.

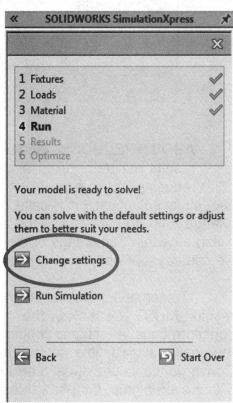

For larger, more complex models there would be a noticeable difference in computing time. SOLIDWORKS Simulation Professional offers many more options and the ability to analyze entire assemblies; with multiple components, a smaller mesh size will most definitively impact the solution time.

If needed, the element size or the mesh density can be changed. Select "Change settings," then "Change mesh density." Under "Mesh Parameters" we can set the global element size and tolerance or move the mesh density slider to make it coarse or fine. For this example, we'll use the default mesh settings. Click OK to continue.

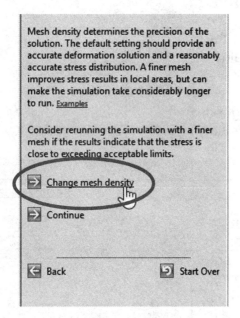

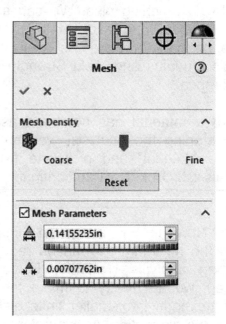

ABOUT REDUCING MESH SIZE: Making the mesh size smaller improves results accuracy up to a certain degree, but there comes a point where reducing the element's size will not significantly improve accuracy in sequential solutions; this is called convergence and is measured in percentage of change from one solution to the next. For example, if the highest stress value is within a certain percentage of the previous study, for example 2%-5% difference, we can say that we have reached convergence and assume our results are valid.

To consider reducing the mesh size and look for convergence, our analysis results MUST be below the yield stress and preferably show no stress concentrations. After these conditions are met, we can re-run the analysis reducing the mesh size a little at a time (10% reduction is a good starting point). Convergence values depend on the accuracy needed and can vary around 2% to 5% or even higher. If these values seem high and/or your design has a low factor of safety you may want to consider analyzing your design with a dedicated analysis package to simulate real world conditions and improve the accuracy of your analysis to be safe.

34.11. - As soon as we click OK in the "Mesh" properties, the mesh is created; in other words, the model is broken into many small pieces (*Finite Elements*). Notice the model icon in the Analysis Manager is meshed, indicating the mesh is complete. To display the mesh, right-mouse-click in the meshed icon and select "Show Mesh" (to hide it, select "Hide Mesh"). Now our model is ready to run the analysis.

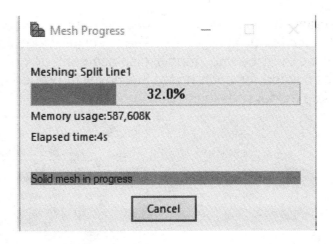

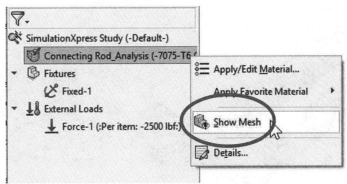

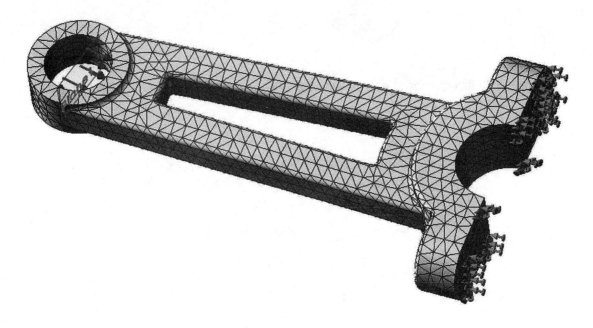

34.12. - Now that the model is meshed, select "**Run Simulation**" from the SimulationXpress wizard. What happens next, in the simplest terms, is that the solver (the part that calculates stresses and displacements) will generate an equation for each element, with an unknown variable for each node. Since adjacent elements share nodes between them this generates a very large matrix of hundreds or thousands of simultaneous equations.

After generating the equations, an equations matrix is assembled and solved for the unknown variables at every node. The resulting value for each unknown value (one per node) will be the node's displacement or deformation. After a displacement is calculated for each node, the physical properties of the material are used to calculate model stresses at each node.

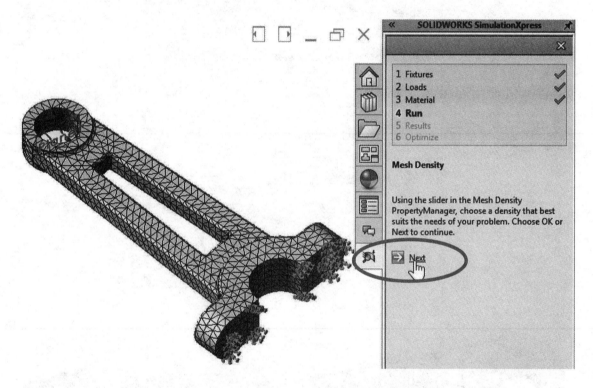

Since this is a small model it only takes a second to complete the analysis.

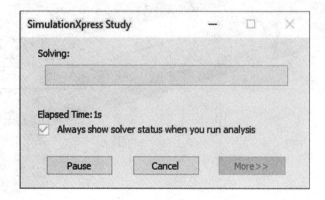

When the solver is finished with the calculations, we see an animation of the deformed model.

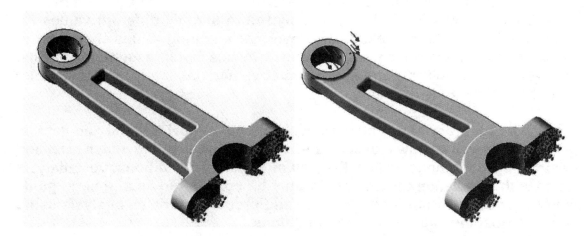

34.13. - The model deforms as we expected in the direction of the force, so we can proceed to view the results. If the model had not deformed as expected, we would have to go back to modify the fixtures and/or loads to make the necessary corrections. Select "**Yes, continue**" to view the analysis results. Select "**Show Von Misses Stress**" to view the resulting stresses calculated using the **von Mises** method, which is the most commonly used criteria for isotropic materials (materials with the same physical properties in all directions; metals are mostly isotropic). The deformation is exaggerated for visual clarity (in this case 13 times).

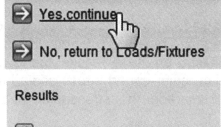

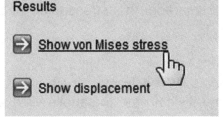

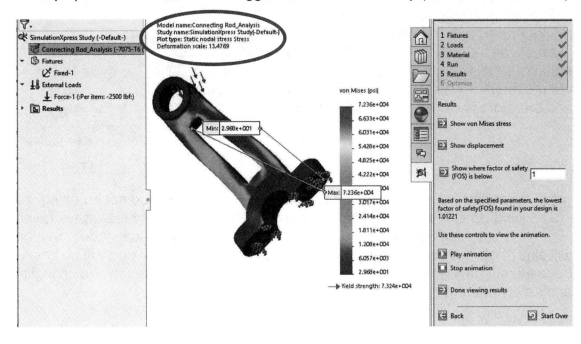

It is important to know that, by the very nature of the Finite Element Analysis process, the results are an *approximation* to the actual physical phenomenon and solving such a large number of simultaneous equations will most likely return slightly different values for stress and displacement results. However, the resulting values should be, in general, very close to the results shown in this book, assuming that the analysis is defined the same way (geometry, material, loads, and restraints) and the mesh size is the same.

The accuracy of any analysis will depend on how well the geometry is modeled, as well as how closely material properties, loading, mesh size, and restraining conditions match the actual physical conditions. An analysis done with SimulationXpress should only be considered as a general guide and never a substitute for a complete and properly modeled analysis using accurate loading and restraining conditions.

34.14. - In these results, we can see the Maximum Stress in the model is approximately 72,000 psi. The Yield Strength of Aluminum 7075-T6 is 73,200 psi. Based on these results, the lowest Factor of Safety in our model is almost 1.01. The color-coded stress scale helps to identify stresses in the rest of the model. Remember that stresses close to fixed faces may not be accurate.

The Factor of Safety is calculated by dividing the material's yield strength by the maximum stress in the model. We are interested in finding the lowest factor of safety value because if any area of the model has a factor of safety below 1, this means the stress is higher than the material's Yield Strength; if this is the case, our model will have permanent deformation and therefore is considered to have failed.

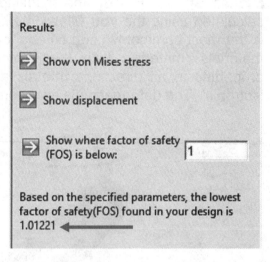

Note that we are talking about permanent deformation and not breaking. To yield does not necessarily mean that it will break; it will break if we reach the Breaking Stress. For most practical purposes, a permanent deformation will almost always be considered as a failed design because the part will no longer have the shape it was supposed to have, compromising the integrity, functionality and safety of the product and/or end user.

The different values for **Factor of Safety** (FOS) mean:

Value	Meaning
> 1 ✓	The stress at this location in the part is less than the yield strength and is therefore safe. Depending on the specific application, we may have to design with a higher FOS to ensure our design is safe and have room for unexpected loads.
= 1 ✗	The stress at this location in the part is exactly the yield strength. This part has started or is about to yield (deform plastically) and will most likely fail if the forces acting on the model are slightly more than the forces used in the analysis.
< 1 ✗	The stress at this location in the part has exceeded the yield strength of the material and the component will fail with the analysis parameters used, either deforming plastically or breaking.

34.15. - To see how safe our design is, type "2" in the Factor of Safety box and click "Show where factor of safety (FOS) is below." Areas under the specified factor of safety (*unsafe*) are shown in red, and the areas over the factor of safety (*safe*) are shown blue. Essentially, if our design requires us to have a minimum factor of safety of 2 (able to carry twice the design load), areas colored in red would fail.

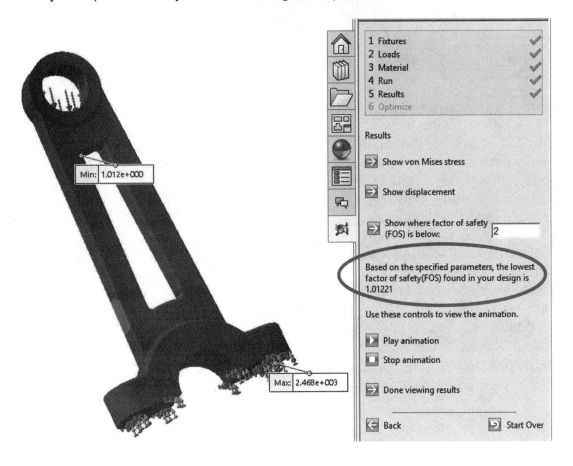

34.16. - Now activate "**Show Displacement**" to see how much our model deforms using the current conditions. The maximum deformation is 0.04″ at the top of the part, as expected since the bottom is rigidly supported. (The values in the scale are listed using scientific notation.)

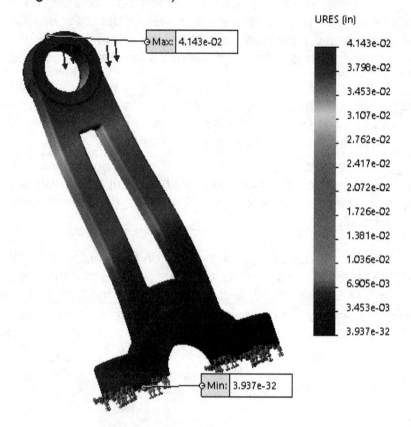

34.17. - Select "Done viewing results" to move to the next screen where we can generate reports either in Word or eDrawings format. Selecting "Generate report" will generate the files needed for the report and allow us to fill in additional information before creating it. It will include information as well as screen captures of results and details of the analysis including loads, fixtures, and material properties.

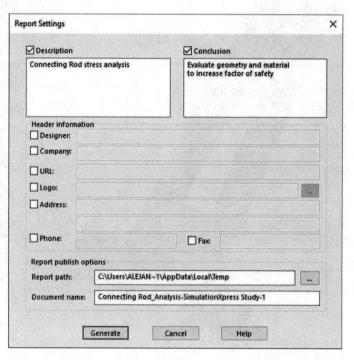

34.18. - Selecting "Generate eDrawings file" will make an eDrawings file with the analysis results. The main difference from the Microsoft Word report is that, in eDrawings, we can rotate the models, turn the mesh on or off, zoom in or out, and easily share the results with other people.

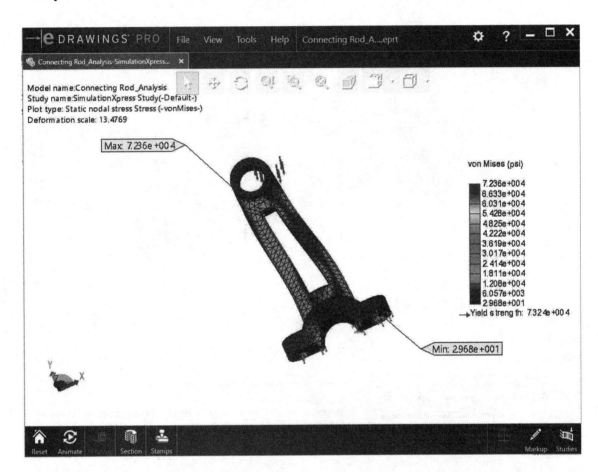

34.19. - What we are going to do now is to change a dimension in the model and re-run the analysis to see how much the results change. Close the SimulationXpress wizard. When asked if we want to save the results select "Yes" from the dialog box.

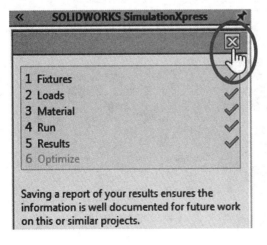

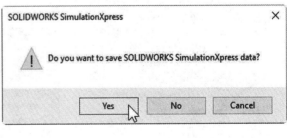

34.20. – To increase the factor of safety and reduce the stress concentration, increase the inside fillet's radius to 0.2" and rebuild the model.

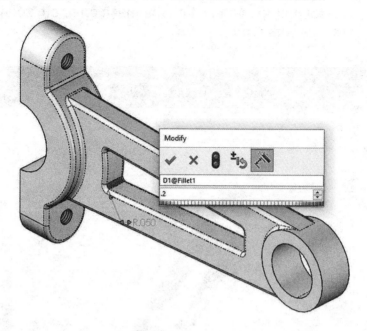

Change the material to AISI 4340

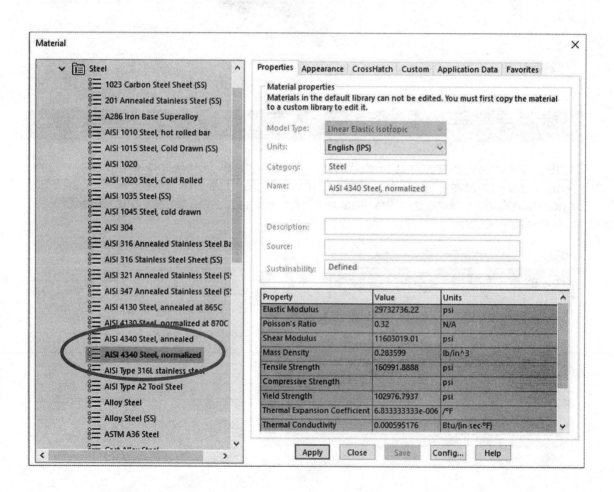

After rebuilding the model and changing the material, return to the SimulationXpress analysis.

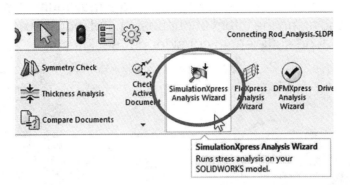

Notice the warning icons in the SimulationXpress study. These icons indicate the results obtained earlier are no longer valid because the geometry was changed (Fillet radius) and now the mesh no longer represents the model's geometry. Also, by changing the material, the deformation of the model under the specified loads will be different.

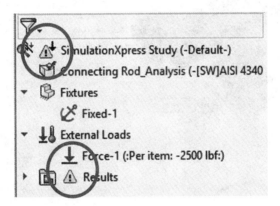

To update the analysis results, right-mouse-click at the top of the analysis tree and select "**Run**" from the pop-up menu to update the results. SimulationXpress will re-mesh the model to account for the geometric changes and run the analysis with the updated geometry and material properties.

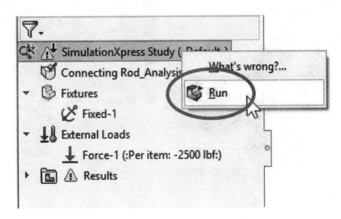

34.21. - After updating the analysis with the modified geometry and material, review the new results. Now we have a maximum von Mises stress of about 51,000 psi, giving us a minimum factor of safety of 2 and a maximum displacement of just over 0.01″. Depending on the intent of our design and industry accepted practices, this may or may not be sufficient to consider our design safe.

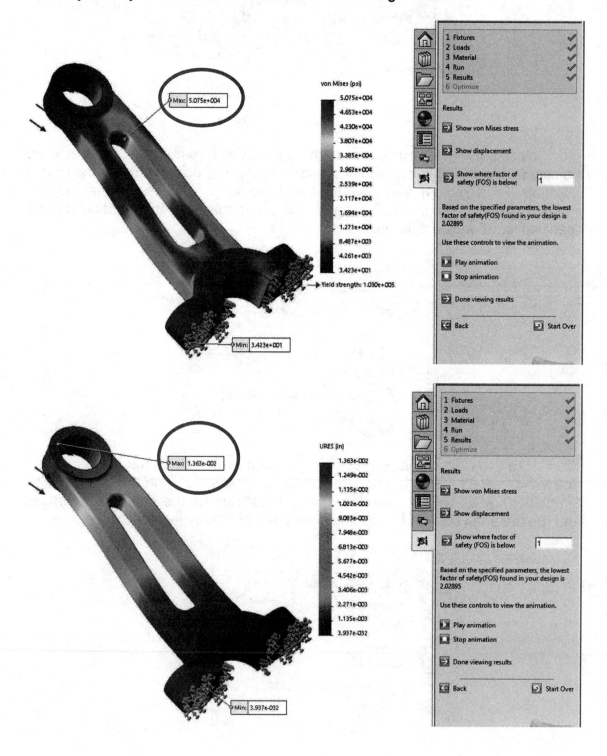

34.22. - SimulationXpress includes an Optimization wizard that can be accessed after reviewing the results and reports. The Optimization wizard can help us find the lightest model (minimum mass) without exceeding the constraints defined. Select "Done viewing results" and click "Next" to continue to the optimization wizard. Select "Yes" when asked to optimize your model.

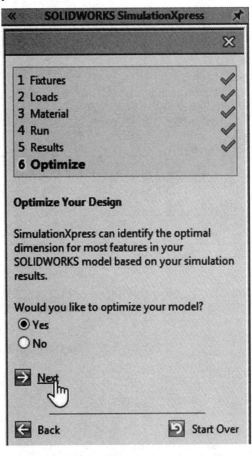

A limitation of SimulationXpress is that it only allows one dimension to be changed within a user defined range of values. The selected dimension will change multiple times within the given range, analyzed for each value and the results compared to the constraint value. The constraints of the optimization can be a minimum factor of safety, a maximum von Mises stress, or a maximum displacement.

For our exercise, we want to run the optimization analysis to find a better design that has a minimum factor of safety of 1.5 by varying the part's thickness. To run the optimization, we must define a range of values for the selected dimension; in our case, vary the connecting rod's thickness between 0.250" and 0.325". The optimization wizard will calculate five values in this range. After rebuilding the model a new analysis will be automatically run and the results analyzed. After the studies are completed, we will be shown the solution that meets the optimization requirements. Select the part's thickness dimension (currently 0.375") to add it to the parameters window. Click OK to continue.

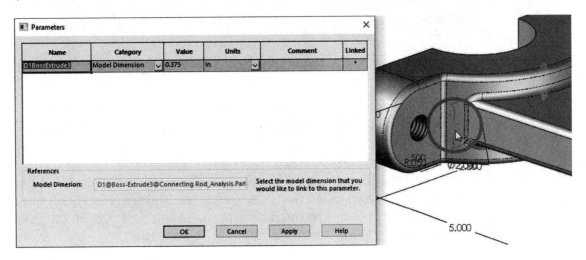

34.23. – In the "Variables" section enter a range between 0.250" and 0.325" for the dimension. In the "Constraints" area, select "Factor of Safety" and "is greater than" and enter 1.5. In the "Goals" section select "Mass" and "Minimize" to optimize for the lightest part that meets a factor of safety of *at least* 1.5. When finished click Run to optimize the model.

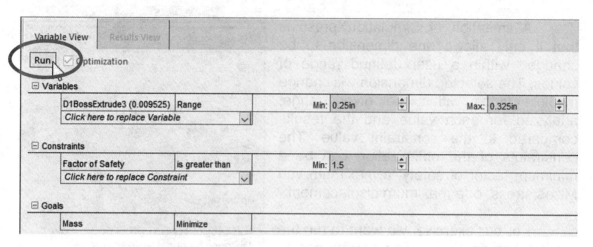

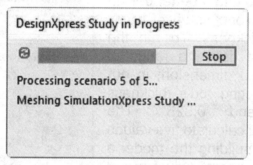

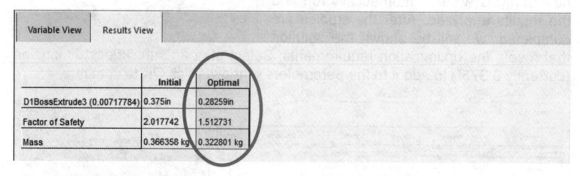

After running the optimization, we can see the "Results View" with the initial value and the optimal value. Selecting a column will update the model with those values.

For optimization to work, we need to have a study with a factor of safety higher than what we need, basically *over designed*. If the study has a high factor of safety, we can optimize the model to make the part lighter, and still have a factor of safety that meets our design requirements.

34.24. - After the optimization is completed we can choose to keep the original dimension or the optimized value.

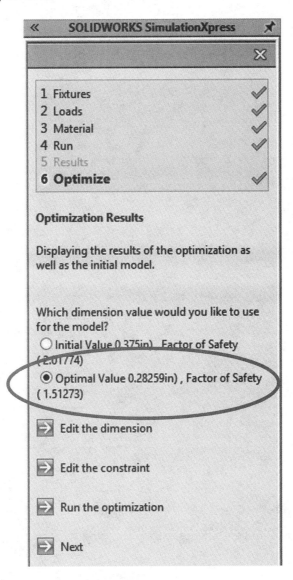

Notes:

A final word on analysis

The analysis world is very fascinating; it has highly sophisticated and powerful tools to predict how machines, products, or designs will perform in any given scenario, and allow us to anticipate the best way to assure their safe operation. However, it is extremely important to make sure the correct assumptions for restraints, loads and external conditions are applied to a model to correctly simulate the actual physical phenomena, as the accuracy and reliability of the results are directly tied to them. Talking about results, it's just as important to understand what the results obtained from an analysis mean to correctly interpret them and avoid over designing or putting our designs at risk for failure.

As we explained before, SimulationXpress is a *first-pass-analysis* tool. This means that its purpose is to give us a general idea of how the design will perform within the set of applied conditions. If our design is not *comfortably* safe, and/or we know this is a critical component of the design, it may be better to perform a more complete analysis using the full suite of SOLIDWORKS Simulation software, where we can add more realistic loading, restraining and external conditions. Besides, SimulationXpress is only capable of performing linear static analysis; therefore, if we anticipate that a model will be exposed to temperature, vibrations, large deformations, uses non-linear materials, or other conditions that cannot be correctly modeled with SimulationXpress, those analyses should be done using a version of SOLIDWORKS Simulation (or any other analysis software package) capable of properly modeling those situations.

Analysis and simulation can be a very complex subject depending on the task being analyzed, and depending on the specifics of it, a higher understanding of mechanics of materials, physics, vibrations, heat transfer and many other areas will be likely needed to properly set up, analyze and interpret the results. An analysis can be as simple as a linear stress analysis in a small part like our example, or as complex as a space ship re-entering the atmosphere considering wind force, temperature, vibration, pressure, radiation, phase changes, aerodynamics, etc., and based on our analysis and results, astronauts would be returning home safely, or not.

Ultimately, safety is the main reason why we make an analysis, and why it must be done right, with the right set of tools and more importantly, the necessary knowledge.

Exercise: From the included files, open *'Crankshaft Analysis'*. Add a split line to divide the indicated face vertically and run an analysis with SimulationXpress using the information given for forces and restraints.

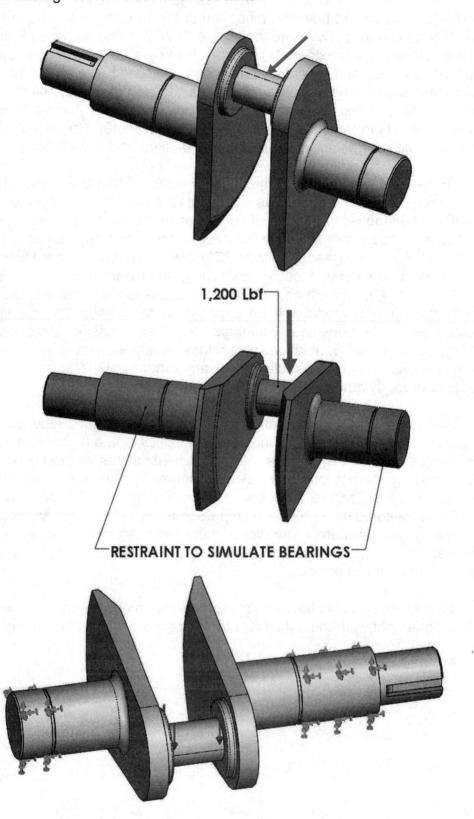

1,200 Lbf

RESTRAINT TO SIMULATE BEARINGS

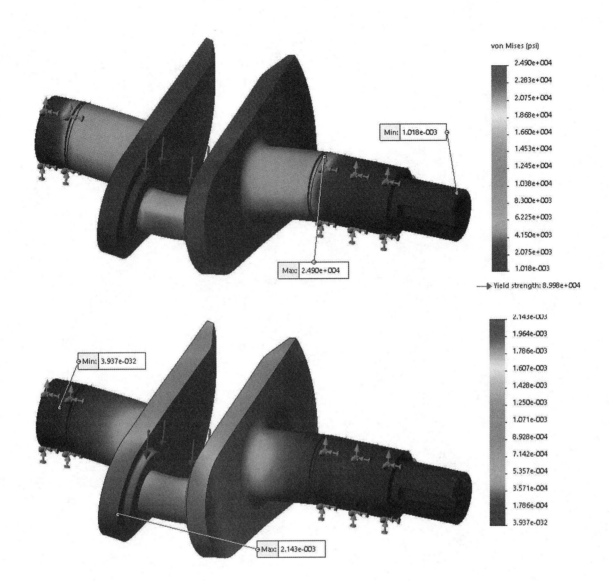

Notes:

Chapter 9: Collaboration: eDrawings
How to collaborate with non-SOLIDWORKS users

Sometimes we need to collaborate with other members of the design team in the same room, same building, or even across the world. With the convenience of e-mail, it's easier than ever to share our designs using eDrawings. We can send our extended design team, customers or suppliers a file by e-mail with all the design information needed for them to review and send back feedback to the designer. A big advantage of eDrawings is that it is available for Windows and Apple, iPad, iPhone, and Android devices.

We can generate eDrawings files from Parts, Assemblies, Drawings, and simulation studies. The only thing we need to do is to press the Publish eDrawings File icon from the "Save" fly-out toolbar, or go to the menu "**File, Publish to eDrawings**" while the document we want to share is active. The eDrawings viewer will be loaded with the same document. After we generate an eDrawings file, it becomes a *read-only* file; we cannot modify it, but we can print it, zoom in, or out, section it, etc. For this example, we'll publish an eDrawings file from the '*Gear Box*' assembly drawing. Depending on your installation of eDrawings, your screen will look something like this:

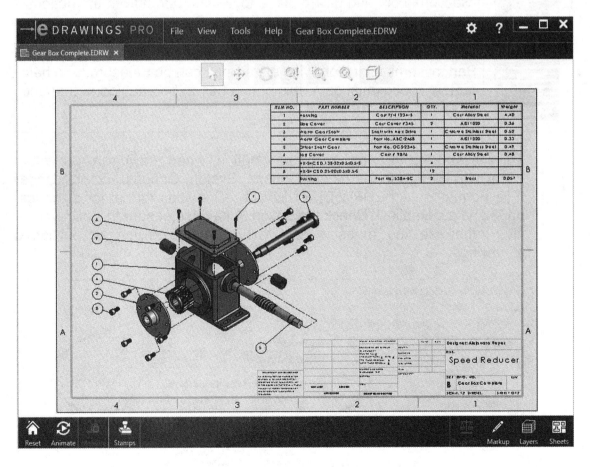

If the eDrawings file is generated using SOLIDWORKS Office Professional or the educational edition, the eDrawings Professional version is loaded. eDrawings Professional includes additional options, including tools to move the components in assemblies, measure, add markups and stamps and, if it's a part or assembly, make section views of the models.

Selecting the "Animate" command will allow us to navigate the different views in the eDrawings file. Pressing "Play" animates the views and will change from one to the next. At any time, the animation can be stopped, and the view zoomed in, rotated and panned.

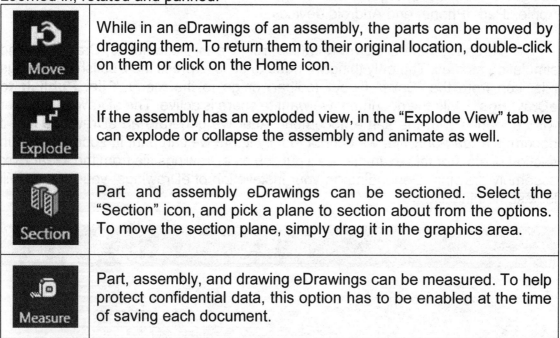

Move	While in an eDrawings of an assembly, the parts can be moved by dragging them. To return them to their original location, double-click on them or click on the Home icon.
Explode	If the assembly has an exploded view, in the "Explode View" tab we can explode or collapse the assembly and animate as well.
Section	Part and assembly eDrawings can be sectioned. Select the "Section" icon, and pick a plane to section about from the options. To move the section plane, simply drag it in the graphics area.
Measure	Part, assembly, and drawing eDrawings can be measured. To help protect confidential data, this option has to be enabled at the time of saving each document.

 To activate the "Enable measure" option when saving a published eDrawings file, go to the menu "Options, System Options, Export." Under "File Format" select EDRW/EPRT/EASM (eDrawings format for drawings, parts and assemblies). Check the option "Okay to measure this eDrawing's file," otherwise the measure tool will be disabled for all published eDrawings.

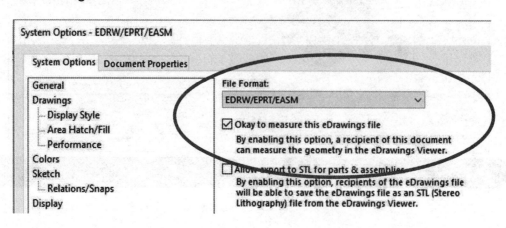

A powerful option in eDrawings Professional is the markup tools to allow anyone to add annotations to a file and send it back to the generator to make the required modifications.

The markup tools allow us to add:
- Dimensions
- Notes
- Add images
- Clouds with text
- Lines
- Rectangles
- Circles
- Arcs and
- Splines, etc.

After creating the eDrawings file, we must save it. eDrawings allows us to save the files in different ways.

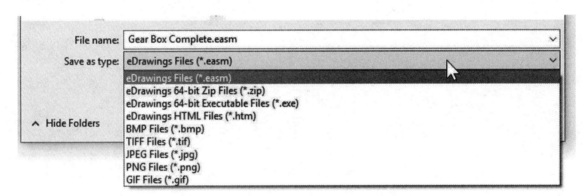

Saving as an eDrawings File (*.edrw) will result in a very small file size that can be easily emailed. The only thing the recipient needs to do is to install the free eDrawings viewer in their computer.

Saving as an HTML file (*.html) will create a web page that can be e-mailed. eDrawings will be loaded automatically if it is already installed on the computer; if it is not installed, the web browser will ask to install and run the corresponding plug-in. Optionally, a link to download the eDrawings viewer is listed at the bottom of the web page.

Notes:

Final Comments

By completing the exercises in this book, we learned how to apply many different SOLIDWORKS features and the various options to common design tasks. As we stated at the beginning, this book is meant to be an introduction to SOLIDWORKS, and as the reader could see, the breadth of options and possibilities available to the user after completing this book are enough to accomplish many different tasks from part and assembly modeling to detailing in a short time, using the most commonly used commands, including a stress analysis with SimulationXpress.

We hope this book serves as a stepping stone for the reader to learn more. A curious reader will be able to venture into more advanced features and take advantage of the similarities and consistency of the user interface to his/her advantage. We tried very hard to make the content as understandable and easy to follow as possible, as well as to get the reader working on SOLIDWORKS almost immediately, maximizing the "hands-on" time.

After completing all parts, drawings, an assembly including the fasteners, exploded view and the assembly drawing with a Bill of Materials, the reader is ready to apply the learned concepts in different design projects.

As a follow-up, be sure to continue learning SOLIDWORKS with **_Beginner's Guide to SOLIDWORKS 2019 – Level II._** Go further and learn about Sheet Metal, Surfacing, Mold Making and more.

One final tip: If you get lost or can't find what you are looking for while working in SOLIDWORKS, click with the right mouse button. Chances are, what you are looking for is in that pop-up menu. ☺

Notes:

Appendix

Document templates

One of the main reasons to have multiple templates is to have different settings, especially units. We can have millimeter and inch templates, pre-defined materials for part templates, dimensioning standards, etc. and every time we make a new part based on that template, the new document will have the same settings of the template. A good idea is to have a folder to store our templates and add it to the list of SOLIDWORKS templates.

Using *Windows Explorer*, make a new folder to store our new templates. For this example, we'll make a new folder in the Desktop called *"MySWTemplates."* In SOLIDWORKS, go to the menu "**Tools, Options, System Options, File Locations**." From the drop-down menu select "Document Templates."

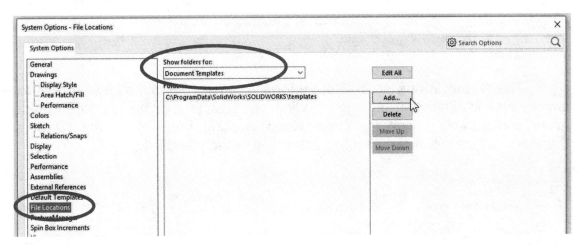

Click the "Add" button and locate the folder we just made to save the templates. Click OK when done. You will now see the additional template folder listed.

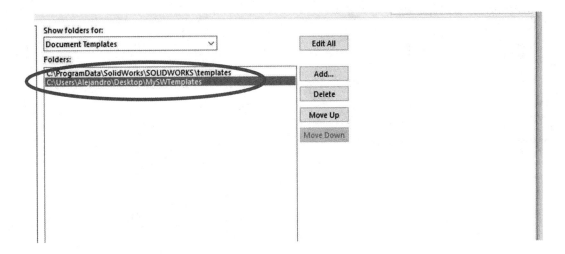

To create a new template, make a new Part, Assembly, or Drawing file, and go to the menu "**Tools, Options, Document Properties**." Everything we change in this tab is saved with the template. The most common options changed in a template are Units, Sketch Grid, Detailing Standards, and Annotation Fonts. Let's review them in the order that they appear in the options.

The first section is "Drafting Standard"; here we can change the dimensioning standard to ANSI, ISO, DIN, JIS, etc.; by changing the standard, all the necessary changes will be made to arrows, dimensions, annotations, etc., per the selected standard.

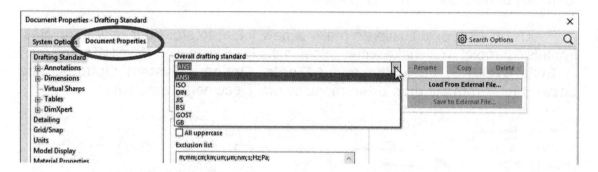

The "**Annotations**" and "**Dimensions**" sections contain all the options for annotation and dimension settings including font, size, arrow type, appearance and every configurable option for them. Keep in mind that setting the "Drafting Standard" will modify these sections according to the standard.

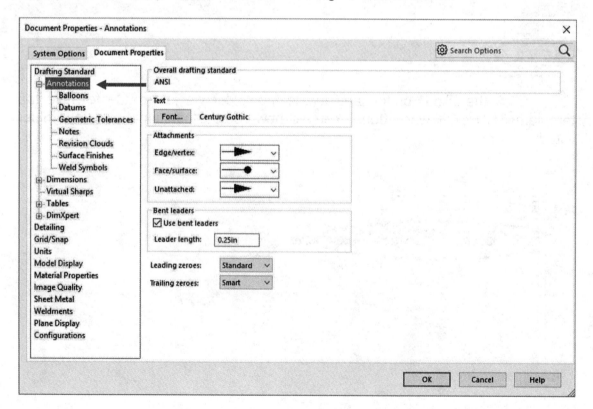

In the "Detailing" area we can change what is displayed in parts and assemblies; for drawings, we have more options including which annotations are automatically inserted when a new drawing view is created.

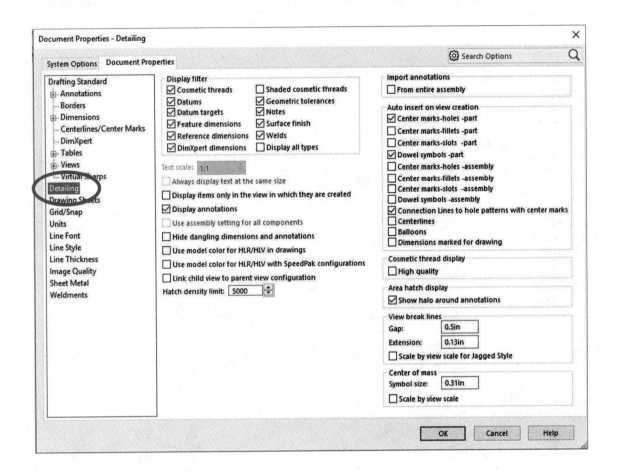

In the "Units" section, select the units that we want to use in the template, number of decimal places or fractional values, angular units, length, mass, volume and force, etc. Selecting a unit system will change all the corresponding units to that system, saving us time. Selecting "Custom" will allow us to mix unit systems if so desired.

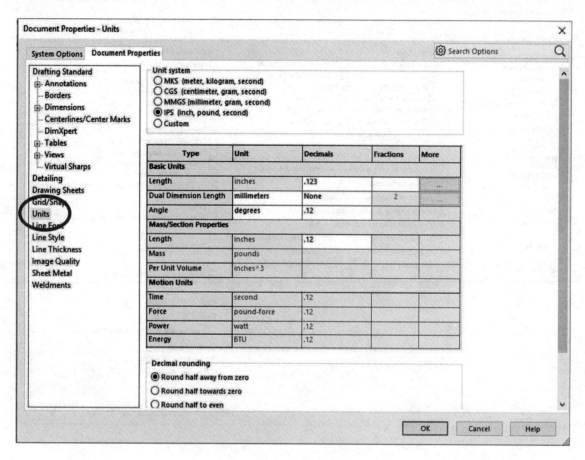

Click OK to set the new document properties to the selected options. Another option saved with the template is the material; for this part template select the aluminum alloy 6061-T6. To save the document as a new template, go to the menu "**File, Save As**." From the "Save As Type" drop down box, select "Part Template (*.prtdot)," "Assembly Template (*.asmdot)" or "Drawing Template (*.drwdot)" (depending on the type of document we are working on). SOLIDWORKS will automatically change to the first folder listed under the "File Locations" list for templates; if needed, browse to the "*MySWTemplates*" folder, give it a name and click "Save." As a suggestion, give the template a name that tells you something about the options set in it, like material, standard, units, etc. or whatever works for you.

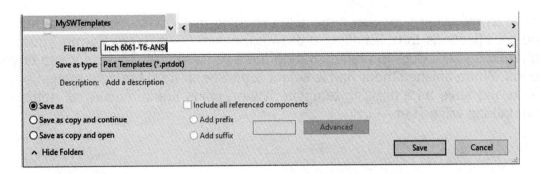

After saving the template, when we select the "**New Document**" icon and use the "**Advanced**" option, we'll see a new "Templates" tab with our new templates listed in it.

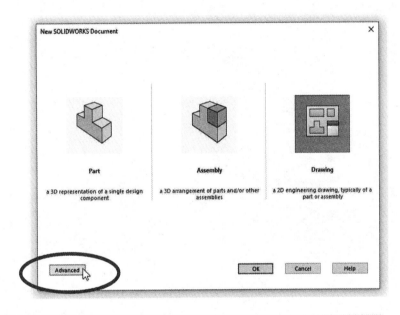

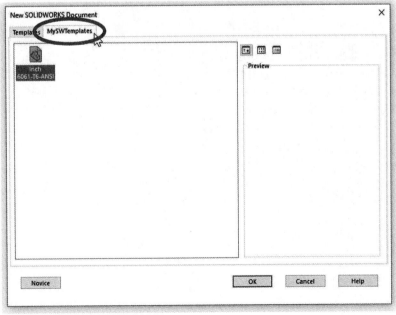

Drawing templates will include the sheet format, title block with any changes and modifications to it. If editing a drawing, select "Edit Sheet Format" from the right-mouse-click pop-up menu; change the annotations and title block to fit your needs. When finished modifying it, select "Edit Sheet" from the right mouse button menu and save as a drawing template. When using this template, all notes and annotations will be set.

Index

Notes: